Food Bioactive Ingredients

Series Editor

Seid Mahdi Jafari, Department of Food Materials and Process Design Engineering, Gorgan University of Agricultural Sciences and Natural Resources, Gorgan, Iran

The Food Bioactive Ingredients Series covers recent advances and research on the science, properties, functions, technology, engineering and applications of food bioactive ingredients and their relevant products. The series also covers health-related aspects of these bioactive components, which have been shown to play a critical role in preventing or delaying different diseases and to have many health-improving properties. The books in this series target professional scientists, academics, researchers, students, industry professionals, governmental organizations, producing industries and all experts performing research on functional food development, pharmaceuticals, cosmetics and agricultural crops.

More information about this series at https://link.springer.com/bookseries/16307

Seid Mahdi Jafari • Ana Sanches Silva
Editors

Releasing Systems in Active Food Packaging

Preparation and Applications

Springer

Editors
Seid Mahdi Jafari
Department of Food Materials and Process Design Engineering
Gorgan University of Agricultural Sciences and Natural Resources
Gorgan, Iran

Ana Sanches Silva
National Institute for for Agrarian and Veterinary Research (INIAV)
Vila do Conde, Portugal

ISSN 2661-8958 ISSN 2661-8966 (electronic)
Food Bioactive Ingredients
ISBN 978-3-030-90298-8 ISBN 978-3-030-90299-5 (eBook)
https://doi.org/10.1007/978-3-030-90299-5

This Springer imprint is published by the registered company Springer Nature Switzerland AG
The registered company address is: Gewerbestrasse 11, 6330 Cham, Switzerland

To Elham, AmirReza, and Elina, with love
Seid Mahdi Jafari

To Maria Inês, João, and Ricardo, with love
Ana Sanches Silva

Preface

Active food packaging represents a revolution for food industry, allowing to extend food shelf life, which results into a postponed expiry date, additional distribution from the point of origin, and decrease in food waste. The book *Releasing Systems in Active Food Packaging: Preparation and Applications* focuses on emitters and temperature systems. Active food packaging with temperature systems allow the heating or cooling of food, while emitters release different compounds with key function in food quality.

Active food packaging had an impressive development in the last two decades; however, all advances in this area have to obey to food safety principles, including requirements for food contact materials and authorized substances. These features are also addressed in this book.

Releasing Systems in Active Food Packaging: Preparation and Applications ensures an interdisciplinary approach to provide a complete understanding of active food packaging releasing systems from some of the most recognized international experts in the field of food packaging. This state-of-the-art guide provides information on different types of emitters, exploring the pros and cons of each one, preparation and effectiveness of releasing systems in active food packaging, and applications of these systems in different food categories.

The book firstly overviews the evolution of active food packaging concept and its current and future perspectives (Chap. 1). The second part is dedicated to different releasing systems including emitters of antimicrobials (Chap. 2); antioxidants (Chap. 3); essential oils (Chap. 4); flavors, colorants, other food ingredients (Chap. 5); and temperature control systems (Chap. 6).

The third part of the book is dedicated to the preparation and effectiveness of releasing systems in active food packaging. In this frame, different approaches for the inclusion of bioactive compounds are addressed (Chap. 7) and the impact of included bioactive compounds in barrier and mechanical properties of active packaging is focused (Chap. 8). In addition, the effectiveness and release studies of bioactive releasing systems is discussed (Chap. 9) as well as the preparation of edible active coating systems for food purposes (Chap. 10).

The fourth part of the book is dedicated to the application of releasing active packaging in different food categories, including meat products (Chap. 11), dairy products (Chap. 12), beverages (Chap. 13), cereals and cereal based products (Chap. 14), fruits and vegetables (Chap. 15), and oils and fats (Chap. 16).

This book aims to be an indispensable tool for professionals in the food packaging field and a key instrument for students and researchers in this area. We look forward that this edition will be well received and that in the future other active food systems can be addressed.

Herein, the editors would like to thank all the contributors of the book for their collaboration and efforts in bringing together different subjects dealing with releasing active food packaging in a comprehensive way with ultimate advances in the field. Their acceptance of our invitation in these critical and pandemic times is highly appreciated. Also, it is necessary to express our sincere thanks to the editorial staff at Springer for their help and support throughout the project. Finally, special acknowledgment is to our family for their understanding and encouragement during the editing of this great project.

Gorgan, Iran Seid Mahdi Jafari
Vila do Conde, Portugal Ana Sanches Silva

Contents

About the Editors

Seid Mahdi Jafari received his PhD in food process engineering from the University of Queensland (Australia) in 2006. Now, he is an academic member of GUASNR (Iran) and adjunct professor at UVigo (Spain). He has published more than 350 papers in international journals (h-index=64 in Scopus) and 60 book chapters and 36 books with Elsevier, Springer, and Taylor & Francis. Selected achievements:

- One of the top 1% world scientists by Thomson Reuters (2015)
- One of the top national researchers by the Iranian Ministry of Science, Research, and Technology (2017)
- One of the world's highly cited researchers by Clarivate Analytics (Web of Science), in 2018, 2019, and 2020.
- A top reviewer in the field of agricultural and biological sciences by Publons (2018 and 2019).

Ana Sanches Silva obtained her degree in pharmaceutical sciences from the Pharmacy Faculty at the University of Coimbra (FFUC), Portugal, and received her European PhD in pharmacy from the University of Santiago de Compostela (USC), Spain, with honors. In addition, she was presented with two awards for best PhD thesis. She is a member of the executive board of Animal Science Studies Center and invited professor at the FFUC. Ana has a remarkable track record, namely as co-author of papers in peer-reviewed journals with high impact factor, book chapters, and as co-editor of scientific books in the food science field. Ana's research focuses on the development of novel packaging, namely bioactive, edible, and nanopackaging. In addition, she has special interest in the development and validation of analytical methodologies, especially mass spectrometry related, to determine food and food packaging components and contaminants.

Part I
Overview and Future Perspectives of Active Food Packaging

Chapter 1
The Evolution of Food Packaging, the Active Food Packaging Concept and Its Current and Future Trends

Ana Sanches Silva and Seid Mahdi Jafari

Abstract Food packaging is responsible for a multitude of benefits for food items, since the beginning of humankind. These include containment, protection, preservation, transportation, identification and convenience of food. In this chapter, the evolution of the food packaging from pre-history will be addressed as well as the latest advancements, namely regarding the introduction of the 'active packaging' concept. The legal frame of food contact materials, in particular of active packaging, will be discussed as well as the features concerning the migration from these specific materials. Finally, current and future trends will be exploited.

Keywords Food packaging · Active food packaging · Food safety · Food quality · Food shelf life · Bioactive · Biopackaging

1 Packaging Since Pre-history

The use of packaging started in Pre-history with the basic need of contain/protect, transport, and store foods (Risch 2009). Pre-historic man started by using materials "given" by nature such as animal skins, leaves, gourds and nuts to protect and transport food. Late on, in Ancient Egypt other materials were used such as glass (Rasmussen 2012). In fact, Egyptians invented glass blowing and started to produce

A. S. Silva (✉)
National Institute for Agrarian and Veterinary Research (INIAV), I.P., Vairão, Vila do Conde, Portugal

Center for Study in Animal Science (CECA), University of Oporto, Oporto, Portugal

S. M. Jafari
Department of Food Materials and Process Design Engineering, Gorgan University of Agricultural Sciences and Natural Resources, Gorgan, Iran

S. M. Jafari, A. S. Silva (eds.), *Releasing Systems in Active Food Packaging*, Food Bioactive Ingredients, https://doi.org/10.1007/978-3-030-90299-5_1

Table 1.1 Relevant packaging achievements along history

Year	Achievement
1795	Food Preservation Prize offered by Napoleon Bonaparte
1810	Invention of canning by Nicolas Appert who also claimed Napoleon's prize
1810	Peter Durand patented tin-coating iron cans
1817	Invention of the first cardbox box by Sir Malcolm Thornhill
1852	Francis Wolle invented the first bag making machine
1906	Kellogg brothers started to distribute and market their cereal using cardboard
1908	Invention of cellophane by Jacques Brandenberger
1957	Invention of Bubble warp by Al Fielding and Marc Cavannes
1973	Invention of polyethylene terephthalate bottles by Nathaniel Wyeth

glass containers (although these were not transparent) for both food and beverages. On the other hand, in Ancient China, the developments in paper-making allowed to start using paper to wrap foods (Cartwright 2017). In fact, this is considered to be the first flexible packaging.

In medieval ages, wood had a major role as food packaging material. Wood boxes and wooden barrels were used to transport different items such as water, wine and dried food. The Industrial revolution (1760–1840) was rich in many technological developments, also applied to packaging industry (Mauer and Ozen 2004). However, the new products were expensive and mainly applied to high quality goods. Table 1.1 summarizes some of the most relevant achievements on food packaging along history.

Nowadays, in order to reduce the impact of packaging industry in the environment, there is a tendency to develop eco-friendly solutions. Therefore, there is a tendency to move from linear economy, where disposals are not re-used, to circular economy which aims to reutilized disposals in order to obtain new added-value products.

In this chapter, the evolution of the food packaging and of the concept 'active packaging' will be addressed. The legal frame of food contact materials, specially the one applied to of active packaging, will be discussed as well as the features concerning the migration from these specific materials. Future trends of active packaging will be also focused.

2 Legal Frame of Food Contact Materials

Materials already in contact with food, those intended to be in contact with food or those that can reasonably be brought into contact with food or transfer their constituents to the food under normal or foreseeable use are recognized as Food Contact

Materials and Articles (FCM). They comprise a vast number of materials including paper, paperboard, wood, metal and plastics (Regulation (EC) No 1935/2004 2004).

In addition, they can be comprised of a single material, so-called monolayer material or they can be comprised of different layers, so-called multilayer material, and the layers can be of different materials (so-called multilayer multi material). The formulation of these materials is quite variable and any of the components of the formulation can move from packaging to food surface or to the headspace/internal atmosphere of the packaging. They can be used as a single material or in combination (e.g. multilayer multimaterials). A broad number of substances can be used in the manufacture of these materials and they can constitute a possible source of food contamination during any of the following steps: manufacturing, packaging, processing, and storage. The legislation of FCM is compulsory, although can present differences among different parts of the globe. It is an obligation of food industry to meet all the conditions required by the legislation, in order to assure Food Safety and indirectly to contribute to consumers Health.

FCM are not intended to be added directly to food, therefore they are also known as "indirect additives". Title 21 of the U.S. Code of Federal Regulations (21CFR) (21CFR, 2017) includes information on more than 3 thousand substances used as food additives. Indirect additives are included in Parts 175 to 178 of 21CFR. In fact, Part 175 dedicates to adhesives and components of coatings, Part 176 to paper and paperboard components, Part 177 to polymers while Part 178 is focused on adjuvants and production aids.

US FDA frequently updates the list of food contact materials ("indirect additives") which includes information on their identity, intended use, and conditions of use of each authorized substance (US FDA 2018).

The European Union has a set of regulations regarding FCM which are also continually being updated. Some of the FCM legal documents are regulations applied to all FCM, while others are specific to groups of materials or to groups of materials (e.g. (Commission Regulation (EC) No 1895/2005 on epoxy derivatives in FCM). The Framework Regulation No 1935/2004 defines the general safety principles for FCM in order to make possible a standardization among different countries of EU (Regulation (EC) No 1935/2004 2004). According to this regulation, FCM should be as inert as possible so that their components do not influence food quality or consumer health (Regulation (EC) No 1935/2004 2004).

In the case of absence of specific rules at EU level, this legislation can be complemented by the national legislation of the Member States. Moreover, the European Food Safety Authority (EFSA 2021) carries out the assessment of the safety of FCMs and publishes scientific opinions while, the European Reference Laboratory for Food Contact Materials (EURL-FCM 2021) is responsible for the scientific knowledge and technical competence of the methods for testing the safety of FCM.

According to Article 3 of Regulation (EC) No 1935/2004, on general requirements, FCM, under normal or foreseeable conditions of use, shall not "transfer their constituents to food in quantities which could: (a) endanger human health; or (b) bring about an unacceptable change in the composition of the food; or (c) bring about a deterioration in the organoleptic characteristics thereof." (Regulation (EC) No 1935/2004 2004).

Regulation (EC) No 1935/2004 specifies that individual measures should be established for seventeen groups of materials. However, up to now , there are only specific measures to: ceramic materials (Directive No 84/500/EEC 1984), regenerated cellulose film (Commission Directive 2007/42/EC 2007), recycled plastic materials (Commission Regulation (EC) No 282/2008 2008), active and intelligent materials (Commission Regulation (EC) No 450/2009 2009) and plastics (Commission Regulation (EU) No 10/2011 2011). Certain starting substances used in the production of FCM also have specific rules (Commission Regulation (EC) No 1895/2005 2005; Vilarinho et al. 2019)

Two types of FCM that should be distinguish from those traditionally used are Active food contact materials and articles and Intelligent packaging and they are defined in Regulation No 1935/2004 (Regulation (EC) No 1935/2004 2004). Intelligent packaging monitors a condition of the food, e.g. the historical of temperatures of the packaged item (Ghaani et al. 2016). Active FCM incorporate active components which are deliberately released into the food (releasing active FCM) or absorbed from food or their surrounding environment (absorbing active FCM). Therefore, active FCM and articles interact with food with the aim of increasing their shelf life, therefore they are not inert as conventional materials (Fig. 1.1).

Regulation No 1935/2004 defines specific requirements for active and intelligent materials and articles (Regulation (EC) No 1935/2004 2004). These include the following: (i) active materials and articles "may bring about changes in the composition or organoleptic characteristics of food on condition that the changes comply with the Community provisions applicable to food"; (ii) Intelligent materials and articles shall not mislead consumers when giving information about the condition of

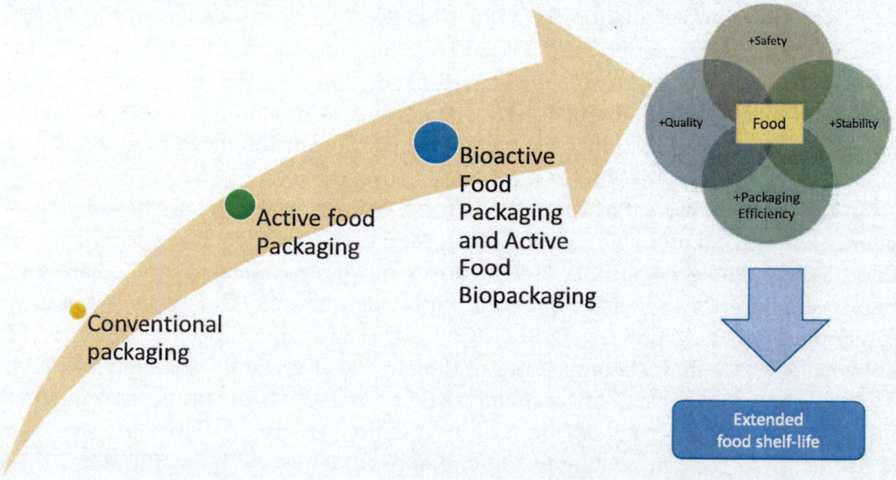

Fig. 1.1 Evolution from "inert" packaging to "active" packaging concept and its main consequences for packaged food

the food; (iii) Active and intelligent materials and articles shall be correctly labelled to allow the consumers to easily identify possible non-edible parts and to indicate the type of materials they constitute.

Another regulation that applies to all FCM is Regulation (EC) No 2023/2006 (Regulation (EC) No 2023/2006 2006) which concerns good manufacturing practices. According to this regulation, Good manufacturing practices must be applied during the whole process, although other legislation covers the production of the starting substances.

In this chapter, the evolution of the food packaging and of the concept 'active packaging' will be addressed. The legal frame of food contact materials, specially the one applied to of active packaging, will be discussed as well as the features concerning the migration from these specific materials. Future trends of active packaging will also be focused.

3 Migration

The process of mass transfer from FCM to food is so called "migration". This is a phenomenon affected by different factors, including: (1) the temperature of contact (higher temperatures increase migration rate), (2) the composition of food and its physical properties (e.g. the fat content of a food item can greatly influence the migration rate), (3) the type of packaging and its formulation (e.g. ingredients of food packaging formulation), (4) the type of contact and surface area between the packaging and the food item (Sanches Silva et al. 2008, 2009).

Ideally the evaluation of migration should be include all known intentionally added substances (IAS) (e.g. additives) as well as non-intentionally added substances (NIAS), which are reaction and degradation products (Koster et al. 2015). Regulation No 1935/2004 defines the conditions to carry out the migration tests (Regulation (EC) No 1935/2004 2004). The labelling, advertising and presentation of a material or article shall not mislead the consumers.

Regulation (EU) No 10/2011 dedicates specifically to plastic materials and articles and lays down specific rules for their safe (Commission Regulation (EU) No 10/2011 2011). On the opposite to the former Commission Directive on plastic materials, Directive 2002/72/EC (Commission Directive 2002/72/EC 2002), which was applied only to materials and articles comprising exclusively of plastics and plastic liner seals, Regulation (EU) No 10/2011 applies also to those with multiple layers of plastic (multilayer) bound together by adhesives or other means; and those with a set of two or more layers of different types of materials, being one of them of plastic (Commission Regulation (EU) No 10/2011 2011).

Regulation (EU) No 10/2011 establishes rules on the composition of plastic FCM and establishes a list which includes all substances that are permitted for use in the manufacture of plastic FCM, which is continually being updated (Commission Regulation (EU) No 10/2011 2011). This list of permitted substances is also known as Positive list and includes monomers and other starting substances; additives other than colorants; polymer production aids, excluding solvents; and macromolecules obtained from microbial fermentation. Regulation (EU) No 10/2011 has already been amended 15 times since it was first published in 2011, most of these amendments were related with the positive list.

This Regulation defines two migration limits, the Overall Migration Limit (OML), which is the maximum permitted amount of non-volatile substances released from a material or article into food simulants, and the Specific Migration Limit (SML) which constitutes the maximum permitted amount of a given substance released from a material or article into food or food simulants.

The food simulants are test media representing foods and their behavior mimics the migration from the plastic FCM. OML is 10 mg/dm^2 plastic surface area while SML is defined for each permitted substance. Besides SML, Regulation (EU) No 10/2011 also defines the Total Specific Migration Limit (SML (T)) as the maximum permitted sum of substances released in food or food simulants expressed as total of moiety of the substances indicated (Commission Regulation (EU) No 10/2011 2011). There are substances from the positive that still do not have a SML or SML(T) due to the lack of information (namely toxicological) that allows to define this limit. Regulation (EU) No 10/2011 also sets out the food simulants that can be used and indicates which simulants represent the food by food category. In sum, Simulant A- Ethanol 10% (v/v) represents Foods that have a hydrophilic character; Simulant B- Acetic acid 3% (w/v) represents Foods which have a pH below 4.5; Simulant C- Ethanol 20% (v/v) mimics. Alcoholic foods with an alcohol content of up to 20% and those foods which contain a relevant amount of organic ingredients that render the food more lipophilic; Simulant D1- Ethanol 50% (v/v) represents foods that have a lipophilic character and for alcoholic foods with an alcohol content of above 20% and for oil in water emulsions; Simulant D2- Vegetable oil represents foods that have a lipophilic character and foods which contain free fats at the surface and Simulant E- Poly(2,6-diphenyl-p-phenylene oxide), particle size 60–80 mesh, pore size 200 nm mimics dry foods.

Additives authorized as food additives by Regulation (EC) No 1333/2008 (Regulation (EC) No 1333/2008 2008) or by Regulation (EC) No 1334/2008 (Regulation (EC) No 1334/2008 2008) as flavorings shall not migrate from plastic FCM to food in quantities which have a technical effect on the final food and shall not exceed the restrictions defined in these two Regulations or in Annex I of Regulation (EU) No 10/2011 for foods for which they are authorized as food additives or flavoring substances; or exceed the restrictions defined in Annex I of this Regulation in foods for which they are not authorized as food additives or flavouring substances.

Regulation (EU) No 10/2011 also allows to predict migration through the use of diffusion models based on scientific evidence. These diffusion models are intended to overestimate the actual migration in order to assure food safety and consequently Human Health (Silva et al. 2008).

4 Actual and Future Trends in Active Food Packaging

In the last few years we have been experiencing a considerable change regarding the replacement of conventional packaging materials by active food packaging, able to intentionally interact with food (Fig. 1.1). These changes are driven by different factors. Figure 1.2 compiles the main driving forces to move from conventional food packaging to active food packaging while Fig. 1.3 resumes the main driving forces to move from these type of packaging to bioactive food packaging and food biopackaging.

Nowadays there is also the tendency for using bioactive compounds, i.e., active compounds obtained from natural sources, such as plants (Ribeiro-Santos et al. 2018), microbes, fungi, among others, that have biological activity (e.g. antimicrobial and antioxidant capacity). Another tendency is the use of biopackaging, i.e., biodegradable packaging and/or packaging obtained from renewable resources. Also, the application of the circular economy to food packaging is a reality, being considerable the number of materials already developed based on food by-products such as whey protein films (Andrade et al. 2018; Brink et al. 2019) and pectin-based films (Espitia et al. 2014). This tendency is linked with the environmental concerns related with the disposal of plastics in nature.

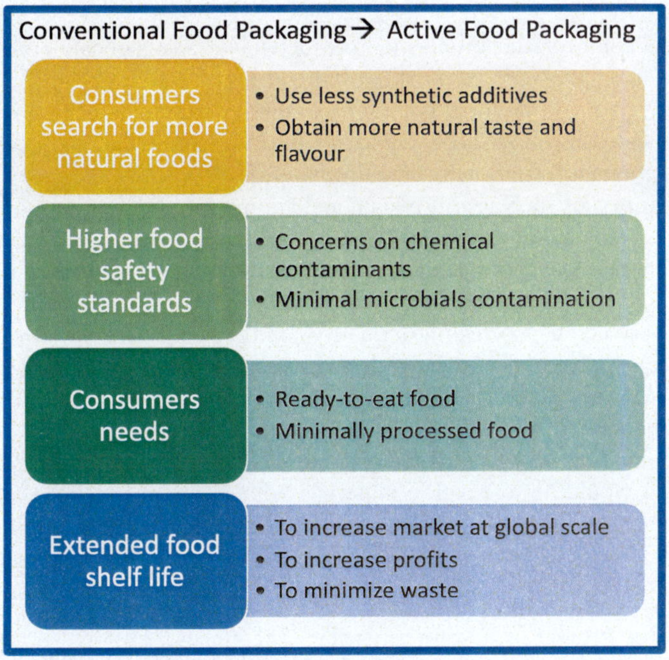

Fig. 1.2 Main driving forces to move from conventional ("inert") food packaging to active food packaging

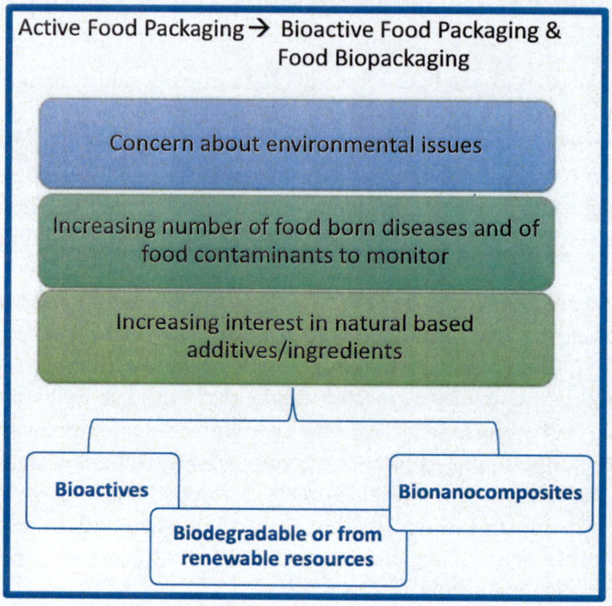

Fig. 1.3 Main driving forces to move from active food packaging to bioactive food packaging and food biopackaging

However, some of the bioplastics present lower mechanical, optical, barrier or thermal

properties than conventional polymers. In these cases, nanotechnology can help to surpass these constraints, allowing to insert nanofillers in the matrices of the biopackaging (bionanocomposites). Nanoforms can also allow active packaging to have their properties such as antioxidant and antimicrobial properties (Hoseinnejad et al. 2018). Some nanoforms have already been evaluated by EFSA and some are authorized in the EU and included in the positive list (Commission Regulation (EU) No 10/2011 2011). Nanotechnology can also be used to manufacture intelligent packaging based on nanosensors for quality control purposes.

The most valuable pros of these tendencies are the increment of food safety, food quality, food stability and the efficiency of the packaging that contributes for the extension of food shelf life.

In the future there will be a tendency to merge active and intelligent FCM materials. The idea would be to develop materials with different functions whose main objective is to simultaneously increase the shelf life of foods and monitor their quality, using the lowest amount of materials and having the lowest negative impact in the environment.

Acknowledgments The work was supported by UIDB/00211/2020 with funding from FCT/MCTES.

References

Andrade MA, et al. Characterization of rosemary and thyme extracts for incorporation into a whey protein based film. LWT Food Sci Technol. 2018;92. https://doi.org/10.1016/j.lwt.2018.02.041.

Brink I, Šipailienė A, Leskauskaitė D. Antimicrobial properties of chitosan and whey protein films applied on fresh cut turkey pieces. Int J Biol Macromol Elsevier. 2019;130:810–7. https://doi.org/10.1016/j.ijbiomac.2019.03.021.

Cartwright M. Paper in ancient China. World history encyclopedia. 2017. Available at https://www.worldhistory.org/article/1120/paper-in-ancient-china/. Accessed 7 July 2021.

Commission Directive 2002/72/EC. Commission directive 2002/72/EC of 6 August 2002 relating to plastic materials and articles intended to come into contact with foodstuffs. Off J Eur Commun. 2002;L220/18-L2.

Commission Directive 2007/42/EC. Commission Directive 2007/42/EC of 29 June 2007 relating to materials and articles made of regenerated cellulose film intended to come into contact with foodstuffs. Off J Eur Union. 2007;L172/71-L1.

Commission Regulation (EC) No 1895/2005. Commission Regulation (EC) No 1895/2005 of 18 November 2005 on the restriction of use of certain epoxy derivatives in materials and articles intended to come into contact with food. Off J Eur Union. 2005;L 302/28-3.

Commission Regulation (EC) No 282/2008. Commission Regulation (EC) No 282/2008 of 27 March 2008 on recycled plastic materials and articles intended to come into contact with foods and amending Regulation (EC) No 2023/2006. Off J Eur Union. 2008;L86/9-L86/.

Commission Regulation (EC) No 450/2009. Commission Regulation (EC) No 450/2009 of 29 May 2009 on active and intelligent materials and articles intended to come into contact with food. Off J Eur Union. 2009;L135/3-L13.

Commission Regulation (EU) No 10/2011. Commission Regulation (EU) No 10/2011 of 14 January 2011 on plastic materials and articles intended to come into contact with food. Off J Eur Union. 2011;L12/1-L12/.

Directive No 84/500/EEC. Council Directive 84/500/EEC of 15 October 1984 on the approximation of the laws of the Member States relating to ceramic articles intended to come into contact with foodstuffs. Off J Eur Commun. 1984;L 277-12/L.

EFSA. European Food Safety Authority. 2021. Available at https://www.efsa.europa.eu/en. Accessed 1 July 2021.

Espitia PJP, et al. Edible films from pectin: physical-mechanical and antimicrobial properties – a review. Food Hydrocolloids Elsevier. 2014;35:287–96. https://doi.org/10.1016/j.foodhyd.2013.06.005.

EURL-FCM. EURL food contact materials. 2021. Available at https://ec.europa.eu/jrc/en/eurl/food-contact-materials. Accessed 20 June 2021.

Ghaani M, et al. An overview of the intelligent packaging technologies in the food sector. Trends Food Sci Technol Elsevier. 2016;51:1–11. https://doi.org/10.1016/j.tifs.2016.02.008.

Hoseinnejad M, Jafari SM, Katouzian I. Inorganic and metal nanoparticles and their antimicrobial activity in food packaging applications. Crit Rev Microbiol. 2018;44(2):161–81.

Koster S, Bani-Estivals MH, Bonuomo M, Bradley E, Chagnon MC, Garcia ML, Godts F, Gude T, Helling R, Paseiro-Losada P, Pieper G. Guidance on best practices on the risk assessment of non intentionally added substances (NIAS) in food contact materials and articles. ILSI Europe Report Series. 2015.

Mauer LF, Ozen BF. Food packaging. In: Smith, Scott J, Hu YH, editors. Food processing principle and applications. Iowa: Blackwell; 2004. p. 151–62.

Rasmussen SC. Origins of glass: myth and known history. In: How glass changed the world, Springer briefs in molecular science, vol. 3. Berlin/Heidelberg: Springer; 2012. https://doi.org/10.1007/978-3-642-28183-9_2.

Regulation (EC) No 1333/2008. Regulation (EC) No 1333/2008 of the European Parliament and of the Council of 16 December 2008 on food additives. Off J Eur Union. 2008;L354/16-L354/33.

Regulation (EC) No 1334/2008. Regulation (EC) No 1334/2008 of the European Parliament and of the Council of 16 December 2008 on flavourings and certain food ingredients with flavouring properties for use in and on foods and amending Council Regulation (EEC) No 1601/91, Regulations (EC). Off J Eur Union. 2008;L 354/34–L 354/50.

Regulation (EC) No 1935/2004. Regulation (EC) No 1935/2004 of the European Parliament and of the Council of 27 October 2004 on materials and articles intended to come into contact with food and repealing Directives 80/590/EEC and 89/109/EEC. Off J Eur Union. 2004.

Regulation (EC) No 2023/2006. Commission Regulation (EC) No 2023/2006 of 22 December 2006 on good manufacturing practice for materials and articles intended to come into contact with food. Off J Eur Union. 2006;L384/75-L3.

Ribeiro-Santos R, et al. Essential oils for food application: natural substances with established biological activities. Food Bioprocess Technol. 2018;11(1). https://doi.org/10.1007/s11947-017-1948-6.

Risch SJ. Food packaging history and innovations. J Agric Food Chem. 2009;57(18):8089–92. https://doi.org/10.1021/jf900040r.

Sanches Silva A, et al. Time-temperature study of the kinetics of migration of diphenylbutadiene from polyethylene films into aqueous foodstuffs. Food Res Int. 2008; https://doi.org/10.1016/j.foodres.2007.10.007.

Sanches Silva A, et al. Migration and diffusion of diphenylbutadiene from packages into foods. J Agric Food Chem. 2009;57(21). https://doi.org/10.1021/jf901666h.

Silva AS, et al. Mass transport studies of model migrants within dry foodstuffs. J Cereal Sci. 2008;48(3). https://doi.org/10.1016/j.jcs.2008.02.006.

US FDA. Code of Federal Regulations – Title 21 – Food and drugs. 2018. Available at https://www.fda.gov/medical-devices/medical-device-databases/code-federal-regulations-title-21-food-and-drugs. Accessed 21 June 2021.

Vilarinho F, et al. Bisphenol A in food as a result of its migration from food packaging. Trends Food Sci Technol. 2019;91. https://doi.org/10.1016/j.tifs.2019.06.012.

Part II
Releasing Systems in Active Food Packaging

Chapter 2
Emitters of Antimicrobials

Jaime López-Cervantes, Dalia Isabel Sánchez-Machado, Ana Aglahe Escárcega-Galaz, Benjamín Ramírez-Wong, Gabriela Servín de la Mora-López, and María del Rosario Martínez-Macias

Abstract Among the modern active packaging systems are the emitters of antimicrobial agents that are intended to improve storage life and food safety. Currently, there is a tendency to study the continuous release of volatile antimicrobial agents from sachets attached to the container, incorporated directly into films or applied as coatings to the surfaces of the containers. Among the active agents are SO_2, CO_2, ClO_2, ethanol and essential oils. The various research studies reviewed have demonstrated their effectiveness in minimizing the deterioration of food of animal and plant origin caused by microbial contamination. By selecting the ideal antimicrobial emitter, it is possible to control the quality of food and improve the safety of foods along all steps of food chain.

Keywords Sulfur dioxide · Carbon dioxide · Chlorine dioxide · Ethanol · Essential oils

J. López-Cervantes · D. I. Sánchez-Machado (✉) · A. A. Escárcega-Galaz (✉)
M. d.R. Martínez-Macias
Instituto Tecnologico de Sonora, Ciudad Obregon, Sonora, Mexico
e-mail: dalia.sanchez@itson.edu.mx; dsanchez@itson.edu.mx

B. Ramírez-Wong
Universidad de Sonora, Hermosillo, Sonora, Mexico

G. S. de la Mora-López
Universidad Autonoma de Occidente, Culiacan Rosales, Sinaloa, Mexico

S. M. Jafari, A. S. Silva (eds.), *Releasing Systems in Active Food Packaging*, Food Bioactive Ingredients, https://doi.org/10.1007/978-3-030-90299-5_2

1 Introduction

In the food industry, it is a priority to reduce spoilage and prolong the shelf life of various food products (Mu et al. 2017). This has favored the development of active packaging systems. These are systems where packaging interactions with food or with the headspace to maintain sensory and nutritional quality and product safety (Ayala-Zavala et al. 2008). Some are designed to contain, protect and preserve foods with their characteristic properties (Altan et al. 2018). Others are developed to incorporate substances that are intended to be released in the food or their environment. Some of them are able to absorb substances that the same food generates and release beneficial substances for food (Demitri et al. 2016). Therefore, Higueras et al. (2014) reported that the inhibition or delay in the deterioration of food occurs due to the interaction among the food, the headspace and the container walls.

Microbial activity is one of the main reactions in food and promotes the loss of 25% of food before consumption (Zemljič et al. 2013). During the deterioration of food by microbial activity, unpleasant tastes and odors are produced, as well as compounds that might be harmful to health (Souza et al. 2015). Antimicrobial packaging is intended to extend the shelf life of food by preventing or slowing the growth of microorganisms (Mlalila et al. 2018). Thus, polymeric sheets have been designed for the packaging of foods that release the added components in a controlled manner without affecting the taste and smell of the food itself (Ramos et al. 2012).

In active packaging systems intended to have antimicrobial action, the antimicrobial agent can be released in the headspace of the container or directly in the surface of the packaged food and it can, (i) be mixed in the formulation of the packaging material, (ii) be fixed on the surface of the container/packaging material (Ayala-Zavala et al. 2008).

The most common volatile antimicrobial components used in this type of packaging systems are sulfur dioxide (SO_2), carbon dioxide (CO_2), chlorine dioxide (ClO_2), ethanol and essential oils, which have been applied to control the growth of bacteria and fungi. Volatile antimicrobials have the advantage of reaching the entire food surface with a greater effect than solutions applied by contact (Pereira de Abreu et al. 2012). However, some of them have limitations, such as their sensory perception and stability during storage.

SO_2 has traditionally been used as a preservative for wines, nuts, fruits and vegetables. CO_2 emitters have been used to extend the shelf life of fresh meats and fruits. ClO_2 has been recommended for the active packaging of fruits and vegetables. Ethanol is used to increase the shelf life of bakery products with high humidity due to its activity against molds, yeasts and bacteria.

Essential oils are an alternative to synthetic additives that are incorporated into food for preservation. Due to their natural origin, volatility, antimicrobial action and antioxidant activity, essential oils have been widely studied for as active packaging ingredients (Zhou et al. 2020). The most active and major components of essential oils are of varied chemical nature. Among the most popular are allicin (garlic, *Allium sativum*), carvacrol (oregano, *Origanum vulgare*), menthol (peppermint,

Mentha piperita), thymol (thyme, *Thymus vulgaris*), and cinnamaldehyde (cinnamon, *Cinnamomun zeylanicum*).

The main focus of this chapter is to present an in-depth review of the most recent and relevant publications on the release of antimicrobial agents from active packaging systems. The active components and polymers used in the formulation of emitters of antimicrobials were analyzed from the reviewed release systems. Moreover, the applications proposed for different food groups were addressed.

2 Antimicrobial Agents in Emitters

The antimicrobial compounds used in active packaging can be volatile and nonvolatile. As nonvolatile antimicrobials, metallic nanoparticles, organic acids, enzymes and bacteriophages have been studied (Otoni et al. 2016), in which the release occurs by direct contact between the food and the packaging material (Almasi et al. 2020). Volatile antimicrobial agents are released from the carrier materials in gaseous form and occupy all the free space within the packaging. This characteristic of gases favors the compound to envelop all food, improving its antimicrobial activity (Lucera et al. 2016).

Bisulfite salts in contact with moisture generate sulfur dioxide (SO_2), which is an active gaseous component with antimicrobial and antioxidant effects. For the design of SO_2 releasers, bisulfite salts are incorporated in the inner layer of the packages or in sachets (Chen et al. 2020a). SO_2 has been used to inhibit the growth of pathogenic fungi in packaged fruits and to prolong their shelf life (Xu et al. 2013a). Specifically, to extend the storage of grapes, less than 60 ppm SO_2 is required.

Carbon dioxide (CO_2) is used to inhibit the development of bacteria such as *Listeria* and aerobic fungi on the surface of food (Lucera et al. 2016). As reported by Wyrwa and Barska (2017), CO_2 emitters decrease the metabolic rate of microorganisms, affect the development of microorganisms, maintain the quality of food and prevent the swelling of packaging. The development of microorganisms in food is delayed by CO_2, and its antimicrobial effect is a function of its concentration in the headspace (Tsironi et al. 2019). Packaging in atmospheres with high levels of CO_2 has been recommended for the preservation of cheese, fish and fresh meat. Additionally, to prolong the storage life of certain fruits and vegetables, CO_2 emitters have been used. To remove CO_2 from the containers, porous sachets filled with calcium hydroxide embedded in silica gel are placed (Wyrwa and Barska 2017).

Chlorine dioxide (ClO_2) is a green-yellow gas very soluble in water that must be produced *in situ* because its storage is not possible due to the risk of explosion (Otoni et al. 2016). Traditionally, this gas has been applied as a washing agent in food processing industries due to its antimicrobial activity against bacteria, fungi, spores and viruses (Chen et al. 2020b). Additionally, it has been used in fresh products and meat. ClO_2 has potent bactericidal effects because gas molecules are small and easily approach microbial cells to react with their cellular components (Chai et al. 2020). However, it does not react with phenolic or nitrogenous compounds, and therefore chloramines are not formed, which are dangerous and impart a bad odor to food. ClO_2 gas is commercially produced from sodium chlorite and an acid when both dry precursors are in contact with moisture. Another method to produce chlorine dioxide with food application is based on the oxidation of sodium hypochlorite with chlorine gas (Chen et al. 2020a).

Ethanol is an effective antimicrobial agent to inhibit the development of molds, although it can act against yeasts and bacteria. Ethanol vapors inhibit *Aspergillus* and *Penicillium* species (Hempel et al. 2013). Ethanol inhibits or kills microorganisms because it damages their cell wall and causes the coagulation of proteins in microbes (Mu et al. 2017). The activity of ethanol as an antimicrobial agent is due to the decrease in water activity (a_w), so it is recommended to increase the storage life of foods with high a_w (Janjarasskul et al. 2016). Ethanol as an agent to stop the development of molds has been applied by spraying in bakery products (Pasqualone 2019). In bakery products sprayed with 95% food grade ethanol, concentrations of 0.5–1.5 (w/w) are recommended (Haghighi-Manesh and Azizi 2017). However, now the trend is to use membranes or sachets that emit ethanol, and most are patented in Japan (Day 2003). In commercial emitters, food-grade ethanol is released from a carrier material such as silicon dioxide powder contained in sachets made of paper and ethylvinyl acetate (EVA) (Pasqualone 2019). It has been found that if a food is packaged with an ethanol emitter, the food absorbs moisture, and ethanol vapors are released into the headspace of the container (Majid et al. 2018). Some authors have documented that ethanol affects the taste of food products, but during baking, ethanol is naturally generated by yeasts, as occurs in some Italian breads where the amount of ethanol is 2% (Dantigny et al. 2005). Ethanol-emitting sachets are expensive and are only used for very special foods. Additionally, as commercial sources of ethanol, brandy or whiskey have been added, specifically, to add different flavors in bakery products (Hempel et al. 2013).

Essential oils (EOs) are low molecular weight volatile compounds of the secondary metabolism of various plants (Llama-Ruiz et al. 2015). These are extracted by distillation or mechanical pressing of flowers, roots, leaves, seeds and bark of plants (Atarés and Chiralt 2016; Ribeiro-Santos et al. 2017). Essential oils are divided into two groups, terpenes and aromatics. Monoterpenes ($C_{10}H_{16}$) are the most recognized terpenes and correspond to 90% of essential oils. Aromatic compounds are found in lower quantities than terpenes, and some of them are aldehydes, alcohols, phenols, and compounds of the methylene dioxy group (Llana-Ruiz-Cabello et al. 2015). The composition of the EO depends mainly on the source and the extraction method, as well as the climatic and geographical conditions where the plant material inhabits (Akrami et al. 2015). Essential oils are of interest due to the amount and biological properties of the bioactive components they contain.

The active components of garlic oil are allyl disulfide, allyl trisulfide and allyl tetrasulfide. All of them are volatile, and their sulfur atoms bind to the electronegative atoms of the proteins, altering the properties of the biomolecules of the microorganisms (Ayala-Zavala and González-Aguilar 2010).

Since ancient times, EO has been incorporated into food. Therefore, their potential application is now as antimicrobial agents in active packaging to improve the shelf life of food (Becerril et al. 2007). Ramos et al. (2012) reported that terpenes and phenolic acids of essential oils have antimicrobial properties. Zanetti et al. (2018) reported that essential oils act directly on the membrane and cytoplasm and sometimes completely alter the cell. Gram-positive bacteria are more sensitive to essential oils than Gram-negative bacteria. The antimicrobial action of carvacrol is attributed to the fact that phenols affect the permeability of cell membranes (Altan et al. 2018). In addition, essential oils in foods act as antioxidants and flavoring agents (Ribeiro-Santos et al. 2017). Asdagh et al. (2021) discovered a synergistic effect between coconut essential oil and paprika extracts against *Staphylococcus aureus* and *Escherichia coli* when studying membranes made with whey protein and copper nanoparticles. Table 2.1 details some methods applied for the

Table 2.1 Antimicrobial activity and stabilization of essential oils for application in active packaging

Sources of essential oil	Stabilization	Microorganisms evaluated	References
Cinnamon	Films of corn loaded with emulsions made by mixing cinnamon oil and chia seed mucilages.	*Botrytis cinerea*	Díaz-Galindo et al. (2020)
Common rose seeds (*Rosa canina* L.)	Encapsulation in chitosan and Tween 80 films produced by solvent evaporation	*Escherichia coli* *Salmonella typhymurium and Bacillus cereus*	Butnaru et al. (2019)
Garlic (*Allium sativum* L.)	Nanocomposite films based on plasticized banana flour to make sachets	*Aspergillus flavus*	Orsuwan and Sothornvit (2018)
Rosemary and cinnamon	Bilayer film based on low density polyethylene	Total viable count Enterobacteriaceae Psychrotrophic bacteria Bacteria producing H_2S	Dong et al. (2018)
Cinnamon bark and ginger rhizomes	Films of carboxymethyl cellulose, chitosan and oleic acid made by drying	*Aspergillus niger*	Noshirvani et al. (2017)
Rosemary and oregano	Cellulose acetate nanofiber pads produced by electrospinning	*Staphylococcus aureus* *Escherichia coli* *Candida albicans*	Liakos et al. (2017)
Cloves (*Syzygium aromaticum*) and oregano (*Origanum vulgare*)	Methylcellulose films	*Aspergillus niger* *Penicillium* sp.	Otoni et al. (2014)
Cloves	Films formulated with carboxymethyl cellulose and polyvinyl alcohol made by solvent evaporation	*Staphylococcus aureus* *Bacillus cereus*	Muppalla et al. (2014)

stabilization of essential oil extracts and their antimicrobial activity when incorporated and proposed as active packaging.

3 Materials Carrying Antimicrobial Agents

Essentially, in active food packaging, two components can be identified: the antimicrobial agent and its carrier material. The carrier materials trigger and control the release rate of the active compound, and by their origin, they can be synthetic or natural polymers. The characteristics of these polymers determine the release rate of antimicrobial agents. To control the release of antimicrobial agents, various formulations of the various polymers have been manipulated. In addition, the porosity of the nanofiber meshes prepared by electrospinning favors the addition of higher concentrations of active components (Almasi et al. 2020).

Among the factors to consider when incorporating an active antimicrobial component in a material is the physical and chemical affinity between them, as well as the production process, distribution and costs of the release system (Chen et al. 2020a). The materials of the carriers favor and control the release of the antimicrobial components to the headspace of the container. Bags containing the antimicrobial system must be permeable to the active compound to modify the composition of the headspace of the packaging (Otoni et al. 2016).

Currently, as a basis for the production of antimicrobial active packaging, conventional synthetic materials and biopolymers have been used. There are many formulations of synthetic carriers, but natural polymers are preferred (Demitri et al. 2016). Among the biopolymers of protein nature are corn zein, casein, soy isolates, collagen, gelatin and wheat gluten. Additionally, polysaccharides such as alginates, chitosan, starch, and celluloses have been investigated for the same purpose.

Chitosan is a linear polysaccharide formed by D-glucosamine and N-acetyl-D-glucosamine molecules linked by β-(1,4) bonds (Escárcega-Galaz et al. 2018). In active packaging, its functionality is double; because of its antimicrobial activity, it is selected as an active compound and can participate as a carrier due to its biocompatibility, zero toxicity and antioxidant properties (Peng et al. 2013; Pereira et al. 2015). In addition, it has the property of film formation and low oxygen permeability (Kanatt et al. 2012). The antimicrobial activity of chitosan is mainly against yeasts, molds and Gram-positive bacteria and less against Gram-negative bacteria. The properties of chitosan depend on its molecular weight, degree of deacetylation and pH, with antimicrobial activity being better when the molecular weight is low (Lago et al. 2014). Chitosan films are flexible and resistant (Hosseini et al. 2013); however, their barrier and mechanical properties are limited. Additionally, chitosan is hydrophilic, so it is combined with other polymers (Sogut and Seydim 2018).

Cyclodextrins (CDs) are widely used for the development of microcapsules. Cyclodextrins have a toroid form and are cyclic oligosaccharides formed of alpha-D-(1,4) glucose. Its toroid structure has a hydrophilic outer surface and a hydrophobic hollow interior where it can trap lipophilic compounds. Regularly, six (α-CD), seven (β-CD) or eight (γ-CD) D-glucopyranose units are formed. β-CD is used in encapsulation processes to protect active compounds and increase their solubility (Cai et al. 2019).

Polylactic acid (PLA) obtained from corn is preferred for food applications due to its biodegradability. However, its application in food packaging is limited by its permeability to gas and vapor. To solve this problem, PLA has been mixed with other polymers (Huang et al. 2019). In addition, it is biocompatible, hydrophilic, nontoxic, soluble in water and chemically stable.

Starch is a reserve polymer in vegetables. The starch granules are formed from two polysaccharides, the linear structure called amylose and the branched structure called amylopectin. Starch is used for microencapsulation due to its high efficiency and low cost since it generates spherical porous aggregates (Esquivel-Chávez et al. 2021).

Gum arabic is an exudate of the acacia tree composed mainly of arabinogalactans and fractions of arabinogalactan protein and glycoprotein. The arabinogalactan protein forms hydrophobic blocks of the polypeptide chain and hydrophilic carbohydrates, determining its emulsifier property (Cai et al. 2019). Specifically, the polypeptide fractions are absorbed on the surface of the oil droplets, and the polysaccharide fractions protrude from the aqueous phase.

Polyvinyl alcohol (PVA) is a synthetic, emulsifying, adhesive, nontoxic and biodegradable polymer. This polymer is characterized by forming films and provides excellent mechanical properties when mixed with other materials. Its use has been reported in the packaging of meat and poultry (Kanatt et al. 2012; Pereira et al. 2015; Liu et al. 2017).

Polyethylene (PE) is one of the most commonly used polymers as a matrix in the manufacture of active packaging due to its low cost and high impact resistance (Al-Naamani et al. 2016). In particular, low-density polyethylene (LDPE) is used in food packaging due to its flexibility, transparency, easy processing and thermal stability. In addition, it is compatible with a large number of antimicrobial agents, such as natural extracts and silver nanoparticles (Jokar et al. 2012).

Cellulose is a structural polymer of plant cell walls. It is biodegradable, ecological and very abundant in corn cobs (Huang et al. 2019). Cellulose membranes are sensitive to water and expensive, which limits their applications.

4 Design of Antimicrobial Emitter Systems

Commercial emitters of antimicrobial agents use innovative technologies to design packaging that meets the specific needs of consumers and food processors. Table 2.2 describes recent research related to the design of antimicrobial emitters for food packaging.

Gaseous active agent release systems are applied to improve the quality and safety of fresh foods. Chen et al. (2020a) documented the applications, purposes and regulatory limitations of commercial antimicrobial releasers (sulfur dioxide,

Table 2.2 Research on the design of antimicrobial emitters for food packaging

Release system	Results	References
Nonwoven PET pads containing β-cyclodextrin and cinnamaldehyde complexes.	In fatty food simulants, cinnamaldehyde is released in a sustained and uniform manner. The release rate of cinnamaldehyde is different in each simulant.	Zhou et al. (2020)
Corrugated cardboard boxes with two-layer internal coating to release ClO_2. The first was a mixture of polyvinyl alcohol-$NaClO_2$ -diatomite and the second of chitosan in acetic acid.	The amount of ClO_2 gas released depends on the concentration of $NaClO_2$ in the cardboard and the diatomite stabilizes and prolongs the release of ClO_2.	Li et al. (2020)
Films based on LDPE in mixture with carvacrol or natural trans-cinnamaldehyde or encapsulated in β-cyclodextrins and compared with pure LDPE films.	The release of carvacrol is greater than that of trans-cinnamaldehyde, and both increase with time. In encapsulated agents, release is slow and controlled.	Canales et al. (2019)
PLA membranes added with microcapsules of ClO_2 and tartaric acid made by solvent melting.	The release of ClO_2 gas is stimulated by moisture. Membranes with 20% microcapsules release ClO_2 for up to 240 hours and inhibit *Escherichia coli* and *Staphylococcus aureus*	Huang et al. (2018)
Fibrous films of zein and polylactic acid in mixture with carvacrol obtained by electrospinning.	The release of carvacrol depends on the amount of carvacrol and the surface morphology of the fibers of both polymers.	Altan et al. (2018)
White gelled ethanol powder. Which is formed when the alcohol reacts with sodium stearate, and the gelled is mixed with diatomite to improve its mechanical strength.	The ethanol release process is of the first order and can be controlled by the concentration of sodium stearate.	Mu et al. (2017)
Chitosan films in mixture with graphene particles added with cinnamaldehyde.	When the films are in humid environments, the release of cinnamaldehyde is promoted and antifungal activity is favored.	Demitri et al. (2016)
Perforated film formed of two layers of LDPE with zeolite. Between the two layers there is a mixture of citric acid, sodium metabisulfite and calcium sulfite that when reacted produce sulfur dioxide.	SO_2 is released by reacting the water of respiration with the mixture based on sodium metabisulfite. SO_2 increases the storage life of grapes due to its antibacterial and antifungal activity.	Xu et al. (2013b)
Sachets with garlic oil encapsulated in beta-cyclodextrins	The relative humidity of 100% favors the release of volatile compounds up to 70% and reduces microbial growth.	Ayala-Zavala and González-Aguilar (2010)

chlorine dioxide and essential oils), growth regulators (1-methyl cyclopropene), and maturation agents (ethylene). Recently, Zhou et al. (2020) designed antimicrobial pads with sustained release using small amounts of additives and maintained their long-term effect.

Recently, Vasile and Baican (2021) pointed out that volatile release systems are applicable for essential oils and low molecular weight active compounds such as linalol, menthol and carvacrol. Additionally, the latter authors document the influence of external factors such as humidity and temperature on the emission of the active agent. Additionally, the intermolecular forces between the components of the carrier material and the active agent determine its release rate.

Otoni et al. (2016) analyzed various active sachet-type containers as releasers of materials with antimicrobial activity. In this context, they refer to two types of sachets, one of which generates and releases antimicrobial compounds. Other sachets transport and release antimicrobial agents.

To keep the quality and safety of food products, the release rate of the antimicrobial active agent must be considered. When the antimicrobial agent is released rapidly, its availability decreases to inhibit microbial growth on the surface of the food. In contrast, microorganisms can grow when the release rate of the antimicrobial agent is very slow (Altan et al. 2018).

According to Chen et al. (2020a), the release profiles of the active components stabilized in various materials can be manipulated as a function of time to increase the storage life of food products. Sometimes the release is required in a short time, while in other situations, the release is required for medium or long periods of time.

Sulfites have been used to prevent darkening and decomposition of fruits such as table grapes. Chen et al. (2016) studied the evolution of the quality of table grapes stored at low concentrations of SO_2 (0, 10 and 20 $cm^3/m^3$of space) and low temperatures (0, 10, 20 and 25 °C) to predict their shelf life based on the content of soluble solids and pH. However, sulfites in high concentrations are dangerous to human health.

Xing et al. (2011) encapsulated sulfite powders in ethylcellulose to study the SO_2 release profile of microparticles. The authors optimized the production process, characterized the microparticles and evaluated the release of SO_2 at relative humidity similar to the transport and storage conditions of the fruits. They found that the release rate accelerates with increasing relative humidity.

In recent decades, the development of packaging where SO_2 release is a function of moisture from fruit respiration and organic acids generated by fungi has increased. To control the release of SO_2, Xu et al. (2013a) formulated several flexible three-layer packages, an external polypropylene and an internal ethylene vinyl acetate (EVA) mixed with low-density polyethylene (LDPE). Thus, sodium metabisulfite and calcium sulfite were trapped in the intermediate layer so that water and organic acids gradually penetrated. With these designs, it was possible to release SO_2 for up to 50 days with ranges from 10 to 40 ppm.

Many commercial CO_2 emitters have a double function by emitting CO_2 and absorbing O_2, but CO_2 must be generated continuously to maintain the required concentration because it is very permeable in various plastic membranes (Haghighi-Manesh and Azizi 2017). To achieve this, sachets containing ferrous carbonate or a mixture with ascorbic acid and sodium bicarbonate have been prepared. However, there are fruits and vegetables that, due to their respiration rate, generate excessive amounts of CO_2, affecting their color and odor. To reduce these losses in quality,

CO_2 absorbers based on activated carbon or calcium hydroxide have been designed. At the commercial level, there are CO_2 emitters that are used in the form of absorbent pads and are mixtures of sodium bicarbonate and ascorbic acid or ferrous carbonate. In other commercial emitters, pads absorb exudates from food, and then these liquids react with sodium bicarbonate and citric acid to generate CO_2 (Majid et al. 2018).

Chen et al. (2020b) produced ClO_2 and α-cyclodextrin complexes by coprecipitation until optimizing the ratio of both components. The agglomerates formed were characterized, and ClO_2 release was evaluated by UV-vis spectrophotometry. They reported that ClO_2 oxidizes α-cyclodextrin, that its transport occurs by diffusion and that it is retained for up to three months at −20 °C. Because the ClO_2 release rate is slow, these agglomerates have potential applications in prolonged emission periods.

Abbasi et al. (2019) designed polymeric films of polyvinyl alcohol (PVA) or polyolefin (POD) loaded with sodium chlorite, which are activated with ultraviolet light to produce ClO_2 gas after contact with moisture. The authors found that the generation of ClO_2 is greater when small crystals of sodium chlorite are used. These films are a new class of food packaging materials because the amount of ClO_2 gas released is sufficient to inhibit pathogenic bacteria and improve the storage life of food products.

Bai et al. (2016) reported the release of ClO_2 gas from a two-component system. One of the acrylate films contains sodium chlorite, and the other film contains polyvinyl alcohol charged with tartaric acid. When both films are united, chlorite and acid react in a wet medium, thereby releasing ClO_2. The release of ClO_2 gas is rapid and can be controlled depending on the thickness and composition of the films. This system was proposed for use in food packaging.

The consumption of fresh foods has been associated with foodborne infections. Therefore, at the beginning of industrial processing, food is washed with aqueous chlorine to remove dirt and destroy pathogenic microorganisms. Saade et al. (2018) designed labels for containers made of alternating layers of gelatin with sodium chlorite and pectin with citric acid to produce chlorine dioxide (ClO_2). In the labels, the ClO_2 precursors were separated by multilayers to avoid gas loss and to control the release rate until the total consumption of the precursors. These labels generate sufficient amounts of ClO_2 gas to inactivate pathogenic bacteria in fresh packaged foods.

Essential oils are usually liquid at room temperature but are used as vapors due to their high volatility and easy release. To control their release, they must be stabilized by encapsulation, emulsification or extrusion (Chen et al. 2020a). The design of an essential oil release system in an antimicrobial packaging material should maximize the effect of the bioactive compounds of the oil (Tongnuanchan and Benjakul 2014).

Canales et al. (2019) added carvacrol or *trans*-cinnamaldehyde in LDPE-based films to assess their effect against *Botrytis cinerea,* which causes gray rot in grapes. Carvacrol or *trans*-cinnamaldehyde were used in their natural form or encapsulated in β-cyclodextrins. The release systems with added natural active agents showed that the fungal activity was 99%. When carvacrol or *trans*-cinnamaldehyde is encapsulated, the release rate is continuous and controlled. The authors confirm that their preparations have applications in the design of active packaging.

5 Applicability of Antimicrobial Emitters in Food

Active packaging technologies with antimicrobial emitters are designed to solve problems of deterioration of food quality. Thus, each food requires specific packaging conditions to prolong its edible quality and commercial value. Here, the food applications of the new proposals for active packaging and the experiments carried out by the researchers that support their effectiveness as antimicrobial emitters are documented.

5.1 *Sulfur Dioxide (SO_2) Emitters*

The most traditional practice for preserving table grapes is based on the release of SO_2 due to its activity against the growth of molds and aerobic bacteria. Chen et al. (2017) reported that 28 mg/m^3 SO_2 is required to control mold deterioration in grapes and twice as much SO_2 when stored at room temperature.

Grapes, like other fruits, are susceptible to microbial deterioration and quality loss from harvest to consumption. Therefore, Xu et al. (2013b) developed bilayer LDPE films filled with a combination of sodium metabisulfite, calcium sulfite and citric acid as SO_2 generators. In their test with grapes packed in perforated bags, they found that SO_2 was continuously released for 48 days. In addition, at 18 days of storage, the decomposition was only 5%, and the changes in the overall quality of the grapes were minimal.

At room temperature, strawberries have a storage life of less than 3 days. Fu et al. (2015) developed a multilayer active packaging as an SO_2 releaser as a fungicide to package strawberries and reduce their decomposition during storage. The outer layer was polypropylene, the central layer had sodium metabisulfite and calcium sulfite as SO_2 generators, and the inner layer had LDPE-EVA. From the tests, it was informed that the shelf life can be extended up to 8 days at 2 °C or up to 5 days at room temperature. SO_2 released from active packaging slows the development of fungi and slows the respiration of packaged fruit.

5.2 *Carbon Dioxide (CO_2) Emitters*

Hansen et al. (2016) packed fresh cod loins (*Gadus morhua*) under vacuum and in modified atmospheres complemented with a CO_2 emitter. The purpose of the investigation was to verify whether the CO_2 emitter compensates for the low availability of CO_2 in both containers and with little headspace. The initial freshness of the cod is best preserved in the tests where the CO_2 emitter was added. The storage life in the packages with modified atmospheres and the CO_2 emitter was 6 days longer than that in the vacuum and CO_2 emitter packages. At the end of storage life in all samples, *Photobacterium* predominated. The authors point out that the packaging of cod using CO_2 emitters improves its preservability with applicability in the retail market of fishery products.

Tsironi et al. (2019) studied and mathematically modeled the effect of packaging in modified atmospheres (50% CO_2/40% N_2/10% O_2) supplemented with CO_2 emitters on microbial growth and refrigerated storage life of gutted bass. The gutted bass were packed in low-density polyethylene bags in modified atmospheres with and without CO_2 emitters and then stored at 0, 5 and 10 °C. The growth of *Pseudomonas* spp. and *Enterobacteriaceae* spp. It was lower in the combination of a modified atmosphere with CO_2 emitters, where the CO_2 concentration stabilized at 60% and the shelf life was extended to 7 days.

5.3 *Chlorine Dioxide Emitters*

Fruits and vegetables need to be protected from contamination by microorganisms after harvest and during processing. To avoid foodborne diseases of microbial origin, safe active components should be applied in food. Based on this, Singh et al. (2020) reviewed the most recent research on the use of chlorine dioxide (ClO_2) in the packaging of fresh agricultural products. In the documents, they found that ClO_2 has antimicrobial effects against yeasts, fungi, *Escherichia coli*, *Listeria monocytogenes* and *Salmonella* spp., which grow on the surface of fresh meats, fruits and vegetables. In addition, they identified that ClO_2 gas has been applied in the conservation of strawberries, mangoes, spinach leaves, carrots, tomatoes, among other foods. Likewise, ClO_2 gas release packaging systems can be pads, perforated sachets or plastic films.

Spice contamination with *Salmonella enterica* can be the cause of foodborne diseases. Golden et al. (2019) studied the efficacy of three concentrations of chlorine dioxide gas (ClO_2) to reduce the number of *Salmonella* in black pepper, cumin and sesame seeds. During the test, envelopes were prepared from the commercial precursor of ClO_2. This consisted of sodium chlorite and an activator of ferric chloride. The number of *Salmonella* was reduced by increasing the concentration of ClO_2 gas and by increasing the storage time up to 30 days. In this test, the effects of ClO_2 gas on the volatile components of the spices were not studied.

Chlorine dioxide (ClO_2) is approved as an antimicrobial agent for application in fruits and vegetables. Saade et al. (2018) confirmed the efficacy of multilayer labels that release ClO_2 gas against *Salmonella*. In the test, mung bean seeds artificially inoculated with *Salmonella* were exposed to multilayer labels. All labels released ClO_2 gas in sufficient quantity for the complete inactivation of *Salmonella* cells.

Fresh walnuts (*Juglans regia* L.) are very susceptible to mold deterioration during storage. With this background, Ma et al. (2020) studied the effect of chlorine dioxide (ClO_2) and sodium diacetate during storage in controlled atmospheres and in the air of nuts. In the tests with air and ClO_2, the growth of molds in the shell and grains was delayed. The combination of controlled atmospheres plus ClO_2 gas was selected as optimal storage, finding a 5% mold incidence at 135 days.

Mangoes after harvesting exhibit yellowing and softening because they are climacteric fruits. Likewise, mangoes are susceptible to microbial contamination and cold storage. To preserve edible and market value, Zhang et al. (2019) packed mangoes in a PLA film coated with continuous release ClO_2 microcapsules. These antimicrobial films delayed the decomposition of the mango and increased its useful life while maintaining its physical and chemical indicators of quality. At the end of the test, the antimicrobial films showed gaps due to the release of ClO_2 that is generated when the humidity of the respiration activates the microcapsules and the tartaric acid.

Li et al. (2020) designed a corrugated cardboard box to package fresh strawberries. The boxes had two coating layers inside, one of them made with polyvinyl alcohol (PVA) mixed with $NaClO_2$ and diatomite and another of chitosan in acetic acid. ClO_2 gas is released when $NaClO_2$, water vapor and acetic acid react, which determines its potential as an antimicrobial container for fruits and vegetables. Mold growth was inhibited, and weight loss was reduced in strawberries packed in boxes with 9 g/L $NaClO_2$. In addition, the surface color, firmness and nutritional quality of fresh strawberries were preserved.

Chai et al. (2020) studied the generation of ClO_2 at the pilot level to decontaminate diced tomatoes, blueberries and carrots. Increasing the cumulative exposure to gas to 1600 ppm/h increased the efficacy against pathogens such as *Salmonella* and *Listeria monocytogenes.*

To disinfect minimally processed fruits and vegetables, chlorine dioxide (ClO_2) is recommended due to its broad antimicrobial spectrum. Based on this, Chiabrando et al. (2018) used ClO_2 emitters during the storage of strawberries. With ClO_2 gas, the quality parameters were preserved for short and prolonged periods of up to 8 days at 2 °C. Likewise, with the gas, the development of yeasts and molds was inhibited in the unwashed strawberries until the emitters were exhausted at 12 days of storage.

5.4 *Ethanol Emitters*

The availability of additives for the preservation of fruits and vegetables is very diverse. Among them, ethanol stands out for its broad-spectrum antimicrobial activity (Mu et al. 2017). However, its commercial applicability in antimicrobial packaging is limited by its volatility and uncontrolled release (Mu et al. 2017).

Latou et al. (2010) evaluated the effect of packaging with an ethanol emitter and an oxygen absorber on the shelf life of wheat bread slices. At 15 days of storage, in the breads with ethanol emitter, little development of molds and yeast was found due to the loss of ethanol through the packaging material. Meanwhile, at 15 days for the breads with the combination of ethanol emitter and oxygen absorber, no microbial growth was observed. For the test with the combination of both active packaging, the storage life was close to 30 days without microbiological or sensory changes.

Hempel et al. (2013) applied an intelligent technology combined with ethanol in ciabatta bread packed in 10% CO_2 and 90% N_2 and complemented it with an optical oxygen sensor. To provide the slow release of ethanol, commercial alcohol gel sachets were used, and for rapid vaporization, the breads were sprayed with ethanol. With this test, the authors found that ethanol emitters extend the storage life of flatbread up to 16 days without requiring modified atmospheres of gases that increase production costs for manufacturers of very prebaked bread products.

In reference to steamed Chinese bread, there is little research on its preservation. Sheng et al. (2015) studied the effect of an oxygen absorber in combination with an ethanol emitter on the quality and preservability of steamed Chinese bread at room temperature. The development of molds and yeasts is inhibited during storage, although by increasing the concentration of ethanol, unpleasant aromas and flavors are produced. This combination of active packaging is innovative and applicable to steamed Chinese bread.

To improve the storage life of packaged biscuits, Janjarasskul et al. (2016) used an O_2 scavenger and an ethanol emitter, both commercial. Their theory was that the ethanol emitter inhibits the reproduction of anaerobic microorganisms that proliferate when the O_2 scavenger decreases the O_2 concentration in the headspace. With this active packaging system, the storage life of the cake increased from 1 to 42 days without the direct addition of chemical preservatives. In addition, the plasticizing effect of ethanol on the proteins of the biscuit crumb slows its aging.

Mu et al. (2017) designed a controlled-rate gelled ethanol emitter to overcome the disadvantages of ethanol emitters. At the same time, they investigated it effects on the shelf life of Chinese blueberry fruits. With the emission of ethanol, it was possible to inhibit the increase in malondialdehyde, decrease the decomposition rate and maintain the firmness of the blueberry fruits. The increase in malondialdehyde content reveals damage to the structure of the plant tissue and accelerated senescence of the fruit.

5.5 *Essential Oil Emitters*

Ayala-Zavala and González-Aguilar (2010) studied the liberation of volatile compounds from encapsulated garlic oil and found that the most abundant volatile compound was allyl disulfide. In addition, increasing the concentration of encapsulated garlic oil stimulates the inhibition of the development of mesophilic bacteria, yeasts and molds in tomatoes.

Altan et al. (2018) encapsulated carvacrol in zein and polylactic acid to evaluate its effectiveness in the preservation of whole wheat bread. In a preliminary study, after 7 days of storage at 25 °C, it was possible to inhibit between 91.3% and 99.6% of molds and yeasts in whole wheat bread. Thus, the authors reveal that carvacrol films have antimicrobial and antioxidant properties to increase the storage life of bread, but it is necessary to evaluate their effect on organoleptic properties.

The mechanical properties of chitosan membranes with cinnamaldehyde can be modified by the addition of graphene nanoparticles. This was confirmed by Demitri et al. (2016), and the membranes were used to package slices of box bread. The authors found that only small amounts of cinnamaldehyde are required in the membranes to inhibit the natural propagation of the fungi.

Zhou et al. (2020) developed antibacterial pads with polyethylene terephthalate (PET) crosslinked with cinnamaldehyde and β-cyclodextrin to study its effect on the shelf life of fresh and cold pork meat. With the test, they showed that the pads inhibited the development of *Escherichia coli* and *Staphylococcus aureus,* prolonging the shelf life of fresh and cold pork. These antibacterial pads can be used as active packaging in meat products due to the sustained release of cinnamaldehyde.

To control the phytopathogens that cause the decomposition of mangoes (*Mangifera indica L.* Esquivel-Chávez et al. (2021) prepared nylon sacks with microcapsules of thyme oil (*Thymus vulgaris*). The essential oil was encapsulated using modified corn starch mixed with agave fructans by spray drying. Micellar growth was inhibited 100% for *Fusarium pseudocircinatum*, *Alternaria alternata*, *Neofusicocum kwambonambiense*, *Cladosporium pseudocladosporioides* and *Colletotrichum gloeosporioides* with 0.15 g of thyme oil in 6 days. Additionally, the effect of active packaging on the color, firmness and sensory characteristics of the mango was evaluated. The parameters of the color and firmness of the fruits were not affected by the use of active packaging. However, the smell of the mangoes was different after storage.

6 Conclusions

Considering that the main cause of food spoilage is attributed to the growth of microorganisms, an infinite of active packaging technologies has been developed based on the addition of natural or synthetic antimicrobial components. In this chapter, a preference was found to develop antimicrobial emitters based on essential oils due to the demand for naturally preserved products. In addition, research has focused on the use of mixtures of natural and non-natural polymers to make films, capsules and other emitter systems to stabilize the active agents. Likewise, real applications are being carried out that confirm the continuous release of antimicrobials during food storage. All this to bring the research achievements closer to the food industry needs. To counteract the lack of food, it is recommended to accelerate the investigations of these active containers, as well as to mass their production, marketing and application in foods with a high risk of deterioration. Additionally, it is important to promote the updating of regulations to ensure the safety of the active components and their possible interaction with the nutrients of each group of packaged foods.

Acknowledgments This investigation was sponsored by the Technological Institute of Sonora with the PROFAPI project (2021-0010).

References

Abbasi R, Jain R, Nelson K, Busche D, Lynn DM, Abbott NL. Polymeric films containing sodium chlorite that release disinfectant gas upon activation with UV light. Adv Funct Mater. 2019;29(7):1804851.

Akrami F, Rodríguez-Lafuente A, Bentayeb K, Pezo D, Ghalebi SR, Nerín C. Antioxidant and antimicrobial active paper based on Zataria (Zataria multiflora) and two cumin cultivars (Cuminum cyminum). LWT-Food Sci Technol. 2015;60(2):929–33.

Almasi H, Jahanbakhsh Oskouie M, Saleh A. A review on techniques utilized for design of controlled release food active packaging. Crit Rev Food Sci Nutr. 2020:1–21.

Al-Naamani L, Dobretsov S, Dutta J. Chitosan-zinc oxide nanoparticle composite coating for active food packaging applications. Innovative Food Sci Emerg Technol. 2016;38:231–7.

Altan A, Aytac Z, Uyar T. Carvacrol loaded electrospun fibrous films from zein and poly (lactic acid) for active food packaging. Food Hydrocoll. 2018;81:48–59.

Asdagh A, Sani IK, Pirsa S, Amiri S, Shariatifar N, Eghbaljoo-Gharehgheshlaghi H, et al. Production and characterization of nanocomposite film based on whey protein isolated/copper oxide nanoparticles containing coconut essential oil and paprika extract. J Polym Environ. 2021;29(1):335–49.

Atarés L, Chiralt A. Essential oils as additives in biodegradable films and coatings for active food packaging. Trends Food Sci Technol. 2016;48:51–62.

Ayala-Zavala JF, González-Aguilar GA. Optimizing the use of garlic oil as antimicrobial agent on fresh-cut tomato through a controlled release system. J Food Sci. 2010;75(7):M398–405.

Ayala-Zavala JF, del Toro-Sánchez L, Alvarez-Parrilla E, Soto-Valdez H, Martín-Belloso O, Ruiz-Cruz S, González-Aguilar GA. Natural antimicrobial agents incorporated in active packaging to preserve the quality of fresh fruits and vegetables. Stewart Postharvest Rev. 2008;4(3):1–9.

Bai Z, Cristancho DE, Rachford AA, Reder AL, Williamson A, Grzesiak AL. Controlled release of antimicrobial ClO_2 gas from a two-layer polymeric film system. J Agric Food Chem. 2016;64(45):8647–52.

Becerril R, Gómez-Lus R, Goni P, López P, Nerín C. Combination of analytical and microbiological techniques to study the antimicrobial activity of a new active food packaging containing cinnamon or oregano against E. coli and S. aureus. Anal Bioanal Chem. 2007;388(5–6):1003–11.

Butnaru E, Stoleru E, Brebu MA, Darie-Nita RN, Bargan A, Vasile C. Chitosan-based bionanocomposite films prepared by emulsion technique for food preservation. Materials. 2019;12(3):373.

Cai C, Ma R, Duan M, Lu D. Preparation and antimicrobial activity of thyme essential oil microcapsules prepared with gum arabic. RSC Adv. 2019;9(34):19740–7.

Canales D, Montoille L, Rivas LM, Ortiz JA, Yañez-S M, Rabagliati FM, et al. Fungicides films of low-density polyethylene (LDPE)/inclusion complexes (Carvacrol and Cinnamaldehyde) against Botrytis Cinerea. Coatings. 2019;9(12):795.

Chai HE, Hwang CA, Huang L, Wu VC, Sheen LY. Feasibility and efficacy of using gaseous chlorine dioxide generated by sodium chlorite-acid reaction for decontamination of foodborne pathogens on produce. Food Control. 2020;108:106839.

Chen X, Mu W, Peter S, Zhang X, Zhu Z. The effects of constant concentrations of sulfur dioxide on the quality evolution of postharvest table grapes. J Food Nutr Res. 2016;55(2)

Chen X, Zhu Z, Zhang X, Oana Antoce A, Mu W. Modeling the microbiological shelf life of table grapes and evaluating the effects of constant concentrations of sulfur dioxide. J Food Process Preserv. 2017;41(4):e13058.

Chen M, Chen X, Ray S, Yam K. Stabilization and controlled release of gaseous/volatile active compounds to improve safety and quality of fresh produce. Trends Food Sci Technol. 2020a;95:33–44.

Chen M, Chen X, Yam K. Encapsulation complex of chlorine dioxide in α-cyclodextrin: structure characterization and release property. Food Control. 2020b;107:106783.

Chiabrando V, Giuggioli N, Maghenzani M, Peano C, Giacalone G. Improving storability of strawberries with gaseous chlorine dioxide in perforated clamshell packaging. Polish J Food Nutr Sci. 2018;68(2)

Dantigny P, Guilmart A, Radoi F, Bensoussan M, Zwietering M. Modeling the effect of ethanol on growth rate of food spoilage molds. Int J Food Microbiol. 2005;98(3):261–9.

Day BPF. Active packaging. In: Coles R, McDowell D, Kirwan M, editors. Food packaging technologies. Boca Raton. ISBN: 9781444392180: CRC Press; 2003. p. 282–302.

Demitri C, De Benedictis VM, Madaghiele M, Corcione CE, Maffezzoli A. Nanostructured active chitosan-based films for food packaging applications: Effect of graphene stacks on mechanical properties. Measurement. 2016;90:418–23.

Díaz-Galindo EP, Nesic A, Bautista-Baños S, Dublan García O, Cabrera-Barjas G. Corn-starch-based materials incorporated with cinnamon oil emulsion: physico-chemical characterization and biological activity. Foods. 2020;9(4):475.

Dong Z, Xu F, Ahmed I, Li Z, Lin H. Characterization and preservation performance of active polyethylene films containing rosemary and cinnamon essential oils for Pacific white shrimp packaging. Food Control. 2018;92:37–46.

Escárcega-Galaz AA, Sánchez-Machado DI, López-Cervantes J, Sanches-Silva A, Madera-Santana TJ, Paseiro-Losada P. Mechanical, structural and physical aspects of chitosan-based films as antimicrobial dressings. Int J Biol Macromol. 2018;116:472–81.

Esquivel-Chávez F, Colín-Chávez C, Virgen-Ortiz JJ, Martínez-Téllez MÁ, de Jesús Avena-Bustillos R, Peña-Madrigal G, Miranda-Ackerman MA. Control of mango decay using antifungal sachets containing of thyme oil/modified starch/agave fructans microcapsules. Future Foods. 2021;3:100008.

Fu YB, Xu WC, Fu L, Li DL, Wang JQ. Preparation of an active packaging material and its application on strawberry preservation. In: Advanced materials research, vol. 1120. Trans Tech Publications Ltd; 2015. p. 643–7.

Golden CE, Berrang ME, Kerr WL, Harrison MA. Slow-release chlorine dioxide gas treatment as a means to reduce Salmonella contamination on spices. Innovative Food Sci Emerg Technol. 2019;52:256–61.

Haghighi-Manesh S, Azizi MH. Active packaging systems with emphasis on its applications in dairy products. J Food Process Eng. 2017;40(5):e12542.

Hansen AÅ, Moen B, Rødbotten M, Berget I, Pettersen MK. Effect of vacuum or modified atmosphere packaging (MAP) in combination with a CO_2 emitter on quality parameters of cod loins (Gadus morhua). Food Packag Shelf Life. 2016;9:29–37.

Hempel AW, O'Sullivan MG, Papkovsky DB, Kerry JP. Use of smart packaging technologies for monitoring and extending the shelf-life quality of modified atmosphere packaged (MAP) bread: application of intelligent oxygen sensors and active ethanol emitters. Eur Food Res Technol. 2013;237(2):117–24.

Higueras L, López-Carballo G, Hernández-Muñoz P, Catalá R, Gavara R. Antimicrobial packaging of chicken fillets based on the release of carvacrol from chitosan/cyclodextrin films. Int J Food Microbiol. 2014;188:53–9.

Hosseini SF, Rezaei M, Zandi M, Ghavi FF. Preparation and functional properties of fish gelatin–chitosan blend edible films. Food Chem. 2013;136(3-4):1490–5.

Huang C, Zhang B, Wang S, Zhang L, Wang J, Huang X, et al. Moisture-triggered release of self-produced ClO_2 gas from microcapsule antibacterial film system. J Mater Sci. 2018;53(18):12704–17.

Huang T, Qian Y, Wei J, Zhou C. Polymeric antimicrobial food packaging and its applications. Polymers. 2019;11(3):560.

Janjarasskul T, Tananuwong K, Kongpensook V, Tantratian S, Kokpol S. Shelf life extension of sponge cake by active packaging as an alternative to direct addition of chemical preservatives. LWT-Food Sci Technol. 2016;72:166–74.

Jokar M, Rahman RA, Ibrahim NA, Abdullah LC, Tan CP. Melt production and antimicrobial efficiency of low-density polyethylene (LDPE)-silver nanocomposite film. Food Bioprocess Technol. 2012;5(2):719–28.

Kanatt SR, Rao MS, Chawla SP, Sharma A. Active chitosan–polyvinyl alcohol films with natural extracts. Food Hydrocoll. 2012;29(2):290–7.

Lago MA, Sendón R, de Quirós ARB, Sanches-Silva A, Costa HS, Sánchez-Machado DI, et al. Preparation and characterization of antimicrobial films based on chitosan for active food packaging applications. Food Bioprocess Technol. 2014;7(10):2932–41.

Latou E, Mexis SF, Badeka AV, Kontominas MG. Shelf life extension of sliced wheat bread using either an ethanol emitter or an ethanol emitter combined with an oxygen absorber as alternatives to chemical preservatives. J Cereal Sci. 2010;52(3):457–65.

Li Y, Ren D, Xu D. Preparation of coated corrugated box for controlled-release of chlorine dioxide and its application in strawberry preservation. Coatings. 2020;10(3):242.

Liakos IL, Holban AM, Carzino R, Lauciello S, Grumezescu AM. Electrospun fiber pads of cellulose acetate and essential oils with antimicrobial activity. Nanomaterials. 2017;7(4):84.

Liu Y, Wang S, Lan W, Qin W. Fabrication and testing of PVA/Chitosan bilayer films for strawberry packaging. Coatings. 2017;7(8):109.

Llana-Ruiz-Cabello M, Pichardo S, Maisanaba S, Puerto M, Prieto AI, Gutierrez-Praena D, et al. In vitro toxicological evaluation of essential oils and their main compounds used in active food packaging: a review. Food Chem Toxicol. 2015;81:9–27.

Lucera A, Conte A, Del Nobile MA. Volatile compounds usage in active packaging systems. Antimicrob Food Packag. 2016:319–27.

Ma Y, Li P, Watkins CB, Ye N, Jing N, Ma H, Zhang T. Chlorine dioxide and sodium diacetate treatments in controlled atmospheres retard mold incidence and maintain quality of fresh walnuts during cold storage. Postharvest Biol Technol. 2020;161:111063.

Majid I, Thakur M, Nanda V. Innovative and safe packaging technologies for food and beverages: updated review. In: Innovations in technologies for fermented food and beverage industries. Springer; 2018. p. 257–87.

Mlalila N, Hilonga A, Swai H, Devlieghere F, Ragaert P. Antimicrobial packaging based on starch, poly (3-hydroxybutyrate) and poly (lactic-coglycolide) materials and application challenges. Trends Food Sci Technol. 2018;74:1–11.

Mu H, Gao H, Chen H, Fang X, Han Q. A novel controlled release ethanol emitter: preparation and effect on some postharvest quality parameters of Chinese bayberry during storage. J Sci Food Agric. 2017;97(14):4929–36.

Muppalla SR, Kanatt SR, Chawla SP, Sharma A. Carboxymethyl cellulose–polyvinyl alcohol films with clove oil for active packaging of ground chicken meat. Food Packag Shelf Life. 2014;2(2):51–8.

Noshirvani N, Ghanbarzadeh B, Gardrat C, Rezaei MR, Hashemi M, Le Coz C, Coma V. Cinnamon and ginger essential oils to improve antifungal, physical and mechanical properties of chitosan-carboxymethyl cellulose films. Food Hydrocoll. 2017;70:36–45.

Orsuwan A, Sothornvit R. Active banana flour nanocomposite films incorporated with garlic essential oil as multifunctional packaging material for food application. Food Bioprocess Technol. 2018;11(6):1199–210.

Otoni CG, Pontes SF, Medeiros EA, Soares NDF. Edible films from methylcellulose and nanoemulsions of clove bud (Syzygium aromaticum) and oregano (Origanum vulgare) essential oils as shelf life extenders for sliced bread. J Agric Food Chem. 2014;62(22):5214–9.

Otoni CG, Espitia PJ, Avena-Bustillos RJ, McHugh TH. Trends in antimicrobial food packaging systems: emitting sachets and absorbent pads. Food Res Int. 2016;83:60–73.

Pasqualone A. Bread packaging: features and functions. In: Flour and breads and their fortification in health and disease prevention. Academic; 2019. p. 211–22.
Peng Y, Wu Y, Li Y. Development of tea extracts and chitosan composite films for active packaging materials. Int J Biol Macromol. 2013;59:282–9.
Pereira de Abreu DA, Cruz JM, Paseiro Losada P. Active and intelligent packaging for the food industry. Food Rev Int. 2012;28(2):146–87.
Pereira VA Jr, de Arruda INQ, Stefani R. Active chitosan/PVA films with anthocyanins from Brassica oleraceae (Red Cabbage) as time–temperature indicators for application in intelligent food packaging. Food Hydrocoll. 2015;43:180–8.
Ramos M, Jiménez A, Peltzer M, Garrigós MC. Characterization and antimicrobial activity studies of polypropylene films with carvacrol and thymol for active packaging. J Food Eng. 2012;109(3):513–9.
Ribeiro-Santos R, Andrade M, de Melo NR, Sanches-Silva A. Use of essential oils in active food packaging: recent advances and future trends. Trends Food Sci Technol. 2017;61:132–40.
Saade C, Annous BA, Gualtieri AJ, Schaich KM, Liu L, Yam KL. System feasibility: designing a chlorine dioxide self-generating package label to improve fresh produce safety part II: solution casting approach. Innovative Food Sci Emerg Technol. 2018;47:110–9.
Sheng Q, Guo XN, Zhu KX. The effect of active packaging on microbial stability and quality of Chinese steamed bread. Packag Technol Sci. 2015;28(9):775–87.
Singh S, Maji PK, Lee YS, Gaikwad KK. Applications of gaseous chlorine dioxide for antimicrobial food packaging: a review. Environ Chem Lett. 2020:1–18.
Sogut E, Seydim AC. Development of chitosan and polycaprolactone based active bilayer films enhanced with nanocellulose and grape seed extract. Carbohydr Polym. 2018;195:180–8.
Souza MP, Vaz AF, Silva HD, Cerqueira MA, Vicente AA, Carneiro-da-Cunha MG. Development and characterization of an active chitosan-based film containing quercetin. Food Bioprocess Technol. 2015;8(11):2183–91.
Tongnuanchan P, Benjakul S. Essential oils: extraction, bioactivities, and their uses for food preservation. J Food Sci. 2014;79(7):R1231–49.
Tsironi T, Ntzimani A, Gogou E, Tsevdou M, Semenoglou I, Dermesonlouoglou E, Taoukis P. Modeling the effect of active modified atmosphere packaging on the microbial stability and shelf life of gutted sea bass. Appl Sci. 2019;9(23):5019.
Vasile C, Baican M. Progresses in food packaging, food quality, and safety – controlled-release antioxidant and/or antimicrobial packaging. Molecules. 2021;26(5):1263.
Wyrwa J, Barska A. Innovations in the food packaging market: active packaging. Eur Food Res Technol. 2017;243(10):1681–92.
Xing YG, Yun J, Li XH, Xu QL, Li WL. The effect of formulation variables on the encapsulation efficiency and SO_2-release behavior of microparticles containing sulphite. In: Advanced materials research, vol. 152. Trans Tech Publications Ltd; 2011. p. 512–5.
Xu W, Li D, Fu Y, Wei H. Preparation and measurement of controlled-release SO_2 fungicide active packaging at room temperature. Packag Technol Sci. 2013a;26:51–8.
Xu W, Li D, Fu Y, Liu Z, Wang Y, Yu X, Shang W. Extending the shelf life of Victoria table grapes by high permeability and fungicide packaging at room temperature. Packag Technol Sci. 2013b;26:43–50.
Zanetti M, Carniel TK, Dalcanton F, dos Anjos RS, Riella HG, de Araujo PH, et al. Use of encapsulated natural compounds as antimicrobial additives in food packaging: a brief review. Trends Food Sci Technol. 2018;81:51–60.
Zemljič LF, Tkavc T, Vesel A, Šauperl O. Chitosan coatings onto polyethylene terephthalate for the development of potential active packaging material. Appl Surf Sci. 2013;265:697–703.
Zhang B, Huang C, Zhang L, Wang J, Huang X, Zhao Y, et al. Application of chlorine dioxide microcapsule sustained-release antibacterial films for preservation of mangos. J Food Sci Technol. 2019;56(3):1095–103.
Zhou Z, Liu Y, Liu Z, Fan L, Dong T, Jin Y, et al. Sustained-release antibacterial pads based on nonwovens polyethylene terephthalate modified by β-cyclodextrin embedded with cinnamaldehyde for cold fresh pork preservation. Food Packag Shelf Life. 2020;26:100554.

Chapter 3
Emitters of Antioxidants (With Special Focus on Natural Antioxidants)

M. D. Celiz, R. Paseiro-Cerrato, L. DeJager, and T. H. Begley

Abstract One of the main functions of food packaging materials (FPM) is to extend food shelf life. FPM act as a barrier against agents such as microorganisms, oxygen or light that may deteriorate food. Preventing food exposure to these external agents not only results to extended shelf life but also decreases food waste. Antioxidants prevent or reduce the actions of oxygen and other radicals in the food packaging as well as food. In addition, they are one of the most used additives by the industry. In this book chapter, we will review and discuss the use of antioxidants in food and food packaging and antioxidants related to active packaging. An introduction related to FPM and antioxidants, types of antioxidants, uses, methods for analysis and most common used analytical techniques (e.g. HPLC-MS) employed for their determination will be presented and discussed in the chapter.

Keywords Antioxidants · Food packaging materials · Active packaging

1 Introduction

Food contact materials (FCMs) have been an essential part in the development of mankind from a social, cultural, and even from an evolutionary point of view. FCMs can be defined as any material (such as glass, paper, plastic, metal, and ceramics) that are intended to be in contact with food. They can be used for manipulating, transporting, storing, and preserving food. Because food is essential for humans, it is not surprising that humans have been linked to FCMs since the beginning of time.

M. D. Celiz · R. Paseiro-Cerrato (✉) · L. DeJager · T. H. Begley
US FDA, Center for Food Safety and Applied Nutrition, College Park, MD, USA
e-mail: MaryDawn.Celiz@fda.hhs.gov; Rafael.Cerrato@fda.hhs.gov; Lowri.DeJager@fda.hhs.gov; Timothy.Begley@fda.hhs.gov

S. M. Jafari, A. S. Silva (eds.), *Releasing Systems in Active Food Packaging*, Food Bioactive Ingredients, https://doi.org/10.1007/978-3-030-90299-5_3

In the beginning, FCMs were made with materials easily found in nature, such as stone and wood. These were mainly used for food handling such as cutting and/or cooking. Later in time, with the discovery of ceramics, glass, and metals, new FCMs started being used. These FCMs had new characteristics that allowed to facilitate storage and transportation more efficiently. A clear example of this are the Phoenician and Roman ceramic vessels found throughout the Mediterranean Sea region. However, it was not until the discovery of synthetic polymers in the nineteenth century that FCMs took a significant leap forward in terms of its utility, particularly in food shelf life improvements. Among FCMs, food packaging materials (FPM) may have contributed the most to this development.

Most of the FPM used today are made of polymeric materials. Polymeric materials, such as plastics and coatings, have several characteristics that make them unique: they can be produced at low cost, a large volume can be manufactured in a reasonable amount of time, they are lightweight materials making the transport of food easier and more affordable, and they take up little space so it increases the amount of food that can be transported. In addition, they are resistant to deterioration when exposed to external stresses such as mechanical impacts, moisture, and UV light.

Perhaps, the most important characteristic is from the food preservation standpoint; polymers protect food from environmental conditions that contribute to food degradation such as oxygen, moisture, light, and microorganisms. This increases food shelf life and reduces food waste. It is not only the packaging that protects the food from external agents and increases shelf life. There is a group of compounds known as additives, that have different properties to contribute to food protection and to enhance food shelf life. Because of advances in food packaging and food additives, food can now be transported everywhere in the world (even to space), ensuring that everyone can have access to it. Some of the advantages of having access to any type of food is that people can achieve a balanced diet, enhancing health and therefore increasing life expectancy.

In the last quarter of the twentieth century, a new type of FCM with superior functions started being developed, the active packaging. Active packaging is designed to actively interact with its content by increasing and maintaining food shelf life (Otero-Pazos et al. 2014). They are also used to improve safety and enhance sensory properties (Koontz et al. 2010; de et al. 2002). These are achieved by actively releasing and/or absorbing substances in the packaged food. The field of active packaging is still under development, but every year more active materials are investigated, making it of interest to the food industry as well as regulatory agencies.

In this brief book chapter, we will focus on antioxidants, a group of substances commonly used in active packaging. In particular, we will focus on natural antioxidants and review methods to analyze antioxidants and antioxidant activity of antioxidants used in active packaging. The main goal is to make the reader familiar with the basics of this type of substances that maybe used in the development of active packaging.

2 Active Packaging

(A) *Types*

Active packaging is designed to increase food shelf life, improve sensory properties and safety of food and beverages (de et al. 2002; Koontz et al. 2010). This is achieved by releasing active ingredients and /or absorbing undesirable compounds from the packed container. Based on these properties, active packaging can be divided in two main types:

- Emitters: This type of active packaging releases active substances inside the packaging. An appropriate amount of substance is actively released to impart a desired technical effect such as minimizing food deterioration. Examples of substances that can be released by this active material are antioxidants, antimicrobials, colorants, and aromas.
- Absorbers or scavengers: Another type of active packaging that modifies the environment inside the packaging by absorbing substances (e.g. oxygen, free radicals, and moisture) that may cause, for instance, food degradation.

Both types of active packaging contribute not only to protect food from external agents such as microorganisms, but actively contribute in preserving food organoleptic properties (color, flavor, aroma), improving safety of the products and increasing food shelf life.

(B) *Manufacturing*

Several types of polymers have been used in the investigation of active materials. Examples of employed polymeric materials are low- and high-density polyethylene films (LDPE and HDPE, respectively), ethylene vinyl alcohol (EVOH) and polypropylene (PP) among others (Lin et al. 2016; Sanches-Silva et al. 2014b; Cerisuelo et al. 2014; Soares and Hotchkiss 1998; Goddard and Hotchkiss 2008; Goddard et al. 2007). Non-petroleum based biodegradable films are also being investigated for their use in active packaging. A common example of biodegradable films are the ones based on polylactic acid (PLA) (Gontard et al. 2011; Otero-Pazos et al. 2016; Jamshidian et al. 2013; Bahrami et al. 2020).

There are three main procedures where active ingredients can be incorporated into active materials such as films. One is by a *casting procedure*, where the active ingredient is dissolved in a solvent with the polymer. Another process is an *extrusion or blending procedure*, where the active ingredient is incorporated during the production of the polymer at high temperatures. Thirdly, a *coating procedure*, where the active ingredient is incorporated on the surface of the film. A more in depth explanation of these procedures can be found in an extensive review written by Dominguez et al. (2018).

There are several challenges related with the manufacture of active packaging. One is the incorporation of the active ingredient in the packaging. Some ingredients may not be compatible with the packaging components, resulting in poor performance of the packaging. Issues related with the degradation of the active ingredients during the processing can also occur, which may result in lower concentration in the film (Gomez-Estaca et al. 2014; Nerin 2010). Taking these processes into account, manufacturers could design and engineer their materials to achieve their desired final product.

(C) *Legislation*

In order to introduce active materials in the market, manufacturers must follow national legislations. Several countries have regulations on FCMs, food ingredients and food contact substances (FCS). In the United States (US), they are regulated under the Federal Food, Drug and Cosmetic Act, which allows US Food and Drug Administration (US FDA) to pre-approve food additives before being placed on the market. A food additive is defined as "any substance the intended use of which results or may reasonably be expected to result – directly or indirectly – in its becoming a component or otherwise affecting the characteristics of any food." There are exclusions to this term such as substances that are generally recognized as safe (GRAS), color additives, and substances approved for their intended use prior 1958. Briefly, there are three processes for receiving approval from FDA: the food additive petition process, the food contact notification process and the threshold of regulation exemption request. In addition, all manufacturers must follow good manufacturing practices (21 CFR Parts 100-169, 21 CFR Parts 170–199). A more in-depth description related to this topic was described by other authors (Song and Hepp 2005; Bailey et al. 2008). More information can be found on the FDA website (www.fda.gov).

In countries that integrate the European Union, food additives and FCM are regulated at a European level. Food additives are regulated under several regulations, the most relevant ones being EC No 1331/2008, 1333/2008 and 231/2012. FCMs are regulated under the framework regulation No 1935/2004 and in the case of plastic FCM, they have the specific regulation EC 10/2011. Active materials have a specific regulation which is EC No 450/2009 on active and intelligent materials and articles intended to come into contact with food. Besides, producers must also comply with Regulation (EC) 2023/2006 on good manufacturing practices. Following the national legislations is a requirement to ensure safety of any food products as well as to protect consumers heath.

3 Antioxidants

Food additives can be used to protect food from undesirable agents such as oxidation and microorganisms. In addition, they may be used to impart certain properties that may help in manufacturing, and in maintaining and enhancing food organoleptic properties. In this chapter, we are going to focus on antioxidants that can be used in active packaging. Antioxidants are substances that prevent food from oxidation. Oxidation is a process generated by oxygen and free radicals, which reacts with a substrate (food) resulting to food deterioration, generating rancidity, off flavors and off odors, and enhancing microorganisms growth (Pereira de Abreu et al. 2012; Otero-Pazos et al. 2014; Vera et al. 2018; Ozdemir and Floros 2004). The degree of oxidation depends on multiple factors, such as type of matrix (e.g. food), amount of free radicals generated, and presence of agents that promote oxidation or radical production (ex. UV light). A substance that interferes with the oxidation process, for example, reacting with free radicals or removing oxygen, could be considered an antioxidant.

Antioxidants may be classified in different ways. A way of classifying antioxidants is according to their mechanism of action. The antioxidants can be grouped into radical scavengers, peroxide decomposers, quenchers of oxygen singlets, antioxidants that avoid enzymatic activity and inhibitors of oxidations (Nerin 2010). But antioxidants can also be classified attending to their origins:

– Synthetic antioxidants: These are manufactured through synthetic ways. They are widely used because of its high efficacy and low cost. A classic example of a synthetic antioxidant is butylated hydroxytoluene or BHT.
– Natural antioxidants: Natural antioxidants can be obtained from natural products such as plant extracts (i.e. rosemary extract) or seafood (i.e. shrimp) among others (Otero-Pazos et al. 2014; Tovar et al. 2005; Sanches-Silva et al. 2012, 2014b; Bentayeb et al. 2007; Barbosa-Pereira et al. 2014; Gomez-Estaca et al. 2014; Ganiari et al. 2017). Several natural antioxidants have been investigated for use in active packaging. Some examples of natural antioxidants are α- tocopherol, caffeic acid and gallic acid (Fig. 3.1).

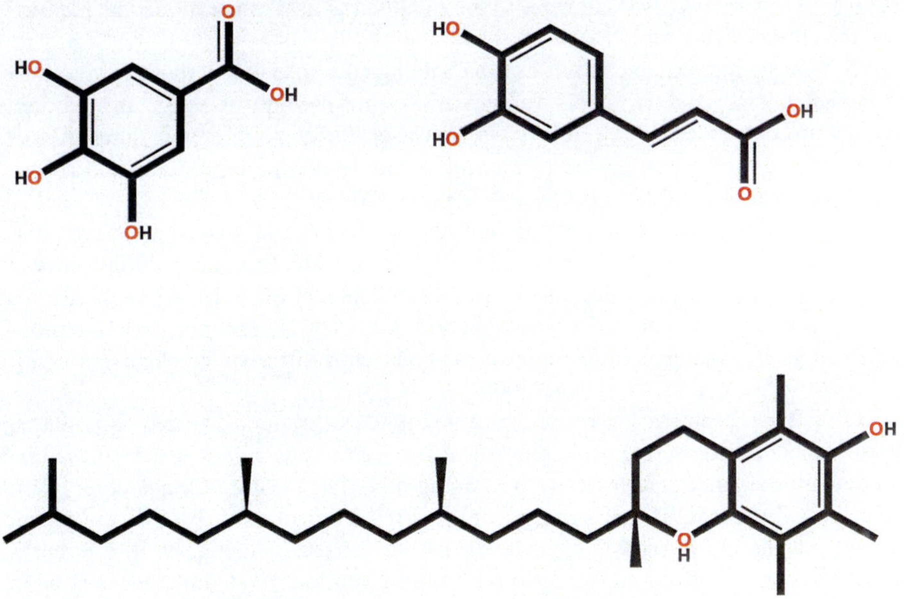

Fig. 3.1 Chemical structure of some common natural antioxidants. Upper left-hand side: Gallic acid; upper right-hand side: Caffeic acid; Bottom: α-tocopherol

In this review chapter, we will focus only on natural antioxidants. Evaluating the activity of these antioxidants and having a good knowledge of the analytical techniques employed in the antioxidant determination are relevant in having a better understanding in the antioxidants uses and applications as well as in their analysis not only in active packaging, but also in other fields of food safety. In the next section, we will focus on some of the main methods to analyze antioxidants in active packaging.

4 Methods for the Analysis of Antioxidants Used in Active Packaging

The purpose of this section is to present to the readers the state-of-the-art review of methods used to determine antioxidant activity of active packaging incorporated with natural antioxidants as emitters.

Different methods are available to evaluate the effectiveness of an antioxidant in an active packaging to protect food from oxidative reactions (Gómez-Estaca et al. 2014; Sanches-Silva et al. 2014a). The methods report the results as antioxidant activity or antioxidant capacity of the material. Antioxidant activity measures the kinetics of the reaction of the antioxidant while antioxidant capacity measures thermodynamic conversion efficiency of reactive species by antioxidants (Apak et al. 2013). However, these two terms have been used interchangeably in the literature. Hence, in this review, we will refer to the methods or assays mentioned as measuring the antioxidant activity.

Comparing antioxidant activities is a challenge because the methods have different mechanisms, and protocols vary for the same method or assay. In addition, results are expressed in different ways and used different reference standards. A standardized method is needed to provide a way to compare products and follow appropriate protocol for the product (Prior et al. 2005).

The chemistries of each method, and its advantages and disadvantages are discussed in different review articles (Antolovich et al. 2002; Prior et al. 2005; Amorati and Valgimigli 2015; Shahidi and Zhong 2015; Apak et al. 2016; Apak 2019). No universal method can show the whole picture of the antioxidant property of a product; hence, it is suggested to employ methods with different mechanisms (Prior et al. 2005; Moon and Shibamoto 2009).

Table 3.1 presents a list of natural compounds or materials tested as potential antioxidants for the active packaging since 2017. Previous review articles have discussed antioxidants used in active packaging materials (Gómez-Estaca et al. 2014; Sanches-Silva et al. 2014a; Domínguez et al. 2018; Yildirim et al. 2018). Antioxidant activity of the active packaging material can be measured using the film directly, film extract by solid-liquid extraction or film dissolution, food simulant, or food in contact with the active packaging material.

Table 3.1 List of natural antioxidants used in active packaging since 2017

Antioxidant	Packaging material	Method of incorporation of antioxidant	Antioxidant activity method/assay[a]	Material tested for antioxidant activity	Food application	References
α-tocopherol	Poly(ε-Caprolactone)	Melt blending and compression moulding	DPPH, ABTS	Food simulant (50% ethanol)	None	Mellinas et al. (2020)
α-tocopherol	Poly(vinyl alcohol)	Casting	DPPH, ABTS	Film extract of 95% ethanol	None	de Carvalho et al. (2019)
α-tocopherol	LDPE	Extrusion	DPPH	Film extract of ethanol	None	Sun et al. (2017)
Apple peel ethanolic extract	Chitosan and gelatin	Casting	DPPH, ABTS	Film	None	Riaz et al. (2020)
Bearberry leaf extract	Multilayer PET/adhesive with antioxidant/LDPE	Coated then laminated	In situ gas-phase hydroxyl radical generation	Film	None	Wrona et al. (2019)
Caffeic acid	Chitosan and gelatin	Casting	DPPH, iron chelating activity, reducing power assay	Film	None	Benbettaieb et al. (2018)
Carob fruit extract	Oriented polyamide/ polyethylene	Coating	Peroxide value, TBARS	Food	Salmon	Goulas et al. (2019)
Chlorophyll	Wheat gluten	Coating	DPPH, peroxide value	Film extract of ethanol and food	Sesame oil	Chavoshizadeh et al. (2020)
Cocoa bean shells	Poly(lactic acid)	Casting	DPPH	Film	None	Papadopoulou et al. (2019)
Curcumin	Carboxymethyl cellulose	Casting	DPPH, ABTS	Film	None	Roy and Rhim (2020a)
Curcumin	Gelatin	Casting	DPPH, ABTS	Film	None	Roy and Rhim (2020b)
Curcumin	Cellulose nanofibril-oil	Casting	DPPH	Film	None	Valencia et al. (2019)

(continued)

Table 3.1 (continued)

Antioxidant	Packaging material	Method of incorporation of antioxidant	Antioxidant activity method/assay[a]	Material tested for antioxidant activity	Food application	References
Curcumin	LDPE	Extrusion and compression moulding	DPPH	Film extract of ethanol	None	Zia et al. (2019)
Curcumin	Bacterial cellulose nanofiber and chitin nanofiber	Antisolvent Precipitation	DPPH, ABTS	Film	None	Yang et al. (2020)
Fungal melanin	Poly(lactic acid)	Extrusion	DPPH, ABTS, Folin-Ciocalteu	Film	None	Łopusiewicz et al. (2018)
Gallic acid	Lentil flour/polyethylene oxide (PEO) nanofibers on poly(lactic acid)	Electrospinning	Peroxide value, TBARS, p-anisidine, Totox value	Food	Walnuts	Aydogdu et al. (2019)
Gallic acid and quercetin	Poly(vinyl alcohol)	Casting	DPPH	Film and food simulant (50% ethanol)	None	Luzi et al. (2019)
Gallic acid and umbelliferone	Poly(vinyl alcohol-co-ethylene)	Casting and extrusion	DPPH	Film	None	Luzi et al. (2018)
Green tea extract	Polyamide	Adsorption	DPPH, ORAC, TBARS	Film extract of methanol, food	Minced beef	Borzi et al. (2019)
Green tea extract	Poly(lactic acid)	Extrusion	Peroxide value, p-anisidine value, TBARS assay and hexanal content	Food	Smoked salmon	Martins et al. (2018)
Green tea extract	Whey protein concentrate	Casting	Peroxide value, p-anisidine, TBARS	Food	Salmon	Castro (2019)
Green tea extract	Polyethylene	Extrusion	DPPH, ORAC	Film extract of methanol or acetone	Minced pork meat	Wrona et al. (2017)
Green tea extract and rosemary extract	Poly(3-hydroxybutyrate-co-3-hydroxyvalerate)	Electrospinning and annealing	DPPH	Film	None	Figueroa-Lopez et al. (2019)

Herba Lophatheri extract	Chitosan	Casting	DPPH	Film extract	None	Wang et al. (2019)
Lignin	Soybean protein	Casting	Peroxide value, headspace oxygen, volatile aldehydes	Food	Soybean oil and fish oil fatty acid ethyl ester	Mohammad Zadeh et al. (2019)
Lycium ruthenicum fruit extract	Starch	Casting	DPPH	Film	Pork (monitor freshness)	Qin et al. (2019a)
Mango kernel extract	Soyprotein isolate or fish gelatin	Casting	DPPH, FRAP, ABTS, Folin-Ciocalteu	Film extract of ethanol	None	Maryam Adilah et al. (2018)
Mangosteen (*Garcinia mangostana* L.) rind	Chitosan	Casting	DPPH, peroxide value, TBARS	Film and food	Soybean oil	Zhang et al. (2020)
Mulberry fruit extract	κ-carrageenan	Casting	DPPH	Film	Milk (monitor freshness)	Liu et al. (2019)
Olive leaf extract	κ-carrageenan	Casting	DPPH, Folin-Ciocalteu	Film	None	da Rosa et al. (2020)
Proanthocyanidins	Chitosan	Casting	DPPH	Film	None	Bi et al. (2019)
Prunus maackii juice	κ-carrageenan	Casting	DPPH, peroxide value, acid value	Film and food	Lard	Sun et al. (2019)
Purple corn extract	Chitosan	Casting	DPPH	Film	None	Qin et al. (2019b)
Purple and black eggplant peel extract	Chitosan	Casting	DPPH	Film	Milk (monitor freshness)	Yong et al. (2019)
Quercetin	Cassava starch–carboxymethyl cellulose	Casting	DPPH, Folin-Ciocalteu, peroxide value	Film extract of methanol	Lard	Tongdeesoontorn et al. (2020)
Quercetin-starch	Chitosan and gelatin	Casting	DPPH, ABTS	Film	None	Yadav et al. (2020)
Red cabbage extract	Oxidized chitin nanocrystals/konjac glucomannan	Casting	DPPH		None	Wu et al. (2020)

(continued)

Table 3.1 (continued)

Antioxidant	Packaging material	Method of incorporation of antioxidant	Antioxidant activity method/assay[a]	Material tested for antioxidant activity	Food application	References
Red pitaya (*Hylocereus polyrhizus*) peel extract	Starch/polyvinyl alcohol	Casting	DPPH	Film	Shrimp (monitor freshness)	Qin et al. (2020)
Rosemary extract	Whey protein	Casting	Hexanal content, TBARS,	Food	Salami	Andrade et al. (2019)
Rosemary extract and chitosan	Poly(lactic acid)	Melt mixing and compression moulding	ABTS	Film extract of ethanol	None	Vasile et al. (2019)
Sage and lemon balm leaves	Poly(lactic acid)	Melt blending	ABTS, DPPH, in situ gas-phase hydroxyl radical generation, reducing power, Folin-Ciocalteu	Film	None	Gavril et al. (2019)
Tea polyphenols	polypropylene/poly(vinyl alcohol)/polypropylene (PP/PVA/PP) multilayer	Casting of tea phenols in PVA, then laminating to make the multilayer film	DPPH, Folin-Ciocalteu	Food simulant (50% ethanol)	None	Chen et al. (2019)
Zein	Chitosan	Casting the chitosan film, then electrospinning zein on chitosan film	DPPH	Film	Apple slices (monitor browning and moisture loss)	Bharathi et al. (2020)

[a] *ABTS* 2,2′-azino-bis(3-ethylbenzothiazoline-6-sulfonic acid, *DPPH* 2,2-diphenyl-1-picrylhydrazyl, *FRAP* Ferric reducing/antioxidant power, *ORAC* oxygen radical absorbance capacity, *TBARS* thiobarbituric acid reactive substances

The next sections are the methods or assays and examples of the methods applied to determine antioxidant activity of the active packaging material.

(A) *Antioxidant activity methods or assays applied on active packaging*

(a) DPPH (2,2-diphenyl-1-picrylhydrazyl) assay

The DPPH assay is one of the most common assays in determining the potential of an active film to generate an antioxidant effect on food as shown in Table 3.1. The antioxidant reacts with DPPH radical (DPPH•),and the decrease in the absorbance of the DPPH radical is monitored at ~517 nm (Blois 1958; Brand-Williams et al. 1995; Kurechi et al. 1980). The DPPH radical is colored purple and loses its color as the radical converts to hydrazine DPPH-H (Blois 1958; Foti 2015).

In active packaging, antioxidant activity for the DPPH assay is expressed in different ways. Most of the studies expressed antioxidant activity as percent inhibition or percent radical scavenging activity (Benbettaieb et al. 2018; Papadopoulou et al. 2019). Another way of expressing antioxidant activity is EC50 (effective concentration) or IC50 (inhibition concentration) which is the concentration of the antioxidant or a reference compound where DPPH radical inhibition is 50% (Rodríguez et al. 2020; Wrona et al. 2019). Antioxidant activity can also be expressed as Trolox equivalents using a Trolox calibration curve (Gavril et al. 2019; Rodríguez et al. 2020).

The reaction mechanisms and kinetics of DPPH radical with antioxidants have been discussed previously and readers are referred to these resources (Foti 2015; Ingold and Pratt 2014; Litwinienko and Ingold 2007; Xie and Schaich 2014).

The succeeding paragraphs give examples on the use of DPPH assay to test antioxidant activity of active films. Chitosan film with apple peel ethanolic extract was mixed with DPPH radical solution in methanol in the dark for 30 min (Riaz et al. 2020). The authors reported increasing antioxidant activity in films with increasing amount of apple peel ethanolic extract. They also found similar observations using Trolox equivalent antioxidant capacity (TEAC) assay.

A papaya film containing ascorbic acid was extracted in methanol and the extract reacted with DPPH radical for 30 min in the dark (Rodríguez et al. 2020). The authors found a higher antioxidant activity for active films with the ascorbic acid compared to the control.

A calcium alginate edible film with tea phenols was extracted with 50% ethanol (Biao et al. 2019). Then, the extract was reacted with DPPH radical solution for 1 h before reading the absorbance. The authors observed increased antioxidant activity as the amount of tea phenols increased in the sample. These observations agree with the results obtained by the authors using TEAC and Folin-Ciocalteau assays.

PLA film containing cocoa bean shell powder was immersed in DPPH radical solution in ethanol (Papadopoulou et al. 2019). Then, absorbance was measured at different reaction times up to 24 h. Although PLA film by itself reacted with DPPH radical, the authors observed improved antioxidant activity of film added with cocoa bean shells.

(b) ABTS (2,2′-azino-bis (3-ethylbenzothiazoline-6-sulfonic acid) or TEAC (Trolox equivalent antioxidant capacity) assay

In this assay, the antioxidant reacts with the ABTS radical cation ($ABTS^{\bullet+}$) and the decrease in absorbance of the $ABTS^{\bullet+}$ is measured at 734 nm (Miller et al. 1993; Re et al. 1999). Different methods have been used to generate $ABTS^{\bullet+}$ (Cano and Arnao 2018). One of the methods uses potassium persulfate to oxidize ABTS and form $ABTS^{\bullet+}$ (Re et al. 1999). In this method, the $ABTS^{\bullet+}$ is generated before reacting with the antioxidant. The advantages and disadvantages of this assay have been discussed by several authors (Prior et al. 2005; Apak et al. 2016; Cano and Arnao 2018).

In testing active films, results may be expressed as percent inhibition or percent free radical scavenging activity (Mellinas et al. 2020; Roy and Rhim 2020a), or against a reference standard. For example, antioxidant activity was reported as Trolox equivalent/g of film (Biao et al. 2019; Gavril et al. 2019) using Trolox as reference standard or gallic acid equivalent/g of film (Maryam Adilah et al. 2018) with gallic acid as reference standard.

Curcumin-containing gelatin films were tested for antioxidant activity using ABTS and DPPH assays (Roy and Rhim 2020b). $ABTS^{\bullet+}$ was generated using potassium persulfate. Then, the films were incubated in the $ABTS^{\bullet+}$ solution for 30 min. Afterwards, the absorbance was measured at 734 nm. The authors observed dose-dependent antioxidant activities in both TEAC and DPPH assays.

PLA films with sage and lemon balm leaves were mixed with the $ABTS^{\bullet+}$ solution generated using potassium persulfate (Gavril et al. 2019). In this study, absorbance at 734 nm was measured at different times up to 24 h. The study showed active films exhibiting higher antioxidant activity compared to the control using TEAC assay, DPPH assay, in situ gas-phase hydroxyl radical generation method, Folin-Ciocalteau assay, and reducing power with ferricyanide. In the TEAC assay, the authors reported no significant difference between sage-containing and lemon balm leaves-containing films after 24 h of incubation.

(c) TBARS (thiobarbituric acid reactive substances) assay

When lipids are oxidized, one of the products formed is malonaldehyde (Frankel 1984). TBARS assay measures the content of malonaldehyde and estimate the degree of lipid oxidation in food (de Koning and Silk 1963). Thiobarbituric acid (TBA) reacts with malonaldehyde and other compounds to form TBA adducts with measurable absorbance at ~532 nm (Hoyland and Taylor 1991). The assay is nonspecific and has been shown to overestimate the content of malonaldehyde in some foods (Papastergiadis et al. 2012). Malonaldehyde concentration is calculated using standards prepared from acid hydrolysis of 1,1,3,3-tetraethoxypropane (Hoyland and Taylor 1991).

TBARS assay was used to test beef stored in LLDPE film containing resveratrol in montmorillonite clay (Busolo and Lagaron 2015). In this work, sampling was done four different times in 17 days. The authors reported lower TBARS values for beef steak wrapped in LLDPE film with resveratrol-montmorillonite clay than the control.

In addition, they found the film exhibited antioxidant behavior when tested with DPPH assay. According to the authors, the results show potential to increase the shelf life of the beef packed in LLDPE film with resveratrol-montmorillonite packaging.

A chitosan film with mangosteen rind powder as antioxidant was tested on preventing oxidation of soybean oil (Zhang et al. 2020). The authors reported lower TBARS and PV for the soybean oil stored in the chitosan film containing mangosteen rind powder than the control throughout the 35-day storage. Moreover, they found dose-dependent antioxidant activity of the film tested by DPPH assay. They concluded that the oxidative stability of soybean oil increased when stored in the active film.

The antioxidant potential of a PLA film containing green tea extract was tested using smoked salmon (Martins et al. 2018). The authors tested the smoked salmon at different days up to 60 days. They reported lower TBARS and p-anisidine values for the smoked salmon stored in the film with 1% green tea extract compared to the control. Hence, according to the authors, the film with 1% green tea extract can protect smoked salmon from oxidation.

(d) In situ gas-phase hydroxyl radical generation method

The method was introduced in 2006 with modification in 2008, and indirectly measures antioxidant activity (Pezo et al. 2006; Pezo et al. 2008). In this method, the antioxidant in the active packaging reacts with hydroxyl radicals generated through UV irradiation of H_2O_2. Afterwards, unreacted hydroxyl radicals are bubbled in salicylic acid solution forming a fluorescent product, 2,5-dihydroxybenzoic acid. The method quantifies the 2,5-dihydroxybenzoic acid by HPLC with fluorescence detector. The authors expressed the results as percentage hydroxylation with lower values indicating higher antioxidant activity.

A multilayer active packaging film made of LDPE/adhesive with burberry extract/PET was tested using the in-situ gas-phase hydroxyl radical generation method (Wrona et al. 2019). The authors observed a higher antioxidant activity for the film with a higher amount of burberry extract.

In another study, the antioxidant property of a multilayer LDPE/acrylic water-based adhesive/PET film with *Pistacia lentiscus* L. leaves extract was investigated (Djebari et al. 2021). In this work, the extract of *Pistacia lentiscus* L. was mixed with the adhesive to make the active film. The authors reported higher antioxidant activity in the film with a higher amount of the *Pistacia lenticus* L. leaf extract.

The antioxidant property of PLA films containing sage and lemon balm leaves was tested (Gavril et al. 2019). In this study, the films either with sage or lemon balm demonstrated antioxidant activity as shown in the results using in-situ hydroxyl radical generation method. In addition, the authors reported presence of antioxidant activity in either films using DPPH, TEAC, reducing powder method with ferricyanide, and Folin-Ciocalteau assays.

(e) Peroxide value (PV)

PV reports the amount of peroxides and can be used as a measure of lipid oxidation during the initial stages of oxidation (Gray 1978). Different methods have been applied to determine PV (Shahidi and Zhong 2015). Iodometric titration is one method wherein peroxides in oil react with iodide to form iodine which is then titrated with sodium thiosulfate using starch as endpoint indicator (Wheeler 1932; Lea 1952). Ferric thiocyanate method is another procedure. Peroxides from the oil oxidizes ferrous ion to ferric ion and, in the presence of ammonium thiocyanate, forms a colored ferric thiocyanate complex (Lips et al. 1943).

PV was used to determine the antioxidant effect of an active packaging film on soybean oil (Colín-Chávez et al. 2014). In this work, the active packaging film was HDPE with marigold extract containing carotenoids as antioxidant and TiO_2 for light protection. The authors sampled the soybean oil for several days up to 16 days. They used the AOCS method that employs iodometric titration to determine PV. They reported a delay in the increase of PV in soybean oil stored in the HDPE film with both marigold extract and TiO_2.

PV of soybean oil was determined using the ferric thiocyanate method (Zhang et al. 2020). In this work, soybean oil was stored in chitosan film with mangosteen rind powder and the oil sampled every 7 days for 35 days. The authors reported lower PV for soybean oil stored in chitosan film with mangosteen rind powder compared to the control. They reported similar behavior in the TBARS assay.

(f) Anisidine test

During lipid oxidation, hydroperoxides are formed which further decomposes to aldehydes (Frankel 1984). Anisidine test measures the amount of aldehydes reacting with *p*-anisidine reagent to form yellow products detected at 350 nm (List et al. 1974). The method was developed using benzidine as reagent (Holm et al. 1957), and later, the reagent was changed to *p*-anisidine.

The lipid oxidation effect of an active film on salmon was determined by *p*-anisidine value test (Castro 2019). In this study, the active film was made of whey protein with green tea extract as antioxidant. The authors extracted fat from the salmon and used for the test. They observed low *p*-anisidine values of salmon stored in the active film and control for 14 days. On the 17th day, they found higher *p*-anisidine values, with the salmon in the active film lower than the control. They also reported lower TBARS value of the salmon in the active film than in the control for the 17 days except on the 10th day.

Anisidine value of walnuts stored in PLA film with gallic acid nanofibers was determined (Aydogdu et al. 2019). In this study, accelerated oxidation of the walnuts was performed at 40 °C for 21 days. Then, the authors extracted the walnut oil to test using *p*-anisidine, PV, and TBARS assays. They reported lower *p*-anisidine and PV for the walnut packaged in the active film than in the control. However, they observed no significant difference in the TBARS results between the active film and the control.

(g) Folin-Ciocalteu assay

The method involves the reaction between phenols and Folin-Ciocalteu reagent to form blue colored products measurable at 765 nm (Singleton et al. 1999; Singleton and Rossi 1965). Folin-Ciocalteu reagent was developed in 1927 to determine tyrosine and tryptophan, and contains sodium tungstate, sodium molybdate, water, phosphoric acid, hydrochloric acid, and lithium sulfate (Folin and Ciocalteu 1927). The method is not specific to phenols with other compounds reacting with Folin-Ciocalteu reagent (Everette et al. 2010).

Folin-Ciocalteu assay was used to determine the phenolics content in a carrageenan film with olive leaf extract as antioxidant (da Rosa et al. 2020). In this work, the film was dissolved in water and the obtained solution used for the test. After reading the absorbance at 765 nm, the authors calculated the concentration of the phenolics using a standard curve of gallic acid. They reported higher values of phenolic content for films with higher content of the olive leaf extract. They also performed DPPH assay with the dissolved film solution. According to the study, higher olive leaf extract content in the film generated higher percent inhibition in the DPPH assay.

Folin-Ciocalteu assay was used to test film made of soy protein isolate and mango kernel extract as antioxidant (Maryam Adilah et al. 2018). In this study, the assay used film extracts obtained by immersing the film in 95% ethanol for 7 days. The authors quantified the results against gallic acid standard. They reported increasing amount of compounds reacting with Folin-Ciocalteu reagent with increasing levels of mango kernel extract in the film.

(h) Hexanal content

During lipid oxidation, hexanal and other aldehydes are formed when lipid hydroperoxides decomposes (Frankel 1980). Hexanal and propanal were found to be the dominant volatiles in cooked pork for the first 6 days and can be used to monitor lipid oxidation and meat flavor deterioration (Shahidi and Pegg 1994).

Hexanal in salami was analyzed using UHPLC (ultrahigh-pressure liquid chromatography) with diode array detector, and quantified at 365 nm (Andrade et al. 2019). In this work, salami was stored in whey protein film with rosemary extract as antioxidant. The authors found that salami stored in the active film took a longer time, around 30 days, than the control to get the highest value of hexanal. They also reported that TBARS values of salami were lower in the active film except on the 30th day which might be the peak of oxidation.

Soybean protein-film incorporated with either alkali lignin or lignosulphonate was investigated as potential active packaging film for soybean oil (Mohammad Zadeh et al. 2019). The authors used GC-MS to quantify the pentanal, hexanal, 2-heptanal, and nonanal of soybean oil containing the active film. They found lower levels of these volatile aldehyde compounds compared to the control. In addition, the authors obtained lower PV for the soybean oil exposed to the active film.

(i) FRAP (Ferric reducing/antioxidant power) assay

The FRAP assay measures the ability of an antioxidant to reduce iron from Fe^{3+}, in the form of ferric tripyridyltriazine (Fe^{3+}-TPTZ) complex, to Fe^{2+} at low pH conditions (Benzie and Strain 1996, 1999). In this assay, the reduction of Fe^{3+}-TPTZ complex produces a blue colored Fe^{2+}-TPTZ complex that has a measurable absorbance at 593 nm (Benzie and Strain 1996).

FRAP assay was used to test the extracts of either soy protein isolate film or fish gelatin film (Maryam Adilah et al. 2018). The authors used mango kernel extract as antioxidant. In this work, the results were quantified against gallic acid standard curve. The authors found a higher antioxidant activity for soy protein isolate films containing a higher concentration of mango kernel extract than fish gelatin films. They also reported higher activity for soy protein isolate films with mango kernel extract in TEAC and Folin-Ciocalteu assays.

(j) ORAC (oxygen radical absorbance capacity) assay

ORAC is a fluorescence-based assay that measures the extent the antioxidant inhibits the fluorescence decay of the fluorescent probe as the probe reacts with peroxy radicals (Cao et al. 1993). The antioxidant activity is measured using the area under the decay curve of the fluorescent probe with and without the antioxidant, and the results quantified against a reference standard such as Trolox (Cao et al. 1993). The fluorescent probe used to be β-phycoerythrin with an excitation wavelength of 540 nm and emission wavelength of 565 nm (Cao et al. 1993). Fluorescein, a more photostable compound, was introduced later as fluorescent probe with an excitation wavelength of 493 nm and emission wavelength of 515 nm (Ou et al. 2001). The peroxy radicals are generated from 2,2′-azobis(2-amidinopropane) dihydrochloride (AAPH) (Cao et al. 1993).

ORAC assay was used to test the extract of a polyamide film with green tea extract as antioxidant (Borzi et al. 2019). The authors used fluorescein as probe and expressed the result as Trolox equivalents. In addition to ORAC, they tested the antioxidant property of the active film on minced beef meat using TBARS assay. According to the study, the TBARS assay indicates that shelf life of the beef can be extended for 23 days with the active film.

(B) *Methods measuring the content and identity of antioxidant components in the active packaging material*

Chromatographic techniques using different detectors can be used to identify and measure the amount of antioxidant in the active packaging material. The antioxidant in the film is extracted by solid-liquid extraction or using a food simulant (López de Dicastillo et al. 2011). Then, the antioxidants in the extract are chromatographically separated, and detected using absorbance (López de Dicastillo et al. 2011), fluorescence (Iñiguez-Franco et al. 2012) or mass spectrometry (MS) (Wrona et al. 2017). High resolution mass spectrometry instruments capable of measuring *m/z* of compounds with high mass accuracy and resolution increase the confidence of identifying antioxidants. For example, time-of-flight MS was used to identify the compounds in the leaf (Wrona et al. 2019) and green tea extracts (López de Dicastillo et al. 2011) incorporated in the active film.

The release of antioxidants from the film to the food can be simulated using food simulants (Castro-López et al. 2014). In this work, catechins and quercetin, from the PP film containing either catechin or green tea extract, were released into the food simulants (10% ethanol and 50% ethanol) and quantified using HPLC with photodiode array detector. A HPLC-PDA-LTQ FT Orbitrap MS was used as confirmatory technique and for detection of other compounds of interest such as methylated catechins. However, the study did not show how the concentration of catechin and quercetin in the food simulant relate to the antioxidant activity of the packaging.

The antioxidants can be quantified, and their antioxidant activities measured in the film extracts or food simulants. The succeeding paragraphs give examples of studies that quantified the antioxidants in the active film by chromatography and determined the antioxidant activity of the film, film extracts, or food simulant.

Catechins and caffeine were quantified from food simulants using UHPLC with triple quadrupole as mass detector (Wrona et al. 2017). In this work, polyethylene film with inorganic capsules containing green tea extract was immersed in 10% or 95% ethanol as food simulants for 10 days at 20 °C. The authors reported epigallocatechin gallate, gallocatechin gallate and epicatechin gallate migrated faster in 95% ethanol. On the other hand, they observed higher migration rate for gallocatechin, epigallocatechin gallate and catechin gallate in 10% ethanol. Moreover, they also found antioxidant activity in the active films using DPPH and ORAC assays.

In a different study, catechins and caffeine were detected in polyamide film containing green tea extract using UPLC with triple quadrupole MS (Borzi et al. 2019). The authors extracted the catechins and caffeine using food simulants with either 10% or 95% ethanol. Moreover, they found that the film showed antioxidant activity using DPPH, ORAC, and TBARS assays.

The concentrations of catechins and caffeine in the ethylene vinyl alcohol copolymer with green tea extract were determined using film extracts and food simulants (water, 3% acetic acid, 10% ethanol and 95% ethanol) by HPLC with diode array detector (López de Dicastillo et al. 2011). Moreover, the authors determined the antioxidant activity of the film extracts and food simulants by DPPH and ABTS assays. They found the ethanolic film extract has reduced antioxidant activity and decreased concentration of the catechin gallates as a result of extrusion. In addition, they observed antioxidant activity in all food simulants tested with 95% ethanol showing the highest. In this study, gallic acid, caffeine, catechin and epigallocatechin gallate were highest in concentration in 95% ethanol food simulant.

HPLC with diode array detector was used to quantify the amount of the citrus extract compounds released from the packaging into cooked turkey meat (Contini et al. 2014). In this work, cooked turkey meat was stored in PET trays coated with citrus extract up to 2 days. Then, the citrus extract compounds were extracted from the turkey meat and analyzed. The authors reported presence of citric acid, salicylic acid, naringin, and hesperdin in the cooked turkey meat. Previous work of the authors showed lower TBARS and hexanal values of the cooked turkey meat stored in PET-citrus packaging for 4 days compared to control (Contini et al. 2012).

Either catechin or epicatechin incorporated in PLA films were diffused into 95% ethanol as food simulant (Iñiguez-Franco et al. 2012). Then, at equilibrium, the catechins in the food simulant were quantified using HPLC with fluorescence detector. In addition, the authors determined the antioxidant activity of the catechins in the food simulant by DPPH assay. They reported positive correlation between concentration of either catechin or epicatechin and antioxidant activity in the 95% ethanol food simulant.

5 Concluding Remarks

Active packaging with antioxidants can extend the shelf-life of food and reduce food waste. In most of the reviewed studies, antioxidants are incorporated into the active packaging material by casting. Numerous natural antioxidants have shown potential as antioxidant in the active packaging materials. Simulating the real storage conditions of the food, then testing the food for spoilage seems to be ideal to demonstrate the performance of the antioxidant in the active packaging material. Different methods are available for use in determining the antioxidant activity of an active packaging material. However, these methods involve different mechanisms and are not standardized. Hence, antioxidant activity is obtained in different ways. In addition, antioxidant activity is expressed in different units which presents a challenge when comparing results from different studies. Majority of the methods used to analyze antioxidants used in active packaging are based in chromatography coupled to several detectors such as PDA and MS.

Briefly, in this book chapter, we have reviewed and discussed the use and analysis of antioxidants in active packaging. The reader will have a knowledge of the basics related to active packaging, natural antioxidants, and will have better understanding of the analytical tools used to analyze antioxidant activity and of the analytical techniques employed in the analysis of antioxidants.

References

COMMISSION REGULATION (EU) No 10/2011 of 14 January 2011 on plastic materials and articles intended to come into contact with food.

COMMISSION REGULATION (EC) No 2023/2006 of 22 December 2006 on good manufacturing practice for materials and articles intended to come into contact with food.

COMMISSION REGULATION (EU) No 231/2012 of 9 March 2012 laying down specifications for food additives listed in Annexes II and III to Regulation (EC) No 1333/2008 of the European Parliament and of the Council.

COMMISSION REGULATION (EC) No 450/2009 of 29 May 2009 on active and intelligent materials and articles intended to come into contact with food.

[FDA] United States Food and Drug Administration. Code of Federal Regulations. Parts 170–199, Title 21. Washington, DC: U.S. Government Printing Office; 2019.

[FDA] United States Food and Drug Administration. Code of Federal Regulations. Parts 100–169, Title 21. Washington, DC: U.S. Government Printing Office; 2020.

REGULATION (EC) No 1331/2008 OF THE EUROPEAN PARLIAMENT AND OF THE COUNCIL of 16 December 2008 establishing a common authorisation procedure for food additives, food enzymes and food flavourings.

REGULATION (EC) No 1333/2008 OF THE EUROPEAN PARLIAMENT AND OF THE COUNCIL of 16 December 2008 on food additives.

REGULATION (EC) No 1935/2004 OF THE EUROPEAN PARLIAMENT AND OF THE COUNCIL of 27 October 2004 on materials and articles intended to come into contact with food and repealing Directives 80/590/EEC and 89/109/EEC.

Amorati R, Valgimigli L. Advantages and limitations of common testing methods for antioxidants. Free Radic Res. 2015;49(5):633–49. https://doi.org/10.3109/10715762.2014.996146.

Andrade MA, Ribeiro-Santos R, Guerra M, Sanches-Silva A. Evaluation of the oxidative status of salami packaged with an active whey protein film. Foods. 2019;8(9):387.

Antolovich M, Prenzler PD, Patsalides E, McDonald S, Robards K. Methods for testing antioxidant activity. Analyst. 2002;127(1):183–98. https://doi.org/10.1039/B009171P.

Apak R. Current issues in antioxidant measurement. J Agric Food Chem. 2019;67(33):9187–202. https://doi.org/10.1021/acs.jafc.9b03657.

Apak R, Gorinstein S, Böhm V, Schaich KM, Özyürek M, Güçlü K. Methods of measurement and evaluation of natural antioxidant capacity/activity (IUPAC Technical Report). Pure Appl Chem. 2013;85(5):957. https://doi.org/10.1351/PAC-REP-12-07-15.

Apak R, Özyürek M, Güçlü K, Çapanoğlu E. Antioxidant activity/capacity measurement. 2. Hydrogen Atom Transfer (HAT)-based, mixed-mode (Electron transfer (ET)/HAT), and lipid peroxidation assays. J Agric Food Chem. 2016;64(5):1028–45. https://doi.org/10.1021/acs.jafc.5b04743.

Aydogdu A, Yildiz E, Aydogdu Y, Sumnu G, Sahin S, Ayhan Z. Enhancing oxidative stability of walnuts by using gallic acid loaded lentil flour based electrospun nanofibers as active packaging material. Food Hydrocoll. 2019;95:245–55. https://doi.org/10.1016/j.foodhyd.2019.04.020.

Bahrami A, Delshadi R, Assadpour E, Jafari SM, Williams L. Antimicrobial-loaded nanocarriers for food packaging applications. Adv Colloid Interf Sci. 2020;278:102140. https://doi.org/10.1016/j.cis.2020.102140.

Bailey A, Batarseh L, Begley T, Twaroski M. US FDA food contact materials regulations. Copyright (C) 2021 American Chemical Society (ACS). All Rights Reserved. Wiley-VCH Verlag GmbH & Co. KGaA; 2008, p. 417–439. https://doi.org/10.1002/9783527621422.ch12.

Barbosa-Pereira L, Aurrekoetxea GP, Angulo I, Paseiro-Losada P, Cruz JM. Development of new active packaging films coated with natural phenolic compounds to improve the oxidative stability of beef. Meat Sci. 2014;97(2):249–54. https://doi.org/10.1016/j.meatsci.2014.02.006.

Benbettaieb N, Nyagaya J, Seuvre A-M, Debeaufort F. Antioxidant activity and release kinetics of Caffeic and *p*-Coumaric acids from hydrocolloid-based active films for healthy packaged food. J Agric Food Chem. 2018;66(26):6906–16. https://doi.org/10.1021/acs.jafc.8b01846.

Bentayeb K, Rubio C, Batlle R, Nerin C. Direct determination of carnosic acid in a new active packaging based on natural extract of rosemary. Anal Bioanal Chem. 2007;389(6):1989–96. https://doi.org/10.1007/s00216-007-1570-y.

Benzie IFF, Strain JJ. The ferric reducing ability of plasma (FRAP) as a measure of "antioxidant power": the FRAP assay. Anal Biochem. 1996;239(1):70–6. https://doi.org/10.1006/abio.1996.0292.

Benzie IFF, Strain JJ. [2] Ferric reducing/antioxidant power assay: direct measure of total antioxidant activity of biological fluids and modified version for simultaneous measurement of total antioxidant power and ascorbic acid concentration. In: Methods in enzymology, vol. 299. Academic; 1999. p. 15–27. https://doi.org/10.1016/S0076-6879(99)99005-5.

Bharathi SKV, Leena MM, Moses JA, Anandharamakrishnan C. Nanofibre-based bilayer biopolymer films: enhancement of antioxidant activity and potential for food packaging application. Int J Food Sci Technol. 2020;55(4):1477–84. https://doi.org/10.1111/ijfs.14492.

Bi F, Zhang X, Bai R, Liu Y, Liu J, Liu J. Preparation and characterization of antioxidant and antimicrobial packaging films based on chitosan and proanthocyanidins. Int J Biol Macromol. 2019;134:11–9. https://doi.org/10.1016/j.ijbiomac.2019.05.042.

Biao Y, Yuxuan C, Qi T, Ziqi Y, Yourong Z, McClements DJ, Chongjiang C. Enhanced performance and functionality of active edible films by incorporating tea polyphenols into thin calcium alginate hydrogels. Food Hydrocoll. 2019;97:105197. https://doi.org/10.1016/j.foodhyd.2019.105197.

Blois MS. Antioxidant determinations by the use of a stable free radical. Nature. 1958;181(4617):1199–200. https://doi.org/10.1038/1811199a0.

Borzi F, Torrieri E, Wrona M, Nerín C. Polyamide modified with green tea extract for fresh minced meat active packaging applications. Food Chem. 2019;300:125242. https://doi.org/10.1016/j.foodchem.2019.125242.

Brand-Williams W, Cuvelier ME, Berset C. Use of a free radical method to evaluate antioxidant activity. LWT Food Sci Technol. 1995;28(1):25–30. https://doi.org/10.1016/S0023-6438(95)80008-5.

Busolo MA, Lagaron JM. Antioxidant polyethylene films based on a resveratrol containing clay of interest in food packaging applications. Food Packag Shelf Life. 2015;6:30–41. https://doi.org/10.1016/j.fpsl.2015.08.004.

Cano A, Arnao MB. ABTS/TEAC (2,2'-azino-bis(3-ethylbenzothiazoline-6-sulfonic acid)/Trolox®-Equivalent Antioxidant Capacity) radical scavenging mixed-mode assay. In: Apak R, Capanoglu E, Shahidi F, editors. Measurement of antioxidant activity & capacity; 2018. p. 117–39. https://doi.org/10.1002/9781119135388.ch7.

Cao G, Alessio HM, Cutler RG. Oxygen-radical absorbance capacity assay for antioxidants. Free Radic Biol Med. 1993;14(3):303–11. https://doi.org/10.1016/0891-5849(93)90027-R.

Castro-López MM, López-Vilariño JM, González-Rodríguez MV. Analytical determination of flavonoids aimed to analysis of natural samples and active packaging applications. Food Chem. 2014;150:119–27. https://doi.org/10.1016/j.foodchem.2013.10.156.

Castro FVRA, Mariana A, Sanches Silva A, Vaz MF, Vilarinho F. The contribution of a whey protein film incorporated with green tea extract to minimize the lipid oxidation of salmon (*Salmo salar* L.). Foods. 2019;8(8):327. https://doi.org/10.3390/foods8080327.

Cerisuelo JP, Gavara R, Hernandez-Munoz P. Natural antimicrobial – containing EVOH coatings on PP and PET films: functional and active property characterization. Packag Technol Sci. 2014;27(11):901–20. https://doi.org/10.1002/pts.2078.

Chavoshizadeh S, Pirsa S, Mohtarami F. Sesame oil oxidation control by active and smart packaging system using wheat gluten/chlorophyll film to increase shelf life and detecting expiration date. Eur J Lipid Sci Technol. 2020;122(3):1900385. https://doi.org/10.1002/ejlt.201900385.

Chen C, Li C, Yang S, Zhang Q, Yang F, Tang Z, Xie J. Development of new multilayer active packaging films with controlled release property based on polypropylene/poly(Vinyl alcohol)/polypropylene incorporated with tea polyphenols. J Food Sci. 2019;84(7):1836–43. https://doi.org/10.1111/1750-3841.14681.

Colín-Chávez C, Vicente-Ramírez EB, Soto-Valdez H, Peralta E, Auras R. The release of carotenoids from a light-protected antioxidant active packaging designed to improve the stability of soybean oil. Food Bioprocess Technol. 2014;7(12):3504–15. https://doi.org/10.1007/s11947-014-1359-x.

Contini C, Katsikogianni MG, O'Neill FT, O'Sullivan M, Dowling DP, Monahan FJ. PET trays coated with Citrus extract exhibit antioxidant activity with cooked turkey meat. LWT. 2012;47(2):471–7. https://doi.org/10.1016/j.lwt.2012.02.009.

Contini C, Katsikogianni MG, O'Neill FT, O'Sullivan M, Dowling DP, Monahan FJ. Mechanism of action of an antioxidant active packaging prepared with Citrus extract. LWT Food Sci Technol. 2014;59(2, Part 1):1082–7. https://doi.org/10.1016/j.lwt.2014.06.025.

da Rosa GS, Vanga SK, Gariepy Y, Raghavan V. Development of biodegradable films with improved antioxidant properties based on the addition of carrageenan containing olive leaf extract for food packaging applications. J Polym Environ. 2020;28(1):123–30. https://doi.org/10.1007/s10924-019-01589-7.

de Carvalho SM, Noronha CM, da Rosa CG, Sganzerla WG, Bellettini IC, Nunes MR, Bertoldi FC, Manique Barreto PL. PVA antioxidant nanocomposite films functionalized with alpha-tocopherol loaded solid lipid nanoparticles. Colloids Surf A Physicochem Eng Asp. 2019;581:123793. https://doi.org/10.1016/j.colsurfa.2019.123793.

de KNN, van BM, Rijk R, Sipilainen-Malm T, Paseiro LP, de MB. Active and intelligent packaging: applications and regulatory aspects. Food Addit Contam. 2002;19(Suppl):144–62.

de Koning AJ, Silk MH. The 2-thiobarbituric acid reagent for determination of oxidative rancidity in fish oils. J Am Oil Chem Soc. 1963;40(5):165–9. https://doi.org/10.1007/BF02632572.

Djebari S, Wrona M, Boudria A, Salafranca J, Nerin C, Bedjaoui K, Madani K. Study of bioactive volatile compounds from different parts of *Pistacia lentiscus* L. extracts and their antioxidant and antibacterial activities for new active packaging application. Food Control. 2021;120:107514. https://doi.org/10.1016/j.foodcont.2020.107514.

Dominguez R, Barba FJ, Gomez B, Putnik P, Bursac Kovacevic D, Pateiro M, Santos EM, Lorenzo JM. Active packaging films with natural antioxidants to be used in meat industry: a review. Food Res Int. 2018;113:93–101. https://doi.org/10.1016/j.foodres.2018.06.073.

Domínguez R, Barba FJ, Gómez B, Putnik P, Bursać Kovačević D, Pateiro M, Santos EM, Lorenzo JM. Active packaging films with natural antioxidants to be used in meat industry: a review. Food Res Int. 2018;113:93–101. https://doi.org/10.1016/j.foodres.2018.06.073.

Everette JD, Bryant QM, Green AM, Abbey YA, Wangila GW, Walker RB. Thorough study of reactivity of various compound classes toward the Folin−Ciocalteu reagent. J Agric Food Chem. 2010;58(14):8139–44. https://doi.org/10.1021/jf1005935.

Figueroa-Lopez KJ, Vicente AA, Reis MAM, Torres-Giner S, Lagaron JM. Antimicrobial and antioxidant performance of various essential oils and natural extracts and their incorporation into biowaste derived poly(3-hydroxybutyrate-co-3-hydroxyvalerate) layers made from electrospun ultrathin fibers. Nanomaterials. 2019;9(2):144.

Folin O, Ciocalteu V. On tyrosine and tryptophane determinations in proteins. J Biol Chem. 1927;73(2):627–50.

Foti MC. Use and abuse of the DPPH• radical. J Agric Food Chem. 2015;63(40):8765–76. https://doi.org/10.1021/acs.jafc.5b03839.

Frankel EN. Lipid oxidation. Prog Lipid Res. 1980;19(1):1–22. https://doi.org/10.1016/0163-7827(80)90006-5.

Frankel EN. Lipid oxidation: Mechanisms, products and biological significance. J Am Oil Chem Soc. 1984;61(12):1908–17. https://doi.org/10.1007/BF02540830.

Ganiari S, Choulitoudi E, Oreopoulou V. Edible and active films and coatings as carriers of natural antioxidants for lipid food. Trends Food Sci Technol. 2017;68:70–82. https://doi.org/10.1016/j.tifs.2017.08.009.

Gavril G-L, Wrona M, Bertella A, Świeca M, Râpă M, Salafranca J, Nerín C. Influence of medicinal and aromatic plants into risk assessment of a new bioactive packaging based on polylactic acid (PLA). Food Chem Toxicol. 2019;132:110662. https://doi.org/10.1016/j.fct.2019.110662.

Goddard JM, Hotchkiss JH. Tailored functionalization of low-density polyethylene surfaces. J Appl Polym Sci. 2008;108(5):2940–9. https://doi.org/10.1002/app.27209.

Goddard JM, Talbert JN, Hotchkiss JH. Covalent attachment of lactase to low-density polyethylene films. J Food Sci. 2007;72(1):E36–41. https://doi.org/10.1111/j.1750-3841.2006.00203.x.

Gomez-Estaca J, Lopez-de-Dicastillo C, Hernandez-Munoz P, Catala R, Gavara R. Advances in antioxidant active food packaging. Trends Food Sci Technol. 2014;35(1):42–51. https://doi.org/10.1016/j.tifs.2013.10.008.

Gómez-Estaca J, López-de-Dicastillo C, Hernández-Muñoz P, Catalá R, Gavara R. Advances in antioxidant active food packaging. Trends Food Sci Technol. 2014;35(1):42–51. https://doi.org/10.1016/j.tifs.2013.10.008.

Gontard N, Angellier-Coussy H, Chalier P, Gastaldi E, Guillard V, Guillaume C, Peyron S. Food packaging applications of biopolymer-based films. Copyright (C) 2020 American Chemical Society (ACS). All Rights Reserved. Wiley; 2011. p. 213–32.

Goulas V, Hadjivasileiou L, Primikyri A, Michael C, Botsaris G, Tzakos AG, Gerothanassis IP. Valorization of carob fruit residues for the preparation of novel bi-functional polyphenolic coating for food packaging applications. Molecules. 2019;24(17):3162.

Gray JI. Measurement of lipid oxidation: a review. J Am Oil Chem Soc. 1978;55(6):539–46. https://doi.org/10.1007/BF02668066.

Holm U, Ekbom K, Wode G. Determination of the extent of oxidation of fats. J Am Oil Chem Soc. 1957;34(12):606–9. https://doi.org/10.1007/BF02638765.

Hoyland DV, Taylor AJ. A review of the methodology of the 2-thiobarbituric acid test. Food Chem. 1991;40(3):271–91. https://doi.org/10.1016/0308-8146(91)90112-2.

Ingold KU, Pratt DA. Advances in radical-trapping antioxidant chemistry in the 21st century: a kinetics and mechanisms perspective. Chem Rev. 2014;114(18):9022–46. https://doi.org/10.1021/cr500226n.

Iñiguez-Franco F, Soto-Valdez H, Peralta E, Ayala-Zavala JF, Auras R, Gámez-Meza N. Antioxidant activity and diffusion of catechin and epicatechin from antioxidant active films made of poly(l-lactic acid). J Agric Food Chem. 2012;60(26):6515–23. https://doi.org/10.1021/jf300668u.

Jamshidian M, Tehrany EA, Desobry S. Antioxidants release from solvent-cast PLA film: investigation of PLA antioxidant-active packaging. Food Bioprocess Technol. 2013;6(6):1450–63. https://doi.org/10.1007/s11947-012-0830-9.

Koontz JL, Moffitt RD, Marcy JE, O'Keefe SF, Duncan SE, Long TE. Controlled release of α-tocopherol, quercetin, and their cyclodextrin inclusion complexes from linear low-density polyethylene (LLDPE) films into a coconut oil model food system. Food Addit Contamin Part A. 2010;27(11):1598–607. https://doi.org/10.1080/19440049.2010.495729.

Kurechi T, Kikugawa K, Kato T. Studies on the antioxidants. XIII. Hydrogen donating capability of antioxidants to 2, 2-Diphenyl-1-picrylhydrazyl. Chem Pharm Bull. 1980;28(7):2089–93. https://doi.org/10.1248/cpb.28.2089.

Lea CH. Methods for determining peroxide in lipids. J Sci Food Agric. 1952;3(12):586–94. https://doi.org/10.1002/jsfa.2740031205.

Lin Z, Roman MJ, Decker EA, Goddard JM. Synthesis of iminodiacetate functionalized polypropylene films and their efficacy as antioxidant active-packaging materials. J Agric Food Chem. 2016;64(22):4606–17. https://doi.org/10.1021/acs.jafc.6b01128.

Lips A, Chapman RA, McFarlane WD. The application of the ferric thiocyanate method to the determination of incipient rancidity in fats and oils. Oil Soap. 1943;20(11):240–3. https://doi.org/10.1007/BF02630882.

List GR, Evans CD, Kwolek WF, Warner K, Boundy BK, Cowan JC. Oxidation and quality of soybean oil: a preliminary study of the anisidine test. J Am Oil Chem Soc. 1974;51(2):17–21. https://doi.org/10.1007/bf02545207.

Litwinienko G, Ingold KU. Solvent effects on the rates and mechanisms of reaction of phenols with free radicals. Acc Chem Res. 2007;40(3):222–30. https://doi.org/10.1021/ar0682029.

Liu Y, Qin Y, Bai R, Zhang X, Yuan L, Liu J. Preparation of pH-sensitive and antioxidant packaging films based on κ-carrageenan and mulberry polyphenolic extract. Int J Biol Macromol. 2019;134:993–1001. https://doi.org/10.1016/j.ijbiomac.2019.05.175.

López de Dicastillo C, Nerín C, Alfaro P, Catalá R, Gavara R, Hernández-Muñoz P. Development of new antioxidant active packaging films based on ethylene vinyl alcohol copolymer (EVOH) and green tea extract. J Agric Food Chem. 2011;59(14):7832–40. https://doi.org/10.1021/jf201246g.

Łopusiewicz Ł, Jędra F, Mizielińska M. New poly(lactic acid) active packaging composite films incorporated with fungal melanin. Polymers. 2018;10(4):386.

Luzi F, Pannucci E, Santi L, Kenny JM, Torre L, Bernini R, Puglia D. Gallic acid and quercetin as intelligent and active ingredients in poly(vinyl alcohol) films for food packaging. Polymers. 2019;11(12):1999.

Luzi F, Puglia D, Dominici F, Fortunati E, Giovanale G, Balestra GM, Torre L. Effect of gallic acid and umbelliferone on thermal, mechanical, antioxidant and antimicrobial properties of poly (vinyl alcohol-co-ethylene) films. Polym Degrad Stab. 2018;152:162–76. https://doi.org/10.1016/j.polymdegradstab.2018.04.015.

Martins C, Vilarinho F, Sanches Silva A, Andrade M, Machado AV, Castilho MC, Sá A, Cunha A, Vaz MF, Ramos F. Active polylactic acid film incorporated with green tea extract: development, characterization and effectiveness. Ind Crop Prod. 2018;123:100–10. https://doi.org/10.1016/j.indcrop.2018.06.056.

Maryam Adilah ZA, Jamilah B, Nur Hanani ZA. Functional and antioxidant properties of protein-based films incorporated with mango kernel extract for active packaging. Food Hydrocoll. 2018;74:207–18. https://doi.org/10.1016/j.foodhyd.2017.08.017.

Mellinas C, Ramos M, Grau-Atienza A, Jordà A, Burgos N, Jiménez A, Serrano E, Garrigós MC. Biodegradable poly(ε-Caprolactone) active films loaded with MSU-X mesoporous silica for the release of α-tocopherol. Polymers. 2020;12(1):137.

Miller NJ, Rice-Evans C, Davies MJ, Gopinathan V, Milner A. A novel method for measuring antioxidant capacity and its application to monitoring the antioxidant status in premature neonates. Clin Sci. 1993;84(4):407–12. https://doi.org/10.1042/cs0840407.

Mohammad Zadeh E, O'Keefe SF, Kim Y-T. Lignin-based biopolymeric active packaging system for oil products. J Food Sci. 2019;84(6):1420–6. https://doi.org/10.1111/1750-3841.14632.

Moon J-K, Shibamoto T. Antioxidant assays for plant and food components. J Agric Food Chem. 2009;57(5):1655–66. https://doi.org/10.1021/jf803537k.

Nerin C. Antioxidant active food packaging and antioxidant edible films. Woodhead Publ Ser Food Sci Technol Nutr. 2010;200 (Oxidation in Foods and Beverages and Antioxidant Applications, Volume 2):496–515.

Otero-Pazos P, Rodriguez-Bernaldo de Quiros A, Sendon R, Benito-Pena E, Gonzalez-Vallejo V, Moreno-Bondi MC, Angulo I, Paseiro-Losada P. Active food packaging based on molecularly imprinted polymers: study of the release kinetics of ferulic acid. J Agric Food Chem. 2014;62(46):11215–21. https://doi.org/10.1021/jf5035042.

Otero-Pazos P, Sendon R, Blanco-Fernandez B, Blanco-Dorado S, Alvarez-Lorenzo C, Concheiro A, Angulo I, Paseiro-Losada P, Rodriguez-Bernaldo de Quiros A. Preparation of antioxidant active films based on chitosan: diffusivity study of α-tocopherol into food simulants. J Food Sci Technol (New Delhi, India). 2016;53(6):2817–26. https://doi.org/10.1007/s13197-016-2256-2.

Ou B, Hampsch-Woodill M, Prior RL. Development and validation of an improved oxygen radical absorbance capacity assay using fluorescein as the fluorescent probe. J Agric Food Chem. 2001;49(10):4619–26. https://doi.org/10.1021/jf010586o.

Ozdemir M, Floros JD. Active food packaging technologies. Crit Rev Food Sci Nutr. 2004;44(3):185–93. https://doi.org/10.1080/10408690490441578.

Papadopoulou EL, Paul UC, Tran TN, Suarato G, Ceseracciu L, Marras S, d'Arcy R, Athanassiou A. Sustainable active food packaging from poly(lactic acid) and cocoa bean shells. ACS Appl Mater Interfaces. 2019;11(34):31317–27. https://doi.org/10.1021/acsami.9b09755.

Papastergiadis A, Mubiru E, Van Langenhove H, De Meulenaer B. Malondialdehyde measurement in oxidized foods: evaluation of the spectrophotometric thiobarbituric acid reactive substances (TBARS) test in various foods. J Agric Food Chem. 2012;60(38):9589–94. https://doi.org/10.1021/jf302451c.

Pereira de Abreu DA, Cruz JM, Paseiro Losada P. Active and intelligent packaging for the food industry. Food Rev Int. 2012;28(2):146–87. https://doi.org/10.1080/87559129.2011.595022.

Pezo D, Salafranca J, Nerín C. Design of a method for generation of gas-phase hydroxyl radicals, and use of HPLC with fluorescence detection to assess the antioxidant capacity of natural essential oils. Anal Bioanal Chem. 2006;385(7):1241–6. https://doi.org/10.1007/s00216-006-0395-4.

Pezo D, Salafranca J, Nerín C. Determination of the antioxidant capacity of active food packagings by in situ gas-phase hydroxyl radical generation and high-performance liquid chromatography–fluorescence detection. J Chromatogr A. 2008;1178(1):126–33. https://doi.org/10.1016/j.chroma.2007.11.062.

Prior RL, Wu X, Schaich K. Standardized methods for the determination of antioxidant capacity and phenolics in foods and dietary supplements. J Agric Food Chem. 2005;53(10):4290–302. https://doi.org/10.1021/jf0502698.

Qin Y, Liu Y, Yong H, Liu J, Zhang X, Liu J. Preparation and characterization of active and intelligent packaging films based on cassava starch and anthocyanins from *Lycium ruthenicum* Murr. Int J Biol Macromol. 2019a;134:80–90. https://doi.org/10.1016/j.ijbiomac.2019.05.029.

Qin Y, Liu Y, Yuan L, Yong H, Liu J. Preparation and characterization of antioxidant, antimicrobial and pH-sensitive films based on chitosan, silver nanoparticles and purple corn extract. Food Hydrocoll. 2019b;96:102–11. https://doi.org/10.1016/j.foodhyd.2019.05.017.

Qin Y, Liu Y, Zhang X, Liu J. Development of active and intelligent packaging by incorporating betalains from red pitaya (*Hylocereus polyrhizus*) peel into starch/polyvinyl alcohol films. Food Hydrocoll. 2020;100:105410. https://doi.org/10.1016/j.foodhyd.2019.105410.

Re R, Pellegrini N, Proteggente A, Pannala A, Yang M, Rice-Evans C. Antioxidant activity applying an improved ABTS radical cation decolorization assay. Free Radic Biol Med. 1999;26(9):1231–7. https://doi.org/10.1016/S0891-5849(98)00315-3.

Riaz A, Lagnika C, Abdin M, Hashim MM, Ahmed W. Preparation and characterization of chitosan/gelatin-based active food packaging films containing apple peel nanoparticles. J Polym Environ. 2020;28(2):411–20. https://doi.org/10.1007/s10924-019-01619-4.

Rodríguez GM, Sibaja JC, Espitia PJP, Otoni CG. Antioxidant active packaging based on papaya edible films incorporated with *Moringa oleifera* and ascorbic acid for food preservation. Food Hydrocoll. 2020;103:105630. https://doi.org/10.1016/j.foodhyd.2019.105630.

Roy S, Rhim J-W. Carboxymethyl cellulose-based antioxidant and antimicrobial active packaging film incorporated with curcumin and zinc oxide. Int J Biol Macromol. 2020a;148:666–76. https://doi.org/10.1016/j.ijbiomac.2020.01.204.

Roy S, Rhim J-W. Preparation of antimicrobial and antioxidant gelatin/curcumin composite films for active food packaging application. Colloids Surf B: Biointerfaces. 2020b;188:110761. https://doi.org/10.1016/j.colsurfb.2019.110761.

Sanches-Silva A, Costa D, Albuquerque TG, Buonocore GG, Ramos F, Castilho MC, Machado AV, Costa HS. Trends in the use of natural antioxidants in active food packaging: a review. Food Additives Contamin Part A. 2014a;31(3):374–95. https://doi.org/10.1080/19440049.2013.879215.

Sanches-Silva A, Costa D, Albuquerque TG, Buonocore GG, Ramos F, Castilho MC, Machado AV, Costa HS. Trends in the use of natural antioxidants in active food packaging: a review. Food Addit Contamin Part A. 2014b;31(3):374–95. https://doi.org/10.1080/19440049.2013.879215.

Sanches-Silva A, Ribeiro T, Albuquerque TG, Paseiro P, Sendon R, de Quiros AB, Lopez-Cervantes J, Sanchez-Machado DI, Valdez HS, Angulo I, Aurrekoetxea GP, Costa HS. Ultra-high pressure LC determination of glucosamine in shrimp by-products and migration tests of chitosan films. J Sep Sci. 2012;35(5–6):633–40. https://doi.org/10.1002/jssc.201100855.

Shahidi F, Pegg RB. Hexanal as an indicator of meat flavor deterioration. J Food Lipids. 1994;1(3):177–86. https://doi.org/10.1111/j.1745-4522.1994.tb00245.x.

Shahidi F, Zhong Y. Measurement of antioxidant activity. J Funct Foods. 2015;18:757–81. https://doi.org/10.1016/j.jff.2015.01.047.

Singleton VL, Orthofer R, Lamuela-Raventós RM. [14] Analysis of total phenols and other oxidation substrates and antioxidants by means of folin-ciocalteu reagent. In: Methods in enzymology, vol. 299. Academic; 1999. p. 152–78. https://doi.org/10.1016/S0076-6879(99)99017-1.

Singleton VL, Rossi JA. Colorimetry of total phenolics with phosphomolybdic-phosphotungstic acid reagents. Am J Enol Vitic. 1965;16(3):144.

Soares NFF, Hotchkiss JH. Naringinase immobilization in packaging films for reducing naringin concentration in grapefruit juice. J Food Sci. 1998;63(1):61–5. https://doi.org/10.1111/j.1365-2621.1998.tb15676.x.

Song YS, Hepp MA. 26 – US Food and Drug Administration approach to regulating intelligent and active packaging components. In: Han JH, editor. Innovations in food packaging. London: Academic; 2005. p. 475–81. https://doi.org/10.1016/B978-012311632-1/50058-9.

Sun G, Chi W, Zhang C, Xu S, Li J, Wang L. Developing a green film with pH-sensitivity and antioxidant activity based on κ-carrageenan and hydroxypropyl methylcellulose incorporating *Prunus maackii* juice. Food Hydrocoll. 2019;94:345–53. https://doi.org/10.1016/j.foodhyd.2019.03.039.

Sun L-n, Lu L-x, Qiu X-l, Tang Y-l. Development of low-density polyethylene antioxidant active films containing α-tocopherol loaded with MCM-41(Mobil Composition of Matter No. 41) mesoporous silica. Food Control. 2017;71:193–9. https://doi.org/10.1016/j.foodcont.2016.06.025.

Tongdeesoontorn W, Mauer LJ, Wongruong S, Sriburi P, Rachtanapun P. Physical and antioxidant properties of cassava starch-carboxymethyl cellulose incorporated with quercetin and TBHQ as active food packaging. Polymers. 2020;12(2):366. https://doi.org/10.3390/polym12020366.

Tovar L, Salafranca J, Sanchez C, Nerin C. Migration studies to assess the safety in use of a new antioxidant active packaging. J Agric Food Chem. 2005;53(13):5270–5. https://doi.org/10.1021/jf050076k.

Valencia L, Nomena EM, Mathew AP, Velikov KP. Biobased cellulose nanofibril–oil composite films for active edible barriers. ACS Appl Mater Interfaces. 2019;11(17):16040–7. https://doi.org/10.1021/acsami.9b02649.

Vasile C, Stoleru E, Darie-Niţa RN, Dumitriu RP, Pamfil D, Tarţau L. Biocompatible materials based on plasticized poly(lactic acid), chitosan and rosemary ethanolic extract I. Effect of chitosan on the properties of plasticized poly(lactic acid) materials. Polymers. 2019;11(6):941.

Vera P, Canellas E, Nerin C. New antioxidant multilayer packaging with nanoselenium to enhance the shelf-life of market food products. Nanomaterials. 2018;8(10):837/831–12. https://doi.org/10.3390/nano8100837.

Wang L, Guo H, Wang J, Jiang G, Du F, Liu X. Effects of *Herba Lophatheri* extract on the physicochemical properties and biological activities of the chitosan film. Int J Biol Macromol. 2019;133:51–7. https://doi.org/10.1016/j.ijbiomac.2019.04.067.

Wheeler DH. Peroxide formation as a measure of autoxidative deterioration. Oil Soap. 1932;9(4):89–97. https://doi.org/10.1007/BF02553782.

Wrona M, Blasco S, Becerril R, Nerin C, Sales E, Asensio E. Antioxidant and antimicrobial markers by UPLC®–ESI-Q-TOF-MS^E of a new multilayer active packaging based on *Arctostaphylos uva-ursi*. Talanta. 2019;196:498–509. https://doi.org/10.1016/j.talanta.2018.12.057.

Wrona M, Nerín C, Alfonso MJ, Caballero MÁ. Antioxidant packaging with encapsulated green tea for fresh minced meat. Innovative Food Sci Emerg Technol. 2017;41:307–13. https://doi.org/10.1016/j.ifset.2017.04.001.

Wu C, Li Y, Sun J, Lu Y, Tong C, Wang L, Yan Z, Pang J. Novel konjac glucomannan films with oxidized chitin nanocrystals immobilized red cabbage anthocyanins for intelligent food packaging. Food Hydrocoll. 2020;98:105245. https://doi.org/10.1016/j.foodhyd.2019.105245.

Xie J, Schaich KM. Re-evaluation of the 2,2-Diphenyl-1-picrylhydrazyl Free Radical (DPPH) assay for antioxidant activity. J Agric Food Chem. 2014;62(19):4251–60. https://doi.org/10.1021/jf500180u.

Yadav S, Mehrotra GK, Bhartiya P, Singh A, Dutta PK. Preparation, physicochemical and biological evaluation of quercetin based chitosan-gelatin film for food packaging. Carbohydr Polym. 2020;227:115348. https://doi.org/10.1016/j.carbpol.2019.115348.

Yang Y-N, Lu K-Y, Wang P, Ho Y-C, Tsai M-L, Mi F-L. Development of bacterial cellulose/chitin multi-nanofibers based smart films containing natural active microspheres and nanoparticles formed in situ. Carbohydr Polym. 2020;228:115370. https://doi.org/10.1016/j.carbpol.2019.115370.

Yildirim S, Röcker B, Pettersen MK, Nilsen-Nygaard J, Ayhan Z, Rutkaite R, Radusin T, Suminska P, Marcos B, Coma V. Active packaging applications for food. Compr Rev Food Sci Food Saf. 2018;17(1):165–99. https://doi.org/10.1111/1541-4337.12322.

Yong H, Wang X, Zhang X, Liu Y, Qin Y, Liu J. Effects of anthocyanin-rich purple and black eggplant extracts on the physical, antioxidant and pH-sensitive properties of chitosan film. Food Hydrocoll. 2019;94:93–104. https://doi.org/10.1016/j.foodhyd.2019.03.012.

Zhang X, Liu J, Yong H, Qin Y, Liu J, Jin C. Development of antioxidant and antimicrobial packaging films based on chitosan and mangosteen (*Garcinia mangostana* L.) rind powder. Int J Biol Macromol. 2020;145:1129–39. https://doi.org/10.1016/j.ijbiomac.2019.10.038.

Zia J, Paul UC, Heredia-Guerrero JA, Athanassiou A, Fragouli D. Low-density polyethylene/curcumin melt extruded composites with enhanced water vapor barrier and antioxidant properties for active food packaging. Polymer. 2019;175:137–45. https://doi.org/10.1016/j.polymer.2019.05.012.

Chapter 4
Emitters of Essential Oils

Regiane Ribeiro-Santos, Victor Gomes Lauriano de Souza, Mariana Alvoco Andrade, Ana Sílvia Boroni de Oliveira, and Joyce Fagundes Gomes Motta

Abstract Essential oils (EOs) are natural bioactive compounds that can be used in active release systems, such as films, coatings or sachets, being gradually transferred from packaging into foods, during storage time. They are a complex mixture of volatile compounds obtained from plants and fruits with biological properties, which can be used as preservatives or flavoring agents aiming to extend the shelf-life and increase the overall quality of foods. Due to their limitations, such as volatility and low solubility in water, and to improve their biological activities, EOs may be encapsulated before being incorporated into the polymeric matrix of a packaging. This technique provides a slow release of the EOs to the food surface or to the headspace of the package. EOs are classified as Generally Recognized as Safe

R. Ribeiro-Santos (✉)
Instituto Federal de Educação, Ciência e Tecnologia de Pernambuco – IFPE campus Vitória de Santos Antão. Propriedade terra preta, Vitória de Santo Antão, Pernambuco, Brasil

V. G. L. de Souza
MEtRICs/Departamento de Ciências e Tecnologia da Biomassa, NOVA School of Science and Technology | FCT NOVA, Universidade NOVA de Lisboa, Caparica, Portugal

INL, International Iberian Nanotechnology Laboratory, Braga, Portugal

M. A. Andrade
Department of Food and Nutrition, National Institute of Health Dr. Ricardo Jorge, Lisbon, Portugal

University of Coimbra, Faculty of Pharmacy, Coimbra, Coimbra, Portugal

REQUIMTE/LAQV, R. D. Manuel II, Oporto, Portugal

A. S. B. de Oliveira
Instituto Federal de Educação, Ciência e Tecnologia do Mato Grosso do Sul - IFMS Campus Coxim, Coxim, Mato Grosso do Sul, Brasil

J. F. G. Motta
Departamento de Ciência dos Alimentos, Universidade Federal de Lavras, Lavras, Minas Gerais, Brasil

S. M. Jafari, A. S. Silva (eds.), *Releasing Systems in Active Food Packaging*, Food Bioactive Ingredients, https://doi.org/10.1007/978-3-030-90299-5_4

(GRAS) by the Food and Drug Administration (FDA); however, the ingestion of higher doses can cause of serious problems of oral toxicity. It is necessary to find an equilibrium between the effective EO dose and the risk of toxicity to consider their use insurance for food purposes. Thus, several studies about EO emitters from packaging to food have been realized and the new technologies have contributed to improve food security, which will be discussed in this chapter.

Keywords Antimicrobial · Antioxidant · Nanotechnology · Emulsion · Encapsulation · Active food packaging · Migration

1 Introduction

Essential oils (EOs) are highlighted as possible natural food additives and have been addressed in numerous research to reduce or replace the use of the traditional synthetic ones. This current trend is related to a more consciousness consumers about potential health problems correlated with synthetic additives (Santana et al. 2013). In addition, EOs are attracting great interest due to their potential use in foods as antimicrobial and antioxidant agents, besides, they can also be used as flavoring. They are natural compounds, extracted from plants and fruits (Ebani et al. 2016; Chouhan et al. 2017; Catarino et al. 2017; Fadel et al. 2019; Ksouda et al. 2019; Goudjil et al. 2020).

EOs have also been used in the formulation of active food packaging, being responsible for the active properties (antimicrobial or antioxidant), aiming the food shelf-life extension or the maintenance or improvement on the condition of packaged food. They are used as emitters (active releasing systems) where may be contained in a separate container as sachet or directly incorporated into the packaging material as films and coating. EOs are released during food transportation and storage from the packaged to the food or the environment surrounding the food (Espitia et al. 2012; Yeddes et al. 2019; Ksouda et al. 2019). Due to this, a large variety of EOs have been indirectly applied to foods of both vegetable and animal origin (Botre et al. 2010; Muriel-Galet et al. 2012). In addition, Food and Drug Administration (FDA) consider EOs as Generally Recognized as Safe (GRAS), which permits their use as food additives. However, reproducibility, organoleptic acceptance and allergic reactions of the EOs are limiting factors for use in real packaging conditions (Ribeiro-Santos et al. 2017a, b; Dvir et al. 2019).

Despite this, EO volatility, low solubility in water and undesirable changes in the flavor are drawbacks which may be overcome for encapsulated techniques that have been used to decrease these impacts and to enhance their biological activities. In this technique, droplets or particles of active substances (core material) are coated into the polymeric matrix (wall material) by small capsules (Souza et al. 2019a).

The interaction between the packaging material and the EOs compounds determines the release rate, which have a direct effect on the food quality (Uzunku and Var 2018; Dvir et al. 2019). Besides, the migration can be influenced by the packaged food(s) and the conditions to which the package will be exposed (Ribeiro-Santos et al. 2017d; Khaneghah et al. 2018). Several materials derived from non-renewable source, such as polypropylene, or from biomaterials such as starch, can be used as polymeric matrices in the production of active food packaging (Zhang et al. 2019; Dvir et al. 2019).

2 Essential Oils: An Overview

EOs are aromatic liquids, volatile, extracted from different parts of plants, such as flowers, buds, seeds, leaves, fruits, roots, stem, bark, and herbs and constituted by several components. These oils have been widely studied due to their active functions that can be exploited by the food industry, replacing synthetic compounds (Burt 2004; Bakkali et al. 2008; Pandey et al. 2017). EOs are extracted by different methods such as hydrodistillation, microwave extraction, solvent extraction, steam distillation, supercritical fluid extraction and ultrasound. At laboratory scale, the most frequently used methods to extract EOs are the conventional steam distillation and hydrodistillation, for example. Depending on the method of extraction, chemical profile of EOs may differ both in the yield and composition (Stratakos and Koidis 2016).

2.1 Biological Properties of Essential Oils

Currently, the research on EOs as natural food additives has increased due to their active functions that include antimicrobial, antioxidant and aromatic properties and which are well regarded for preserving food (Valdivieso-Ugarte et al. 2019). The high content in phenolic compounds confers to EOs activities capable to decrease the growth of spoilage and pathogenic microorganisms and retard lipid oxidation in food products (Ribeiro-Santos et al. 2017c, d, 2018a).

2.1.1 Antimicrobial Activity

The antimicrobial capacity of EOs is considered as one of their main advantages and reasons for their research and application to maintain the quality and safety of foods and, consequently, to extend the shelf-life of these products. In addition, EOs effectively contribute to impede the growth of pathogenic and spoilage microorganisms and consequently assist in the prevention of foodborne diseases (Taylor 2018).

Yet, the EO antimicrobial mechanism of action is not fully understood and therefore cannot be attributed to a single mechanism (Khorshidian et al. 2018). However, one of the plausible justifications is related to their hydrophobic characteristics that tends to promote disorder of cellular structures, making them more permeable due to the lipid partition in the cell membrane and mitochondria. This allows the leakage of ions and other components of the cell which, when in large quantities, cause the cell death of microorganisms (Burt 2004).

Therefore, EOs are effective against molds, yeasts and bacteria (Gavahian et al. 2020). Studies demonstrate that, because to the existence of an external membrane in Gram-negative bacteria which does not exist in Gram-positive bacteria, and that it is defined as a lipid bilayer, composed internally by phospholipids and externally by lipopolysaccharide, hinders the active permeation of the EOs into the plasma membrane (Malanovic and Lohner 2016). So, Gram-positive bacteria are, in general, more susceptible to EOs than Gram-negative bacteria and, bacteria themselves are more resistant to EOs than fungi (Smith-Palmer et al. 1998; Rodriguez-Garcia et al. 2016; Chouhan et al. 2017; Patterson et al. 2019).

The antimicrobial effectiveness of EOs are evaluated by many methods, such as agar diffusion, time-kill (time-dependent antimicrobial effect on EOs), checkerboard test (identification of possible synergistic effects of EOs) and, dilution method (Rao et al. 2019). This dilution method consists of determining the minimum concentration of EOs necessary to inhibit the growth of microorganisms through wells or microdilution tubes (Chouhan et al. 2017). This concentration is known as the minimum inhibitory concentration (MIC) and varies according to the EO, microorganisms, time of exposure of the microorganisms to these EOs, solubility and the way in which the EOs are solubilized or emulsified (Oussalah et al. 2006).

In vitro analysis of oregano EOs (Ebani et al. 2016; Ribeiro-Santos et al. 2018b), cinnamon cassia (Khorram et al. 2018; Vijayan and Mazumder 2018), lemon (Dao et al. 2019; Ilango et al. 2019), mentha (Brahmi et al. 2016), rosemary (Medeiros Barbosa et al. 2016), *Lippia alba* (Souza et al. 2017; Peixoto et al. 2018), Brazilian rose pepper (D'Sousa Costa et al. 2015; Uliana et al. 2016), among others have been summarized in Table 4.1, which demonstrate the vast antimicrobial potential of EOs against different types of microorganisms.

2.1.2 Antioxidant Activity

One more advantage to be pointed out regards the EOs' antioxidant capacity (Dhifi et al. 2016). In food, as lipid oxidation so as enzymatic browning are concerns in food technology. The lipid oxidation reduces the nutritional safety and quality as it causes unpleasant odors and flavors in products. While enzymatic browning in damaged vegetables results in the formation of dark compounds caused, mainly, by the polyphenol oxidase enzyme which is malfunctioning and oxidizes phenolic compounds (Del Ré and Jorge 2012; Chen et al. 2017).

Therefore, EOs can also be used as antioxidant agents due to the presence of phenolic compounds in their composition (Koroch et al. 2007; Amarowicz and Pegg 2019). Phenolics are organic compounds in which hydroxyl groups (OH) are bonded to aromatic rings with oxidizing and redox properties, acting as a reducing agent and hydrogen donor. High molecular weight phenolic compounds have the ability to

Table 4.1 Essential oil *in vitro* antimicrobial activity against different target microorganisms (fungi and bacteria)

Essential oils/ sources	Target microorganisms	References
Brasilian rose pepper	*Enterococcus faecalis* *Staphylococcus aureus*, *Escherichia coli* and *Candida albicans*	D'Sousa Costa et al. (2015) Uliana et al. (2016)
Cinnamon cassia	*Penicillium digitatum* *Staphylococcus aureus, Salmonella typhimurium, Proteus vulgaris, Psudomonas aeruginosa, Escherichia coli* and *Klebsiella pneumoniae*	Khorram et al. (2018) Vijayan and Mazumder (2018)
Lemon	*Bacillus cereus, Escherichia coli, Pseudomonas aeruginosa* and *Staphylococcus aureus* *Lactobacillus, Staphylococcus epidermidis* and *Streptococcus mutans*	Dao et al. (2019) Ilango et al. (2019)
Lippia alba	*Aeromonas* spp. *Fusarium pallidoroseum, Fusarium solani* and *Lasiodiplodia theobromae*	Souza et al. (2017) Peixoto et al. (2018)
Mentha	*Aspergillus flavus, Bacillus subtillis, Klebsiella pneumoniae, Staphylococcus aureus and others*	Brahmi et al. (2016)
Oregano	*Salmonella typhimurium, Yersinia enterocolitica, Listeria monocytogenes, Enterococcus faecium, durans* and *faecalis* *Staphylococcus aureus, Escherichia coli* and *Penicillium* sp.	Ebani et al. (2016) Ribeiro-Santos et al. (2018b)
Rosemary	*Listeria monocytogenes, Escherichia coli* and *Salmonella* Enteritidis	Medeiros Barbosa et al. (2016)

inhibit free radicals. Their main sources are fruits, vegetables and plant extracts (Rappoport 2004; Ivana and Marija 2019; Hasheminya and Dehghannya 2020).

Different *in vitro* assays are reported in the literature to quantify the antioxidant potential of EOs. The most common assays are the Oxygen Radical Absorbance Capacity (ORAC) (Bentayeb et al. 2014); photochemiluminescence (PCL) (Noriega et al. 2019); Ferric Reducing Antioxidant Power (FRAP) (Goudjil et al. 2020); Cupric Reducing Antioxidant Capacity (CUPRAC) (Chemsa et al. 2018); Trolox Equivalent Antioxidant Capacity (TEAC) (Basmacıoğlu-Malayoğlu et al. 2018); Inhibition of the DPPH (2,2′-diphenyl-1-picrylhydrazyl) radical (Liaquat and Ali 2019), and the β-carotene-linoleic acid (linoleate) bleaching assay (Hichri et al. 2019). These tests are distinguished in the intensity of simplifications performed in relation to real oxidative processes. Generally, the DPPH radical inhibition is the most commonly used method (Amorati et al. 2013; Valdivieso-Ugarte et al. 2019).

Studies providing the *in vitro* antioxidant effect of several EOs in different concentrations are also reported, namely the antioxidant capacity of oregano (Stanojević et al. 2018), rosemary, thyme (Zeid et al. 2019), cinnamon (Kim et al. 2018; Tepe and Ozaslan 2020), orange (Tores-Alvarez et al. 2017), basil (Ademiluyi et al. 2016), chilean boldo (Souza et al. 2019b), clove (Ugalde et al. 2017), eucalyptus

(Hafsa et al. 2016), lavender (Jamróz et al. 2018), aloe vera (El Fawal et al. 2019), Salvia (Pirouzifard et al. 2019), ginger (Noori et al. 2018), and lemon (Hsouna et al. 2017) EOs among others.

2.1.3 Flavoring Agent

EOs can also be used to intensify the aroma or flavor of several products besides foods, such as cosmetics, soaps, plastic resins and perfumes (Pandey et al. 2017). For this, two prerequisites are required: adequate production technology and appropriate selection of the raw material (Rios 2016). Some studies reported the use of EOs modified pleasantly the flavor of foods. Chandran et al. (2017) studied coconut oil oxidative and thermal stability when black pepper EOs and ginger were incorporated in their formulation, in addition, the sensory acceptance of these mixtures was also evaluated. It was observed that as for the sensory evaluation, the oils were inserted in a vegetable salad and consumers preferred the combination of black pepper and ginger EOs with the coconut oil more than the control oil (pure coconut oil).

Other EOs such as mint, eucalyptus and those extracted from herbs and fruits are extensively used in the food and cosmetic industries, mainly because of the flavor that these volatile products impart (Hüsnü and Buchbauer 2015). In this sense, EOs improve the flavor of products, enhancing the brand's image and, consequently, gaining consumers' trust by giving primacy to natural products. However, the presence of the flavors generated by the EOs is not always desirable, especially when EOs produce a strong aroma, but undesirable effects can be counterbalanced by meticulous selection of EOs according to the type of food. To overcome this problem, EOs can be incorporated into active packaging, encapsulated or used in smaller amounts combined with other active agents (synergistic effect) (Mariod 2016; Perdones et al. 2016; Souza et al. 2019a).

2.2 *Active Compounds in Essential Oils*

The main active functions of EOs are the result of the components present in each oil (Burt 2004). Thus, the composition of EOs differs according to the type of plant species, geographical origin of the plant, soil, climate, stage of the vegetative cycle, part of the plant used and the method of extraction (Bakkali et al. 2008; Khorshidian et al. 2018). The technique used to identify and quantify the EOs' compounds, in general, gas chromatography coupled with mass spectrometry detector (Ribeiro-Santos et al. 2018a).

Studying cinnamon EOs, Ribeiro-Santos et al. (2017a) found that eugenol is the major component of cinnamon (*Cinnamomum zeylanicum* L.) leaf EOs, while cinnamaldehyde is the major compound of cinnamon cassia (*Cinnamomum cassia* L.) EOs extracted from bark. In addition, Elzaawely et al. (2007) reported that *Alpinia zerumbet* EOs have different main components depending on the part of the plant used to extract the EO (flower fraction was different from the seed fraction).

EOs are constituted by a mixture of compounds, such as terpenoids, alcoholic compounds, acids, aldehydes and phenolics from secondary metabolites of herbs, used as raw materials for their extraction. Among these, terpenes (such as pinene, mircene, limonene, terpinene, *p*-cymene), terpenoids (for example oxigen-containing hydrocarbons), and aromatic phenols (carvacrol, thymol, safrole, eugenol, etc.) are considered, in theory the main constituents of EOs. Each EO contains two or three of these compounds in higher concentrations (20–95%) and other compounds in minor levels. Thus, the biological properties of EOs mainly come from their major compounds (Koul et al. 2008; Shaaban et al. 2012; Pandey et al. 2017). Some EOs and their main compounds are mentioned in Table 4.2.

Thymol is known as the main monopertene phenol found in EOs extracted from plants that belong to the *Lamiaceae* family (Marchese et al. 2016) and carvacrol is a phenolic monoterpenoid, also present in several plants (Sharifi-Rad et al. 2018). Terpenes (*p*-cymene and limonene), terpenoids (thymol and carvacrol), phenylpropenes (eugenol, vanillin and cinnamaldehyde), and other compounds such as allicin and isothiocyanates are components present in EOs associated with antimicrobial activity (Hyldgaard et al. 2012).

Despite the fact that biological activities of the EOs are associated to their major compounds, minor compounds (present in small amounts) have a significant role in the biological activities, acting in synergism (interactions) with the others constituents (Chouhan et al. 2017). This interaction was observed in the study by Reyes-Jurado et al. (2019). The authors tested the antimicrobial activity of Mexican EOs which major compounds are the *p*-cymene followed by oregano and carvacrol. The *p*-cymene is a monopertene present in more than 100 plant species with medicinal and food properties. Reyes-Jurado et al. (2019) showed that the Mexican EOs exhibit lower antimicrobial activity against 18 microorganisms than other evaluated EOs. According to Marchese et al. (2017a), *p*-cymene when used alone does not have an antimicrobial effect, being necessary for other compounds to be used together to potentialize its action. The synergism can also occur when different EOs are combined, as shown by Purkait et al. (2020). The authors assessed the antimicrobial, antifungal and antioxidant properties of black pepper, cinnamon and clove EOs. Combined EOs exhibited higher biological effects than the EOs alone. Among the tested combinations, the best was cinnamon/clove EO.

However, interactions between the compounds (major and minor) that decrease the biological activities of the EOs can also occur, creating a counter-synergistic effect, known as antagonism (Bassolé and Juliani 2012). Antagonism effect on the MIC of *E. coli* was reported by Goñi et al. (2009) when combined cinnamon and clove EOs were tested. On the other hand, at more concentrated levels, these individual EOs and their combinations presented synergistic effect against *E. coli*. Thus, a synergistic or antagonistic effect could be concentration dependent. In another study, Ciesla et al. (2016) studied the synergistic and antagonistic antioxidant interactions for binary mixtures of common monoterpenes. The authors found that there was a significant antioxidant synergism between *p*-cymene and one of the

Table 4.2 Main constituents of some ssential oils

Source of essential oils	Main components	References
Basil (*Ocimum basilicum*)	Methyl chavicol and linalool	Ribeiro-Santos et al. (2017a)
Black cumin (*Nigella sativa*)	p-Cymene, thymoquinone, tricyclene, α-thujene	Viuda-Martos et al. (2011)
Black pepper	β-Caryophyllene	Chandran et al. (2017)
Brazilian rose pepper (*Schinus terebinthifolius* Raddi)	δ-3-carene	Uliana et al. (2016)
Cinnamon cassia (*Cinnamomum cassia*)	*trans*-Cinnamaldehyde	Ribeiro-Santos et al. (2017a)
Cinnamon leaf (*Cinnamomum zeylanicum*)	Eugenol	Ribeiro-Santos et al. (2017a)
Clove	Eugenol	Purkait et al. (2020)
Eucalyptus	*p*-cymene	Hafsa et al. (2016)
Fennel (*Foeniculum vulgare*)	Trans-anethole, methyl chavicol	Viuda-Martos et al. (2011)
Ginger	α-Zingiberene	Noori et al. (2018)
Lavender (*Lavandula officinalis*)	Linalool, linalyl acetate, camphor	Viuda-Martos et al. (2011)
Lemon (*Citrus aurantifolia*)	α-pinene and β citral	Dao et al. (2019)
Mentha (*Mentha pulegium* L.)	Pulegone	Brahmi et al. (2016)
Mentha (*Mentha rotundifolia* L.)	*trans*-Piperitone epoxide	Brahmi et al. (2016)
Mexican oregano (*Lippia berlandieri*),	*p*-cymene and carvacrol	Reyes-Jurado et al. (2019)
Mustard (*Brassica nigra*)	Allyl isothiocyanate	Reyes-Jurado et al. (2019)
Orange	D-limonene	de Andrade et al. (2020)
Oregano (*Origanum Vulgare* L.)	Thymol, carvacrol and β-terpineol	Salvo et al. (2019)
Parsley (*Petroselinum crispum*)	Caryophyllene oxide apiole, α-pinene, β-pinene	Viuda-Martos et al. (2011)
Pennyroyal (*Mentha pulegium*)	Menthone, pulegone	Teixeira et al. (2012)
Rosemary (*Rosmarinus officinalis*)	1,8-Cineole and Camphor	Ribeiro-Santos et al. (2017a)
Summer savory (*Satureja hortensis*)	Carvacrol carvacrol and γ-terpinene	Khorram et al. (2018)
Thyme (*Thymus spathulifolius*)	Thymol, borneol and carvacrol	Ceylan et al. (2016)
Thyme (*Thymus vulgaris*)	O-cymol, linalool and thymol	Reyes-Jurado et al. (2019)

Adapted from Ribeiro-Santos et al. (2018a)

compounds, i.e. (±) -citronellal, α-phellandrene or α-terpinene, characterized by moderate antioxidant activity.

Limonene – a cyclic monopertene, is another EO compound with antimicrobial activity, which is present in the peels of citrus fruits, such as lemon and orange, and in mint and fir (Jongedijk et al. 2016). de Andrade et al. (2020) showed that limonene was the major compound of orange EO and, *in vitro* studies showed that this oil had a great antibacterial property against Gram-negative bacteria including *E. coli*.

Regarding phenylpropenes, eugenol is a hydroxyphenylpropene, found in plants of the families *Lamiaeceae*, *Lauraceae*, *Myrtaceae*, and *Myristicaeae* and cinnamaldehyde is a phenylpropene aldehyde, commonly found in cinnamon. Both compounds have a broad antimicrobial potential reported in the literature (Manu 2016; Marchese et al. 2017b). Ribeiro-Santos et al. (2017a) observed that cinnamaldehyde and cinnamon cassia (*C. cassia* L.) EOs showed a higher antimicrobial effect against Gram-positive *S. aureus*, Gram-negative *E. coli* and the fungus *Penicillium* spp. than eugenol and cinnamon leaf (*C. zeylanicum* L.) EOs. Another example of phenylpropene is vanillin, being the main component of the vanilla aroma, giving it a sweet and creamy aroma. It is found, above all, in the vanilla orchid grains; however, it is extracted in small quantities due to its low levels in the plant (Furuya et al. 2017). Not many studies are found regarding the antimicrobial property of vanillin and/or vanilla EOs; however, due to its pleasant aroma, it can be used in mixtures with other EOs, providing an improvement in aroma (Gutiérrez et al. 2009).

Thus, as well as the antimicrobial effect, the antioxidant activity is also related to the major compounds in EOs (Dhifi et al. 2016). The antioxidant action of EOs can be associated with terpenoids and phenylpropenes, which comprise phenolic compounds with antioxidant characteristics because they react with the peroxyl radicals which are eliminated by hydrogen atom transfer, with thymol and carvacrol as two of the main antioxidant agents precisely because of their phenolic structure (Dhifi et al. 2016). The *in vitro* studies carried out by Ceylan et al. (2016) investigated the chemical composition and antioxidant properties by different methods, on the oil extracted from the aerial part of *Thymus spathulifolius*. EOs exhibited good antioxidants due to the main compound found, i.e., thymol (50.5%), borneol (16.7%) and carvacrol (7.7%). Abou Baker et al. (2020) evaluated several properties of the summer savory EOs (*Satureja hortensis* L.) and identified that carvacrol was the major compound (48.51%), followed by γ-terpinene (36.63%). In addition, the results showed high antioxidant activity of this EO, associated with the major compounds and high levels of phenols and flavonoids.

2.3 Micro- and Nano-Encapsulation of Essential Oils

In order to overcome limitations of EOs, as strong aroma and taste that can change the food's organoleptic characteristic and interfere on their acceptance consequently restricting their direct use in foods, several incorporation techniques into polymeric matrices have been tested. Through this strategy, due to the controlled release towards foods, a lower amount of EOs is necessary to achieve the same preservation results with diminished changes in the food intrinsic characteristics (Gómez-Estaca et al. 2014; Andrade et al. 2018; Hassan et al. 2018; Souza et al. 2019a). Such phenolic compounds, responsible for the bioactivity of EOs, as other bioactive substances (e.g. tocopherol or ascorbic acid), are sensitive to oxygen, heat and light (i.e. they are chemically degradated/modified in the presence of such conditions), and also some of them have a hydrophobic nature (Brewer 2011; Hadian et al. 2017; Yeh et al. 2014), i.e. incompatible with some hydrophilic polymers. Moreover, as most of conventional packaging structures are manufactured by extrusion process at high temperatures, the most critical issue is, therefore, whether the active substance is degraded by the severe thermo-mechanical treatment submitted during the extrusion (Gómez-Estaca et al. 2014).

Thus, encapsulation of EOs may be one alternative to preserve their active properties, control their diffusion and ensure their effectiveness. Besides, other positive aspects are worth mentioning with the encapsulation of EOs: enhancement of the contact area with water-soluble foods; easier dispersion on the surface of food; maintenance of their biological activity; reduction of their volatility and the negative adverse effects due to their strong odor and taste, which may increase their effective utilization (Dias et al. 2014; Ju et al. 2019). The term 'encapsulated active compounds' is applied when droplets or particles of a substance (the core material) are coated within polymeric or lipid-based materials by extremely small capsules (the wall material). According to their size, they are divided into two groups: microencapsulated particles, with size ranging from 1 to 1000 μm, or nanoencapsulated particles, with size from 1 to 100 nm (Becerril et al. 2020; Ribeiro-Santos et al. 2017c).

The list of substances that can be used as wall material in the encapsulation process is vast however, due to the food additives regulations (e.g. Regulation (EC) No 1935/2004 (European Union (EU) 2004), only those certified for food application are allowed to be used in the encapsulation of EOs (Nedovic et al. 2011). Thus, most materials used are biodegradable biomolecules with food-grade status and, with the ability to form a barrier between the encapsulated core material and its surroundings. In food applications, the most widely used wall compounds are polysaccharides, such as starch, dextrins, maltodextrins, cellulose, gums (e.g. Arabic, mesquite), galactomannans, pectins, carrageenans, sodium alginate, chitosan, xanthan and gellan. Proteins (e.g. milk, whey proteins, caseins, gelatine and gluten) and lipids (e.g. fatty acids, beeswax, carnauba wax, glycerides and phospholipids) are also suitable to be used as wall material (Nedovic et al. 2011; Ribeiro-Santos et al. 2017c). The type of active compound, its characteristics and the target application of the encapsulated material are the main criteria when selecting the proper wall material for encapsulation (Nedovic et al. 2011).

Spray drying is the most used technique to encapsulate active substances; however, different techniques can be used, such as melt injection, extrusion, ionic gelation, fluid bed coating, vacuum and freezing-drying, spray chilling, emulsification and molecular inclusions in cyclodextrins and liposomal vesicles (Nedovic et al. 2011; Ribeiro-Santos et al. 2017c). Spray drying is basically used to encapsulate liquid substances that, after being emulsified with the wall material, are pulverized as small droplets into a chamber with hot air circulation, resulting in an almost instantaneous evaporation of solvent, turning the system into powder (solid particles) (Ribeiro-Santos et al. 2017c). Molecular inclusion using cyclodextrins or liposomal vesicles is a less exploit method due to the superior cost involved in them (Nedovic et al. 2011).

Currently, the interest on nano-sized formulation of EOs has expanded once their subcellular size confers significant advantages over conventional encapsulation (microencapsulation), especially in terms of their larger surface area, which enables a higher intracellular uptake and carrying capacity with an extended biological activity (Ghaderi-Ghahfarokhi et al. 2017). Throughout the recent development in the nanotechnology field (i.e. use of nanoencapsulation), with applications in the food science, it is possible to improve the bioavailability and enhance solubility, while protecting EOs during processing, storage and distribution (Neethirajan and Jayas 2010; Souza and Fernando 2016). Despite the novelty of such approach on the application of EOs, several research studies are found in the literature, most of them using nanoencapsulation techniques. Some studies of encapsulated EOs and their encapsulation techniques can be seen in Table 4.3.

As an example, rosemary EO was successfully nanoencapsulated into chitosan benzoic acid nanogels with enhancement on the antimicrobial and antioxidant activities of the EO, as well as improved stability (Hadian et al. 2017). The gel nanoparticles obtained were of spherical shape, diameter <100 nm and uniform size distribution. Pure EOs and the nanoencapsulated ones were also applied in beef cutlet samples inoculated with *Salmonella typhimurium*. The encapsulated EOs presented a promising effect on reducing the pathogens population and, also led to a lower impact on the increasing of pH and on changing the color of the samples. The authors attributed the enhancement on the EO bioactivity to the protecting effect of the chitosan benzoic acid nanocapsules, which reduced the volatility and instability of the EOs when exposed to environmental factors.

Nanogels of chitosan-cashew gum were used to encapsulate peper-rosmarin (*Lippia sidoides*) EOs (Abreu et al. 2012). *Lippia sidoides* is a plant native to Northeast Brazil, and its EO is rich in thymol with strong antimicrobial activity. Nanogels are the designation for hydrogel particles in the nanometer range, being usually formed by physical interaction of oppositely charged ions (Abreu et al. 2012). After the formation of polymeric complex of chitosan: gum, the emulsion containing the EOs was incorporated and dried using a spray dryer. Nanogels were found to show average sizes in the range of 335–558 nm, and encapsulation efficiency of 70%. It is woth mentioning that dispites the average particle size of the gel produced was >100 nm, the authors still considered the gel as a nanogel. *In vitro* release studies demonstrated that nanogel presented slower and sustained release, while the larvicide efficacy was enhanced with the encapsulation reaching mortality rates superior to that obtained by the *L. sidoides* pure oil (Abreu et al. 2012).

Thyme EOs and its major component thymol were microencapsulated in β-cyclodextrin using two different methods: kneading (KD) or freeze-drying (FD) (Tao et al. 2014). The entrapment efficiency was superior for thymol and FD particles. Particles produced using FD showed larger particle sizes than the KN method. The FD method was the best encapsulation method since enhanced antimicrobial activity for both compounds. The differences on the microbiological activity may

Table 4.3 Some selected studies on encapsulation of essential oils

Source of essential Oil	Micro/ Nano-encapsulation	Wall materials	Encapsulation technique	References
Chia	Microencapsulation	WPC with mesquite gum; WPC + gum Arabic	Spray dried	Rodea-González et al. (2012)
Cardamom	Nanoencapsulation	Chitosan nanoparticles	Ionic gelation	Jamil et al. (2016)
Cinnamon	Nanoencapsulation	Chitosan nanoparticles	Emulsion-gelation method	Ghaderi-Ghahfarokhi et al. (2017)
Cinnamon	Nanoencapsulation	Nanoliposomes	Thin film dispersion method	Wu et al. (2015)
Lemon ironbark	Nanoencapsulation	Cashew gum	Spray dried	Herculano et al. (2015)
Rosemary	Nanoencapsulation	Chitosan-benzoic acid	Nanogel formation with 1-Ethyl-3-(3-dimethylaminopropyl) carbodiimide (EDC) mediated reaction and ultrasound	Hadian et al. (2017)
Pepper-rosmarin	Microencapsulation	Chitosan and cashew gum nanogel	Spray dried	Abreu et al. (2012)
Clover, cinnamon bark, and lemongrass	Microencapsulation	β-chitosan	Ionic gelation	Zhang et al. (2017a)
Eugenol, cinnamon, and clove (extracts)	Microencapsulation	Beta-cyclodextrin	Freeze- drying method	Hill et al. (2013)
Thymol and thyme essential oil	Microencapsulation	Beta-cyclodextrin	Freeze-drying and kneading	Tao et al. (2014)
Walnut	Microencapsulation	Skim milk powder (SMP), SMP + Tween 80, and SMP + maltodextrin	Spray dried	Shamaei et al. (2017)
Zataria multiflora	Nanoencapsulation	Chitosan nanoparticles	Ionic gelation	Mohammadi et al. (2015b)

depend on the different steric conformation of the guest molecules, inclusion complex synthesis method and the release rate from the cavity of the β-cyclodextrin (Tao et al. 2014).

EO encapsulation can be used for its protection and, also, to improve the flavor of foods. Fadel et al. (2019) reported that the flavor of biscuits is lost during storage and, to address this problem, the authors encapsulated cinnamon EO in maltodextrin, as an exogenous flavor. The authors observed that this EO was responsible for sensorially pleasing the evaluators and maintaining the biscuit quality. These examples highlight the remarkable potential application for the encapsulation of EOs as antimicrobial and antioxidant complexes in food systems as a valuable alternative method to preserve the food quality by the inhibition of pathogens and/or deteriorative microorganisms and oxidative deteriorative processes in a more sustainable approach. Nevertheless, such technology matches the current trend of consumers to look for greener and more environmentally friendly products (Souza et al. 2020).

2.4 Nanoemulsions of Essential Oils

Nanoemulsions are characterized by droplets of oil, < 100 nm in diameter, surrounded by emulsifier molecules to form a thin layer. The reduction in the droplet size in nanoemulsions significantly reduces gravity and Brownian motion, thus significantly decreasing the separation phase than conventional emulsions, which results in more stable systems (Gharenaghadeh et al. 2017). The mixture between the disperse and continuous phase is generally done using a high-speed homogenizer (Alexandre et al. 2016; Zhang et al. 2017b), however it can be enhanced with the assistance of an ultrasonicator (Otoni et al. 2014b) as it was performed by Gharenaghadeh et al. (2017) to emulsify *Salvia multicaulis* EOs. The authors produced the nanoemulsions with Tween® 80 and Span™ 80 surfactants and investigated the size and physical stability, as well as the antimicrobial and antioxidant activities. The produced nanoemulsions were stable for 60 days, and the antioxidant activity, measured by DPPH• scavenging assay, of the EO nanoemulsion was significantly higher than the antioxidant activity of free *Salvia multicaulis* solubilized in methanol for all concentrations assessed. Also, there was an enhancement in the antibacterial effects of the encapsulated EOs in nanoemulsions, which was attributed to: the system enabled the transference of the antibacterial compound across the cell membrane layer, acting as a carrier agent, bringing considered concentrations of compounds within the aqueous phase to microorganisms. Moreover, the smaller nanoparticles could be smoothly incorporated into the small holes/cavities present on the cell membranes of bacteria, which function as the entrance and exit for nutrients and metabolites, and then release their bioactive (Gharenaghadeh et al. 2017).

Rutin, the glycoside of quercetin, is a flavonoid present abundantly and distributed in plants such as in buckwheat seed, fruits (especially citric fruits), and fruit rinds (Almeida et al. 2010). This compound is present in the extracts of such plants and also in EOs as minor component (Brewer 2011). This phenolic compound has a remarkable antioxidant property (high scavenging ability on oxidizing specimens), however, its poor solubility in aqueous media limits its use. Thus, nanoemulsions and nanoencapsulation are strategies to overcome this drawback (Almeida et al. 2010). Almeida et al. (2010) assessed the activity of rutin in both forms, as a nanoemulsion or nanoencapsulated in poly-ε-caprolactone (PCL). The difference between the nanoemulsion and the nanoencapsulated rutin was only the presence of the polymer, that was added to form the vesicle. The surfactants used were sorbitan monostearate (Span™ 60) and polysorbate 80 (Tween® 80). Results demonstrated that the particles size obtained did not differ regarding the method used, as well as the antioxidant properties, however the presence of the polymeric layer in nanocapsules was essential to obtain a continued antioxidant activity (Almeida et al. 2010). Thus, the use of wall materials (polymers) to encapsulate active compounds has advantages over emulsion systems.

To obtain active packaging incorporated with EOs, based on hydrophilic polymers (e.g. proteins and polysaccharides), it is necessary to apply emulsification or homogenization techniques, where fine emulsions of EOs are achieved containing polymers at the continuous aqueous phase (Atarés and Chiralt 2016). Thus, for this type of material, an emulsion will always be necessary.

This is one more example of how nanotechnology can be applied in the food science to develop more stable emulsified systems. Due to such improvements, several examples can be found in literature, and few are summarized on Table 4.4.

3 Releasing Systems for Essential Oils

The volatile nature of EOs makes them suitable for designing packaging systems as active films (Ju et al. 2019). The theoretical advantage is that because they are volatile, they can penetrate deeper in the food matrix, and the polymer does not necessarily need to be in direct contact with food (Lucera et al. 2016). Exploitation of EOs as additives in active release packaging can be carried out through their inclusion in emission sachets within the packaging space, or directly incorporated into the structure of films or coating materials. (Limbo and Khaneghah 2015; Marinello et al. 2018; Huang et al. 2019; Ju et al. 2019). It is important to note that in order to exercise activity through their release in the product, the EOs must not be toxic to human health (Surwade and Chand 2017). Their migration can be achieved through the diffusion in gas phase of the sachets or inner layer of the packaging onto the surface of the food (Lucera et al. 2016).

3.1 Sachets

EOs are released from the packaging into the headspace. The volatile concentration of these substances in the free space needs to be balanced and their release rate is highly dependent on the volatility that relates to the chemical interactions between the substance and the packaging material. In addition, the absorption rate depends on the composition of the food, such as lipid and water content (Limbo and

Table 4.4 Nanoemulsions loaded with essential oils and their biocompounds

Essential oils/ sources	Surfactants	Polymers incorporated	Main conclusions	References
Salvia multicaulis	Tween 80 and span 80	Not incorporated	Nanoemulsions were stable for 60 days, with enhancement on the antimicrobial and antioxidant activities	Gharenaghadeh et al. (2017)
Ginger essential	Tween 20	Fish skin gelatine + ZnO nanoparticles	Microemulsions with diameter between 0.1 and 0.2 μm were produced. Films with GEO demonstrated enhancement on antimicrobial and antioxidant properties. Reductions on lipid oxidation, on the release of total volatile bases nitrogen were observed for fresh meat storage protected with active films.	Zhang et al. (2017b)
Ginger	1% Tween 20 + 4% Span 80	Pigskin gelatin + montmorillonite (MMT)	Size of droplets obtained were around 127-150 nm, MMT and GEO improved significantly the elongation at break, puncture force and puncture deformation of films, acting as a plasticizer. Films only presented antioxidant property; no antimicrobial activity was observed.	Alexandre et al. (2016)
Thyme, lemongrass or sage	Tween 80	Alginate	SEO nanoemulsions revealed a superior transparency, water vapor resistance and flexibility than films formed from TEO or LEO. Edible films containing TEO were those with the strongest antimicrobial effect against inoculated *Escherichia coli*.	Acevedo-Fani et al. (2015)

(continued)

Table 4.4 (continued)

Essential oils/ sources	Surfactants	Polymers incorporated	Main conclusions	References
Clove Bud and Oregano	Tween 80	Methylcellulose	Films incorporated with coarse emulsion (size of 1.3-1.9 μm) were less active than those with nanoemulsion (size of 180–250 nm), specially for CEO, which were more efficient than OEO. The nanocomposites were used to protect bread and, extended its shelf life compared to samples without films or without EO or the commercial bread with synthetic additive. In terms of mechanical properties, the nanoemulsion only statistically enhanced the elongation at break of films, when compared to coarse emulsion.	Otoni et al. (2014b)
Cinnamaldehyde	Tween 80	high or low methylester pectin film	Pectin film incorporated with cinnamaldehyde nanoemulsions presented increased rigidity and decreased extensibility and permeability to water vapor. Active compounds conferred antimicrobial properties against several foodborne bacteria. The diminishment on droplet diameter did not influence the physical-mechanical properties of the films, but remarkably improved bacterial inhibition.	Otoni et al. (2014a)
α-ocopherol, garlic EO and cinnamaldehyde nanoemulsion	Tween 20 + Span 60	gelatin, gelatin-sodium caseinate and gelatin-chitosan blends	The nanoemulsions provided antioxidant activities to the films, with more intensity for the films of gelatin-sodium caseinate. The incorporation of the nanoemulsions resulted in an increase on the hydrophilicity for the films of gelatin and gelatin sodium-caseinate, while for gelatin-chitosan changed to a more hydrophobic surface.	Pérez Córdoba and Sobral (2017)

Khaneghah 2015). The sachet system, among the controlled release packages, is a technology that allows the diffusion of the volatile compounds contained in its interior by the free space for food components. The use of EOs in sachets represents an interesting approach for the food industry to overcome the limitations of EOs' direct application (Marques et al. 2019). According to Jung and Zhao (2016), their application currently occurs spatially in minimally processed fruits and vegetables, however it is not only limited to them.

The principle of carrier sachets or emitters involves the adsorption of EOs in a carrier that permits their controlled release in a closed packaging system. The sachet material must be permeable to allow release of EOs, in addition to supporting handling and transportation to prevent failures and leaks from the conveyor. The sachet is placed next to the food in a closed system, surrounded by an external packaging with specific permeability requirements according to the application, providing the modification of the internal gas composition of the packaging (Otoni et al. 2016). Thus, the active sachet for use in food is a type of pillow that contains volatile compounds, such as EOs, closed inside a package to allow their release and interaction with its free space (Jung and Zhao 2016). According to Soares et al. (2009), sachets are characterized by being composed of porous and sealable materials, preferably non-woven, containing a high absorption polymer, such as a high absorption polyethylene resin, incorporated with EOs and packed inside the packaging, whose material has a high gas barrier. Their preparation with EOs, therefore, basically comprises the steps of incorporating the agent into a carrier, adding the carrier inside the sachet and sealing the sachet (Wieczyńska and Cavoski 2018).

The incorporation and subsequent release of EOs, as well as other active volatiles, has led to a recent expansion in the use of the sachet system, which has long been used as a gas eliminator, for other functions (Otoni et al. 2016). This technology draws attention because it does not require contact between the EOs and the food product, not impacting its sensory properties (Dıblan and Kaya 2018). According to Otoni et al. (2016), sachets emitting EOs with flavoring and antioxidant activities are continually being explored, however, their most recurrent use in food is for antimicrobial purposes. In addition, being the first product commercially available that antimicrobial agents were incorporated into the packaging, the sachet containing volatile agents that allows the control of microorganisms is the commercially successful and best known application of antimicrobial packaging. Their potential use with the incorporation of EOs is demonstrated in studies that show them as possible substitutes for commercial antimicrobials (Montes et al. 2013; Lucera et al. 2016).

Sachets containing EOs of cinnamon, oregano and lemongrass were used in papaya packaging (Espitia et al. 2012). The authors found that the presence of the developed sachets reduced the growth of aerobic mesophilic bacteria, yeasts and molds, with emphasis on greater effectiveness for those incorporated with cinnamon oil. Sachets containing activated carbon with peppermint oil adsorbed and coated with paraffin showed antifungal activity against *Aspergillus flavus* and had potential for commercial application in brown rice to facilitate long-term storage (Chaemsanit et al. 2019). According to Passarinho et al. (2014), sachets containing oregano EOs proved to be effective against contaminating pathogenic microorganisms in food, evaluated on agar medium, and against fungal deterioration in sliced bread, being indicated for application in this product. Bill et al. (2016) found that sachets of polylactic acid (PLA) impregnated with thyme oil are suitable for application in avocado packaging, as they promoted a significant reduction in the incidence and severity of anthracnose in artificially inoculated fruits. Continuing this study, Bill et al. (2017) suggested the incorporation of 10% thyme oil in low density polyethylene (LDPE) pellets in PLA sachets for commercial avocado packaging as a natural option for preserving fruits and satisfying consumption.

According to Otoni et al. (2016), there are several materials used to pack the carriers in sachets. In a bibliographic study, they highlighted what type and size of material used to transport volatile compounds affects the loading and releasing abilities of antimicrobial sachets. Some of them include semipermeable polymer films, perforated barrier films, porous non-woven fabrics and papers, in addition to ethylene-vinyl acetate (EVA) film, cellulose fiber paper, and filter paper. For the external packaging, metallized polypropylene, paper bag and polyethylene were mentioned. The authors also highlighted that the application of bio-based and/or biodegradable materials is recommended as an ecological alternative for carriers, sachets and external packaging materials.

Despite presenting several advantages and commercial applications, the use of sachets in food packaging systems also has some inherent disadvantages. According to Upasen and Wattanachai (2018), sachets are unsuitable for use in liquids or tight-fitting films, as their functionality would be suffocated, in addition to the risks of accidental ingestion being a serious concern. The concern about accidental consumption or possible confusion with a new sauce, sticker, toy or discount card, caused by the presence of a foreign object inside the package is also reported by Surwade and Chand (2017). Considering the sensory aspect, Marques et al. (2019) reported that the presence of the EOs-loaded sachet in the packaging caused a reduction in global acceptance and in the intention to purchase the product, after conducting a study of its use in ready-to-eat salads. These disadvantages can be mitigated by incorporating the functional components into the plastic film itself. In addition, recent studies indicate that combinations of various packaging techniques to extend shelf-life are more effective (Upasen and Wattanachai 2018).

3.2 *Films*

The incorporation of EOs into packaging materials can be done either by their immobilization or directly incorporation into the polymer matrix (Jung and Zhao 2016; Otoni et al. 2016). According to Ibarra et al. (2016), the active release packaging system has its action achieved by emitting the agent of a packaging material to act directly on food. Those based on immobilization in the polymeric matrix, on the other hand, have action on the surface and require direct contacts between the packaging materials and the products. Due to the convenience and ease of mass production, mixing EOs with polymers is the most commonly used method for manufacturing emitter packaging for food applications (Huang et al. 2019).

Polymeric materials called films are thin sheets previously prepared to be applied to the product as a wrapper or between its layers (Sharmeen et al. 2020). The diversity of studies and applications of films emitting EOs in food products is quite large and involves their use as packaging materials for both foods of animal origin, such as dairy products, meat, fish and their derivatives, and for foods of plant origin especially minimally processed vegetables. According to Botre et al. (2010), the use of films incorporated with oregano EO arouses great interest in the application in pizzas. Mulla et al. (2017) concluded in a study that low density linear polyethylene films containing clove EOs can be used for active poultry packaging and meat products to ensure the food safety of fresh meat. Muriel-Galet et al. (2012) developed films containing oregano EOs and citral for application in minimally processed salads, aiming at their better conservation. Also seeking to preserve and ensure the provolone cheese safety, Lima et al. (2020) produced films incorporated with oregano EOs.

The volatility and thermal sensitivity of EOs is the main factor that, however, limits their direct application in the processing of active emitting films, especially when using the extrusion process, as it implies a reduction in their effectiveness and action during the term of validity (Limbo and Khaneghah 2015; Mulla et al. 2017; Khaneghah et al. 2018). In fact, Valencia-Sullca et al. (2018) found that the heat treatment applied to obtain films in starch-chitosan bilayers containing EOs, reduced its antimicrobial properties. The authors associated probable loss of EOs and their antimicrobial action to the processing of the films. Suppakul et al. (2008) attributed reductions in the concentrations of basil EOs in blown films of LDPE and partial loss of antimicrobial activity to its volatilization during extrusion. Sung et al. (2014), however, did not find a loss effect on the antimicrobial activity of *Allium sativum* EOs in blown films of LDPE and EVA. In addition, Dias et al. (2013) produced flavoring active films of LDPE with EOs and/or aroma of orange and, applied them to sugar biscuits. The sensory evaluation showed that consumers preferred the sugar biscuits packaged by the films which contained 10% EOs and 5% aroma.

Thus, although the incorporation of active agents in a polymer by the total or partial extrusion process is well known, the aggressiveness that the thermomechanical treatment causes with potential lose of the EO biological activity, explains the preference for a chemical or physical method to incorporate EOs (Limbo and Khaneghah 2015). For this reason, according to Shojaee-Aliabadi et al. (2013) and Atarés and Chiralt (2016), films incorporated with EO are generally obtained by the casting method, which is based on the formation of dispersions followed by the evaporation of the solvent on a support. This is because, according to Sung et al. (2013), the molten film is generally produced at a relatively lower temperature and is not subjected to mechanical shear forces.

Due to the growing interest in preserving food using more healty products that also cause less damage to the environment, researchers have sought to explore various types of raw materials for the development of biodegradable films incorporated with EOs. Among them, Wu et al. (2017) used fish gelatine to produce films, Lima et al. (2020) resorted to the use of modified starch and gelatin, and Jouki et al. (2014) explored the quince seed mucilage. Cellulose (Botre et al. 2010), whey protein isolated (Oliveira et al. 2017), as well as other raw materials are also widely used for this purpose.

Despite the several advantages, studies have shown that incorporation of EOs in emitter films can negatively affect their physical-chemical properties, which may imply the modification of various properties of the system, such as optics and traction, which in turn would affect consumer acceptability (Atarés and Chiralt 2016; Rao et al. 2019). According to Atarés and Chiralt (2016), the continuity of the polymer matrix is influenced due to the incorporation of EOs, which can lead to physical changes depending on the interactions of the polymer-oil components. Generally, with the addition of EOs, the film structure is weakened, water barrier properties are improved, and transparency is reduced. Thus, a more in-depth study involving the physical-chemical modifications of materials caused by the addition of EOs, as well as the migration profile and eventual modifications of the systems during storage, in addition to its active profile, is of fundamental importance for the development of these migrant films.

3.3 Coatings

Coatings are defined as a thin layer of edible materials applied directly onto the food surface, commonly by immersion or spraying, forming thin membranes that are imperceptible to the naked eye. What differentiates them from edible films is that they are preformed in molds and dried to then be applied to food (Assis and Britto 2014; Jung and Zhao 2016; Oliveira et al. 2017). Thus, coatings remain adhered to the surface of products, while edible films wrap the product without adhering to them (Paul 2019).

The application of EOs in food coatings is also an important alternative to their incorporation during film extrusion, since EOs are highly volatile and sensitive to light and heat (Realini and Marcos 2014; Assis and Britto 2014). It overcomes adversities of this method, making the EOs available without subjecting them to high temperatures or shear forces. Thus, the coating serves as a carrier for EOs and their bioactive property is based on migration or release by evaporation (Lucera et al. 2016). Studies aiming at the development of edible coatings with EOs have been widely disseminated in order to obtain active emission systems that allow to achieve better food conservation.

According to Oriani et al. (2014), edible cassava starch-based coatings containing EOs can be applied in minimally processed fruits to build a protective barrier against water and gas permeability, despite the improvement on the antimicrobial properties due to the phenolic compounds presents in the EOs. The addition of edible coating based on alginate and incorporated with rosemary and oregano EOs in Ricotta cheese was proposed by Tavares et al. (2014) as an option for improvements in conservation and product quality. It was verified by Shokri et al. (2020)

that chitosan coating containing EOs of *Ferulago angulata* can successfully delay the bacterial growth of rainbow trout fillets stored under refrigeration. The inhibition of bacterial growth in fish fillets was also achieved using a chitosan coating combined with isolated whey protein and tarragon EOs (Farsanipoor et al. 2020). Behbahani et al. (2020) tested Shahri Balangu seed mucilage and cumin oil to developed an active coating that improved the beef's shelf life, suppressing deterioration and inhibition of lipid oxidation.

Thus, several structural characteristics of these coatings are dependent on the formulation of the precursor film-forming solution (Assis and Britto 2014). In general, due to their hydrophobic character, the addition of EOs suggests an increase in hydrophobicity and a consequent reduction in water vapor permeability of the coatings (Atarés and Chiralt 2016). According to Ju et al. (2019), the controlled release of EOs achieved in the coating, has short effect on the mechanical properties of packaging materials, however, their barrier properties have yet to be considered.

4 Essential Oils Emitters from Active Food Packaging

The application of active packaging incorporated with EOs is vast, in a wide types of foods: strawberries, cheese, beef, pork, chicken and meat, to mention a few, and they have exhibited excellent activity (Wen et al. 2016b Shao et al. 2017; Noori et al. 2018; Pabast et al. 2018; Lin et al. 2019). However, the effectiveness (in terms of antimicrobial and antioxidant properties) of active packaging may be different from *in vitro* to *in situ* applications, i.e. some strong activities may be found for *in vitro* studies, but equivalent activity is not demonstrated when applied in real food products. This can be explained due to the interactive effects of a number of factors related to the specific characteristics of the model food such as pH, water activity, fat and protein content, the presence of additives, salt and, also, extrinsic factors such as the storage conditions (temperature and atmosphere composition) (Limbo and Khaneghah 2015). In Table 4.5 several examples of antioxidant and antimicrobial active packaging with EOs as additive are summarized.

5 Essential Oils Migration from Packaging to Food

In view of the active functions of EOs and their properties, they are highly studied as potent food preservatives and, for this function, it is extremely important to know the interaction between EOs and the matrix where it will be incorporated, the minimum concentration to present the antioxidant and/or antimicrobial effects, and possible combination with other preservatives (Hyldgaard et al. 2012). Migration can be described as the transference of components from the package to the packaged food or the other way around by diffusion. This process is directly dependent on the packaging constituents, the packaged food(s) and the conditions (for example,

Table 4.5 Essential oils in different active packaging system with antioxidant and antimicrobial effects applied in food

Essential oils/ sources	Formulation technologies	Active function	Microorganisms	Polymeric matrix	Type of food	References
Active Food Packaging: Films						
Bergamot	–	Antimicrobial	*Penicillium italicum.*	Chitosan	NA	Sanchez-Gonzalez et al. (2010)
Orange	–	Flavoring	NA	Low-Density Polyethylene	Sugar biscuits	Dias et al. (2013)
Lemon, thyme and Cinnamon	–	Antimicrobial	*E. coli* and *S. aureus*	chitosan	NA	Peng and Li (2014)
Cinnamon	Encapsulation	Antimicrobial	*E. coli* and *S. aureus*	Polyvinyl alcohol (PVA) nanofibers film	Strawberry	Wen et al. (2016b)
Cinnamaldehyde	Encapsulation	Antimicrobial	*E. coli* and *Pseudomonas aeruginosa*	Chitosan/poly(ethylene oxide	NA	Rieger and Schiffman (2014)
Cinnamon	Encapsulation	Antimicrobial	*Staphylococcus aureus* and *Escherichia coli*	polylactic acid (PLA) nanofibers	Pork	Wen et al. (2016a)
Oregano and Bergamot	–	Antimicrobial	*Escherichia coli*	Hydroxypropyl methylcellulose	Formosa' plum	Choi et al. (2016)
Eucalyptus	Emulsion	Antioxidant	NA	Chitosan	NA	Hafsa et al. (2016)
Morinda citrifolia	Emulsion	Antioxidant	NA	Fish gelatin	NA	Maryam Adilah and Nur Hanani (2016)
Oregano	–	Antioxidant	NA	Poly (lactic acid)/ poly(trimethylene carbonate)	NA	Liu et al. (2016)

Cinnamon	Encapsulation	Antimicrobial	*Bacillus cereus*	poly(ethylene oxide nanofibers	Beef	Lin et al. (2017).
Clove	NA	Antioxidant	NA	Corn starch	Sausages	Ugalde et al. (2017)
cinnamaldehyde	Encapsulation	Antimicrobial	Bacterial	poly(vinyl alcohol)/ permutite fibrous film	Strawberries	Shao et al. (2018)
Cinnamon bark	Emulsion	Antioxidant	NA	Hagfish skin gelatin	NA	Kim et al. (2018)
Lavender	Emulsion	Antioxidant	NA	Furcellaran, starch and gelatin	NA	Jamróz et al. (2018)
Moringa	Encapsulation	Antimicrobial	*Listeria monocytogenes* and *S. aureus*	Chitosan nanofibers	Cheese	Lin et al. (2019)
Peppermint and chamomile	Encapsulation	Antimicrobial	*E. coli* and *S. aureus*	Gelatin nanofibers	NA	Tang et al. (2019)
Rosemary	Emulsion	Antioxidant	NA	Gelatin	NA	Yeddes et al. (2019)
Rosemary, thyme and oregano	Emulsion	Antioxidant	NA	PLA (Polylactic acid resin)	NA	Zeid et al. (2019)
Rosemary and aloe vera	–	Antioxidant	NA	Cellulose acetate	NA	El Fawal et al. (2019)
Salvia	–	Antioxidant	NA	Potato starch with zeda gum	NA	Pirouzifard et al. (2019)
kaffir lime oil	–	Flavoring	NA	Edible carboxymethylcellulose–alginate film	Tom-Yum soup cooking	Sabphon et al. (2019)

(continued)

Table 4.5 (continued)

Essential oils/ sources	Formulation technologies	Active function	Microorganisms	Polymeric matrix	Type of food	References
Active Food Packaging: Coating						
Cinnamon and Mustard	–	Antimicrobial	*Salmonella enterica* serovar Typhimurium	Zein	Tomatoes	Yun et al. (2015)
Oregano	–	Antimicrobial	*Escherichia coli*	Polypropylene (PP), polyethylene and terephthalate (PET)	Ripened, sheep, cheese, model	Otero et al. (2014)
Oregano and clove	–	Antimicrobial	Total aerobic mesophilic bactéria, *Enterobacteriaceae*, total aerobic psycrotrophic bacteria, lactic acid bacteria and *Pseudomonas* spp	Whey Protein Isolate	Chicken breast	Fernandez-Pan et al. (2014)
Oregano	–	Antioxidant	NA	Whey protein	Portugueses sausages (*paínhos* and *alheiras*)	Catarino et al. (2017)
Ginger	Nanoemulsion	Antioxidant	NA	Sodium caseinate	Chicken breast fillets	Noori et al. (2018)
Satureja	Nanoencapsulation	Antioxidant	NA	Chitosan	Lamb meat	Pabast et al. (2018)
Paulowinia tomentosa	Nanoencapsulation	Antioxidant	NA	Chitosan	Ready-to-cook pork chops	Zhang et al. (2019)
Pimpinella saxifraga	–	Antioxidant	NA	Sodium alginate	Cheese	Ksouda et al. (2019)
Black pepper	–	Flavoring	NA	Mix of flour, vegetable oil, lard and β-cyclodextrin	Dry cured maet	Bi et al. (2019)

Active Food Packaging: Sachets						
Oregano cinnamon or lemongrass	–	Antimicrobial	Fungi (*Alternaria alternata, Fusarium semitectum, Lasiodiplodia theobromae* and *Rhizopus stolonifer)*	Not mentioned	Papaya	Espitia et al. (2012)
Eugenol, carvacrol, anathole and sweet almond oil	–	Antioxidant	NA	Non-woven PLA materials	Organic ready-to eat iceberg lettuce	Wieczyńska and Cavoski (2018)
Chilean boldo	–	Antioxidant	NA	Not mentioned	NA	Souza et al. (2019b)

NA Not Apllied

temperature and radiation) to each the package will be submitted (Arvanitoyannis and Bosnea 2004; Sanches Silva et al. 2008). Migration from packages to foods is a health concern and, therefore, it is regulated with legal limits and limitations.

In the European Union, migration limits are regulated by the Commission Regulation No. 10/2011. According to this Regulation, the overall migration limit (OML) refers to "(...) *the maximum permitted amount of non-volatile substances released from a material or article into food simulants*"; the specific migration limit (SML) refers to the "(...) *maximum permitted amount of a given substance released from a material or article into food or food simulants*", being a food simulant "(...) *a test medium imitating food; in its behaviour the food simulant mimics migration from food contact materials*", and the total specific migration limit (SML(T)) refers to the "*maximum permitted sum of particular substances released in food or food simulants expressed as total of moiety of the substances indicated*" (European Commission 2011).

In a conventional or traditional package, the migration of substances is not desirable, being its main purpose the protection of the packaged food(s) from physical damages that may occur on the transportation chain, changes in temperature and radiation, prolonging food' shelf-life. On the other hand, the main objective of active packaging presupposes the interaction between the food and the package. The emission active packaging implies the emission of certain substances present in the matrix of the package or present in a sachet inside the package. Since EOs have some biological activities that can increase food' shelf life, they are interesting substances to include in the package matrix and/or sachets (Espitia et al. 2012; Ribeiro-Santos et al. 2017d).

Since food packaging migration is a diffusion process, Fick's laws of diffusion can explain how this migration occurs. Briefly, the first Fick's law states that the diffusing flux goes from a matrix with a high concentration of some compound(s) to a matrix with a lower concentration. This flux is proportional to the concentration gradient. The Fick's second law predicts how the diffusion rate evolves or changes over time (Miltz et al. 1997; Arvanitoyannis and Bosnea 2004; Silva et al. 2009; Reinas et al. 2012). For active food packaging, a slow and gradual migration of the EOs from the package (or an emission device) is the most desirable so, the choice of the used polymer in which the EOs will be incorporated and/or encapsulated is also very important. Depending on the purpose of the final product, the polymer choice will have a key impact on the success of the final product. Therefore, the manufacturer must take into account the final objective of the package, since the migration rate can vary according to the food composition, such as fat and moisture content (Ribeiro-Santos et al. 2017d). The temperature to which the packaging will be exposed has to be considered since high temperatures can influence the diffusion rate between the package and the food (Kuorwel et al. 2013).

However, the use of these packaging and the activity they promote is conditioned by the type of EOs used, their concentration in the films and their possible interactions with the biopolymer matrix (Wu et al. 2015; Rao et al. 2019). In addition, Ribeiro-Santos et al. (2017d) pointed out that its diffusion effectiveness also depends on the material of the packed food, on the time and contact temperature of the packaging with it. Alginate is a naturally occurring polysaccharide formed by the monomers of α-L-glucuronic acid and β-D-mannuronic acid that can be found in brown seaweeds and some bacteria, namely *Pseudomonas aeruginosa* (Asbahani et al. 2015; Nesic and Seslija 2017). The chemistry of alginates is determined by the monomer's ratio, which can differ depending on the extraction source. If, on one hand, the chemical composition of alginates extracted from seaweeds is variable and dependent on the edaphoclimatic conditions to which the seaweeds were exposed, the same does not apply to alginates extracted from bacteria since the ratio of the monomers can be controlled through the controlled growth of the bacteria (Brownlee et al. 2009; Karlsen 2010). The ratio between the two monomers in the alginate composition is so important because this ratio will define the diffusion rate of EOs inside the alginate capsules (Karlsen 2010). Alginic acid (E400), sodium alginate (E401), potassium alginate (E402), ammonium alginate (E403), calcium alginate (E404) and propane-1, 2-diol alginate (E405) are authorized to be used as food additives in the European Union (The European Parliment and the Council of the European Union 2008a).

Oussalah et al. (2007) tested the antimicrobial activity of *Corydothymus capitatus* L., *Cinnamomum cassia* L. and *Satureja montana* L. EOs incorporated in alginate-based edible films. The authors also measured the migration of bioactive compounds over 5 days. The films were composed by sodium alginate (3%, w/v), polycaprolactone diol (1%, w/v) and glycerol (2%, w/v) and the EOs were incorporated at a final concentration of 1% (w /v). Then, the films suffered a pretreatment with $CaCl_2$, and their antimicrobial activity was tested, for 5 days, in bologna and ham slices inoculated with *Salmonella enterica* subsp. *enterica* serovar Typhimurim and *L. monocytogenes*. Regarding their antimicrobial activity, all EOs showed antimicrobial activity against the two bacterial strains, although their effect was more powerful against *L. monocytogenes* on bologna slices and against *Salmonella typhimurium* in ham slices (Oussalah et al. 2007). In the *C. capitatus* L. films, the main active compound was carvacrol, while in the *S. montana* L. films, the main active compound was thymol and finally in the *C. cassia* L. films, the main active compounds were *trans*- and methoxy-cinnamaldehyde. Regarding the migration of the active compounds from the films to the ham and bologna slices, regardless of the pre-treatment with $CaCl_2$, the films with *C. cassia* L. presented the highest migration rate of active compounds on the 1st day. The films with *C. capitatus* L. oil with 2% of $CaCl_2$ showed the highest migration of active compounds in both ham and bologna slices. The authors concluded that the presence of $CaCl_2$ permitted a controlled migration of the active compounds from the films to the foods. The release of the active compounds differs with the food matrix, since the films with 2% of $CaCl_2$ showed a lower release of compounds over time and, in the ham slices, the amount of $CaCl_2$ seemed to interfere less with the release of compounds (Oussalah et al. 2007).

Mohammadi et al. (2015a) prepared chitosan nanoparticles loaded with *Cinnamomum zeylanicum* L. EOs. Chitosan is a biodegradable polymer, recognized as GRAS by the FDA, and the most abundant natural polysaccharide after cellulose (Lee 2005; Vunain et al. 2017). It has proven antimicrobial activity and, can be used as a thickening agent and for film formation (Vunain et al. 2017). The chitosan + *C. zeylanicum* L. EOs (1:0.25) were prepared through ionic gelation. Also known as ionotropic gelation, it is a non-toxic and organic solvent free method based on the crosslink capacity of polyelectrolytes in the presence of counter ions (Giri 2016; Mohammadi et al. 2015a). The *in vitro* release kinetics of the *C. zeylanicum* L. EOs were measured over 6 weeks using a spectrophotometer. In the first nine days, the nanoparticles released up to 28.4% of the EOs but, after this period, the EO release abruptly slowed being almost zero. At the end of the 40th day, the released concentration of the EOs was 31.65%. The authors attributed the first burst release to the close to the surface desorption of the *C. zeylanicum* L. EOs during the nanoparticles preparation (Mohammadi et al. 2015a).

As chitosan has low water resistance, poor barrier and mechanical properties, due to its hydrophilic nature, Pires et al. (2018) incorporated two types of montmorillonite (MMT) nanoparticles, Cloisite®Na^+ and Cloisite®Ca^{+2}, to overcome these problems. To improve the antioxidant and antimicrobial properties of chitosan, the authors incorporated rosemary and ginger EOs. Minced poultry meat was used to test the biological properties of the films for a time period of 15 days. The results showed that the unwrapped samples, when compared with the packaged samples, the meat deteriorated more rapidly. Although there were no significant differences between ginger and rosemary EOs, both presented antioxidant activity. Regarding the antimicrobial activity, both EOs seemed to not have an influence, positive or negative, on the antimicrobial activity of chitosan (Pires et al. 2018).

Souza et al. (2018) prepared chitosan films with MMT Cloisite®Na^+ and ginger EOs and tested its antioxidant and antimicrobial properties *in vitro* and *in vivo* with fresh poultry meat. Regarding the *in vitro* migration studies, the authors compared the content of total phenolic compounds in chitosan films containing 0.5%, 1% and 2% of ginger EOs with and without MMT nanoparticles during 250 h. The samples reached a migration equilibrium within the first 24–48 h. The chitosan loaded with 1% ginger EOs and the MMT nanoparticles showed the highest migration of phenolic compounds (Souza et al. 2018). The release of the oregano EOs was found from *in vitro* migration assay to be dependend on the content of oregano EOs in the chitosan nanoparticles (Hosseini Zandi and Rezaei 2013). Shao et al. (2017) reported that EOs release behaviors in polymeric nanofiber films were found to be diffusion controlled. The active packaging with encapsulated EOs showed better activity compared to non-encapsulated EOs (Wen et al. 2016a, b). Abreu et al. (2012) encapsulated pepper-rosemary EOs in a chitosan and cashew gum nanogel packaging and the nanoparticles presented a slow and sustained release.

6 Legal Aspects of the Use of EOs in Food Packaging

In the European Union "*any material or article intended to come into contact directly or indirectly with food must be sufficiently inert to preclude substances from being transferred to food in quantities large enough to endanger human health or to bring about an unacceptable change in the composition of the food or a deterioration in its organoleptic properties*" (European Union 2004). Following this reasoning, the Regulation No 10/2011 states that the plastic materials intent to be in contact with foods should not transfer their constituents to food simulants above 10 mg/dm^2 of food contact surface. The active and intelligent packages have, by design, the main goal, to interact with the packaged food. According to the Regulation No 1935/2004, "*Active food contact materials and articles are designed to deliberately incorporate 'active' components intended to be released into the food or to absorb substances from the food. They should be distinguished from materials and articles which are traditionally used to release their natural ingredients into specific types of food during the process of their manufacture, such as wooden barrels*". This regulation also states that the "*active food contact materials and articles may change the composition or the organoleptic properties of the food only if the changes comply with the Community provisions applicable to food (…)*". The authorized food additives in the EU and their conditions, are listed in the Annex II of the Regulation (EC) No 1333/2008 and should be labeled as ingredients in accordance to the Directive 2000/12/EC.

Regarding food additives, the Regulation (EC) 1333/2008 of the EU states that food additives are not consumed as foods but can be intentionally added to foods for technological purposes, including "*preparations obtained from foods and other natural source material that are intended to have a technological effect in the final food and which are obtained by selective extraction of constituents (e.g. pigments) relative to the nutritive or aromatic constituents, should be considered additives*". This regulation also states that the substances used to provide flavor and/or taste and for nutritional purposes are not considered food additives (EU 2008a). EOs are not on the list of authorized food additives in the Regulation (EC) No 1333/2008 and on the list of authorized flavorings in the Regulation (EC) No 1334/2008 (EU 2008a, b). Regardless of this, most of the major compounds in EOs are present and authorized in both regulations. For instance, the use of anthocyanins (E163) is authorized as a food additive in all kinds of foodstuff excluding the red marbled products (EU 2008a). The use of myrcene, camphene, α-terpinene, limonene, linalool, menthol, citronellol, geraniol, carvacrol and other EOs are authorized by the Regulation (EC) No 1334/2008 (EU 2008b).

Regarding the Code of Federal Regulations (CFR) of the FDA, there are > 150 EOs categorized as GRAS, including basil (*Ocimum basilicum* L.), cinnamon (*Cinnamomum zeylanicum* Nees. and *Cinnamomum cassia* Blume.), grapefruit (*Citrus paradisi* Macf.), lemon peel (*Citrus limon* (L.) Burm. f.) and sage (*Salvia officinalis* L.) (FDA 2020). In the FDA regulation, there is not any EOs categorized as Substances prohibited from use in human food (FDA 2020).

Because EOs are products of natural origin used by human civilization for several centuries, their consumption is considered safe by the general population and also registering as safe food additives; although they have some undesirable effects such as toxicity and hypersensitivity reactions associated with high consumption doses which are usually forgotten (Adams and Taylor 2010). Although the official recommendations and information regarding the safety, efficacy and use of EOs in foods is still ambiguous, their use as flavoring agents or fragrances are subjected to governmental regulations.

7 Concluding and Perspectives Future

The application of EOs in food systems as natural inhibitors or biopreservatives has received high attention mainly due to potential alternatives to chemical preservatives. Their application for shelf-life extension in foods and consequently to ensure food safety, preventing foodborne illness, is due to their antimicrobial and antioxidant properties, already well elucidated, on the control of microbial growth and lipid oxidation. Their biological activity is associated as their major compounds but, minor constituents can contribute to their high action. In spite of benefits, EOs are very sensitive to processing and storage conditions and can lose their nutritional value. They have special use limitations that must be outdated before their application in food, such as low water solubility, high volatility, and strong odor. Thus, noval forms of applications are currently being studied to overcome such limitations. In this scenario, EOs obtained from plants and fruits present great perspectives as natural additives in active food packaging systems such as coatings, films and sachets. Besides, their biological activities can be enhanced using new technologies, as encapsulation and nanotechnology, which are promising strategies to improve the application of EOs in food packaging.

In view of the encapsulation of these compounds, different methodologies can be applied in order to: increase the compatibility with water soluble foods; to facilitate their dispersion on the surface of food; to have controlled release; to preserve their biological activity, which may also increase their effective utilization once they reduce the volatility and instability of EOs when exposed to environmental factors.

There is a large use of EOs in active-realising systems applied to several foods, however, the migrations of EOs to foods and the consequent effectiveness of the active packaging is directly dependent on: the polymeric matrix, the packaged food(s) and the conditions to which the package will be submitted, the choice of the EO(s), the wall material and the applied encapsulation technique. The migration of EOs from packaging to food is a health concern and, therefore, should be regulated within legal limits. Being a natural product, EOs are considered safe, although some undesirable effects are associated with high doses of their consumption. However, it is important to ensure if EOs and their compounds does not compromise the food and the consumer's health at doses able to produce efficiency. In addition, standardization of their composition would be needed for their safe use in the food industry. Despite this, EOs in food packaging systems have a wide range of applications that can strongly benefit the consumers. Therefore, it is predicted that, in the coming years, new products and applications will emerge based on the exploitation of EOs.

References

Abou Baker DH, Al-Moghazy M, ElSayed AAA. The *in vitro* cytotoxicity, antioxidant and antibacterial potential of Satureja hortensis L. essential oil cultivated in Egypt. Bioorg Chem. 2020;95:103559. https://doi.org/10.1016/j.bioorg.2019.103559.

Abreu FOMS, Oliveira EF, Paula HCB, et al. Chitosan/cashew gum nanogels for essential oil encapsulation. Carbohydr Polym. 2012;89:1277–82. https://doi.org/10.1016/j.carbpol.2012.04.048.

Acevedo-Fani A, Salvia-Trujillo L, Rojas-Graü MA, et al. Edible films from essential-oil-loaded nanoemulsions: physicochemical characterization and antimicrobial properties. Food Hydrocoll. 2015;47:168–77. https://doi.org/10.1016/j.foodhyd.2015.01.032.

Adams TB, Taylor SV. Safety evaluation of essential oils: a constituent-based approach. In: Baser KHC, Buchbauer G, editors. Handbook of essential oils: science, technology and applications. CRC Press; 2010. p. 185–208.

Ademiluyi AO, Oyeleye SI, Oboh G. Biological activities, antioxidant properties and phytoconstituents of essential oil from sweet basil (*Ocimum basilicum* L.) leaves. Comp Clin Pathol. 2016;25:169–76.

Alexandre EMC, Lourenço RV, Bittante AMQB, et al. Gelatin-based films reinforced with montmorillonite and activated with nanoemulsion of ginger essential oil for food packaging applications. Food Packag Shelf Life. 2016;10:87–96. https://doi.org/10.1016/j.fpsl.2016.10.004.

Almeida JS, Lima F, Da Ros S, et al. Nanostructured systems containing rutin: in vitro antioxidant activity and photostability studies. Nanoscale Res Lett. 2010;5:1603–10. https://doi.org/10.1007/s11671-010-9683-1.

Amarowicz R, Pegg RB. Chapter One – natural antioxidants of plant origin. In: Ferreira ICFR, Barros LBT-A in F and NR, editors. Functional food ingredients from plants. Academic; 2019, p. 1–81

Amorati R, Foti MC, Valgimigli L. Antioxidant activity of essential oils. J Agric Food Chem. 2013;61:10835–47. https://doi.org/10.1021/jf403496k.

Andrade MA, Ribeiro-Santos R, Costa Bonito MC, et al. Characterization of rosemary and thyme extracts for incorporation into a whey protein based film. LWT Food Sci Technol. 2018;92:497–508. https://doi.org/10.1016/j.lwt.2018.02.041.

Arvanitoyannis IS, Bosnea L. Migration of substances from food packaging materials to foods. Crit Rev Food Sci Nutr. 2004;44:63–76. https://doi.org/10.1080/10408690490424621.

Asbahani AE, Miladi K, Badri W, et al. Essential oils: from extraction to encapsulation. Int J Pharm. 2015;483:220–43. https://doi.org/10.1016/j.ijpharm.2014.12.069.

Assis OBG, Britto D. Review: edible protective coatings for fruits: fundamentals and applications. Braz J Food Technol. 2014;17:87–97. https://doi.org/10.1590/bjft.2014.019.

Atarés L, Chiralt A. Essential oils as additives in biodegradable films and coatings for active food packaging. Trends Food Sci Technol. 2016;48:51–62. https://doi.org/10.1016/j.tifs.2015.12.001.

Bakkali F, Averbeck S, Averbeck D, et al. Biological effects of essential oils–a review. Food Chem Toxicol. 2008;46:446–75.

Basmacıoğlu-Malayoğlu H, Özdemir P, Aktaş B. *In vitro* antioxidant activities, total phenolic contents and main phenolic compounds of essential oil blend and grape seed extract. Hayvansal Üretim. 2018;59:43–7.

Bassolé IHN, Juliani HR. Essential oils in combination and their antimicrobial properties. Molecules. 2012;17:3989–4006. https://doi.org/10.3390/molecules17043989.

Becerril R, Nerín C, Silva F. Encapsulation systems for antimicrobial food packaging components: an update. Molecules. 2020; 25. https://doi.org/10.3390/molecules25051134

Behbahani BA, Noshad M, Jooyandeh H. Improving oxidative and microbial stability of beef using Shahri Balangu seed mucilage loaded with Cumin essential oil as a bioactive edible coating. Biocatal Agric Biotechnol. 2020;101563. https://doi.org/10.1016/j.bcab.2020.101563

Bentayeb K, Vera P, Rubio C, Nerín C. The additive properties of Oxygen Radical Absorbance Capacity (ORAC) assay: the case of essential oils. Food Chem. 2014;148:204–8.

Bi Y, Zhou G, Pan D, et al. The effect of coating incorporated with black pepper essential oil on the lipid deterioration and aroma quality of Jinhua ham. Food Measure. 2019;13:2740–50. https://doi.org/10.1007/s11694-019-00195-4.

Bill M, Obianom PC, Korsten L, et al. Using plant volatiles in biodegradable sachets and exposure of volatiles under pallet covers to control postharvest decay in avocados. South African Avocado Growers' Assoc Yearb. 2016;39:44–9.

Bill M, Pillai SK, Tinyane P, et al. The effect of thyme oil low-density polyethylene impregnated pellets in polylactic acid sachets on storage quality of ready-to-eat avocado. Food Bioprocess Technol. 2017;11:141–51. https://doi.org/10.1007/s11947-017-2001-5.

Botre DA, Soares NFF, Espitia PJP, et al. Avaliação de filme incorporado com óleo essencial de orégano para conservação de pizza pronta. Revista Ceres. 2010;57:283–91. https://doi.org/10.1590/S0034-737X2010000300001.

Brahmi F, Abdenour A, Bruno M, et al. Chemical composition and *in vitro* antimicrobial, insecticidal and antioxidant activities of the essential oils of Mentha pulegium L. and Mentha rotundifolia (L.) Huds growing in Algeria. Ind Crop Prod. 2016;88:96–105. https://doi.org/10.1016/j.indcrop.2016.03.002.

Brewer MS. Natural antioxidants: sources, compounds, mechanisms of action, and potential applications. Compr Rev Food Sci Food Saf. 2011;10:221–47. https://doi.org/10.1111/j.1541-4337.2011.00156.x.

Brownlee IA, Seal CJ, Wilcox M, et al. Applications of alginates in food. In: Rehm BHA, editor. Alginates: biology and applications. Berlin/Heidelberg: Springer; 2009. p. 212–28.

Burt S. Essential oils: their antibacterial properties and potential applications in foods—a review. Int J Food Microbiol. 2004;94:223–53.

Catarino MD, Alves-Silva JM, Fernandes RP, et al. Development and performance of whey protein active coatings with Origanum virens essential oils in the quality and shelf life improvement of processed meat products. Food Control. 2017;80:273–80. https://doi.org/10.1016/j.foodcont.2017.03.054.

Ceylan R, Zengin G, Uysal S, et al. GC-MS analysis and *in vitro* antioxidant and enzyme inhibitory activities of essential oil from aerial parts of endemic Thymus spathulifolius Hausskn. et Velen. J Enzyme Inhib Med Chem. 2016;31:983–90.

Chaemsanit S, Sukmas S, Matan N, et al. Controlled release of peppermint oil from paraffin-coated activated carbon contained in sachets to inhibit mold growth during long term storage of brown rice. J Food Sci. 2019;84:832–41. https://doi.org/10.1111/1750-3841.14475.

Chandran J, Nayana N, Roshini N, Nisha P. Oxidative stability, thermal stability and acceptability of coconut oil flavored with essential oils from black pepper and ginger. J Food Sci Technol. 2017;54:144–52.

Chemsa AE, Zellagui A, Öztürk M, et al. Chemical composition, antioxidant, anticholinesterase, antimicrobial and antibiofilm activities of essential oil and methanolic extract of Anthemis stiparum subsp. sabulicola (Pomel) Oberpr. Microb Pathog. 2018;119:233–40. https://doi.org/10.1016/j.micpath.2018.04.033.

Chen X, Ren L, Li M, et al. Effects of clove essential oil and eugenol on quality and browning control of fresh-cut lettuce. Food Chem. 2017;214:432–9. https://doi.org/10.1016/j.foodchem.2016.07.101.

Choi WS, Singh S, Lee YS. Characterization of edible film containing essential oils in hydroxypropyl methylcellulose and its effect on quality atributes of "Formosa" plum (*Prunus salicina* L.). LWT Food Sci Technol. 2016;70:213–22. https://doi.org/10.1016/j.lwt.2016.02.036.

Chouhan S, Sharma K, Guleria S. Antimicrobial activity of some essential oils—present status and future perspectives. Medicines. 2017;4:58.

Ciesla LM, Wojtunik-Kulesza KA, Oniszczuk A, et al. Antioxidant synergism and antagonism between selected monoterpenes using the 2,2-diphenyl-1-picrylhydrazyl method. Flavour Fragr J. 2016;31:412–9. https://doi.org/10.1002/ffj.3330.

D'Sousa Costa CO, Ribeiro PR, Loureiro MB, et al. Phytochemical screening, antioxidant and antibacterial activities of extracts prepared from different tissues of Schinus terebinthifolius

Raddi that occurs in the coast of Bahia, Brazil. Pharmacogn Mag. 2015;11:607–14. https://doi.org/10.4103/0973-1296.160459.

Dao TP, Tran TH, Nguyen PTN, et al. Chemical composition and evaluation of antibacterial activities of essential oil from lemon (Citrus aurantifolia L.) leaves growing Tien Giang Province, Vietnam. Asian J Chem. 2019;31:2284–6. https://doi.org/10.14233/ajchem.2019.22144.

De Andrade MF, de Lima D, Silva I, Alves da Silva G, et al. A study of poly (butylene adipate-co-terephthalate)/orange essential oil films for application in active antimicrobial packaging. LWT Food Sci Technol. 2020;125:109148. https://doi.org/10.1016/j.lwt.2020.109148.

Del Ré PV, Jorge N. Especiarias como antioxidantes naturais: aplicações em alimentos e implicação na saúde. Rev Bras Plantas Med. 2012;14:389–99.

Dhifi W, Bellili S, Jazi S, et al. Essential oils' chemical characterization and investigation of some biological activities: a critical review. Medicines. 2016;3 https://doi.org/10.3390/medicines3040025.

Dias MV, Melo NR, Soares NFF, et al. Application of flavoring active films of low-density polyethylene in sugar biscuits. J Food Process Eng. 2013;36:656–62. https://doi.org/10.1111/jfpe.12029.

Dias MV, Machado Azevedo V, Borges SV, et al. Development of chitosan/montmorillonite nanocomposites with encapsulated α-tocopherol. Food Chem. 2014;165:323–9. https://doi.org/10.1016/j.foodchem.2014.05.120.

Dıblan S, Kaya S. Antimicrobials used in active packaging films. Food Health. 2018;4:63–79.

Dvir IM, Weizman O, Lewitus D, et al. Antimicrobial active packaging combining essential oils mixture: migration and odor control study. Polym Adv Technol. 2019;30:2558–66. https://doi.org/10.1002/pat.4642.

Ebani VV, Nardoni S, Bertelloni F, et al. Antibacterial and antifungal activity of essential oils against some pathogenic bacteria and yeasts shed from poultry. Flavour Fragr J. 2016;31:302–9. https://doi.org/10.1002/ffj.3318.

El Fawal GF, Omer AM, Tamer TM. Evaluation of antimicrobial and antioxidant activities for cellulose acetate films incorporated with Rosemary and Aloe Vera essential oils. J Food Sci Technol. 2019;56:1510–8. https://doi.org/10.1007/s13197-019-03642-8.

Elzaawely A, Xuan T, Koyama H, et al. Antioxidant activity and contents of essential oil and phenolic compounds in flowers and seeds of Alpinia zerumbet (Pers.) B.L. Burtt. & amp; R.M. Sm. Food Chem. 2007;104:1648–53. https://doi.org/10.1016/j.foodchem.2007.03.016.

Espitia PJP, Soares NFF, Botti LCM, et al. Assessment of the efficiency of essential oils in the preservation of postharvest papaya in an antimicrobial packaging system. Braz J Food Technol. 2012;15:333–42. https://doi.org/10.1590/S1981-67232012005000027.

European Union (EU). Regulation (EC) No 1333/2008 of The European Parliment and the Council of 16 December 2008 on food additives. Off J Eur Union L 354;2008a.

European Union (EU). Regulation (EC) No 1334/2008 of the European Parliament and of the Council of 16 December 2008 on flavourings and certain food ingredients with flavouring properties for use in and on foods and amending Council Regulation (EEC) No 1601/91, Regulations (EC) No 2232/96 and (EC) No 110/2008 and Directive 2000/13/EC. Off J Eur Union, 2008b.

European Commission (EU). Commission regulation no 10/2011. Off J Eur Union; 2011.

European Union (EU). Regulation (EC) no 1935/2004 of the european parliament and of the council of 27 October 2004 on materials and articles intended to come into contact with food and repealing Directives 80/590/EEC and 89/109/EEC. Off J Eur Union (Vol. L338), 2004.

Fadel HHM, Hassan IM, Ibraheim MT, et al. Effect of using cinnamon oil encapsulated in maltodextrin as exogenous flavouring on flavour quality and stability of biscuits. J Food Sci Technol. 2019;56:4565–74. https://doi.org/10.1007/s13197-019-03931-2.

Farsanipoor A, Khodanazary A, Hosseini SM. Effect of chitosan-whey protein isolated coatings incorporated with tarragon *Artemisia dracunculus* essential oil on the quality of Nemipterus japonicus fillets at refrigerated condition. Int J Biol Macromol. 2020;155:766–71. https://doi.org/10.1016/j.ijbiomac.2020.03.228.

FDA, Food And Drug Administration (FDA). Part 189 – Substances prohibited from use in human food. In: Code of Federal Regulations (CFR); 2020.

Fernandez-Pan I, Carrion-Granda X, Mate JI. Antimicrobial efficiency of edible coatings on the preservation of chicken breast fillets. Food Control. 2014;36:69–75. https://doi.org/10.1016/j.foodcont.2013.07.032.

Furuya T, Kuroiwa M, Kino K. Biotechnological production of vanillin using immobilized enzymes. J Biotechnol. 2017;243:25–8.

Gavahian M, Chu Y-H, Lorenzo JM, et al. Essential oils as natural preservatives for bakery products: understanding the mechanisms of action, recent findings, and applications. Crit Rev Food Sci Nutr. 2020;60:310–21. https://doi.org/10.1080/10408398.2018.1525601.

Ghaderi-Ghahfarokhi M, Barzegar M, Sahari MA, et al. Chitosan-cinnamon essential oil nano-formulation: application as a novel additive for controlled release and shelf life extension of beef patties. Int J Biol Macromol. 2017;102:19–28. https://doi.org/10.1016/j.ijbiomac.2017.04.002.

Gharenaghadeh S, Karimi N, Forghani S, et al. Application of salvia multicaulis essential oil-containing nanoemulsion against food-borne pathogens. Food Biosci. 2017;19:128–33. https://doi.org/10.1016/j.fbio.2017.07.003.

Giri TK. Alginate containing nanoarchitectonics for improved cancer therapy. In: Holban AM, Grumezescu AM, editors. Nanoarchitectonics for smart delivery and drug targeting. Oxford: William Andrew; 2016. p. 565–88.

Gómez-Estaca J, López-de-Dicastillo C, Hernández-Muñoz P, et al. Advances in antioxidant active food packaging. Trends Food Sci Technol. 2014;35:42–51. https://doi.org/10.1016/j.tifs.2013.10.008.

Goñi P, López P, Sánchez C, et al. Antimicrobial activity in the vapour phase of a combination of cinnamon and clove essential oils. Food Chem. 2009;116:982–9. https://doi.org/10.1016/j.foodchem.2009.03.058.

Goudjil MB, Zighmi S, Hamada D, et al. Biological activities of essential oils extracted from *Thymus capitatus* (Lamiaceae). South African J Bot. 2020;128:274–82.

Gutiérrez L, Escudero A, Batlle R, Nerín C. Effect of mixed antimicrobial agents and flavors in active packaging films. J Agric Food Chem. 2009;57:8564–71. https://doi.org/10.1021/jf901459e.

Hadian M, Rajaei A, Mohsenifar A, et al. Encapsulation of *Rosmarinus officinalis* essential oils in chitosan-benzoic acid nanogel with enhanced antibacterial activity in beef cutlet against Salmonella typhimurium during refrigerated storage. LWT Food Sci Technol. 2017;84:394–401. https://doi.org/10.1016/j.lwt.2017.05.075.

Hafsa J, Smach M, Ben Khedher MR, et al. Physical, antioxidant and antimicrobial properties of chitosan films containing Eucalyptus globulus essential oil. LWT Food Sci Technol. 2016;68:356–64. https://doi.org/10.1016/j.lwt.2015.12.050.

Hasheminya S-M, Dehghannya J. Composition, phenolic content, antioxidant and antimicrobial activity of Pistacia atlantica subsp. kurdica hulls' essential oil. Food Biosci. 2020;34:100510. https://doi.org/10.1016/j.fbio.2019.100510.

Hassan B, Chatha SAS, Hussain AI, et al. Recent advances on polysaccharides, lipids and protein based edible films and coatings. A review. Int J Biol Macromol. 2018;109:1095–107. https://doi.org/10.1016/j.ijbiomac.2017.11.097.

Herculano ED, de Paula HCB, de Figueiredo EAT, et al. Physicochemical and antimicrobial properties of nanoencapsulated Eucalyptus staigeriana essential oil. LWT Food Sci Technol. 2015;61:484–91. https://doi.org/10.1016/j.lwt.2014.12.001.

Hichri F, Omri Hichri A, Maha M, et al. Chemical composition, antibacterial, antioxidant and *in vitro* antidiabetic activities of essential oils from Eruca vesicaria. Chem Biodivers. 2019;16:e1900183. https://doi.org/10.1002/cbdv.201900183.

Hill LE, Gomes C, Taylor TM. Characterization of beta-cyclodextrin inclusion complexes containing essential oils (trans-cinnamaldehyde, eugenol, cinnamon bark, and clove bud extracts) for antimicrobial delivery applications. LWT Food Sci Technol. 2013;51:86–93. https://doi.org/10.1016/j.lwt.2012.11.011.

Hosseini Zandi M, Rezaei MSF. Two-step method for encapsulation of oregano essential oil in chitosan nanoparticles: preparation, characterization and in vitro release study. Carbohydr Polym. 2013;95:50–6.

Hsouna AB, Halima NB, Smaoui S, et al. Citrus lemon essential oil: chemical composition, antioxidant and antimicrobial activities with its preservative effect against *Listeria monocytogenes* inoculated in minced beef meat. Lipids Health Dis. 2017;16:146.

Huang T, Qian Y, Wei J, et al. Polymeric antimicrobial food packaging and its applications. Polymers. 2019;11:560. https://doi.org/10.3390/polym11030560.

Hüsnü CB, Buchbauer G. Handbook of essential oils: science, technology, and applications. Handb Essent Oils Sci Technol Appl; 2015.

Hyldgaard M, Mygind T, Meyer RL. Essential oils in food preservation: mode of action, synergies, and interactions with food matrix components. Front Microbiol. 2012;3:12.

Ibarra VG, Sendón R, Rodríguez-Bernaldo de Quirós A. Antimicrobial food packaging based on biodegradable materials. In: Antimicrobial food packaging; 2016. p. 363–84. https://doi.org/10.1016/B978-0-12-800723-5.00029-2.

Ilango P, Suresh V, Vummidi AV, et al. Evaluation of antibacterial activity of lemongrass oil against oral clinical isolates – an *in vitro* study. Pharm J. 2019;11:1023–8. https://doi.org/10.5530/pj.2019.11.161.

Ivana B, Marija L. 8 – Role of phenols in energy and functional beverages. In: Grumezescu AM, Holban AMBT-S and ED, editors. Woodhead Publishing; 2019, p. 229–268.

Jamil B, Abbasi R, Abbasi S, et al. Encapsulation of cardamom essential oil in chitosan nanocomposites: in-vitro efficacy on antibiotic-resistant bacterial pathogens and cytotoxicity studies. Front Microbiol. 2016;7:1–10. https://doi.org/10.3389/fmicb.2016.01580.

Jamróz E, Juszczak L, Kucharek M. Investigation of the physical properties, antioxidant and antimicrobial activity of ternary potato starch-furcellaran-gelatin films incorporated with lavender essential oil. Int J Biol Macromol. 2018;114:1094–101. https://doi.org/10.1016/j.ijbiomac.2018.04.014.

Jongedijk E, Cankar K, Buchhaupt M, et al. Biotechnological production of limonene in microorganisms. Appl Microbiol Biotechnol. 2016;100:2927–38.

Jouki M, Yazdi FT, Mortazavi SA, et al. Effect of quince seed mucilage edible films incorporated with oregano or thyme essential oil on shelf life extension of refrigerated rainbow trout fillets. Int J Food Microbiol. 2014;174:88–97. https://doi.org/10.1016/j.ijfoodmicro.2014.01.001.

Ju J, Chen X, Xie Y, et al. Application of essential oil as a sustained release preparation in food packaging. Trends Food Sci Technol. 2019;92:22–32. https://doi.org/10.1016/j.tifs.2019.08.005.

Jung J, Zhao Y. Antimicrobial packaging for fresh and minimally processed fruits and vegetables. Antimicrobial Food Packaging. 2016;243–256 https://doi.org/10.1016/B978-0-12-800723-5.00018-8.

Karlsen J. Encapsulation and other programmed release techniques for essential oils and volatile terpenes. In: Baser KHC, Buchbauer G, editors. Handbook of essential oils: science, technology and applications. CRC Press Taylor & Francis Group; 2010. p. 855–61.

Khaneghah AM, Hashemi SMB, Limbo S. Antimicrobial agents and packaging systems in antimicrobial active food packaging: an overview of approaches and interactions. Food Bioprod Process. 2018;111:1–19. https://doi.org/10.1016/j.fbp.2018.05.001.

Khorram F, Ramezanian A, Saharkhiz MJ. *In vitro* activity of some essential oils against penicillium digitatum. Adv Hortic Sci. 2018;32:487–94. https://doi.org/10.13128/ahs-22569.

Khorshidian N, Yousefi M, Khanniri E, et al. Potential application of essential oils as antimicrobial preservatives in cheese. Innov Food Sci Emerg Technol. 2018;45:62–72.

Kim H, Beak S-E, Song KB. Development of a hagfish skin gelatin film containing cinnamon bark essential oil. LWT. 2018;96:583–8. https://doi.org/10.1016/j.lwt.2018.06.016.

Koroch AR, Rodolfo Juliani H, Zygadlo JA. Bioactivity of essential oils and their components BT – flavours and fragrances: chemistry, bioprocessing and sustainability. In: Berger RG, editors. Berlin/Heidelberg: Springer; 2007, p. 87–115.

Koul O, Walia S, Dhaliwal GS. Essential oils as green pesticides: potential and constraints. Biopestic Int. 2008;4:63–84.

Ksouda G, Sellimi S, Merlier F, et al. Composition, antibacterial and antioxidant activities of Pimpinella saxifraga essential oil and application to cheese preservation as coating additive. Food Chem. 2019;288:47–56. https://doi.org/10.1016/j.foodchem.2019.02.103.

Kuorwel KK, Cran MJ, Sonneveld K, et al. Migration of antimicrobial agents from starch-based films into a food simulant. LWT Food Sci Technol. 2013;50:432–8. https://doi.org/10.1016/j.lwt.2012.08.023.

Lee DS. Packaging containing natural antimicrobial or antioxidative agents. In: Han JH, editor. Innovations in food packaging. Elsevier Science & Technology Books; 2005. p. 108–22.

Liaquat SM, Ali R. Development of antioxidant rich functional dairy product and study their total antioxidant status measured by DPPH assay. Int J Food Nutr Sci. 2019;4:65–73.

Lima JV, Moreira MN, Santos AMB, et al. Production of added active film for essential oil for cheese conservation. Revista Ingi. 2020;4:754–68.

Limbo S, Khaneghah AM. Active packaging of foods and its combination with electron beam processing. In: Electron beam pasteurization and complementary food process technologies. Cambridge: Woodhead Publishing; 2015. p. 195–217. https://doi.org/10.1533/9781782421085.2.195.

Lin L, Dai Y, Cui H. Antibacterial poly (ethylene oxide) electrospun nanofibers containing cinnamon essential oil/beta-cyclodextrin proteoliposomes. Carbohydr Polym. 2017;178:131–40. https://doi.org/10.1016/j.carbpol.2017.09.043.

Lin L, Gu Y, Cui H. Moringa oil/chitosan nanoparticles embedded gelatin nanofibers for food packaging against Listeria monocytogenes and Staphylococcus aureus on cheese. Food Packag Shelf Life. 2019;19:86–93. https://doi.org/10.1016/j.fpsl.2018.12.005.

Liu D, Li H, Jiang L, et al. Characterization of active packaging films made from poly (lactic acid)/poly(trimethylene carbonate) incorporated with oregano essential oil. Molecules. 2016;21 https://doi.org/10.3390/molecules21060695.

Lucera A, Conte A, Del Nobile MA. Volatile compounds usage in active packaging systems. Antimicrobial Food Packaging. 2016:319–27. https://doi.org/10.1016/B978-0-12-800723-5.00025-5.

Malanovic N, Lohner K. Gram-positive bacterial cell envelopes: the impact on the activity of antimicrobial peptides. Biochim Biophys Acta (BBA)-Biomembranes. 2016;1858:936–46.

Manu DK. Antimicrobial activity of cinnamaldehyde or geraniol alone or combined with high pressure processing to destroy *Escherichia coli* O157: H7 and *Salmonella enterica* in juices. 2016.

Marchese A, Orhan IE, Daglia M, et al. Antibacterial and antifungal activities of thymol: a brief review of the literature. Food Chem. 2016;210:402–14.

Marchese A, Arciola CR, Barbieri R, et al. Update on monoterpenes as antimicrobial agents: a particular focus on p-cymene. Materials (Basel). 2017a;10:947..

Marchese A, Barbieri R, Coppo E, et al. Antimicrobial activity of eugenol and essential oils containing eugenol: a mechanistic viewpoint. Crit Rev Microbiol. 2017b;43:668–89.

Marinello F, La Storia A, Mauriello G, et al. Atomic force micrscopy techniques to investigate activated food packaging materials. Trends Food Sci Technol. 2018;87:84–93. https://doi.org/10.1016/j.tifs.2018.05.028.

Mariod AA. Effect of essential oils on organoleptic (smell, taste, and texture) properties of food. In: Essential oils in food preservation, flavor and safety. Elsevier; 2016. p. 131–7.

Marques CS, Carvalho SG, Bertoli LD, et al. β-Cyclodextrin inclusion complexes with essential oils: obtention, characterization, antimicrobial activity and potential application for food preservative sachets. Food Res Int. 2019;119:499–509. https://doi.org/10.1016/j.foodres.2019.01.016.

Maryam Adilah ZA, Nur Hanani ZA. Active packaging of fish gelatin films with Morinda citrifolia oil. Food Biosci. 2016;16:66–71. https://doi.org/10.1016/j.fbio.2016.10.002.

Medeiros Barbosa I, da Costa Medeiros JA, de Oliveira KÁR, et al. Efficacy of the combined application of oregano and rosemary essential oils for the control of Escherichia coli, *Listeria monocytogenes* and *Salmonella Enteritidis* in leafy vegetables. Food Control. 2016;59:468–77. https://doi.org/10.1016/j.foodcont.2015.06.017.

Miltz J, Ram A, Nir MM. Prospects for application of post-consumer used plastics in food packaging. Food Addit Contam. 1997;14:649–59. https://doi.org/10.1080/02652039709374577.

Mohammadi A, Hashemi M, Hosseini SM. Chitosan nanoparticles loaded with *Cinnamomum zeylanicum* essential oil enhance the shelf life of cucumber during cold storage. Postharvest Biol Technol. 2015a;110:203–13. https://doi.org/10.1016/j.postharvbio.2015.08.019.

Mohammadi A, Hashemi M, Hosseini SM. Nanoencapsulation of *Zataria multiflora* essential oil preparation and characterization with enhanced antifungal activity for controlling Botrytis cinerea, the causal agent of gray mould disease. Innov Food Sci Emerg Technol. 2015b;28:73–80. https://doi.org/10.1016/j.ifset.2014.12.011.

Montes SS, Santana-Neta LG, Cruz RS. Óleos essenciais em embalagens para alimentos - Revisão de literatura de 2000 a 2012. Perspectivas da Ciência e Tecnologia. 2013;5(1–2).

Mulla M, Ahmed J, Al-Attar H, et al. Antimicrobial efficacy of clove essential oil infused into chemically modified LLDPE film for chicken meat packaging. Food Control. 2017;73:663–71. https://doi.org/10.1016/j.foodcont.2016.09.018.

Muriel-Galet V, Cerisuelo JP, López-Carballo G, et al. Development of antimicrobial films for microbiological control of packaged salad. Int J Food Microbiol. 2012;157:195–201. https://doi.org/10.1016/j.ijfoodmicro.2012.05.002.

Nedovic V, Kalusevic A, Manojlovic V, et al. An overview of encapsulation technologies for food applications. Procedia Food Sci. 2011;1:1806–15. https://doi.org/10.1016/j.profoo.2011.09.265.

Neethirajan S, Jayas DS. Nanotechnology for the Food and Bioprocessing Industries. Food Bioprocess Technol. 2010;4:39–47. https://doi.org/10.1007/s11947-010-0328-2.

Nesic AR, Seslija SI. The influence of nanofillers on physical–chemical properties of polysaccharide-based film intended for food packaging. In: Grumezescu AM, editor. Food packaging. Elsevier Inc.; 2017. p. 637–97. https://doi.org/10.1016/b978-0-12-804302-8.00019-4.

Noori S, Zeynali F, Almasi H. Antimicrobial and antioxidant efficiency of nanoemulsion-based edible coating containing ginger (Zingiber officinale) essential oil and its effect on safety and quality attributes of chicken breast fillets. Food Control. 2018;84:312–20.

Noriega P, Guerrini A, Sacchetti G, et al. Chemical composition and biological activity of five essential oils from the Ecuadorian Amazon rain forest. Molecules. 2019;24:1637.

Oliveira SPLF, Bertan LC, Rensis CMVB, et al. Whey protein-based films incorporated with oregano essential oil. Polímeros. 2017;27:158–64.

Oriani VB, Molina G, Chiumarelli M, et al. Properties of cassava starch-based edible coating containing essential oils. J Food Sci. 2014;79:E189–94. https://doi.org/10.1111/1750-3841.12332.

Otero V, Becerril R, Santos JA, et al. Evaluation of two antimicrobial packaging films against *Escherichia coli* O157: H7 strains *in vitro* and during storage of a Spanish ripened sheep cheese (Zamorano). Food Control. 2014;42:296–302. https://doi.org/10.1016/j.foodcont.2014.02.022.

Otoni CG, De MMR, Aouada FA, et al. Antimicrobial and physical-mechanical properties of pectin/papaya puree/cinnamaldehyde nanoemulsion edible composite films. Food Hydrocoll. 2014a;41:188–94. https://doi.org/10.1016/j.foodhyd.2014.04.013.

Otoni CG, Pontes SFO, Medeiros EAA, et al. Edible films from methylcellulose and nanoemulsions of clove bud (*Syzygium aromaticum*) and oregano (*Origanum vulgare*) essential oils as shelf life extenders for sliced bread. J Agric Food Chem. 2014b;62:5214–9. https://doi.org/10.1021/jf501055f.

Otoni CG, Espitia PJP, Avena-Bustillos RJ, et al. Trends in antimicrobial food packaging systems: emitting sachets and absorbent pads. Food Res Int. 2016;83:60–73. https://doi.org/10.1016/j.foodres.2016.02.018.

Oussalah M, Caillet S, Saucier L, Lacroix M. Antimicrobial effects of selected plant essential oils on the growth of a Pseudomonas putida strain isolated from meat. Meat Sci. 2006;73:236–44. https://doi.org/10.1016/j.meatsci.2005.11.019.

Oussalah M, Caillet S, Salmiéri S, et al. Antimicrobial effects of alginate-based films containing essential oils on *Listeria monocytogenes* and *Salmonella Typhimurium* present in Bologna and Ham. J Food Prot. 2007;70:901–8. https://doi.org/10.4315/0362-028X-70.4.901.

Pabast M, Shariatifar N, Beikzadeh S, Jahed G. Effects of chitosan coatings incorporating with free or nano-encapsulated Satureja plant essential oil on quality characteristics of lamb meat. Food Control. 2018;91:185–92. https://doi.org/10.1016/j.foodcont.2018.03.047.

Pandey AK, Kumar P, Singh P, et al. Essential oils: sources of antimicrobials and food preservatives. Front Microbiol. 2017;7:2161.

Passarinho ATP, Dias NF, Camilloto GP, et al. Sliced bread preservation through oregano essential oil-containing sachet. J Food Process Eng. 2014;37:53–62. https://doi.org/10.1111/jfpe.12059.

Patterson JE, McElmeel L, Wiederhold NP. *In vitro* activity of essential oils against gram-positive and gram-negative clinical isolates, including carbapenem-resistant enterobacteriaceae. Open Forum Infect Dis. 2019;6 https://doi.org/10.1093/ofid/ofz502.

Paul SK. Edible films and coatings for fruits and vegetables. Reference module in materials science and materials engineering. 2019. https://doi.org/10.1016/B978-0-12-803581-8.11509-7.

Peixoto MG, Blank AF, Arrigoni-Blank MF, et al. Activity of essential oils of lippia alba chemotypes and their major monoterpenes against phytopathogenic fungi. Biosci J. 2018;34:1136–46. https://doi.org/10.14393/BJ-v34n5a2018-39385.

Peng Y, Li Y. Combined effects of two kinds of essential oils on physical, mechanical and structural properties of chitosan films. Food Hydrocoll. 2014;36:287–93. https://doi.org/10.1016/j.foodhyd.2013.10.013.

Perdones Á, Escriche I, Chiralt A, Vargas M. Effect of chitosan–lemon essential oil coatings on volatile profile of strawberries during storage. Food Chem. 2016;197:979–86.

Pérez Córdoba LJ, Sobral PJA. Physical and antioxidant properties of films based on gelatin, gelatin-chitosan or gelatin-sodium caseinate blends loaded with nanoemulsified active compounds. J Food Eng. 2017;213:47–53. https://doi.org/10.1016/j.jfoodeng.2017.05.023.

Pires JRA, Souza VGL, Fernando AL. Chitosan/montmorillonite bionanocomposites incorporated with rosemary and ginger essential oil as packaging for fresh poultry meat. Food Packag Shelf Life. 2018;17:142–9. https://doi.org/10.1016/j.fpsl.2018.06.011.

Pirouzifard M, Yorghanlu RA, Pirsa S. Production of active film based on potato starch containing Zedo gum and essential oil of Salvia officinalis and study of physical, mechanical, and antioxidant properties. J Thermoplast Compos Mater. 2019;0892705718815541.

Purkait S, Bhattacharya A, Bag A, Chattopadhyay RR. Synergistic antibacterial, antifungal and antioxidant efficacy of cinnamon and clove essential oils in combination. Arch Microbiol. 2020:1–10.

Rao J, Chen B, McClements DJ. Improving the efficacy of essential oils as antimicrobials in foods: mechanisms of action. Annu Rev Food Sci Technol. 2019;10:365–87.

Rappoport Z. The chemistry of phenols. Wiley; 2004.

Realini CE, Marcos B. Active and intelligent packaging systems for a modern society. Meat Sci. 2014;98:404–19. https://doi.org/10.1016/j.meatsci.2014.06.031.

Reinas I, Oliveira J, Pereira J, et al. Migration of two antioxidants from packaging into a solid food and into Tenax®. Food Control. 2012;28:333–7. https://doi.org/10.1016/j.foodcont.2012.05.023.

Reyes-Jurado F, Cervantes-Rincón T, Bach H, et al. Antimicrobial activity of Mexican oregano (*Lippia berlandieri*), thyme (*Thymus vulgaris*), and mustard (*Brassica nigra*) essential oils in gaseous phase. Ind Crop Prod. 2019;131:90–5. https://doi.org/10.1016/j.indcrop.2019.01.036.

Ribeiro-Santos R, Andrade M, Mela NR, et al. Biological activities and major components determination in essential oils intended for a biodegradable food packaging. Ind Crop Prod. 2017a;97:201–10. https://doi.org/10.1016/j.indcrop.2016.12.006.

Ribeiro-Santos R, Andrade M, Melo NR, et al. Use of essential oils in active food packaging: recent advances and future trends. Trends Food Sci Technol. 2017b;61:132–40. https://doi.org/10.1016/j.tifs.2016.11.021.

Ribeiro-Santos R, Andrade M, Sanches-Silva A. Application of encapsulated essential oils as antimicrobial agents in food packaging. Curr Opin Food Sci. 2017c;14:78–84. https://doi.org/10.1016/j.cofs.2017.01.012.

Ribeiro-Santos R, Melo NR, Andrade M, et al. Potential of migration of active compounds from protein-based films with essential oils to a food and a food simulant Essential oils for food application: natural substances with established biological activities. Food Bioprocess Technol. 2017d;11:43–71.

Ribeiro-Santos R, Andrade M, Sanches-Silva A, et al. Essential oils for food application: natural substances with established biological activities. Food Bioprocess Technol. 2018a;11:43–71. https://doi.org/10.1007/s11947-017-1948-6.

Ribeiro-Santos R, Ventura LAF, Santos DC, et al. Effects of oregano, cinnamon, and sweet fennel essential oils and their blends on foodborne microorganisms. Int Food Res J. 2018b;25:540–4.

Rieger KA, Schiffman JD. Electrospinning an essential oil: cinnamaldehyde enhances the antimicrobial efficacy of chitosan/poly(ethylene oxide) nanofibers. Carbohydr Polym. 2014;113:561–68. https://doi.org/10.1016/j.carbpol.2014.06.075

Rios J-L. Essential oils: what they are and how the terms are used and defined. In: Essential oils in food preservation, flavor and safety. Elsevier; 2016. p. 3–10.

Rodea-González DA, Cruz-Olivares J, Román-Guerrero A, et al. Spray-dried encapsulation of chia essential oil (*Salvia hispanica* L.) in whey protein concentrate-polysaccharide matrices. J Food Eng. 2012;111:102–9. https://doi.org/10.1016/j.jfoodeng.2012.01.020.

Rodriguez-Garcia I, Silva-Espinoza BA, Ortega-Ramirez LA, et al. Oregano essential oil as an antimicrobial and antioxidant additive in food products. Crit Rev Food Sci Nutr. 2016;56:1717–27. https://doi.org/10.1080/10408398.2013.800832.

Sabphon C, Srichoosilp A, Wanichwecharungruang S, et al. Dissolvable and edible film for long-lasting kaffir lime aroma in food. Food Sci Technol. 2019; https://doi.org/10.1111/ijfs.14426.

Salvo A, La Torre GL, Rotondo A, et al. Multiple analytical approaches for the organic and inorganic characterization of Origanum vulgare L. samples. Nat Prod Res. 2019;33:2815–22. https://doi.org/10.1080/14786419.2018.1503269.

Sanches Silva A, Cruz Freire JM, Franz R, et al. Time-temperature study of the kinetics of migration of diphenylbutadiene from polyethylene films into aqueous foodstuffs. Food Res Int. 2008;41:138–44. https://doi.org/10.1016/j.foodres.2007.10.007.

Sanchez-Gonzalez L, Chafer M, Chiralt A, et al. Physical properties of edible chitosan films containing bergamot essential oil and their inhibitory action on Penicillium italicum. Carbohydr Polym. 2010;82:277–83. https://doi.org/10.1016/j.carbpol.2010.04.047.

Santana MCCB, Machado BAS, Silva TN, et al. Incorporação de urucum como aditivo antioxidante em embalagens biodegradáveis a base de quitosana. Ciênc Rural. 2013;43:544–50.

Shaaban HAE, El-Ghorab AH, Shibamoto T. Bioactivity of essential oils and their volatile aroma components: review. J Essent Oil Res. 2012;24:203–12. https://doi.org/10.1080/10412905.2012.659528.

Shamaei S, Seiiedlou SS, Aghbashlo M, et al. Microencapsulation of walnut oil by spray drying: effects of wall material and drying conditions on physicochemical properties of microcapsules. Innov Food Sci Emerg Technol. 2017;39:101–12. https://doi.org/10.1016/j.ifset.2016.11.011.

Shao P, Yan Z, Chen H, et al. Electrospun poly (vinyl alcohol)/permutite fibrous film loaded with cinnamaldehyde for active food packaging. J Appl Polym Sci. 2017;135:46117. https://doi.org/10.1002/app.46117.

Sharifi-Rad M, Varoni EM, Iriti M, et al. Carvacrol and human health: a comprehensive review. Phyther Res. 2018;32:1675–87. https://doi.org/10.1002/ptr.6103.

Sharmeen JB, Suroowan S, Rengasamy K, et al. Chemistry, bioactivities, mode of action and industrial applications of essential oils. Trends Food Sci Technol. 2020;111:89–105. https://doi.org/10.1016/j.tifs.2020.04.025.

Shojaee-Aliabadi S, Hosseini H, Mohammadifar MA, et al. Characterization of antioxidant-antimicrobial κ-carrageenan films containing *Satureja hortensis* essential oil. Int J Biol Macromol. 2013;52:116–24. https://doi.org/10.1016/j.ijbiomac.2012.08.02.

Shokri S, Parastouei K, Taghdir M, Abbaszadeh S. Application an edible active coating based on chitosan- Ferulago angulata essential oil nanoemulsion to shelf life extension of Rainbow trout fillets stored at 4 °C. Int J Biol Macromol. 2020;15:846–54. https://doi.org/10.1016/j.ijbiomac.2020.03.080.

Silva AS, Cruz Freire JM, Sendón R, et al. Migration and diffusion of diphenylbutadiene from packages into foods. J Agric Food Chem. 2009;57:10225–30. https://doi.org/10.1021/jf901666h.

Smith-Palmer A, Stewart J, Fyfe L. Antimicrobial properties of plant essential oils and essences against five important food-borne pathogens. Lett Appl Microbiol. 1998;26:118–22. https://doi.org/10.1046/j.1472-765X.1998.00303.x.

Soares NFF, Silva WA, Pires ACS, et al. New developments and applications in food packaging. Revista Ceres. 2009;56:370–78.

Souza VGL, Fernando AL. Nanoparticles in food packaging: biodegradability and potential migration to food – a review. Food Packag Shelf Life. 2016;8:63–70. https://doi.org/10.1016/j.fpsl.2016.04.001.

Souza RC, da Costa MM, Baldisserotto B, et al. Antimicrobial and synergistic activity of essential oils of Aloysia triphylla and Lippia alba against Aeromonas spp. Microb Pathog. 2017;113:29–33. https://doi.org/10.1016/j.micpath.2017.10.013.

Souza VGL, Pires JRA, Vieira ET, et al. Shelf life assessment of fresh poultry meat packaged in novel bionanocomposite of chitosan/montmorillonite incorporated with ginger essential oil. Coatings. 2018;8:177. https://doi.org/10.3390/coatings8050177.

Souza VGL, Rodrigues C, Ferreira L, et al. In vitro bioactivity of novel chitosan bionanocomposites incorporated with different essential oils. Ind Crop Prod. 2019a;140:111563. https://doi.org/10.1016/j.indcrop.2019.111563.

Souza WFM, Mariano XM, Isnard JL, et al. Evaluation of the volatile composition, toxicological and antioxidant potentials of the essential oils and teas of commercial Chilean boldo samples. Food Res Int. 2019b;124:27–33. https://doi.org/10.1016/j.foodres.2018.12.059.

Souza VGL, Pires JRA, Rodrigues C, et al. Chitosan composites in packaging industry—current trends and future challenges. Polymers. 2020;12:417. https://doi.org/10.3390/polym12020417.

Stanojević LP, Stanojević JS, Cvetković DJ, Ilić DP. Antioxidant activity of oregano essential oil (*Origanum vulgare* L.). Biol Nyssana. 2018;7:131–9. https://doi.org/10.5281/zenodo.200410.

Stratakos AC, Koidis A. Methods for extracting essential oils. Essential oils in food preservation, flavor and safety; 2016, p. 31–38. https://doi.org/10.1016/b978-0-12-416641-7.00004-3.

Sung S-Y, Sin LT, Tee T-T, et al. Antimicrobial agents for food packaging applications. Trends Food Sci Technol. 2013;33:110–23. https://doi.org/10.1016/j.tifs.2013.08.001.

Sung S-Y, Sin L T, Tee T-T, et al. Effects of Allium sativum essence oil as antimicrobial agent for food packaging plastic film. Innov Food Sci Emerg Technol. 2014;26:406–14. https://doi.org/10.1016/j.ifset.2014.05.009

Suppakul P, Sonneveld K, Bigger S. W, et al. Efficacy of polyethylene-based antimicrobial films containing principal constituents of basil. LWT. 2008;41:779–788. https://doi.org/10.1016/j.lwt.2007.06.006.

Surwade SA, Chand K. Antimicrobial food packaging: an overview. Eur J Biotechnol Biosci. 2017;5:85–90.

Tang Y, Zhou Y, Lan X, et al. Electrospun gelatin nanofibers encapsulated with peppermint and chamomile essential oils as potential edible packaging. J Agric Food Chem. 2019;67:2227–34.

Tao F, Hill LE, Peng Y, et al. Synthesis and characterization of β-cyclodextrin inclusion complexes of thymol and thyme oil for antimicrobial delivery applications. LWT Food Sci Technol. 2014;59:247–55. https://doi.org/10.1016/j.lwt.2014.05.037.

Tavares FO, Pieretti GG, Antigo JL, et al. Edible coating with oregano and rosemary essential oils for use in ricotta cheese. Rev Inst Laticinios Candido Tostes. 2014;69:249–57. https://doi.org/10.14295/2238-6416.v69i4.309.

Taylor TM. Natural food antimicrobials: recent trends in their use, limitations, and opportunities for their applications in food preservation. In: Natural and bio-based antimicrobials for food applications. ACS Publications; 2018. p. 25–43.

Teixeira B, Marques A, Ramos C, et al. European pennyroyal (Mentha pulegium) from Portugal: chemical composition of essential oil and antioxidant and antimicrobial propertiesof extracts and essential oil. Ind Crop Prod. 2012;36:81–7. https://doi.org/10.1016/j.indcrop.2011.08.011.

Tepe AS, Ozaslan M. Anti-Alzheimer, anti-diabetic, skin-whitening, and antioxidant activities of the essential oil of Cinnamomum zeylanicum. Ind Crop Prod. 2020;145:112069.

Tores-Alvarez C, Núñez González A, Rodríguez J, et al. Chemical composition, antimicrobial, and antioxidant activities of orange essential oil and its concentrated oils. CyTA-J Food. 2017;15:129–35.

Ugalde ML, de Cezaro AM, Vedovatto F, et al. Active starch biopolymeric packaging film for sausages embedded with essential oil of Syzygium aromaticum. J Food Sci Technol. 2017;54:2171–5. https://doi.org/10.1007/s13197-017-2624-6.

Uliana MP, Fronza M, da Silva AG, et al. Composition and biological activity of Brazilian rose pepper (Schinus terebinthifolius Raddi) leaves. Ind Crop Prod. 2016;83:235–40. https://doi.org/10.1016/j.indcrop.2015.11.077.

Upasen S, Wattanachai P. Packaging to prolong shelf life of preservative-free white bread. Heliyon. 2018;4:e00802. https://doi.org/10.1016/j.heliyon.2018.e00802.

Uzunku S, Var I. Active antimicrobial food packaging. 2018; https://doi.org/10.5772/intechopen.82429.

Valdivieso-Ugarte M, Gomez-Llorente C, Plaza-Díaz J, Gil Á. Antimicrobial, antioxidant, and immunomodulatory properties of essential oils: a systematic review. Nutrients. 2019;11 https://doi.org/10.3390/nu11112786.

Valencia-Sullca C, Vargas M, Atarés L, Chiralt A. Thermoplastic cassava starch-chitosan bilayer films containing essential oils. Food Hydrocoll. 2018;75:107–15. https://doi.org/10.1016/j.foodhyd.2017.09.008.

Vijayan V, Mazumder A. *In vitro* inhibition of food borne mutagens induced mutagenicity by cinnamon (*Cinnamomum cassia*) bark extract. Drug Chem Toxicol. 2018;41:385–93. https://doi.org/10.1080/01480545.2018.1439056.

Viuda-Martos M, Mohamady MA, Fernández-López J, et al. In vitro antioxidant and antibacterial activities of Essentials oils obtained from Egyptian aromatic plants. Food Control. 2011;22:1715–22. https://doi.org/10.1016/j.foodcont.2011.04.003.

Vunain E, Mishra AK, Mamba BB. Fundamentals of chitosan for medical applications. In: Jennings JA, Bumgardner JD, editors. Chitosan based biomaterials, Volume 1: Fundamentals. Woodhead Publishing Limited; 2017. p. 3–30.

Wen P, Zhu D-H, Feng K, et al. Fabrication of electrospun polylactic acid nanofilm incorporating cinnamon essential oil/β -cyclodextrin inclusion complex for antimicrobial packaging. Food Chem. 2016a;196:996–1004. https://doi.org/10.1016/j.foodchem.2015.10.043.

Wen P, Zhu D-H, Wu H, et al. Encapsulation of cinnamon essential oil in electrospun nanofibrous film for active food packaging. Food Control. 2016b;59:366–76.

Wieczyńska J, Cavoski I. Antimicrobial, antioxidant and sensory features of eugenol, carvacrol and trans-anethole in active packaging for organic ready-to-eat iceberg lettuce. Food Chem. 2018;259:251–60. https://doi.org/10.1016/j.foodchem.2018.03.137.

Wu J, Liu H, Ge S, et al. The preparation, characterization, antimicrobial stability and in vitro release evaluation of fish gelatin films incorporated with cinnamon essential oil nanoliposomes. Food Hydrocoll. 2015;43:427–35. https://doi.org/10.1016/j.foodhyd.2014.06.017.

Wu J, Sun X, Guo X, et al. Physicochemical properties, antimicrobial activity and oil release of fish gelatin films incorporated with cinnamon essential oil. Aquac Fish. 2017;2:185–92. https://doi.org/10.1016/j.aaf.2017.06.004.

Yeddes W, Nowacka M, Rybak K, et al. Evaluation of the antioxidant and antimicrobial activity of rosemary essential oils as gelatin edible film component. Food Sci Technol Res. 2019;25:321–9.

Yeh HY, Chuang CH, Chen HC, et al. Bioactive components analysis of two various gingers (*Zingiber officinale* Roscoe) and antioxidant effect of ginger extracts. LWT Food Sci Technol. 2014;55:329–34. https://doi.org/10.1016/j.lwt.2013.08.003.

Yun J, Fan X, Li X, et al. Natural surface coating to inactivate *Salmonella enterica* serovar Typhimurium and maintain quality of cherry tomatoes. Int J Food Microbiol. 2015;193:59–67. https://doi.org/10.1016/j.ijfoodmicro.2014.10.013.

Zeid A, Karabagias IK, Nassif M, Kontominas MG. Preparation and evaluation of antioxidant packaging films made of polylactic acid containing thyme, rosemary, and oregano essential oils. J Food Process Preserv. 2019;43 https://doi.org/10.1111/jfpp.14102.

Zhang H, Jung J, Zhao Y. Preparation and characterization of cellulose nanocrystals films incorporated with essential oil loaded β-chitosan beads. Food Hydrocoll. 2017a;69:164–72. https://doi.org/10.1016/j.foodhyd.2017.01.029.

Zhang L, Liu A, Wang W, et al. Characterisation of microemulsion nanofilms based on Tilapia fish skin gelatine and ZnO nanoparticles incorporated with ginger essential oil: meat packaging application. Int J Food Sci Technol. 2017b;52:1670–9. https://doi.org/10.1111/ijfs.13441.

Zhang H, Li X, Kang H. Chitosan coatings incorporated with free or nano-encapsulated Paulownia Tomentosa essential oil to improve shelf-life of ready-to-cook pork chops. LWT. 2019;116 https://doi.org/10.1016/j.lwt.2019.108580.

Chapter 5
Emitters of Flavours, Colorants and Other Food Ingredients

Ana Luísa Guimarães

Abstract Active packaging is defined as a packaging system that involves interactions between the package and its content, in order to extend shelf-life and/or enhance the organoleptic properties of the packaged food. There are various ways by which active packaging may enrich the consumer experience, being the incorporation of diverse kinds of additives the most common one. This chapter focuses on packaging systems engineered to emit flavours, colorants and other food ingredients expected to improve the five-sense experience of eating. The legal framework regulating the usage of these substances and their kinetics of release and diffusion are explored in the first half of the chapter. In the second part of the chapter, various systems for the incorporation of bioactive substances in the package are delved into, and some of the existing products and patents are presented. The emission of flavours, colorants and other food ingredients by the food package is a field with numerous applications of market interest. However, there are still a few aspects to be improved, which, together with expected market trends for the near future, are briefly explored in the finishing part of this chapter.

Keywords Packaging · Food · Technology

1 Introduction

Changes in the food packaging industry nearly escorted human kind lifestyle changes. For many centuries, human beings would simply eat what they could find in their surroundings, with no need for storing or packaging. When the nomadic

A. L. Guimarães (✉)
Colab4Food – Collaborative Laboratory, Vila do Conde, Portugal
e-mail: ana.guimaraes@colab4food.com

S. M. Jafari, A. S. Silva (eds.), *Releasing Systems in Active Food Packaging*,
Food Bioactive Ingredients, https://doi.org/10.1007/978-3-030-90299-5_5

habits were abandoned, the practice of store gathered foods in containers arose. At this time, however, food containers were mainly made out of wood, shells, leaves, glass, ceramics and/or paper, and were not very sophisticated. It was not until the nineteenth century that the packaging industry significantly evolved, due to the new materials and manufacturing methods brought by the industrial revolution. Metal cans, for example, were initially used for storing snuff, avoiding the flavour to be lost and the moisture to enter in contact with the product, and only later, were they used to store food. Paperboard boxes started only to be used as storage and protection units by the same time, and plastic was only used for packaging in the twentieth century, during World War II (Risch 2009). Therefore, historically, packaging systems for food or otherwise, have been used mainly to contain and protect the product and, only more recently, to communicate and offer convenience to the consumer. These rather simple packaging items provided the basis for the development of modern food preservation and packaging systems, as the focus on the safety and quality of food increased after World War II. Several packaging improvements allowed the consumers to have access to a wider range of quality food all year round (Risch 2009).

Currently, packaging plays a central role, not only in the protection of the packaged food, but also in enhancing products' freshness, nutritional value and organoleptic properties, in the case of active packaging. Active packages, as defined by the European regulation (EC) No 450/2009, are intended to extend shelf-life or to preserve and/or improve the condition of packaged food. They are produced to release or absorb compounds into or from the food and/or the external environment. The features which can be included in a package are incredibly diverse. Barriers for gases and UV-light, moisture absorbers and flavour emitters are just some examples of recent concepts of active packaging, designed/engineered to bring convenience and improved food quality for consumers (Lee et al. 2015). This chapter focuses on packaging systems with emitters of flavours, colorants and other ingredients of interest. An emitter is defined as a component that releases substances into or onto the packaged food or the environment around the food (European regulation (EC) No 450/2009).

To achieve this level of active packaging technology, a lot of chemical, physical and biological knowledge was necessary. Taking the aroma-releasing packages as an example, it was necessary to understand the importance of aroma for the eating experience, identify the crucial molecules and compounds for the development of a certain aroma, gain insight into the release kinetics and diffusion dynamics within the system and develop the technology that allows the production of such a package. Thus, several aspects have still to be studied and properly understood before active packages become the norm in grocery stores. Recently, with the rising of sustainability concerns, particularly within the food industry, a large body of research has been dedicated to the development of environmentally friendly packaging, non-petroleum-based and, ideally, edible (Risch 2009).

Inclusion of bioactive substances in the packaging material for controlled release is one of the biggest trends in the active packaging world. Among all the substances one can incorporate in a package, this chapter focuses on flavours, colorants and other food ingredients of interest. Those are essential players in food quality and, above all, in the preservation of organoleptic features during storage, contributing to satisfy the ever-increasing demands for foods with extended shelf-life.

2 Legal Framework

The food industry has to follow a group of restrict legal prerequisites in order to become available to the consumer, which are mainly related with safety and health concerns. The authority responsible for this evaluation depends on the country where the product is labeled. The Food and Drug Administration (FDA) regulates the market in the USA, while the European Food Safety Authority (EFSA) does it in Europe. Every and each ingredient or substance that may become part of a food product must be registered as a food ingredient or additive, and be subjected to a premarket approval requirement, meaning that these should be approved by the competent authority prior to commercialization. The rules applied to ingredients which directly take part of the food preparation process are also valid to any additive that will, or at least is expected to, become a component of the food or otherwise affect its characteristics. This includes any substance used in the production, processing, treatment, packaging, transportation or storage of foods. The most common additives present in food products are sweeteners, colorants, emulsifiers, antioxidants, gelling and thickener agents, preservatives and stabilisers (Burgos et al. 2017).

The FDA classifies additives in two groups: direct and indirect. The former are added to the food with a specific purpose (e.g. xanthan gum is usually used to add texture), while the latter may become part of the food itself, such as minute amounts of packaging substances that were not intended to be directly added to the food. Both kinds of additives have to be proven safe before use and commercialization. All of the authorized additives are defined and listed by FDA which may be found in their website, with some exceptions that are by default accepted as safe (e.g. GRAS ingredients), which have their own list, also available in the FDA website (Food Additive Status List 2019). When it comes specifically to colouring additives, it is important to understand that FDA distinguishes between synthetically produced and natural sources-derived ones. The widely used synthetic compounds, which are much cheaper and deliver more reproducible results, have to be evaluated to become certified colours in order to be used in food products. The natural colouring agents, while generally are more expensive and less trustworthy, including dehydrated beets, beta-carotenes and annatto extract, are exempt from certification, although they still have to be approved by FDA (Food Additive Status List 2019). Furthermore, in the US market, some ingredients can be listed collectively as 'spices' or 'flavours' such as ginger or pepper, without need to name each one individually. If in such a group of additives there is a putative allergenic ingredient, it may be mentioned individually (CPG Sec 525.750 Spices – Definitions 1980).

In the European Union, EFSA is responsible for the evaluation of food additives (including antioxidants, colouring agents, flour treatment agents, preservatives), food enzymes and flavourings, which are submitted to the European Commission. Note that, among the flavourings, the smoking flavours follow a special path of approval by the European commission, which may be found in the Regulation (EU) No 1321/2013 (Regulation (EU) No. 1321/2013). The applications submitted to the European Commission to evaluate a food additive, should include a dossier on the substance, containing scientific data on its proposed uses and usage concentration. The decision of whether or not authorize a certain substance is based on the EFSA's assessment of the application dossier. The application for a new use for a previously approved substance follows the same procedure. In the European Commission website, a list of all the approved food additives may be found, accordingly to the Annex II of Regulation (EC) No 1333/2008 (Regulation (EC) No 1333/2008). Authorized food additives must comply not only with the approved usage purpose, but also with the purity criteria. Accordingly to the European Union legislation, all food additives must be designated by the category of its principle function (e.g. colorant, preservative), followed by their specific name, or, their E number (Food Improvement Agents – European Commission).

3 Eating – A Five Senses Experience

Although one may not think about it very often, every time we eat the five senses are engaged and each and every one has an important contribution to the whole experience. A classic example is how essential are the crunchiness felt when eating chips and the sound produced by each bite, to a full experience. Further, the visuals of a certain food, how colourful it is and how fresh it looks, have a tremendous impact on our experience, as it is the first impression we get and one that will influence everything following. But the biggest contributing component is the flavour of a food, which is a combination of any sensation of heat or cooling in the mouth, the aromas of the food, given by volatile aromatic components, and the taste/gustatory components. There are five essential tastes from which all the complexes tastes are originated, sweet, sour, umami, salty and bitter. Thus, the flavour profile is an attribute of utter importance, which plays a major role in determining the consumer acceptability of a food product (Arabi et al. 2012).

Both aroma and taste compounds make up the flavour profile of a food. Usually, less than 1% of a food is composed by aroma compounds. These molecules are very small, with very low molecular weight (<400 Da) and high vapor pressure, which allows for an easy release into the gas phase and then perceived by the olfactory receptors. In the food database for volatile compounds (Volatile Compounds in Food Database), more than fifteen thousand volatile compounds can be identified. Volatile compounds may belong to several chemical classes, for instance, alcohols, esters or lactones, and may or not be odorants. Taste compounds, on the other hand, may be mineral salts, peptides, nucleotides, organic acids, among others, and different classes present varying thresholds for detection by the taste buds. They are responsible for the five basic tastes mentioned above (Guichard and Salles 2016).

Aroma and taste molecules are released from the food piece into the saliva (taste) or air phase of the oral and nasal cavities (aroma), in order to be detected. Consequently, the releasing process is highly dependent on the structure of the food and it is also influenced by particular features, such as the salt content. An increase in salt concentration leads to an increase in the air/water partitioning of many aroma compounds, due to the salting out effect, the reduction of the available solvent in the liquid phase (Guichard and Salles 2016).

Historically, flavour molecules were extracted from plants or animal parts by enzymatic, microbiological or physical methods. More recently, they have started to be produced synthetically as well. There are several limitations for the use of either natural or synthetic compounds for flavouring foods and every and each one should be registered before use, as outlined in the previous section (Burgos et al. 2017).

From the moment a food is harvest or is out of the cooking pan, its quality and organoleptic properties start to degrade, from the flavour itself to the bright colour of a freshly harvested fruit, which diminish the eating experience. One of the most common strategies to delay any alteration that the food may suffer is packaging and, more specifically, active packaging. This fairly recent concept refers to packages with unique and innovative capacities that prolong the shelf-life and enhance the freshness and quality of the food product up until the moment of consumption, besides the standard protecting properties (European regulation (EC) No 450/2009). Annually, an average of 1.3 billion tonnes of foods are wasted, representing approximately 30% of the global food production. This number include both, the waste at the production level, such as harvesting problems, and the waste at the consumer level. In developed countries, the majority of food waste occurs at the retail and consumer levels, often because the food is not timely consumed, since people tend to buy more than their families can possibly eat. The food that is lost and wasted every year represents an economic burden of one trillion USD and would be enough to feed the 870 million hungry people in the world, four times over (FAO 2020).

4 Release and Diffusion

In order to create effective active packages, which are able to harbor and timely release the substances of interest, it is essential to understand the mechanisms of retention, release and diffusion of these molecules.

4.1 *Kinetics of Release and Diffusion*

The release of any substance from an active package is highly dependent on thermodynamic and kinetic parameters. In the particular case of volatile molecules incorporated in the package, such as aromas, the release is dependent on the volatilization of the compounds, which may occur in different steps, such as formulation, storage, preparation or consumption. The release of volatiles will vary depending on the physical state of the matrix. Liquid matrixes (aqueous or lipidic) are usually the simplest to understand and are often used to study the dynamics of partitioning and release. In solid matrixes though, other parameters need to be considered. Food matrixes are frequently very complex, with heterogeneous composition and convoluted structure, demanding a case-by-case study (Voilley and Souchon 2006).

Furthermore, depending on its type, package structure may differ, for example from amorphous to crystalline, which, together with the characteristics of the molecules to be transferred, affect the extend and the kinetics of the mass transference. In polymeric matrices, which represent the majority of food packaging materials, amorphous and crystalline regions are mixed together. In the amorphous regions, there are small empty spaces which allow for the movement of molecules, creating the conditions to mass transfer. Low molecular weight substances can be adsorbed or absorbed into the spaces and move inside the network. Contrarily, crystalline regions are not permeable to movement and may function as blocking knots, decreasing the capacity for sorption and delaying the diffusion of molecules inside the package. The chemical composition of the package also interferes with the whole process, mainly due to the level of compatibility between the package and the active substance (Gavara and Catalá 2002).

Food constituents, which may be transferred, are very diverse in physical properties. Some are fairly mobile, such as gases or aroma molecules, whereas others are rather static, like polymer molecules. Among the food constituents with high mobility are the flavour molecules, which are volatile and odorous at atmospheric pressure. They are usually present in foodstuff in very low amounts and have to be transferred to the gas phase for their aroma to be felt. Depending on the situation, transference of molecules between different materials and/or phases may occur by migration, permeation, sorption or diffusion processes. The release desired or undesired substances, such as additives or polymeric residues, from the packages to the package content is classified as migration. When molecules are exchanged between internal and external atmospheres, through the package, in response to a concentration gradient, the process is called permeation or permeance. The capacity of a molecule to permeate a package or some sort of membrane is measured by the permeability coefficient. The permeability coefficient is calculated using Darcy's law (Eq. 5.1) after experimentally determining the flow rate of a substance through a material:

$$Q = \frac{k.A\left(P_i - P_o\right)}{\mu.L} \tag{5.1}$$

Where,

Q (m^3/s) is the flow rate
k (m^2) is the permeability of the sample
A (m^2) is the area of the sample
P_ie P_0 (Pa) are the inlet and the outlet fluid pressure, respectively

μ (Pa.s) is the dynamic viscosity of the fluid
L (m) is the length of the sample (McPhee et al. 2015)

Sorption (absorption or desorption) represents the affinity between package and food component, and it is measured by the solubility coefficient (S) for gases, and by the partition coefficient (K) in the case of solutes. In practical terms, sorption represents the capacity of the package to retain a certain substance, being flavour scalping (retention of flavour by the package) a classic example of sorption. Water, oil and fats are the main ingredients most commonly absorbed in the package, although sugars, alcohols, colorants and aromas are also potentially absorbed. Sorption frequently leads to detrimental effects on the appearance of the product and, consequently, a decrease in consumers' acceptance, however it is hardly implicated in food spoilage. The absorption and desorption of a compound to the package depend on certain properties, such as the package material, the solvent and solutes, and the degree of agitation on both sides of the packaging material. All of these processes are expressions of the system tendency towards equilibrium between its, usually four, phases: external environment, package, internal atmosphere and packaged product, food in this particular case (Gavara and Catalá 2002; Han and Scanlon 2013). Lastly, diffusion is the spontaneous movement of molecules within a material, also accordingly to a concentration gradient. Diffusivity (D) is the measure for how well a compound, either a gas or a solute, may diffuse in a certain material, defined by the transfer rate of a given amount of diffusant across a demarcated distance. When the phase is considered homogeneous, such as in gaseous or stirred liquid phases, the kinetics of molecules is assumed to follow Fick's laws (Gavara and Catalá 2002; Han and Scanlon 2013; Teixeira et al. 2007).

Diffusion was firstly described in the nineteenth century by Adolf Fick, a German physician, in the Fick's first law. This law's equation may be written in several forms, being the molar-based formula, presented below, the most common one (Eq. 5.2).

$$J_c = -D_c \frac{dC_c}{dz} \tag{5.2}$$

Where,

J_c is the molar diffusion flux of molecule *c*
D_c ($m^2.s^{-1}$) is the diffusion coefficient of molecule *c*, usually assumed to be exclusively dependent on temperature for a given substance/medium
$\frac{dC_c}{dz}$ ($mol.m^{-3}.m^{-1}$) is the concentration gradient

The minus symbol means that the mass transfer occurs always from the area where molecule *c* is the more concentrated to the area where it is least concentrated (Gavara and Catalá 2002; Teixeira et al. 2007).

In a biphasic mixture, the film theory helps describing mass transfer among both phases. This theory defends that an interface is formed between the two phases and that, a molecule that is to go from one phase to the other has to first be transferred from the first phase to the interphase and then from the interphase to the second phase. The film theory assumes that both phases are perfectly homogenised and that there is no resistance to the movement of molecules in each phase nor in the interphase. Thus, the guiding force is the concentration gradient between the two phases in the vicinity of the interphase. The mass transfer flux in this calculated using Eq. 5.2.

$$J_c = K_1\left(C_1 - C_{1i}\right) = k_2\left(C_2 - C_{2i}\right) \tag{5.3}$$

Where,

k_1 and k_2 are the mass transfer coefficients in the film adjacent to phase one and phase two, respectively
C_1 and C_2 are the concentration of the molecule *c* in phases one and two
C_{1i} and C_{2i} are the concentration of molecule *c* in the interphase of phases one and two.

As mention above the transference of molecules from packaging materials to food may occur by several physical processes, such as diffusion or migration. Furthermore, to fully classify the mass transference between the packaging material and the food, one should verify if there is direct contact between food and package or not. The steps occurring in each situation are briefly described below. Although the process differs in both situations, direct and indirect systems often overlap, given that usually there is some headspace, even if very small.

- Direct contact between food and package: the release of molecules from the package to the package content by direct contact may be applied to volatile and non-volatile molecules, being mainly used with beverages. Typically, a direct system follows the steps below (Fig. 5.1). Using an aroma as example:
  1. The aroma molecule is diffused within the packaging material, usually a polymer.
  2. The aroma molecule is dissolved at the package-food interface or package-headspace interface.
  3. The aroma molecule is dispersed or diffused in the food.
  4. The aroma molecule is release to the food-headspace interface.
  5. The aroma molecule is dispersed in the headspace (Arabi et al. 2012).
- Indirect contact between food and package: the release of molecules from the package to the package content by indirect contact is much simpler. Typically, this occurs by releasing to the headspace the compounds embedded in the packaging materials, following the steps below (Fig. 5.2). Using an aroma as example:
  1. The aroma molecules are diffused within the packaging material, usually a polymer.
  2. The aroma molecules are then released to the interface package-headspace.
  3. The aroma molecules are adsorbed into the food (Arabi et al. 2012).

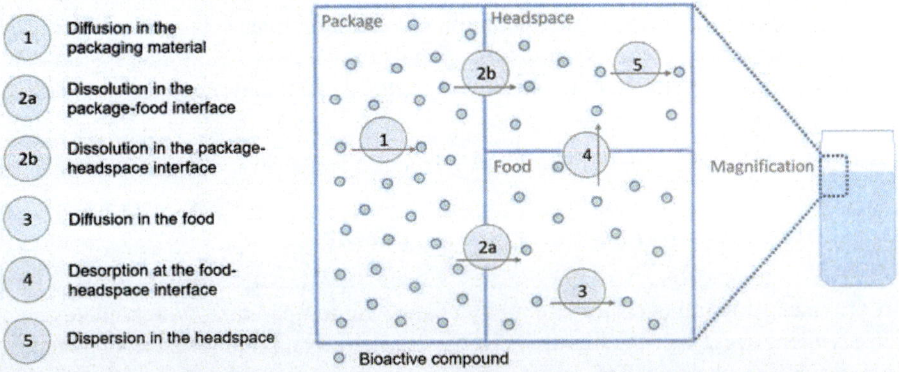

Fig. 5.1 Representation of a bioactive compound release in a food package by direct contact. (Adapted from Arabi et al. 2012)

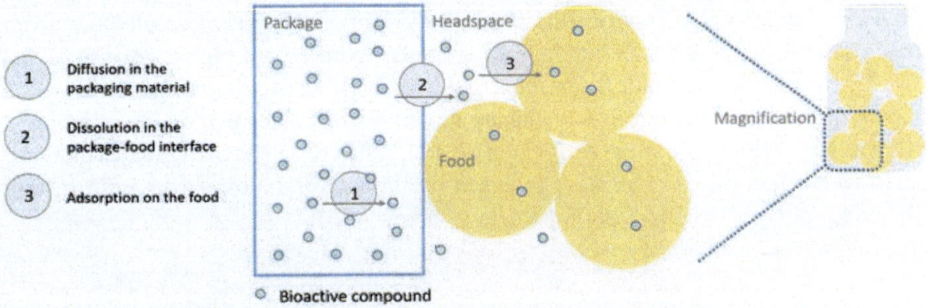

Fig. 5.2 Representation of a bioactive compound release in a food package by indirect contact. (Adapted from Arabi et al. 2012)

One example of such a system is the release of aromas from the wood barrels to the wine during the aging process. Although barrels are not a package, the same principles apply, showing that even without the complete understanding of the physical laws, these principles have been applied for centuries in order to improve the quality of the finished product.

5 Triggering Systems

Volatile compounds, including flavour molecules, due to their susceptibility, are often encapsulated to resist the processing and storage procedures. However, later on, they must be released at the right moment and place and, for this to happen, a sophisticated mechanism of controlled release has to be engineered and optimized. The releasing systems shall stay dormant until the right trigger is activated, which can be temperature or pressure, among others. Once the release is activated, the diffusion characteristics of the system will tell how long it lasts. Triggering systems are discussed in more detail below (Arabi et al. 2012) and some examples are described in the Sect. 5.3.

5.1 *Substance Release Triggered by Heat*

A polymer structure is easily altered by changes in temperature, which makes heat an excellent trigger for the liberation of the bioactive compounds that may be incorporated in the package. This system is often used in packaging for microwavable food products. The heat generated during microwave heating cracks the encapsulating system and the, usually, aroma compounds are released to the headspace and later on, will reach the olfactory system of the consumer, enhancing the eating experience.

Some sealants also rely on this principle to protect the aroma molecules from being released at low temperatures, but once the temperature raises, during microwave heating for example, the sealant melts and allows the release of the aroma molecules. Commonly, these sealants are made of substances that are mostly lipids, such as lard, tallow or butter (Arabi et al. 2012). It has been shown a greater increase in flavour when the aromas were released by this method, comparing with instant flavour spray – US6066347A (Nikhil and Jeffrey 2000).

5.2 *Substance Release Triggered by Mechanical Force*

One of the most important moments for the aromatic molecules to be release is when the consumer opens the food package, as the first aromas released will greatly impact that meal. Knowing this, a clever way of trigger the aroma release is by mechanical force, which is already used to open the package. This is used for instance, in bottles, which have capsules encapsulating aromas as part of the cap screws. The opening of the cap causes the capsules to break and the compounds of interest to be released (Arabi et al. 2012) (Fig. 5.3).

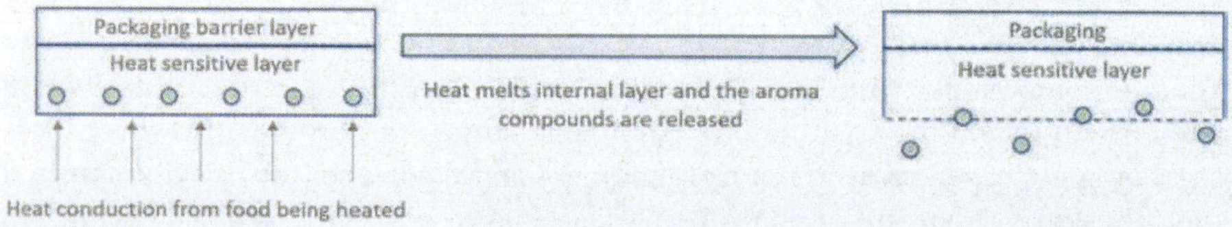

Fig. 5.3 Illustrative example of substance release triggered by heat. (Adapted from Burgos et al. 2017)

5.3 *Other Triggers for Substance Release*

Other triggering systems may be used in active packages, accordingly to specific characteristics and needs. Humidity is one trigger commonly used, namely to initiate the release of active substances which are trapped in crystalline structures. When relative humidity increases in the package, the excess of water molecules may induce conformational changes in the structure, which weakens the interactions between the bioactive molecule and the packaging material, favouring the release of the previously trapped substances. There is another class of releasing systems, where the aroma compartment is separated from the food compartment. AROMA-Can®, for example, presents a packaging design for wine or beer with two compartments, one for the beverage and one for the aroma compounds, that have to be opened individually. This kind of systems is however the target of some criticism, as the aroma-releasing system demands extra steps (opening the aroma compartment) and is thus, less subtle (Arabi et al. 2012).

6 Active Packaging

As previously introduced, during the last centuries, food packaging items, such as carton boxes and plastic containers, were ideally inert structures, used as a way to merely protect the food and make the transportation easier. However, with the technologic advances seen in the recent decades, in particular the onset of nanotechnologies, also the food packaging sector has felt the pressure and need to evolve. Nowadays, traditional packaging is often not enough to the consumer, as it is not able to respond to the current requirements for convenience, quality and freshness. Therefore, there is a constant call for effective means to maintain freshness and extend shelf life of, especially, fresh products, such as meat, fruits and vegetables, through packaging technology, namely active packaging.

Active packaging is defined as a packaging system involving interaction between the package itself and food and/or package and internal atmosphere (European regulation (EC) No 450/2009). These systems may extend the shelf-life of a food, by several means, namely by maintaining the food's nutritional value, inhibiting the growth of microorganisms, providing an easy open-close system and preventing or flagging the migration of compounds. Often, active packaging systems involve the incorporation of compounds of interest into the packaging material, which will timely be released to interact with the food product. Pivotal examples include: ethanol emitters, oxygen scavengers, flavour releasing systems, temperature control systems and antioxidant/antimicrobial-containing materials (Ozdemir and Floros 2004). The last is maybe the one with more diverse commercial applications. One elegant way of incorporating antimicrobials in a package for controlled-release was invented and patented – WO2019FI50770 (Lehtonen et al. 2020). Certain aldehydes and ketones, such as hexanal, a product of lipid oxidation, have shown to reduce the growth of bacteria and fungi in fresh fruits and vegetables. In this work, an active packaging material incorporating a system for production and release of these substances was developed. This material contains a lipid phase and an initiator that enables the controlled release of lipid oxidation products and was approved to be used in food packages.

Many applications may be explored under the scope of active packaging, some of them having even dual-function. Some spices for example, may be incorporated in the packaging for controlled release in order to, both, enhance the flavour, due to its flavouring and aromatic properties, and provide antioxidant/antimicrobial properties, uplifting the product worth in a double way (Gomes et al. 2011). Another example of dual function packaging is the simultaneous absorption of oxygen and release of carbon dioxide, to more efficiently preserve the freshness of the product and extend shelf-life, by retarding the growth of several microorganisms – IN2010MA00166 (Obaiah et al. 2012).

In the USA, incorporation of nano-clays in polymeric or ceramic matrixes for the production of packages is already widely used, however it is yet not common in Europe. The inclusion of this mineral in the packaging material confers antimicrobial and gas barrier functionalities, leading to the extension of the food product shelf-life – KR20117001865 (Cabello et al. 2009). Although there are a lot of ground to cover regarding antimicrobial properties of the packaging systems, incorporation of antimicrobial and antioxidant compounds is out of this chapter's scope.

One of the applications of active packaging with the greatest impact in consumers' experience is the flavour incorporation in the package for controlled release, thus greatly appealing to industries that want to differentiate their products from competitors. Flavour-releasing systems may be applied to volatile or non-volatile compounds, however the major focus is current on volatile compounds (aromas). Various forms of compound-release packaging systems are available, which may be produced by several different processes. One example is the inclusion of flavour molecules in the polymers that constitute the package, as it is explored further, later in this chapter (Arabi et al. 2012).

Nowadays, the majority of active packaging technologies for the release of substances applied in the market are based on sachet systems. Although sachets are a convenient way of including many compounds of interest in a food package, such as moisture absorbers or aroma molecules, they are not very well accepted by consumers. There is the fear of ingestion by children and the fear of accidental consumption, together with the package content (Ozdemir and Floros 2004). Aiming to fight some of these fears and make the system more convenient to the consumer, a sticker using a moisture absorption film, which can be attached to the packaging material, was already patented to replace sachets- KR20110042317 (Gensuke and Yasunari 2012).

Together, the consumers' demand for better solutions and the recent development of innovative techniques, are the perfect stimulus to the packaging industry to try new solutions. Thus, the vulgarization of technologies like thin coating films with active ingredients, is expected in the near future. Active packaging is an evolving concept which will still be the target of many enhancements, aiming to improve food safety, quality and stability (Lee et al. 2015).

Table 5.1 Summary of active packaging classes (Lee et al. 2015)

Features of the package	Brief description	Example
Absorption/ scavenging of compounds	Designed to scavenge any gas or off-flavour developed during storage.	Scavenging of oxygen, ethylene or moisture
Emission of compounds	Designed to release compounds with specific properties, in order to extend the product's shelf life	Release of preservatives and flavours
Removal of food ingredients	Designed to remove some compounds of the food product, in order to increase its nutritional value	Removal of cholesterol or lactose
Temperature control	Designed to change temperature of the packaged food	Self-cooling or self-heating

Presently, there are several types of active packaging, which may be classified accordingly with the following five broad categories (Table 5.1).

There is yet another strategy related with active packaging that, although it may involve engineering the package itself, is of utter importance when it comes to preserve food. In the majority of packages, there is some air contained in the free space between package and food, the so-called headspace. The composition of this space may highly influence the food characteristics and it is commonly customized by producers, creating modified atmosphere packages. Common examples are the removal of oxygen, as it promotes the food degradation, in order to extend the post-harvest life of perishable food products, such as fish, meat, fruits and vegetables (Bakar 2013; López-de-Dicastillo et al. 2010; McMillin 2008; Rojas-Graü et al. 2009; Szente and Fenyvesi 2018).

7 Systems for Inclusion of Bioactive Compounds in the Package

The inclusion of bioactive molecules in food packages may be done using several techniques. With the expected growth of the active packaging field in the forthcoming years, new methods for packaging production will emerge and the existing ones are also expected to evolve. In this section, some of the currently most used ones are briefly explored.

7.1 Cyclodextrin Complexes

Cyclodextrins are an example of a method for the incorporation of bioactive substances is a package. They are cyclic oligosaccharides made of glucose units, which form a ring structure with a hydrophilic surface and a hydrophobic cavity. Cyclodextrins may bind and hold non-polar aromatic compounds and lipophilic molecules, forming complexes that may then be incorporated in a polymer and become part of an active food package. Cyclodextrin may also be used as a sponge to trap unwanted substances, such as odours, lactose and cholesterol (Szente and Fenyvesi 2018). To this end, 'empty' complexes of cyclodextrins may be used. Either way, the cyclodextrin complexes shall be mixed with the thermoplastic polymers and several types of plastic products may be produced (films, laminates, containers…). This method assures a homogeneous distribution of the complexes and consequently a slow release of the molecules of interest, usually triggered by humidity. The release rate may be defined by the affinity of the bioactive substance to the cyclodextrin complexes, its concentration within the polymer, the temperature and the partition coefficient between the packaging material and the cyclodextrin, as well as between the package and the packaged material. Typically, the complexes of cyclodextrin and bioactive molecules are prepared by mixing powdered cyclodextrin with the guest molecule and water, followed by the removal of the water, which may be done by several methods (filtration + drying, spray-drying, freeze-frying) (Szente and Fenyvesi 2018).

Some successful examples of the incorporation of cyclodextrin complexes in polymers are transparent sheets with 3% of lemon oil which released fragrance during 6 months and garbage bags made of animal repellent plastic (Kuwabara et al. 1998; Nakagawa and Ukon 2007). Still in the breakfast cereals domain, volatile substances incorporated in the colour-printed carton boxes were gradually released to the headspace and penetrated the internal polyethylene bags, reaching the food product. Edible films are other example of success of this technique, namely humidity-releasing edible film and antioxidant-containing chewing gum edible film (Szente and Fenyvesi 2018).

Another example of cyclodextrin usage relates to fruits which are commonly harvested unripe and the ripening process has to be tightly controlled during transportation, so the fruit arrives to the consumer in the best conditions for consumption. Cyclodextrins complexes of ethylene were incorporated in polymeric films used in fruit containers to control the postharvest ripening of fruits, instead of ethylene injections (Fuenmayor et al. 2013).

Furthermore, nanoencapsulation of limonene into a polymer of pullulan and β-cyclodextrin has shown advantages when compared to the direct application of limonene in the polymer matrix itself. The highest rate of limonene release was observed in the first three days of contact with the food, having stabilized after seven days and kept a constant releasing rate for 45 days (Kayaci et al. 2013, 2014). Complexes of cyclodextrins with eugenol, a natural phenolic with flavouring properties, or geraniol, a compound with aromatic and insect repellent properties, also proved to be effective in active food packaging (Szente and Fenyvesi 2018).

There are several ways to incorporated desired molecules in a package. Depending on the final aim, cyclodextrin-complexes present some advantages. The losses that usually occur during food processing due to high temperatures are diminished, the chemical stability of the bioactive molecules increases, the amount of required bioactive molecule is small since one can evenly distribute it through the packaging material, avoiding wasting. Moreover, the release of the molecules to the foodstuff is dependent on the hydrophobicity and permeability to water of the polymer, size of the cyclodextrin complexes and temperature, hence it can be partially controlled by the manufacture. When empty cyclodextrins, which work as traps for unwanted substances, are used, the penetration of volatile compounds in the package is significantly decreased in both directions, contributing to the retention of valuable volatiles to the consumer experience inside (Burgos et al. 2017).

Following a similar line of thought, instead of cyclodextrins, hydrotalcite complexed with bioactive molecules may also be incorporated in polymer matrixes, as described in the patent – IT2011MI01921 (Bianchi et al. 2014).

7.2 Microencapsulation

Flavour molecules are essential compounds to the full experience of eating a food. However, due to their volatile and unstable character, it is not always easy to keep them functional until the moment of consumption. The encapsulation of these compounds represents a promising approach to stabilize them during storage and overcome this problem. Thus, food industries often show a great interest towards encapsulation techniques, which allow then for the incorporation of stabilized flavour molecules in polymer matrices. Several compounds, such as cinnamaldehyde and thymol are currently encapsulated and can be incorporated in food packages (Arabi et al. 2012; Blanco-Padilla et al. 2014; Gomes et al. 2011; Zhu et al. 2012).

Microencapsulation had its first commercial application in the twentieth century in the production of carbonless copy paper and, only later, it started to be used in the food industry. Microencapsulation was then applied in pharmaceutical industry for the coating and granulation of pharmaceutical tablets and pills, by Professor Dale E. Wurster, and it still is today. These works and all the related research served as springboards to the many applications of microencapsulation existing nowadays. In the food industry in particular, custom encapsulation formulas and techniques allow for dozens of compounds to be encapsulated and included in the package, or even in the food itself. To select the most appropriated material to form the capsule in each particular case, one shall consider the properties of the active compound to be encapsulated and the target release rate, among other factors. In general terms, all the capsule materials should be hygroscopic at high solid content and have low viscosity and good emulsification properties. They should also have low reactivity with the active compound and the solvent, do not interfere with the organoleptic profile of the food and, ideally, be cheap. As a rule of thumb, capsules for hydrophobic active agents are usually made of hydrophilic compounds (e.g. polysaccharides), whereas capsules for hydrophilic active agents are typically made of hydrophobic materials (e.g. lipids). Regardless of the chosen material, it has to be included in the list of safe ingredients for consumption, accordingly with the current legislation (Regulation (EC) No 1333/2008; Abreu et al. 2012; Madene et al. 2006; Wang et al. 2015).

Recently, with the growing social demand for environmentally friendly materials, biopolymers have been increasingly used in the preparation of capsules. The main advantage of these natural materials are their biocompatibility, low toxicity and biodegradability. A summary of the most commonly used materials for the

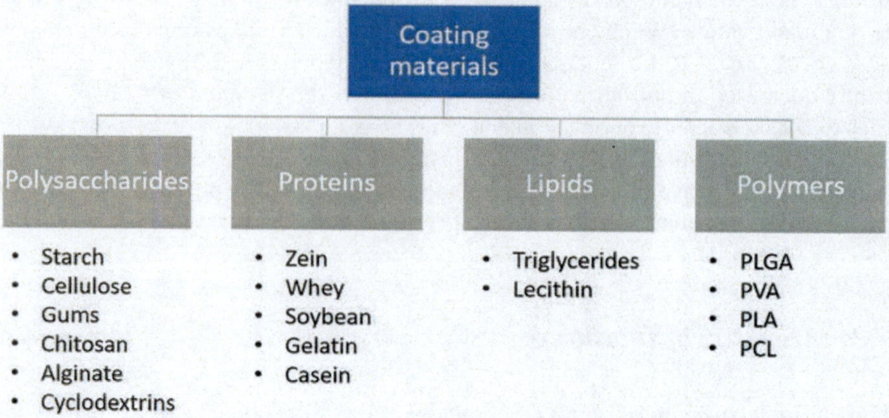

Fig. 5.4 Materials widely used to encapsulate flavours and aromas. (Adapted from Burgos et al. 2017)

encapsulation of flavours and aromas is presented in Fig. 5.4 (Wang et al. 2015; Zhu et al. 2012).

After choosing the capsule material, the selection of the best method to encapsulate the bioactive compound is the next step, which should consider several factors, namely characteristics of the active ingredient itself, the particle size, the desired concentration inside the capsule and the mechanism of release (Nussinovitch 2003).

Some of the considerations that need to be made during the whole process follow:

- The functionality that the encapsulated ingredients must provide to the finished product;
- The non-reactivity of the coating material with, both the encapsulated ingredients and the food matrix.
- The processing conditions that the microcapsules still have to survive before consumption.
- The mechanism by which the ingredients will be released from the capsule and if there is any form of release required (sustained or delayed release).
- Cost constrains for the encapsulation.

Taking all this in consideration, the best method to create microcapsules should be selected, which in more than 90% of the cases is spray drying (Janjarasskul and Krochta 2010).

7.3 Films and Coatings – Edible Packaging

The idea of coating food products to confer protection is not novel, with the first documents mentioning it dating from the nineteenth century and the first patents being published in the twentieth century. In Asian countries, the so-called soy-milk skin has widely been used as an edible film. However, due to several restrictions, this concept is still not commonly applied at the commercial level (Guilbert et al. 1996).

Considering the current concerns about climate change and the huge amount of waste produced by human kind, which is destroying landscapes and affecting life in oceans to a great extent, any and every approach that may help reduce waste is valuable. The food sector is one of the greatest producers of waste, in part because of the packaging materials, which are often excessively used, sometimes even in products that do not require packaging, such as individual pieces of fruits or vegetables. Edible packaging is a novel approach which can improve the product stability, quality and safety, without compromising the convenience for consumers nor the environment (Krochta and Mulder-Johnston 1997).

In general terms, edible packaging consists in films, coatings, sheets or pouches which confer protection to the food product and can be ingested by humans without health concerns. If the layer of edible material is produced directly onto the surface of the food product, it is considered an edible coating. However, if it is a stand-alone structure, produced separately from the food and then applied on the food surface, it is either a film (<254 μM) or a sheet (>254 μM) (Gennadios 2004).

Edible packages are an integral part of the food product, intended to be ingested. Even if the coating is not ingested, it is inherently biodegradable and may be compost with other biological materials, representing a big advantage of this packaging system. Similarly to what happened with capsules, in order to increase the consumers' acceptance of edible packaging, there has been a marked tendency towards the use of proteins, polysaccharides, resins and other edible compounds, derived from renewable sources to produce edible packages (Janjarasskul and Krochta 2010). Besides simply protecting the food, edible packaging makes it easier to incorporate bioactive ingredients and enhance the overall value of the product.

Although edible packaging is a rapidly evolving technology with many very promising applications, the lack of knowledge and data still precludes a wider use of it. Before being commonly commercialised, edible packaging has yet to overcome challenges at various levels: legal requirements, consumer acceptance/concerns and the scaling-up of the production from laboratory to industrial scale (Ackermann 1995). As it serves as both packaging and food, it has to obey to very strict regulations to be considered safe (Gennadios 2004). Reliant on the applications of the edible package, it can be classified and regulated as a food product, a food ingredient, a food additive, a food contact substance, or food packaging material. But, in any of the cases, it must present neutral sensorial properties or be compatible with the sensorial properties of the packaged food, in order to not interfere with the consumer experience.

In the market for years already, soft-gel capsules and tablet coatings may also be considered edible packaging, protecting the active principle inside, which may be several kinds of drugs, mainly antibiotics. Edible packaging materials may also be used in non-edible packages, namely ones with improved protective or barrier function, controlling mass transfer events between food and outside atmosphere. More recently, these materials have also been used as separators inside food packages, delaying the migration between different food components of flavours, aromas or oils, among others, and increasing self-life (Janjarasskul and Krochta 2010).

Edible packaging can make easier the packaging of assorted pieces of a certain food type, satisfying the constant demand for variety. Besides that, edible package can be the carrier for bioactive compounds which may be timely released and increase the market value of a food. Some examples are formulations which carry pigments, flavours, antioxidants, spices and nutrients, among many others. Also, the visual appearance of a food can be improved with the help of edible packaging, by including in the packaging material with colouring or glossing properties, or even texture features, such as substances that make the surfaces smoother or non-greasy. Wax coating and other lipid-based coating have long been used to provide high-gloss properties to fruit pieces. Whey protein films are also in the run as coating materials presenting the big advantage of not having neither colour nor flavour and not interfering with the product properties (Han and Gennadios 2005).

As the consumers are increasingly aware of labels and the food they buy and eat, the marketing of edible films, with or without added functional properties, is of utter importance. Careful and detailed instructions may be presented on how to wash, prepare or even cook the packaged food. On their hand, producers demand more research on large-scale production and cost reduction, prior to invest in coating equipment (Janjarasskul and Krochta 2010). Finally, it is important to keep in mind that edible packaging will have a shorter duration when compared to conventional packaging, due to their inherently biodegradable nature. Thus, the conditions for store edible packages safely have also to be investigated (Ozdemir and Floros 2004; Sacharow 1995).

8 Emitters of Flavours, Colorants and Other Food Ingredients

The concept of active packaging is a very wide one, allowing for the implementation of several ideas and concepts in the name of food quality and consumer convenience. Among all, some emitters of flavours, colorants and other food ingredients are explored below.

Oxygen Scavengers For decades now, we are used to see food product packed in vacuum in order to improve quality and extend shelf life. The rationale behind this is to preclude the contact between food and oxygen, an atmospheric gas that highly contributes to the degradation of food. It turns out that there is a cheaper way of doing this, by simply including oxygen scavengers in the food packages. Typically, the system is based on one of two mechanisms: iron powder available to oxidation that uses up the atmospheric oxygen, before it gets to the food products or, enzymes which react with a substrate, consuming oxygen. These ideas were firstly applied in sachet systems, however recently, it had become clear that in order to solve sachet-related problems, it would be better to simply incorporate these systems on the packaging material itself. Scavenging agents may either be imbedded in a solid or homogeneously distributed in a polymer matrix, performing better the role of barrier and precluding oxygen to contact with the food. Going a step forward, dual function packages were developed that both trap the oxygen from the exterior world, but also the ethylene, delaying the ripening of packed fresh fruits (Ozdemir and Floros 2004).

CO_2 Emitters and Scavengers Contrary to oxygen, the presence of carbon dioxide retards the food deterioration, by delaying microbial growth and decreasing the respiratory rate of vegetables and fruits. A dual function system which traps oxygen and emits carbon dioxide may be of great benefit to extend the shelf-life of highly perishable goods. In the case of coffee packages though, our interest lays on the scavenging of carbon dioxide as roasted coffee beans tend to release carbon dioxide that, if not trapped, may lead to the burst of the coffee bags (Ozdemir and Floros 2004).

Ethanol and Aroma Emitters For decades that food products, especially baked goods, such as breads and cookies, are sprayed with ethanol aiming to delay food spoilage. With the recent advantages in food packaging, the spraying can now be replaced by sachets or coatings that release ethanol, in exchange to water vapor that is absorbed by the carrier material. Besides releasing ethanol, the package often concomitantly releases aromas, such as vanilla, increasing the organoleptic quality of the food. Although sachets present some disadvantages when compared to films, they are still the main system encountered in the market as they are a cheaper system (Ozdemir and Floros 2004).

Flavour and Aroma Emitters or Scavengers There are several reasons why a food may lose some flavour after packaging. Firstly, there is a natural tendency to lose organoleptic properties once the food is processed and packaged. Then, the packaging material itself can interact with the food in a way that traps part of the flavour molecules, known as scalping. The same is true for aroma molecules, which are very powerful as they can travel a direct line to the limbic system, the brain centre for emotions, triggering memories and influencing behaviours. The usage of plastic packages with high barrier properties may help in the preservation of the flavour and aromas, however, in some instances, there is the need for extra help in order to provide a strong aromatic experience to the consumer. A classic example, used for a long time, is the filling of the headspace of instant dry coffee flasks with volatiles captured during the dehydration process, in order to deliver a fresh and strong fragrance the first time the flask is opened. More recently, packaging materials started to incorporate flavours (microcapsules, cyclodextrin, among others) that are then released to contact with food and enhance the organoleptic profile. The advantage of this system relatively to the just including aromas in the headspace, is the sustained release of the bioactive molecules. In opposition, there are other instances when flavour scavengers are incorporated in the package aiming to absorb off-flavours and aromas that may be developed during storage. This raises however some controversy, since many people believe that this system may mask off-flavours, which may jeopardize food quality and safety (Ozdemir and Floros 2004).

Colour Emitters Several foods may lose colour intensity during processing or storage and the reestablishment of this colour is often of interest to the producer and consumer. The most well-known example of these systems refers to surimi, the artificial crab meat sold in stripes. The wrapping paper of surimi contains an edible red pigment which migrates to the food in order to give it the desired colour (Ozdemir and Floros 2004).

Other Agents There are several processes during storage that may interfere with the presentation and quality of the food. One example is the respiration of certain fresh products and the resulting condensation inside the food package which leads to a foggy appearance. This not only makes the product less appealing to the consumer, but also accelerates the spoilage rate of the foods. Fortunately, there are anti-fogging films that may be used to coat the package and prevent this to occur. Similarly, films that incorporate anti-sticking agents also exit and are mainly used to preclude soft candies and sliced cheese to stick to the package (Arabi et al. 2012).

9 Existing Products and Patents

Many of the concepts explored throughout this chapter are already being commercialized and are briefly covered in this section.

9.1 Encapsulated Flavours and Fragrances

Currently, there are several producers of encapsulated flavour or fragrances masterbatches which can be incorporated in the structure of polymeric matrix used in food packaging, at the time of manufacturing, becoming integral parts of the package itself. The incorporation of distinctive flavours and aromas create memorable experiences associated with the food brand. Furthermore, as wellness and healthy products are a growing interest of food companies, the incorporation of flavour in the package is a great way of improving the experience provided by the food, without adding undesired ingredients, sugars, fats or calories.

The incorporated flavours and fragrances may be release at different rates and at different times, accordingly to the producer desire. They can be uniformly and slowly released during the food shelf-life, to counterpart some dilution and/or flavour loss induced by the processing of the food up until then. Alternatively, flavours and fragrances may be released when the package is opened or even during the food preparation (e.g. microwave heating), presenting a great opportunity to enhance the consumer experience. Some producers of flavour and fragrances masterbatches offer concomitantly the option to include features that reduce the flavour scalping extend and inhibit the release of undesirable flavours and/or odours from the package to the product. Integrating aromas and flavours in the packaging system is a step forward towards satisfying the contemporary demands for distinct flavours, for stronger and long-lasting fragrances, as well as for unique and innovative ideas.

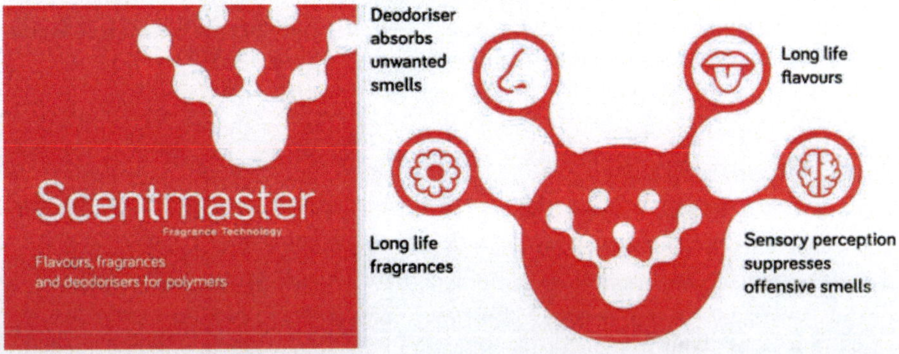

Fig. 5.5 Scentmaster infographic

Addmaster is one company that produce masterbatches (Scentmaster) of flavours and aromas to be incorporated in the packaging material, originating scented products. In Scentmaster masterbatches, flavours or fragrances are supplied in high concentrations and the producer only has to incorporate them at a 1% or lower rate, hence being a highly efficient system. A vast array of compounds may be chosen from their catalogue and personalized ones may be produced upon request. Masterbatches may be used in toys, bin liners, educational products and food and beverages packages. They are compatible with many materials, including the most common ones, polyethylene, polypropylene, polyvinyl chloride and ethylene vinyl acetate. The company ensures the retention of subtle top notes of more sensitive fragrances and the stability of the scents or flavours up to 200 °C. Besides the capacity to release flavours and fragrances, they may also be used to trap and absorb undesired smells, as it is shown in Fig. 5.5 (Add master website). (Add master website; https://www.addmaster.co.uk/scentmaster/scentmaster-technology. Accessed on 31 Aug 2020).

Other companies offering the same kind of product which have a variety of samples to be tested by consumers are ScentSational (ScentSational technologies website). (Add master website; https://scentt.com/scentsationals-compelaroma-technology-scentsational-news/. Accessed 31 Aug 2020) and Wells Plastics (Wells Plastics website. https://wellsplastics.com/. Accessed 31 Aug 2020).

9.2 Scent Prints and Coatings

It is currently possible to paint packaging products with scented ink. These paints have incorporated microcapsules designed to be ruptured by a stimulus as gentle as a finger swipe. This makes the scent to be release constantly when the package is being handled, representing a powerful marketing tool.

These paints may be printed directly onto a variety of packaging material, the aromas may then stay dormant up to twelve months, before being released by touching. Each activation of scent release will last a few minutes and may be reactivated over and over again. ENCAPSCENT® is one of the companies offering this kind of ink (ScentSational technologies website). (Scent website; https://scentt.com/about-our-technologies/#!/encapScent-coatings. Accessed 31 Aug 2020).

9.3 Flavoured Straws

Straws are widely used by beverages' consumers and, although they are not part of the package itself, but simply a complement, they can be engineered to enhance the flavour and/or nutritional value of beverages. If a beverage producer wants to give an extra bit of flavour to their drinks, they are now able to do so without changing the beverage formulation. Instead, they can simply incorporate a flavour-releasing system in the straws, constituted by three parts, a transparent straw, beads containing flavour compounds attached to the inside straw wall and filters, one in each end of the straw to keep the beads inside. The beverage flowing through the straw leads to the release of flavour molecules and allows for the enhancement of flavour without manipulating the beverage formulation. Unistraw™ is the original manufacture and has available many flavours to satisfy every taste. Each straw contains half a teaspoon of sugar and different flavours may be combined in one straw accordingly to the consumer desire (Unistraw website; http://www.unistraw.com/straws.html. Accessed 31 Aug 2020 and Arabi et al. 2012).

More recently and after the success of flavoured straws, other kind of molecules and bioactive compounds started to be incorporated in the straws in order to obtain functional beverages. Some examples are straws that release collagen, vitamins, caffeine, probiotics and even energy beads in order to make your drink an energetic one (Unistraw website. http://www.unistraw.com/straws.html. Accessed 31 Aug 2020 and Arabi et al. 2012).

9.4 Beverages Foamy Heads

A beverage can incorporate the widget technology and will automatically create a foamy head on the beverage, once the can is opened, such as the ones produced by Ball (Ball website; https://www.ball.com/na/solutions/markets-capabilities/capabilities/beyond-beverage-cans/widget-technology. Accessed 31 Aug 2020). The technology relies on the inclusion of capsules containing nitrogen in the can material, which are released upon opening. Widget technology may be used whether the beverage is alcoholic or non-alcoholic, being especially promising for coffee drinks and beers. It is expected that this concept will be soon adapted to combine pressurized gas and aromas or flavours inside the same widget (Arabi et al. 2012). Alternatively, the beverage may be injected with a gas, such as nitrous oxide, and then liquid nitrogen is used to fill part of the head space, before crimping and sterilizing. This process also leads to the formation of a foamy-head when the beverage is poured and was developed especially for milk-based beverages – US6669973B1 (Jolivet 2013).

9.5 *Anti-fogging Coating*

The importance and convenience of internally coating food packages with anti-fogging films was already pointed out, previously in this chapter. Nowadays, there are available masterbatches of anti-fogging agents that can be incorporated in the food packaging plastics. One of the products commercially available are the Ampacet masterbatches, which reach peak performance in 24 h and last at least two weeks. Ampacet is recommended for PET laminated products, and it is an FDA-approved additive for cold and hot food packaging applications. Ampacet also offers masterbatches to odor scavenging, absorving a wide range of odors when used in polyethylene material. This can be used not only for food and beverages packages, but also trash bags and barrier films (Ampacet website; https://www.ampacet.com/masterbatch-products/ampacet-additives/antifog-concentrates/. Accessed 3 Aug 2020).

9.6 *Edible Natural Apple Essence Sustained-Release Liposomes*

This process to incorporate apple essence in food packages relies on the production of liposomes containing apple essence. The liposome walls are made of soybean lecithin and cholesterol, whereas the apple essence is in the core embedded into vesicles formed by the liposomes, which allow for its slow release. This method is simple and adapted to large-scale production and the apple essence-containing liposomes are stable during long storage and transportation. This invention is particularly suited for the preparation of freshly-cut apple packages, as it can effectively make up for the flavour loss that occurs during processing of the fresh pieces – CN201811232285 (Yun et al. 2019).

9.7 *Apparatus for Improving the Aroma and Flavour of Beverages*

Vitamins and certain flavours, such as citrus-flavoured beverages are not very stable, especially the ones made with essential citrus oils or other water-dispersible liquid flavours such as cranberry, passion fruit, and carambola. Caps may include a reservoir containing a separate component, which may be mixed with the unflavoured beverage in the bottle, before consumption. In the special case of highly susceptible beverages, this would, not only enhance the flavour profile, but also extend the shelf-life of the product – US20020157970A1 (Masayuki and Takahashi 2003). The InCap is currently commercialized and freshly releases several compounds, such as vitamins from the cap, which blends into the beverage at the time of opening (InCap website; https://www.incap.hk/. Accessed 1 Sep 2020).

9.8 Controlled Release of Fragrance

The first tried solution to incorporate fragrance in the food packages was the incorporation of tapes outside the containers that exude aromas, known as "scratch and sniff". However, this system, once used may present an untidy, undesired appearance and it is not able to retain the aroma over repeated uses. Moreover, it does not entirely fit the consumers desire for an aromatic packaged product, as there is the need to open an extra container in the package and the fragrances are not coming directly from the food.

Trying to achieve the perfect solution, a system of waterproof embodiments, able to comprise and release aroma compounds held in a polymer matrix, was invented. These embodiments are not in direct contact with the food, but rather in an area under the container cover that breaks or peels off when the container is opened, releasing the aroma. Several sorts of containers may incorporate this technology, such as screw cap bottles, as long as the polymer matrix is destroyed, releasing the aromatic compound – RU2533362C2 (Zhang and Given 2014).

9.9 Packaging for Delaying Spoilage and Maintaining Organoleptic Qualities

Recently, Ferrero has announced the inclusion of active substances in their packages in order to delay the spoilage and maintain the organoleptic qualities of product for longer and the weight losses to a maximum of 0.5% (https://www.foodweb.it/2020/07/ferrero-brevetta-un-packaging-anti-batteri-e-muffe/). However, much work of optimization at the production level of active packages is still needed in order to make it economically viable.

10 A Promising Future for Active Packaging – Concluding Remarks

The concept of packaging has come a long way in the last century, starting with major improvements during World War II, up until the active packaging solutions used nowadays.

The consumers demand for packaged food products that retain freshness as well as nutritional and organoleptic properties for longer was, and continue to be, the main driver for active package improvements. Active packaging solutions promise new ways to prolong shelf-life of food products, such scavenging oxygen and emitting flavours, colorants and antioxidants, besides what is already offered by the traditional passive packaging. The evolution of active packaging is however far from concluded, as many aspects have still to be optimized before it becomes the norm in grocery stores.

Different active packaging systems present different readiness levels to be commonly applied. Active coating films, for example, are mainly found at laboratory scale and still need scaling up experiments to be commercialized. Other types of active packages, such as the incorporation of fragrances in polymers used for packaging, are already used and may be found in commercialized products. Independently of the readiness level, the concept of active packaging is not yet completely accepted by the consumers. One main concern is the possibility of the flavours released by the package will mask off-flavours and thus some level of deterioration. In this regard, some ethical norms should be established in order to guarantee that active packages will never risk food safety.

Active packaging solutions have proved themselves very useful to extend shelf-life and increase quality of food products, hence significant developments in this field are expected in the near future. The first step is the recognition of active packaging benefits by the food industry, which is already happening. The last challenge will be gaining the confidence of the consumers, something that will happen progressively once these solutions acquire their place in the market and consumers witness their benefits. Continued innovation at this level towards a better quality, safety and stability is expecting, aligned with the consumers' convenience needs.

References

Abreu F, Oliveira E, Paula H, de Paula R. Chitosan/cashew gum nanogels for essential oil encapsulation. Carbohydr Polym. 2012;89(4):1277–82. https://doi.org/10.1016/j.carbpol.2012.04.048.

Ackermann P. Foods and packaging materials: chemical interactions. Cambridge: Royal Society of Chemistry; 1995.

Addmaster website. https://www.addmaster.co.uk/scentmaster/scentmaster-technology. Accessed on 31 Aug 2020.

Ampacet website. https://www.ampacet.com/masterbatch-products/ampacet-additives/antifog-concentrates/. Accessed 3 Aug 2020.

Arabi S, Chen X, Shen L, Lee D. Flavor-release food and beverage packaging. In: Emerging food packaging technologies. Elsevier; 2012. p. 96–108. https://doi.org/10.1533/9780857095664.1.96.

Bakar A. Effect of modified atmosphere packaging on microbial flora changes in fishery products. Int Food Res J. 2013;20(1):17–26.

Ball website. https://www.ball.com/na/solutions/markets-capabilities/capabilities/beyond-beverage-cans/widget-technology. Accessed 31 Aug 2020.

Bianchi R, Bugatti V, Tammaro L et al. Process for preparing a polymer composition comprising hydrotalcites intercalated with active molecules, composition thus obtained and shaped articles comprising the same. IT1408555B1. 2014.

Blanco-Padilla A, Soto K, Hernández Iturriaga M, Mendoza S. Food antimicrobials nanocarriers. Sci World J. 2014; https://doi.org/10.1155/2014/837215.

Burgos N, Mellinas A, García-Serna E, Jiménez A. Nanoencapsulation of flavor and aromas in food packaging. In: Food packaging. Alicante: Elsevier; 2017. p. 567–601. https://doi.org/10.1016/b978-0-12-804302-8.00017-0.

Cabello J, Pons M, Calzado M. Active nanocomposite materials and production method thereof. KR20110044981A. 2009.

CPG Sec 525.750 Spices – Definitions. Food and Drug Administration. 1980. https://www.fda.gov/regulatory-information/search-fda-guidance-documents/cpg-sec-525750-spices-definitions. Accessed 14 Aug 2020.

FAO Food and Agriculture Organization of the United Nations (2020) Seeking end to loss and waste of food along production chain. 2020; http://www.fao.org/in-action/seeking-end-to-loss-and-waste-of-food-along-production-chain/en/. Accessed 28 Aug 2020.

Ferrero brevetta un packaging anti batteri e muffe. Food web website. https://www.foodweb.it/2020/07/ferrero-brevetta-un-packaging-anti-batteri-e-muffe/. 2020.

Food Additive Status List. Food and drug administration. 2019. https://www.fda.gov/food/food-additives-petitions/food-additive-status-list. Accessed 15 Aug 2020.

Food Improvement Agents. European Commission. https://ec.europa.eu/food/safety/food_improvement_agents_en. Accessed 13 Aug 2020.

Fuenmayor C, Mascheroni E, Cosio M, et al. Encapsulation of R-(+)-limonene in edible electrospun nanofibers. Chem Eng Trans. 2013;32:1771–6. https://doi.org/10.3303/CET1332296.

Gavara R, Catalá R. Mass transfer in food/plastic packaging systems. In: Engineering and food for the 21st century. CRC Press; 2002. https://doi.org/10.1201/9781420010169.

Gennadios A. Edible films and coatings from proteins. In: Proteins in food processing. Elsevier Ltd; 2004. p. 442–67. https://doi.org/10.1533/9781855738379.3.442.

Gensuke N, Yasunari S. Sticker using Water-absorbent Film. KR20120124586A. 2012.

Gomes C, Moreira R, Castell-Perez E. Poly (DL-lactide-co-glycolide) (PLGA) nanoparticles with entrapped trans-Cinnamaldehyde and eugenol for antimicrobial delivery applications. J Food Sci. 2011;76(2):N16–24. https://doi.org/10.1111/j.1750-3841.2010.01985.x.

Guichard E, Salles C. Retention and release of taste and aroma compounds from the food matrix during mastication and ingestion. In: Flavor: from food to behaviors, wellbeing and health. Elsevier Inc.; 2016. p. 3–22. https://doi.org/10.1016/B978-0-08-100295-7.00001-3.

Guilbert S, Gontard N, Gorris L. Prolongation of the shelf-life of perishable food products using biodegradable films and coatings. Lebensm Wiss Technol. 1996;29:10–7.

Han J, Gennadios A. Edible films and coatings: a review. In: Innovations in food packaging. Elsevier Ltd.; 2005. p. 239–62. https://doi.org/10.1016/B978-012311632-1/50047-4.

Han J, Scanlon M. Mass transfer of gas and solute through packaging materials. In: Innovations in food packaging. 2nd ed. Elsevier Ltd; 2013. p. 37–49. https://doi.org/10.1016/B978-0-12-394601-0.00003-5.

InCap website. https://www.incap.hk/. Accessed 1 Sept 2020.

Janjarasskul T, Krochta J. Edible packaging materials. Annu Rev Food Sci Technol. 2010;1:415–48. https://doi.org/10.1146/annurev.food.080708.100836.

Jolivet E. Assembly comprising a container and a ready-to-drink beverage. AU774227B2. 2013.

Kayaci F, Ertas Y, Uyar T. Enhanced thermal stability of eugenol by cyclodextrin inclusion complex encapsulated in electrospun polymeric nanofibers. J Agric Food Chem. 2013;61(34):8156–65. https://doi.org/10.1021/jf402923c.

Kayaci F, Sen H, Durgun E, Uyar T. Functional electrospun polymeric nanofibers incorporating geraniol-cyclodextrin inclusion complexes: high thermal stability and enhanced durability of geraniol. Food Res Int. 2014;62:424–31. https://doi.org/10.1016/j.foodres.2014.03.033.

Krochta J, Mulder-Johnston C. Edible and biodegradable polymer films: challenges and opportunities. Food Technol. 1997;51(2):61–73.

Kuwabara N, Takaku H, Oku S, Kogure Y. Production of transparent perfumed plastic article. JP Pat Appl. JPS51132296A. 1998.

Lee S, Lee S, Choi D, et al. Current topics in active and intelligent food packaging for preservation of fresh foods. J Sci Food Agric. Wiley. 2015;95(14):2799–810. https://doi.org/10.1002/jsfa.7218.

Lehtonen M, Mikkonen K, Tenkanen M. Composition and method for active packaging and storage of fresh plant products. WO2020089525A1. 2020.

López-de-Dicastillo C, Gallur M, Catalá R, et al. Immobilization of β-cyclodextrin in ethylene-vinyl alcohol copolymer for active food packaging applications. J Membr Sci. 2010;353(1–2):184–91. https://doi.org/10.1016/j.memsci.2010.02.049.
Madene A, Jacquot M, Scher J, Desobry S. Flavour encapsulation and controlled release – a review. Int J Food Sci Technol. 2006;41(1):1–21. https://doi.org/10.1111/j.1365-2621.2005.00980.x.
Masayuki M, Takahashi M. Bottle cap with a chamber for raw material and pressure gas. CA2374538A1. 2003.
McMillin KW. Where is MAP going? A review and future potential of modified atmosphere packaging for meat. Meat Sci. 2008;80(1):43–65. https://doi.org/10.1016/j.meatsci.2008.05.028.
McPhee C, Reed J, Zubizarreta I. Routine core analysis. In: Developments in petroleum science, vol. 64. Elsevier; 2015. p. 181–268. https://doi.org/10.1016/B978-0-444-63533-4.00005-6.
Nakagawa S, Ukon M. Garbage bag protected from Crow. JP Pat Appl. JP2007223781. 2007.
Nikhil P, Jeffrey W. Aromatized food package. US6066347A. 2000.
Nussinovitch A. Water-soluble polymer applications in foods. Blackwell; 2003.
Obaiah M, Nagarajarao R, Krishnaswamy S. Dual action active packaging system combining carbon dioxide emitter and oxygen scavenger for extending the shelf life of refrigerated fish products. 2012.
Ozdemir M, Floros J. Active food packaging technologies. Crit Rev Food Sci Nutr. 2004;44(3):185–93. https://doi.org/10.1080/10408690490441578.
Regulation (EC) No 1333/2008; Regulation (EC) No 1333/2008 of the European Parliament and of the Council of 16 December 2008 on food additives – Annex II (2008) European Commission, European Union.
Regulation (EU) No. 1321/2013; Regulation (EC) No 1333/2008 for establishing the Union list of authorised smoke flavouring primary products for use as such in or on foods and/or for the production of derived smoke flavourings. (2013) European Commission, European Union.
Risch S. Food packaging history and innovations. J Agric Food Chem. 2009;57(18):8089–92. https://doi.org/10.1021/jf900040r.
Rojas-Graü MA, Oms-Oliu G, Soliva-Fortuny R, et al. The use of packaging techniques to maintain freshness in fresh-cut fruits and vegetables: a review. Int J Food Sci Technol. 2009;44(5):875–89. https://doi.org/10.1111/j.1365-2621.2009.01911.x.
Sacharow S. Commercial applications in North America. In: Active food packaging. Chapman & Hali; 1995. p. 203–14.
Scentt website. https://scentt.com/scentsationals-compelaroma-technology-scentsational-news/. Accessed 31 Aug 2020.
Szente L & Fenyvesi ÉCyclodextrin-enabled polymer composites for packaging. Molecules. 2018;23(7). https://doi.org/10.3390/molecules23071556.
Teixeira J, Vicente A, Belo I. Transferência de massa. In: Reactores Biológicos – Fundamentos e aplicações. Lidel; 2007.
Unistraw website. http://www.unistraw.com/straws.html. Accessed 31 Aug 2020.
Voilley A, Souchon I. Flavour retention and release from the food matrix: an overview. In: Flavour in food. Elsevier Ltd; 2006. p. 117–32. https://doi.org/10.1533/9781845691400.2.117.
Volatile Compounds in Food Database. http://www.vcf-online.nl/VcfHome.cfm. Accessed 13 Aug 2020.
Wang X, Yuan Y, Yue T. The application of starch-based ingredients in flavor encapsulation. Starch/Staerke. 2015;67(3–4):225–36. Wiley-VCH Verlag. https://doi.org/10.1002/star.201400163.
Wells Plastics website. https://wellsplastics.com/. Accessed 31 Aug 2020.
Yun D, Zhongqi S, Danfeng W. Edible natural apple essence sustained-release liposome and preparation method thereof. 2019.
Zhang N, Given P. Controlled release of fragrance, held retained with application of polymer matrix. RU2533362C2. 2014.
Zhu G, Xiao Z, Zhou R, Yi F. Fragrance and flavor microencapsulation technology. Adv Mater Res. 2012;535–537:440–5. https://doi.org/10.4028/www.scientific.net/AMR.535-537.440.

Chapter 6
Temperature Control Emitters

Bruno Joaquim Henriques

Abstract Temperature control emitters consist of a diverse family of chemical compounds that are used to control the surrounding temperature of the food when in use. These chemicals can be organic or inorganic and are mostly found as liquids and solids. The range of temperatures to be controlled is quite vast, requiring an equally vast list of chemical compounds with different properties to choose from. Phase-change materials are amongst the most used compounds as temperature control emitters, whose phase-change occurs at a specific temperature, allowing to maintain the surrounding temperature during a longer period. Another well-known concept both in the literature and in practice are chemical reactions: exothermic and endothermic, to heat or cool a product, respectively. These are mostly used in more convenient products (e.g., self-heating or self-cooling cans). These technological solutions usually involve chemical reactions between two or more reactants. The combination of these materials with good insulation properties helps to prevent energy losses from the system, while in transport, storage or in the shelf. Due to sheer convenience of use, these products will surely appear more and more throughout the world as time goes by and new developments and improvements are made in insulation materials and chemical reactant systems.

Keywords temperature · PCM · heat · energy

B. J. Henriques (✉)
Colab4Food – Collaborative Laboratory, Vila do Conde, Portugal
e-mail: bruno.henriques@colab4food.com

S. M. Jafari, A. S. Silva (eds.), *Releasing Systems in Active Food Packaging*,
Food Bioactive Ingredients, https://doi.org/10.1007/978-3-030-90299-5_6

1 Introduction

The continuous and increasing growth in cross-border trade of perishables across long distances incorporating multi-modal supply routes requires optimal temperature control and monitoring for longer periods. This growth is fuelled by the increasing health consciousness of consumers worldwide opting to acquire and consume healthier goods: fresh biological fruits, vegetables, meat, fish, dairy alternatives, prebiotics, and probiotics while avoiding products rich in fat (both saturated and unsaturated), (added) sugars, and food preservatives (especially synthetic additives).

Among these niche food products, some have relatively high unit values when compared to staple food products, which grants them a considerable share in the world trade. As such, the protection of the cold chain is key in safeguarding the safety and quality of many high-value food products. Temperature fluctuation during cold chain transport has undesirable effects on food products. Temperatures from 2 to 8 °C aid in avoiding or delaying bacteriological, physiological, and chemical changes in these products. Meat and fish, for instance, are transported in a temperature range of 1–2 °C as temperature fluctuation by 2–7 °C often leads to decay and foul-smelling produce. These effects result from inadequate thermal insulation and meagre thermal buffering capacity of packaged products (Likar and Jevšnik 2006; Laguerre et al. 2013; Křížek et al. 2014; Singh et al. 2016).

Packaging plays an important part in the protection of food products against a variety of external influences (e.g., environmental oxygen and water vapour), controlling undesirable fluctuations in ambient temperature and keeping its contents at a suitable temperature for longer periods, thus extending their shelf life.

On account of consumers' growing interest in consumption of fresh foodstuffs with prolonged shelf life and controlled quality, manufacturers must provide modern and safer packaging solutions. This proves to be both a challenge and chief motive for the food packaging industry to develop new and enhanced concepts of technology for packaging (Dainelli et al. 2008a). Owing to its scale in the creation of new products, the food industry has created its own set of demands on the packaging market.

Packaging production is an international industry branded by its diversity and each subdivision affects the market. There are systematically more demands towards packaging and materials that are to contact directly with food products.

A range of parameters, varying from product characteristics to consumer requirements and trends, affect the selection of materials used for the package, and are presented in Fig. 6.1.

Packaging can be defined as either passive or active. Passive packaging, by far the most common type of packaging present in every branch of the world trade market, plays the role of a physical barrier which protects its contents from physical damage and keeps them in an orderly fashion for easier transport and helps in maintaining the inside temperature during transportation and storage, minimizing temperature fluctuations through the use of insulating materials in its construction. Standard materials for product packaging (plastic, cardboard, wood, etc.) usually

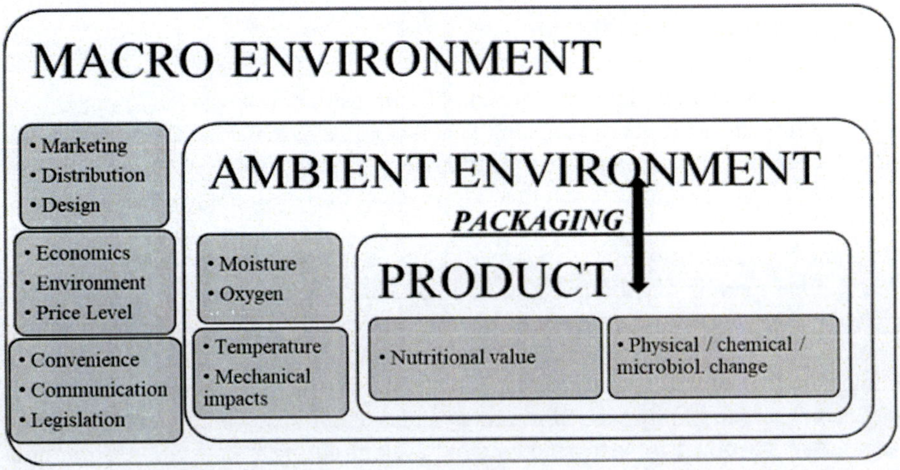

Fig. 6.1 Interaction of the packaged product with the environment. (Sonneveld 2000)

have quite limited thermal buffering capacities, leading the industry in continuous search for better options to preserve temperature and reduce fluctuations along the cold chain. To this end, packaging can be designed to play an active role in maintaining food temperature within optimal limits by using phase-change materials (PCMs) (Gaikwad et al. 2016) which are usable during the transport, distribution, and storage stages to maintain the cold chain for foods and beverages.

Farmer (2013) defined active packaging as a passive container and protection for the product as well as performing additional actions whereas smart packaging or intelligent packaging trace and inform. Active packaging is an upgrade to the conventional packaging purposes such as providing physical protection, comfort and ease of use, longer shelf life and storage periods. Farmer additionally stresses that in the future besides acting as a barrier, packaging will interact with the packaged products themselves (Farmer 2013). Further definitions of packaging comprise a harmonised system of readying goods for transport, distribution, storage, retailing and end-use a means of ensuring safe delivery to the ultimate consumer in sound condition at optimum cost and a techno-commercial function targeting cost of delivery optimization and sales maximization (and, hence, profits) (Coles 2011).

The designations active packaging, intelligent packaging, and smart packaging apply to packaging systems used for food and beverages (besides with pharmaceuticals and numerous other products) to extend shelf life, monitor product freshness, displaying quality information, and improve their safety and convenience (Dainelli et al. 2008b). Despite being correlated to one another they differ however, in actual function: intelligent packaging and smart packaging encompass the capability to sense or measure a characteristic of the product and the internal and/or external atmosphere of the package. On the other hand, active packaging has dynamic functions beyond the inert/passive containment and protection of the product such as the scavenging of oxygen from the atmosphere to prevent oxidation or having cold sources that absorb external thermal energy before it reaches the cold product.

2 Phase-Change Materials (PCMs)

A PCM is a substance that undergoes a phase transition at a specific temperature, enabling it to absorb and release latent heat when isothermal conditions are altered (Jin et al. 2010). Usually, a phase transition between one fundamental state of matter – solid, liquid, or gas - to another. However, it may also happen between non-classical states of matter, for instance the conformity of crystals (the material goes from one crystalline structure to another of higher or lower energy state).

By either melting or solidifying at the phase change temperature, a PCM is capable of both storing and releasing large quantities of energy in comparison to sensible heat storage. Heat is absorbed or released as the material changes from solid to liquid and vice-versa or when the material's internal structure changes.

Latent heat (also known as latent energy and heat of transformation) is the energy released or absorbed by a thermodynamic system during a constant-temperature process, usually a first-order phase transition. It can be understood as energy in a hidden form which is supplied or removed to change the state of a substance without varying its temperature (e.g., latent heat of fusion).

Equation 6.1 represents the latent heat for a given mass of a substance

$$Q = m.L \tag{6.1}$$

where:

Q is the amount of energy released or absorbed during the change of phase of the substance (kJ),
m is the mass of the substance (kg), and
L is the specific heat of the particular substance (kJ.kg^{-1}), either L_f for fusion or L_v for vaporization.

The specific latent heat (L) is an intensive property of the substance, as it is independent from the size or extent of the sample, and is commonly quoted and documented in the literature making it readily available for consult.

In choosing the adequate PCM one must consider several aspects and properties of the substance to be used that may pose as challenges (Fig. 6.2).

Regarding the chemical properties in Fig. 6.2 and compatibility wise, permeability is a critical factor. For instance, if the PCM is unsuitable it may permeate the packaging after manufacture, leaking either out and/or into the product leading to contamination.

The key features in the characterization of thermal performance of PCM are thermostability, phase change temperature (melting point, freezing point, flash point, etc.), latent heat, thermal conductivity, and vapour pressure.

Thermostability is the quality of a substance to resist irreversible change in its chemical or physical structure, often by resisting decomposition or polymerization, at high relative temperature (Masiulanis 1984).

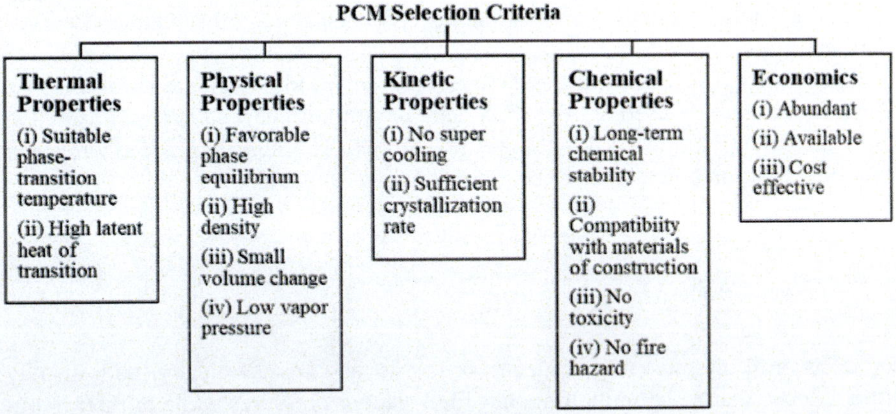

Fig. 6.2 Selection parameters for PCMs. (Kim et al. 2015)

Vapour Pressure is defined as the pressure exerted by a vapour in thermodynamic equilibrium with its condensed phases (solid or liquid) at a given temperature in a closed system. The equilibrium vapour pressure is an indication of a liquid's evaporation rate. It relates to the tendency of particles to escape from the liquid (or a solid) (Gopal et al. 2004). The International System of Units recognizes pressure as a derived unit with the dimension of force per area and designates Pascal (Pa) as its standard unit (1 Pa = 1 $N.m^{-2}$ or $kg.m^{-1}.s^{-2}$)

Thermal Conductivity (k, $W.m^{-1}.K^{-1}$) of a material is a measure of its ability to conduct heat. According to the second law of thermodynamics, heat will flow from the hot environment to the cold one in an attempt to equalize the temperature difference. This is quantified in terms of a heat flux q, which gives the rate, per unit area, at which heat flows in a given direction (Polezhaev 2011).

A simple mathematical definition of heat flux is as being the rate, per unit area, at which heat flows in a given direction (e.g., the x-direction). In many materials, q is observed to be directly proportional to the temperature difference and inversely proportional to their position. This relation can be seen in Eq. 6.2.

$$q = -k\nabla T = -k\frac{T_2 - T_1}{L} \tag{6.2}$$

where T_1 and T_2 are the temperatures of the body at two given points and L is the separation between the two points.

Another important note is that some PCMs are aqueous solutions and are rather safe. Others are either hydrocarbons, other combustible materials, or even toxic. Per se, PCMs must be carefully chosen and applied in compliance with fire and building codes and sound engineering practices (SEP). As there is an augmented fire risk, flame spread, smoke, the potential for explosion when held in containers, and liability, it is wise not to use combustible PCMs in housing or other frequently occupied buildings.

PCMs can be organised in three classes: organic, inorganic, and solid-to-solid. The most commonly used organic PCMs are paraffin, lipids, eutectic mixtures, and sugar alcohols (Farid et al. 2004; Agyenim et al. 2011; Singh et al. 2018) while salt hydrates, salts, and metals are examples of inorganic PCMs.

2.1 *Organic PCMs*

Paraffins fulfil the most requirements for use as they are chemically inert, display little volume change upon melting, have low vapour pressures while in their liquid form, are nontoxic, predictable, reliable, and stable below 500 °C (Kim et al. 2015). Some examples of paraffin used for this purpose are n-dodecane, n-nonadecane, n-eicosane, n-octacosane, n-triacontane, n-tritriacontane, or mixtures. The melting of the (CH_3) chain crystals absorb large amounts of heat however, due to cost considerations, only technical-grade paraffin is usable as PCM in latent heat storage systems.

Fatty acids and respective blends have interesting properties such as absence of supercooling, chemical stability, concurrent melting point, high latent heat, low volume change during phase change, nontoxicity, and self-nucleating behaviour (Hasan and Sayigh 1994; Sari and Kaygusuz 2001, 2002; Kumaresan et al. 2012; Yuan et al. 2014). Their availability is an advantage since they can be renewably obtained from common vegetable oils and animal fat (Cedeño et al. 2001). They are however more expensive than paraffin (2–2.5 times higher than that of technical-grade paraffin), slightly corrosive, and may have or develop unpleasant odours.

Specific examples of fatty acids with PCM aptitude are arachidic acid, behenic acid, capric acid, caprylic acid, cerotic acid, lauric acid, lignoceric acid, myristic acid, palmitic acid, stearic acid, or a combination/mixture of the different fatty acids. There are also PCMs comprising alkyl esters of fatty acids in the form of methyl, ethyl, isopropyl, butyl, or hexyl ester. Few examples of these esters are methyl caprate, ethyl oleate, isopropyl ecosapentanoate, butyl laurate, and hexyl docosahexanoate (to cover all the ester forms mentioned).

2.2 *Inorganic PCMs*

Conversely, inorganic PCMs are affordable, readily available, and have higher thermal conductivities. They are however offset by serious drawbacks, being some of them supercooling, low rate of nucleation, and, the most critical, their corrosivity. Some specific examples of suitable salt hydrates for PCM application are:

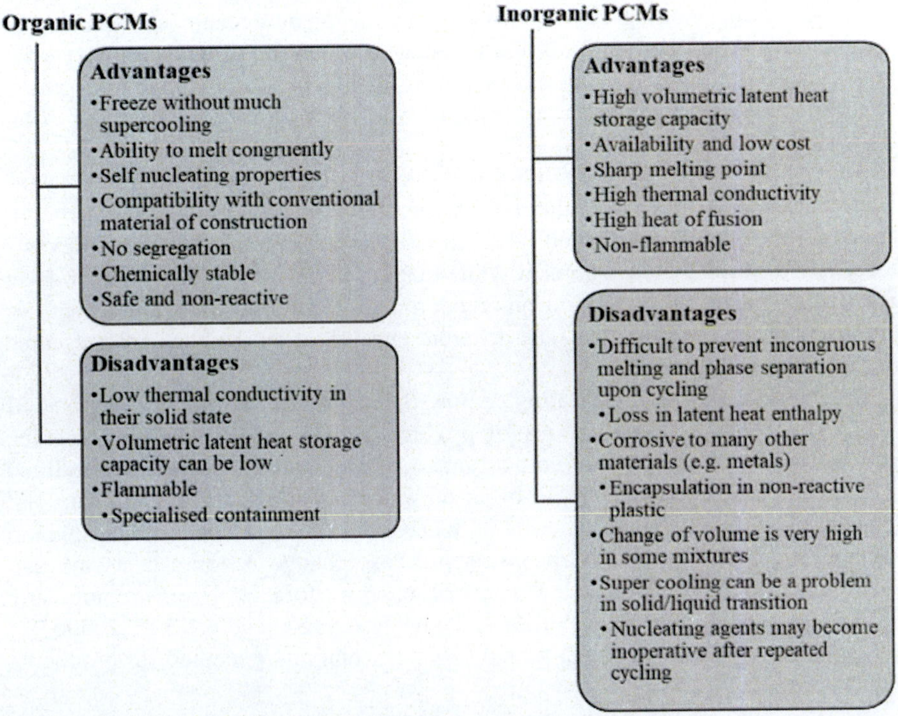

Fig. 6.3 Advantages and drawbacks of both organic and inorganic PCMs

$Ca(NO_3)_2{\cdot}3H_2O$;	$NaNO_3{\cdot}6H_2O$;	$Zn(NO_3)_2{\cdot}2H_2O$;	$Fe(NO_3)_2{\cdot}6H_2O$;
$Ni(NO_3)_2{\cdot}6H_2O$;	$Cd(NO_3)_2{\cdot}4H_2O$;	$Cd(NO_3)_2{\cdot}H_2O$;	$Co(NO_3)_2{\cdot}6H_2O$;
$Na_2HPO_4{\cdot}12H_2O$;	$Na_3PO_4{\cdot}12H_2O$;	$NaOH{\cdot}H_2O$;	$FeSO_4{\cdot}7H_2O$;
$NaAl(SO_4)_2{\cdot}12H_2O$;	$MgCl_2{\cdot}4H_2O$;	$MnCl_2{\cdot}4H_2O$;	$FeCl_3{\cdot}2H_2O$;
$NaCH_3COO{\cdot}3H_2O$;	$Na_2B_4O_7{\cdot}10H_2O$;	$LiCH_3COO{\cdot}2H_2O$;	

or mixtures of the mentioned hydrates.

All PCMs have, understandably, both advantages and disadvantages (Fig. 6.3).

2.3 *Solid-to-Solid PCMs*

Solid-to-solid PCMs are a particular group undergoing solid-to-solid phase transition associated with the absorption and release of large amounts of heat. These materials change their crystalline structure from one lattice configuration to another at a fixed and well-defined temperature, and this transformation can consist of latent heats comparable to the most effective solid-to-liquid PCMs. These materials are advantageous as, unlike solid-to-liquid PCMs, nucleation is unnecessary in avoiding supercooling, there are no noticeable changes in the appearance of the PCM, and difficulties related to handling liquids are non-existent (e.g., containment, potential leakage, etc.). At present, the temperature range of solid-solid PCM solutions spans from −50 up to 175 °C (PhaseStor solutions n.d.).

Due to the difficulty in handling and avoiding leakage of a liquid PCM the application method is of absolute importance. Encapsulation of PCMs is acknowledged as an application method which improves the heat transfer area, prevents reactivity towards the environment while making them suitable for application in both packaging and refrigeration, and rendering their handling more convenient for end-users (Alkan et al. 2011). The different encapsulation methods are presented in Fig. 6.4.

A humble list of patents relating to (micro)encapsulated PCMs is comprised of WO2017043986A1 ("The manner of production of composite with a sandwich panel structure on the basis of aerogel mat, polyurethane or epoxy resin modified with glycolisate obtained on the basis of waste polyethylene terephthalate and encapsulated phase change material"), WO2012175777A1 ("PCM encapsulation method"), US10151542B2 ("Encapsulated phase change material heat sink and method"), US20170254601A1 ("Thermal energy storage systems comprising encapsulated phase change materials and a neutralizing agent"), KR102225007B1 ("Capsule having an outer shell bound to a surfactant and a method for producing the capsule").

Encapsulated PCMs can be directly applied to packaging structures for smart packaging to help control temperature fluctuations. However, this method is not without its disadvantages as it increases the supercooling effect. Plus, during melting, microcapsules may rupture, decreasing the overall thermal buffering capacity of the PCM.

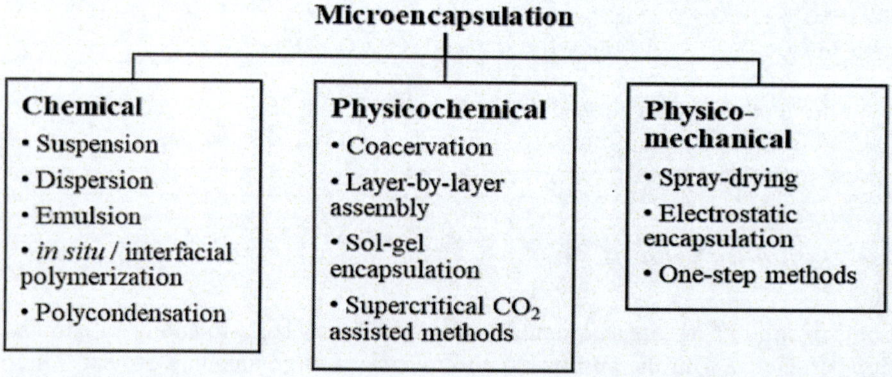

Fig. 6.4 Methods employed in microencapsulation of PCMs. (Kim et al. 2015; Giro-Paloma et al. 2016)

Encapsulating the PCM in a highly porous matrix (thermal composite) is an alternative method to improve applicability and its implementation is simpler. Johnston et al. (2008) studied this method by means of nanostructured calcium silicate as the porous matrix and an alkane as the PCM. The greatest achievement was obtaining a dry, free-flowing powder that could easily be contained in a bubble wrap liner, giving this product an extensive range of possible applications in the packaging industry for the safe transport of temperature-sensitive and perishable products.

Many available PCMs are covering a wide range of temperatures that come in many formats and from various natural and synthetic sources. Table 6.1 lists some commercial products, both PCMs and end-products, used in food packaging for refrigeration and transport.

Table 6.1 Commercial products containing PCMs used in food and pharmaceutical packaging and refrigeration

Product	Trade name	Active compound	Temperature range (°C)	Application
	PureTemp	Palm oil, palm kernel, rapeseed oil, coconut oil, soybean oil	−37 to 151	Cold Storage, refrigeration, air conditioning, heat recovery, heat storage
	MatVesl	Biodegradable organic products	1 to 18	Withstand rigorous long-term use
	BlockVesl	Biodegradable products	−15 to 37	Stackable blocks
	Lava Lunch	Lava rock technology	48 to 65	Containing hot food
	Joulies	PureTemp 60	57 to 63	Cooling hot beverages and maintaining ideal temperature
	savENRG Ice Pack	Hydrated salt	−26 to −10	Keeping specimens frozen
	savENRG Cool Pack	Organic products	2 to 8	Keeping payloads refrigerated
	GreenBox	Vegetable oil-based	2 to 8	Perishable products
	Yoomi	Supersaturated sodium acetate solution	32 to 34	Simulate breast milk temperature

3 Heat/Cold Emitters

Another type of active packaging does not require PCMs. Instead, the release and absorption of heat are achieved by chemical reactions that happen inside the package, in a second compartment containing a chemical reactant or a mixture of chemical reactants. This application relies on the heat of reaction (enthalpy of reaction) of a determined chemical reaction, be it exothermic or endothermic.

An exothermic reaction is a reaction for which the overall standard enthalpy change (ΔH^0) is negative (IUPAC 1997), meaning they release heat and encompass the replacement of weak bonds with stronger ones (Schmidt-Rohr 2015). On the other hand, an endothermic reaction is a reaction for which the overall standard enthalpy change (ΔH^0) is positive (IUPAC 1997), meaning the system absorbs heat and requires a favourable entropy increase ($\Delta S > 0$) to overcome the unfavourable increase in enthalpy to keep $\Delta G < 0$.

The Gibbs free energy ($\Delta G = \Delta H - T\Delta S$ (measured in Joules)) is a thermodynamic potential that can be used to calculate the maximum of reversible work that may be performed by a thermodynamic system at a constant temperature and pressure. Explained simplistically, the Gibbs free energy tells us whether a reaction will occur spontaneously at constant pressure and temperature or not.

While endothermic phase-transitions into more disordered states of higher entropy (e.g., melting and vaporization) are common, spontaneous chemical reactions at moderate temperatures are rarely endothermic. The activation energy for these reactions is usually quite high and must be made available so the system may react.

A rather simple commercial form taken by these self-heating packages is permeable packets containing a reactant or a reactant mixture inside which is activated by the addition of water. These are particularly convenient for military operations, during natural disasters, or when conventional cooking is unavailable since they come with MREs (Meal, Ready-to-Eat). An example of such a reactant is quicklime (calcium dioxide) reacting with water, forming calcium hydroxide as the product of the reaction, by the ensuing equilibrium:

$$CaO(s) + H_2O(l) \leftrightarrow Ca(OH)_2(aq) \tag{6.3}$$

and releasing thermal energy in the process (ΔH_r = −65.21 kJ.mol^{-1} of CaO) (Ebbing 1990). This reaction is known to achieve temperatures of up to 300 °C.

Another example, used in MREs, is finely powdered magnesium metal alloyed with a small amount of iron and table salt. The reaction is then activated by adding a small volume of water to create a myriad of minute short-circuited batteries that quickly burn out, thus producing heat (Scott and Meadows 1992).

By adapting these same principles to different, often more convenient forms and shapes for both supplier and consumer to usufruct of the product, different developments start to appear. Self-heating cans, for instance, have twin chambers; one enclosing the other with a “ring” which acts as a mechanical trigger, that breaks the barrier keeping the reactants separate. This application has, at least, two basic versions.

In one version, the inner chamber holds the food or drink while the outer chamber houses the reactants that shall undergo the exothermic reaction once combined. Once the user wishes to heat the contents of the can, a ring may be pulled to break the barrier separating the reactants in the outer chamber from a small water compartment to start the reaction.

The other version has the reactants in the inner chamber and the beverage surrounding it in the outer chamber. To start the reaction, the user pushes the bottom of the can, breaking a seal and allowing contact between the chemical reactants and water. This design is more advantageous as it is more efficient in heating the product instead of overheating the can's exterior and losing energy to the atmosphere, potentially burning an unsuspecting user.

The way this second version works is quite simple: it is built as a triple-walled container. The container for the beverage surrounds another container for the heating agent which, in turn, is separated from yet another container with water by a thin breakable membrane. By pressing the bottom of the can, a rod pierces the membrane to allow the water and heating agent to mix, resulting in the release of heat, thus warming the beverage surrounding this reaction chamber (Scudder and Berntsen 1995).

Hot-Can North America Inc. is "an innovative smart-packaging company with a revolutionary, self-heating can called Hot-Can". The layout of their packaging is a hybrid of the two versions mentioned above, where the compartment containing the heating element is pierced by a rod upon popping the seal in the underside of the can while the beverage is contained in the middle compartment. Their marketing approach instructs the user to shake for 30 seconds. This action further promotes the mixing of the heating element and the water, increasing the overall heating efficiency of the reactants by creating more surface area for the reaction to occur. This also promotes some thermal energy transfer from the reaction chamber to the product.

HeatGen (previously known as HeatGenie) is a company that has patented technology that heats beverages an efficient, safe, and simple in off-the-shelf aluminum packaging using a solid-state thermal reaction. In 2018 an idea that had been abandoned 15 years prior by Nestlé, was brought to market: thermite.

Thermite has a rather simple composition: metal powder as fuel (e.g. aluminum, magnesium, titanium, zinc, silicon, and boron) and metal oxide as the oxidizer (e.g., bismuth(*III*) oxide, boron(*III*) oxide, silicon(*IV*) oxide, chromium(*III*) oxide, copper(*II*) oxide , iron(*III*) oxide, iron(*II,III*) oxide, lead(*II,IV*) oxide, and manganese(*IV*) oxide) (Hall 2014), a safe and stable mix until given a jolt of energy (e.g. lighting a magnesium metal fuse). In the most common example of aluminum and iron oxide, once ignited the aluminum grabs the oxygen from the rust and produces iron as well as an enormous amount of heat (Eq. 6.4), easily reaching 2500 °C.

$$Fe_2O_3(s) + 2Al(s) \rightarrow 2Fe(s) + Al_2O_3(s) + heat \tag{6.4}$$

This reaction is the perfect example of an extremely exothermic reaction as its energy content is 3956 $J.g^{-1}$ and an energy density of 16516 $J.cm^{-3}$ (Fischer and Grubelich 1996). This massive heat output allows the use of smaller quantities of reactant which amounts to less room taken by the heating element in the can.

HeatGen cans use aluminum and silicon dioxide as a lower-power 'thermite' which can still reach a whopping 1600 °C. However, 'pure' silicon thermite consisting exclusively of silicon dioxide and aluminum is very difficult to ignite and fizzles out quickly. All thermite reactions need proper ignition through locally applied strong heat, usually generated by setting off a small amount of ignition mixture however, silicon thermite blends require extra aluminum and sulphur to guarantee a self-sustaining reaction and to avoid fizzling. An example of a potential formula for the SiO_2:Al:S mixture, in parts by weight, is 9:10:12. Equations 6.5 and 6.6 are two distinct chemical reactions taking place.

$$3S(s) + 2Al(s) \rightarrow Al_2S_3(s) + heat \tag{6.5}$$

$$3SiO_2(s) + 4Al(s) \rightarrow 2Al_2O_3(s) + 3Si(s) + heat \tag{6.6}$$

What HeatGen does is to develop ways to cover safety issues (e.g., burning your hands). These include complex arrangements of insulation elements ('firewalls') to block excessive heat and energy-absorbing heatsinks to ensure heat distribution throughout the drink, as well as safety vents to let off any steam (Lorch 2018). It is claimed that with all this in place only 10% of the packaging is taken by heating elements, which is outstanding in comparison with the usual 50% taken by limestone.

Other solid chemical reactants or mixtures include calcium chloride ($CaCl_2$), calcium oxide (CaO), and/or magnesium chloride ($MgCl_2$), as well as an organic acid (acetic acid, citric acid, or lactic acid) in some cases. These can be in a specified hydration state such as anhydrous, monohydrate, or dihydrate. Upon contacting the aqueous solution, the aqueous solution dissolves and reacts with the solid chemical reactant mixture producing heat. Some of this heat is derived from the heat of solution of the solid chemical reactant mixture and is specific to the exact form of the chemical species present in the mixture (Farone and Palmer 2009, 2013; Ford and Lund 2013).

As mentioned before, examples of endothermic reactions that take place spontaneously at moderate temperatures are scarce. As a result of this, most applications tend to explore different approaches when choosing which endothermic phenomenon to use in their application.

Despite scarce, they are not non-existent: using chemical reactants to refrigerate a product in a self-chilling can according to patent US 8826672B2 (Heat Wave Technology LLC (US)) (Farone and Palmer 2014) "at least one solid chemical reactant of the cooling composition must be selected independently from a group consisting of ammonium nitrate (NH_4NO_3), urea (CH_4N_2O), and mixtures thereof, and a second solid chemical reactant selected independently from the group consisting of ammonium chloride (NH_4Cl), calcium sulphate ($CaSO_4$), borax ($Na_2H_4B_4O_9 \cdot nH_2O$), phosphates ($H_2PO_4^-$, HPO_4^{2-}, PO_4^{3-}), magnesium chloride ($MgCl_2$), and carrageenan". Optionally, there may be a third chemical reactant. The reactants are then allowed contact with a solution in the following order: the first chemical reactant is supposed to react with the liquid enough to form a heat-absorbing solution within a lower temperature than the desired; the second chemical reactant is then allowed to react with this solution to maintain this lower temperature; and the optional third chemical reactant would react to maintain the temperature range of the heat-absorption solution, thereby cooling the substance.

In 2018, Mitchell Joseph, chairman/CEO of the Joseph Co. Intl. and West Coast Chill, launched in the market the two-piece aluminum Chill-Can®, an update to a previous product that streamlines the activation process – from pushing a button to a twist – that uses liquefied CO_2 as the active coolant. The CO_2 chamber ends in a sealed valve that upon twisting, opens the valve allowing the pressurized CO_2 to rush out of the underside of the can and into the surrounding atmosphere. This is an isochoric process as it occurs at constant volume. In an isochoric process, the temperature and pressure are proportional. As the expansion of the CO_2 takes place (pressure drop), heat is absorbed by the now lower temperature CO_2, lowering the liquid's temperature as the result (proportional temperature drop) by 17 °C.

A different example without the use of gases is I.C. Can'sTM, from Tempra Technology and Crown Holdings. These cans have a different layout since they have a centerpiece called an evaporator and a bottom compartment with a desiccant stored in vacuum. The evaporator is filled with a watery gel. Rotating the bottom chamber releases the vacuum, sucking the water vapor that is absorbed in the desiccant granules and heat (isochoric process) from the gel to the bottom of the can, dropping the temperature by 17 °C.

4 Temperature-Controlled Packaging

It goes without saying that to take full advantage of a thermal process it is important to keep the thermal energy either inside or outside of the system, thus the use of materials with good insulant properties.

The Food and Beverages industry holds a majority of the market share in the global temperature-controlled packaging solutions market, closely followed by the healthcare industry. The high demand from this industry is mainly driven by the necessity of insulated containers and bins coupled with a refrigerant for proper transportation of perishables and temperature-sensitive products.

Performance-oriented products can be produced using a variety of insulation materials available to the cold chain protection products industry. Such insulation materials are typically flexible thermoplastics, foam or encapsulated air type of products that are made by employing foil or metalized films to 'hold' or encapsulate panels or are laminates of the like when combined with encapsulated air products such as polyethylene bubble.

A good insulation performance from the building materials (passive refrigeration) greatly aids in preserving the inside temperature of the package and alleviates the thermal load otherwise faced by the PCMs (active refrigeration). An effective, simultaneous, use of these approaches may extend the period at which temperature can be kept under control without the use of electrical refrigeration units both during transport and storage.

Overall, passive temperature-controlled packaging is largely preferred over its active counterpart mainly due to it being relatively more economical, of easier handling, and disposability.

Materials commonly used for packaging are paper (e.g., cardboard, whiteboard prepared from sulphite pulp), cellophane, polyethylene (low-density (LDPE), high-density (HDPE) and linear low-density (LLDPE)), polypropylene ((PP) cast, oriented, heat-set, coated, and composite), polystyrene (PS), Polyester (e.g., PET), polyamides (nylon), polyvinyl chloride (PVC), ionomers (e.g., Surlyn™), copolymers (e.g., EVA) and aluminum foil (Gopal and Shankar 2010).

Plastics contain different additives that give it certain characteristics, that either inherent or deliberately added to change the plastics' properties. These components can be fit into three categories (Gopal and Shankar 2010):

- Polymerization residues (catalyst remnants, polymerization solvent, residual monomers, etc.);
- Processing aids (antioxidants, antistatic agents, lubricants, plasticizers, slip agents, stabilizers, etc.); and
- End-use additives (antioxidants, blowing agents, brighteners, colorants, mold release agents, UV stabilizers, etc.).

Aluminum foil, and other products of its family, is used in packaging as it is highly malleable and acts as a barrier to both light and oxygen, odors and flavors, moisture and germs. Due to its abilities, it is broadly used in food and pharmaceutical packaging, including long-life packs (aseptic packaging) for drinks and dairy products. It acts as a radiation shield due to its reflectivity, minimizing heat transfer to occur through radiative means (Hanlon et al. 1998).

The selection of an appropriate packaging system must obey several criteria:

- Firstly, the stability of the food product itself (proteins, lipids, and certain vitamins may experience detrimental changes due to variation in water activity of the product) since its chemical, biochemical, and physical nature will be markedly influenced by the permeability or barrier properties of the package;
- Secondly, environmental factors such as temperature, relative humidity, oxygen tension, and light intensity, to which the product-package system is exposed during distribution and storage must also be considered when evaluating the barrier properties required for the package; and
- Lastly, the nature and composition of the specific packaging material and its potential effect on the intrinsic quality and safety of the packaged food as a con-

sequence of the migration of components from the packaging material into the food should also be considered.

A recent study by Future Market Insights (FMI), published on the 6th of February of 2020, predicts growth at an impressive Compound Annual Growth Rate (CAGR) of over 8% through 2030 for the global temperature-controlled packaging solutions market alongside a doubled revenue. The global temperature-controlled packaging solutions market is segmented in depth to cover every single facet of the market and display comprehensive market intelligence to readers (Tiwari and Godge 2020), as it is represented in Fig. 6.5.

Table 6.2 lists some types of existing commercial products and solutions for temperature-controlled packaging used both in the Food & Beverages industry and in the Pharmaceutical industry.

Fig. 6.5 List of temperature-controlled packaging solutions provided by the packaging market

System Type

- Active
- Passive

Product Type

- Insulated Shippers
  - Panels and Envelopes
  - EPS Shippers
  - PUR Shippers
  - VIP Shippers
- Insulated Containers
  - Chest Style
  - Upright Style
- Others

Application

- Frozen
- Chilled
- Ambient

End Use

- Healthcare
- Food and Beverages
- Others

Region

- North America, Latin America, Europe, South Asia, East Asia, Oceania, MEA

Table 6.2 Commercial solutions for temperature-controlled packaging

Product	Trade name	Building material	Application
	Cooliner	Bubble/metalized film product	Pouches, box liners, pallet covers, roll stock, pallet shipper products
	GreenLiner	Foam/metalized film product	
	Temprecision® Foam	Open-cell phenolic foam	Passive refrigerant and dry foam insulation
	Thermabrick®	Open-cell phenolic foam	Sealed and delivered passive cooling element
	AEON™	Vacuum insulation panels (VIPs) and PCMs	Reusable parcel shipper, very healthcare oriented
	Skypod	Thermal insulation	LTE/UAV-connected drones for delivery of vaccines
	SilverSkin™	Radiant barrier materials and strong flexible materials	Cost-effective transportation of pharmaceutical products within specified label claim parameters
	Silverpod® MAX	Insulation materials and recyclable PCM coolants	Manufactured to meet all industry established temperature ranges
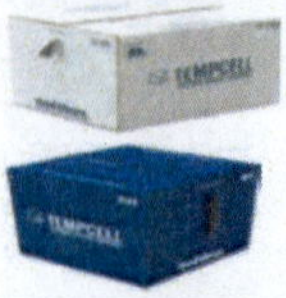	Tempcell™ and Tempcell™ MAX	Water-gel temperature control parcel Recyclable PCM coolants	Single-use parcel shippers to regulate temperature during transit
	Coolimate® Insulated Envelopes	Insulating foil	Reflects up to 97% of radiation heat and 97% of the cold air back into the product

5 Future Development and Prospects

The packaging market will continue to grow and evolve as different and more needs come to be. Already existing solutions tend to develop into new, more efficient ones and to be stepping stones for breakthrough ideas and applications with new and improved features. Active packaging along with intelligent and smart packaging will gain more visibility and their momentum will take them to everyone's shelves for their convenience.

The study and investigation of new compounds may bring about further growth of the list of reactants and PCMs to choose from and, hopefully, green and sustainable. Insulation materials also share this evolution trend, making the packaging market a truly diverse and dynamic system that is everchanging.

References

Agyenim F, Eames P, Smyth M. Experimental study on the melting and solidification behaviour of a medium temperature phase change storage material (Erythritol) system augmented with fins to power a LiBr/H2O absorption cooling system. Renew Energy. 2011;36(1):108–17.

Alkan B, Sari A, Karaipekli A. Preparation, thermal properties and thermal reliability of microencapsulated n-eicosane as novel phase change material for thermal energy storage. Energy Convers Manag. 2011;52:687.

Cedeño FO, Prieto MM, Espina A, García JR. Measurements of temperature and melting heat of some pure fatty acids and their binary and ternary mixtures by differential scanning calorimetry. Thermochim Acta. 2001;369:39.

Coles R. Food and beverage packaging technology, vol. 1. New Jersey: Blackwell Publishing Limited, Hoboken; 2011. p. 1–29.

Dainelli D, Gontard N, Spyropoulos D, Zondervan-van den Beuken E, Tobback P. Active and intelligent food packaging: legal aspects and safety concerns. Trends Food Sci Technol. 2008a;19(1):103.

Dainelli D, Nathalie G, Dimitrios S Z-v, den Beuken E, Tobback P. Active and intelligent food packaging: legal aspects and safety concerns. Trends Food Sci Technol. 2008b;19(1):167–77.

Ebbing DD. General chemistry. 3rd ed. Boston: Houghton Mifflin Company; 1990. p. 225.

Farid MM, Khudhair AM, Razack SAK, Al-Hallaj S. A review on phase change energy storage: materials and applications. Energy Convers Manag. 2004;45:1597.

Farmer N. Trends in packaging of food, beverages and other fast-moving consumer goods (FMCG), vol. 1. Oxford: Woodhead Publishing Limited, Cambridge; 2013. p. 1–21.

Farone W, Palmer T. Self-heating apparatuses using solid chemical reactants. US 20090078711A1, United States Patent and Trademark Office, 26 March 2009. PatSeer.

Farone W, Palmer T. Self-heating systems and methods for rapidly heating a comestible substance. US 8556108B2, United States Patent and Trademark Office, 15 October 2013. PatSeer.

Farone W, Palmer T. Self-cooling compositions, systems and methods. US 8826672B2, United States Patent and Trademark Office, 9 September 2014. PatSeer.

Fischer SH, Grubelich MC. A survey of combustible metals, thermites, and intermetallics for pyrotechnic applications. Sandia National Laboratories, Albuquerque, NM (United States), 1996, August 1st. Retrieved from https://www.osti.gov/servlets/purl/372665.

Ford J, Lund D. Self-heating systems and methods for rapidly heating a comestible substance. US 8578926B2, United States Patent and Trademark Office, 12 November 2013. PatSeer.

Gaikwad KK, Lee JY, Lee YS. Development of polyvinyl alcohol and apple pomace bio-composite film with antioxidant properties for active food packaging application. J Food Sci Technol. 2016;53:1608.

Giro-Paloma J, Martinez M, Cabeza LF, Fernandez AI. Types, methods, techniques, and applications for microencapsulated phase change materials (MPCM): a review. Renew Sust Energ Rev. 2016;53:1059.

Gopal TKS, Shankar CNR. Handbook of seafood quality, safety and health applications, vol. 12. New Jersey: Blackwell Publishing Limited, Hoboken; 2010. p. 139–55.

Gopal H, Kleinsmidt J, Case J, Musonge P. An investigation of tertiary students' understanding of evaporation, condensation and vapour pressure. Int J Sci Educ. 2004;26(13):1597.

Hall AC. Powder safety awareness for additive manufacturing. Sandia National Laboratories, Albuquerque, NM (United States), 2014, October 1st. Retrieved from https://www.osti.gov/servlets/purl/1242062.

Hanlon JF, Kelsey RJ, Forcinio HE. Handbook of package engineering, vol. 3. 3rd ed. Boca Raton: CRC Press; 1998. p. 59–105.

Hasan A, Sayigh AA. Some fatty acids as phase-change thermal energy storage materials. Renew Energy. 1994;4:69.

IUPAC. Compendium of Chemical Terminology, 2nd ed. (the "Gold Book"). Compiled by A.D. McNaught and A. Wilkinson. Oxford: Blackwell Scientific Publications; 1997. Online version (2019) created by S.J. Chalk.

Jin Y, Lee W, Musina Z, Ding Y. A one-step method for producing microencapsulated phase change materials. Particuology. 2010;8:588.

Johnston JH, Grindrod JE, Dodds M, Schimitschek K. Composite nano-structured calcium silicate phase change materials for thermal buffering in food packaging. Curr Appl Phys. 2008;8:508.

Kim D, Park I, Seo J, Han H, Jang W. Effects of the paraffin wax (PW) content on the thermal and permeation of the LDPE/PW composite films. Polym Res; 2015.

Křížek M, Matějková K, Vácha F, Dadáková E. Biogenic amines formation in high-pressure processed pike flesh (Esox lucius) during storage. Food Chem. 2014;151:466.

Kumaresan V, Velraj R, Das SK. The effect of carbon nanotubes in enhancing the thermal transport properties of PCM during solidification. Heat Mass Transf Und Stoffuebertragung. 2012;48:1345.

Laguerre O, Hoang HM, Flick D. Experimental investigation and modelling in the food cold chain: thermal and quality evolution. Trends Food Sci Technol. 2013;29:87.

Likar K, Jevšnik M. Cold chain maintaining in food trade. Food Control. 2006;17:108.

Lorch M. Self-heating drinks cans return – here's how they work, 2018, June 19th. Retrieved from https://theconversation.com/self-heating-drinks-cans-return-heres-how-they-work-98476.

Masiulanis B. Thermoplastic polyurethane elastomers – thermostability in relation to composition. J Appl Polym Sci. 1984;29(2):681.

"PhaseStor solutions". Retrieved from https://phasechange.com/products/phasestor/.

Polezhaev YV. Thermal conductivity, 2011, February 2nd. Retrieved from http://www.thermopedia.com/content/1186/.

Sari A, Kaygusuz K. Thermal energy storage system using stearic acid as a phase change material. Sol Energy. 2001;71:365.

Sari A, Kaygusuz K. Thermal performance of a eutectic mixture of lauric and stearic acids as PCM encapsulated in the annulus of two concentric pipes. Sol Energy. 2002;72:493.

Schmidt-Rohr K. Why combustions are always exothermic, yielding about 418 kJ per Mole of O2. J Chem Educ. 2015;92(12):2094–9.

Scott D. Meadows R. Hot meals. Chem Matters; 1992, February. Retrieved from https://heatermeals.com/wp-content/uploads/2013/07/chemmattersarticle.pdf.

Scudder JA, Berntsen JL. Container with integral module for heating or cooling the contents. US 5461867, United States Patent and Trademark Office, 31 October 1995. Free Patents Online. http://www.freepatentsonline.com/5461867.html.

Singh S, Ho Lee M, Park I, Shin Y, Lee YS. Antimicrobial seafood packaging: a review. J Food Sci Technol. 2016;53:2505.

Singh S, Gaikwad KK, Lee, M.ho Lee, Y.S. Temperature-regulating materials for advanced food packaging applications: a review. J Food Meas Charact. 2018;12:588.

Sonneveld K. What drives (food) packaging innovation? Packag Technol Sci. 2000;13:29.

Tiwari H, Godge A. Temperature controlled packaging solutions market, 2020, February 6th. Retrieved from https://www.futuremarketinsights.com/reports/temperature-controlled-packaging-solutions-market.

Yuan Y, Zhang N, Tao W, Cao X, He Y. Fatty acids as phase change materials: a review. Renew Sust Energ Rev. 2014;29:482.

Part III
Preparation and Effectiveness of Releasing Systems in Active Food Packaging

Chapter 7
Different Approaches for the Inclusion of Bioactive Compounds in Packaging Systems

Amro Shetta, Isra H. Ali, Fatma Elshishiny, and Wael Mamdouh

Abstract There is a growing need for alternative materials that could enhance the food nutritional value, dominating moisture and solute migration, prolong shelf life, and improve the quality of gas exchange and oxidative reaction rates. Besides, making the polymer-based plastics eco-friendly and biodegradable. Several bioactive compounds are implemented for the preparation of food active packaging including herbal extracts, metallic, and biological compounds. For the inclusion of these bioactive molecules, either synthetic polymers or natural polymers have been used. The encapsulation could be classified as microencapsulation and nanoencapsulation according to the carrier final diameter. Distinct methods were used to encapsulate bioactive substances like spray and freeze-drying, fluidized bed technologies, and molecular inclusion. Moreover, solid lipid nanoparticles and liposomes are lipid-based nanosystems that usually applied as a carrier of bioactive molecules and exhibited oxygen scavenging and antimicrobial properties. Hydrophobic phytochemicals molecules such as essential oils can be loaded in nanoemulsions and polymeric nanoparticles using emulsification–solvent evaporation, and supercritical fluid technologies. As well, Nanofibers are considered one of the promising materials in the inclusion of bioactive molecules in food packaging. This chapter sheds light on the methods used for inclusion, limitations, current trends, and regulatory issues, and biosafety in the encapsulation of bioactive components.

A. Shetta · F. Elshishiny · W. Mamdouh (✉)
Department of Chemistry, School of Sciences and Engineering (SSE),
The American University in Cairo (AUC), New Cairo, Egypt
e-mail: wael_mamdouh@aucegypt.edu

I. H. Ali
Department of Chemistry, School of Sciences and Engineering (SSE),
The American University in Cairo (AUC), New Cairo, Egypt

Department of Pharmaceutics, Faculty of Pharmacy, University of Sadat City,
Sadat City, Egypt

S. M. Jafari, A. S. Silva (eds.), *Releasing Systems in Active Food Packaging*,
Food Bioactive Ingredients, https://doi.org/10.1007/978-3-030-90299-5_7

Keywords Active packaging · Antimicrobial · Antioxidant · Bioactive molecules · Encapsulation · Liposomes · Nanocomposites · Nanoemulsions · Nanofibers · Nanostructures

1 Introduction

Food Packaging represents a vital role in the food sector. It maintains the food condition, extends shelf-life, and protects against environmental contamination such as temperature, microorganisms, and humidity (Han et al. 2018). Polymers play a substantial role in this regard, concerning their competitive properties. These include being cost-effective, easy to handle, and ideal mechanical properties, making them the perfect choice for food packaging. Some major type of polymers used as a matrix in the industry of food packaging involves Polyethylene terephthalate (PET), High-density polyethylene (HDPE), Low-density polyethylene (LDPE) (Siracusa and Blanco 2020). However, numerous different bioactive materials have recently been incorporated with those polymers to enhance the characteristics of the developed package material. For example, to improve food quality and safety, scientists have managed to delay oxidation, by creating packaging products with employed antioxidants and antimicrobial active components (Jouki et al. 2014). Others could include moisture absorbers, CO_2 scavengers, oxygen scavengers, and emitters (Crisosto et al. 1993). Besides other packaging passive characteristics including thermal persistence, and mechanical capacity (Fernández-Pan et al. 2014). For instance, factors like discoloration, lack of aroma, and lipid oxidation can be controlled by comprising active additives to modify the surface of Low-density polyethylene (LDPE) and get rid of meat spoilage (Ben-Yehoshua. 1985).

1.1 *Bioactive Compound Sources*

Materials used in encapsulation should meet the safety guidelines according to the Food and Drug Administration (FDA). Most of these materials are biomolecules that are food-grade and biodegradable. The most utilized materials in food packaging industries could be natural or manufactured. For example, natural sources include polysaccharides, starch, and cellulose, besides plant extracts such as soluble soybean polysaccharides, pectins, Arabic gum, and galactomannans. While synthetic materials could be synthetic antioxidants, like ethylenediaminetetraacetic acid (EDTA) and poly(acrylic acid) (PAA) (https://reader.elsevier.com/reader/sd/pii/S0924224418302760?token=022AB9E1FF4BBDFD1F36E75608D69A35F711C9EAD3A6DAD330D36E5DB5D6887A3E7612D769D095AE946C531685EFDA1C) (Table 7.1).

Table 7.1 Types and sources of utilized bioactive compounds used in food packaging

Type	Source	Function	References
Ascorbic, Gallic acids	Ganic acids	Oxygen scavengers	Pant et al. (2017)
Dextran, Chitosan	Polysaccharides	Antimicrobial activity	Lazić et al. (2020)
Silica gel, Alumina	Silicates, Zeolites	Ethylene scavenger	Gaikwad et al. (2020)
Copper, Gold, Zinc oxide, Titanium dioxide	Metals	Antimicrobial activity	Brandelli et al. (2017a)
Carrageenan, Alginate	Marine extracts	Gel forming agent	Bhat et al. (2020)
Asmonoterpenes, Flavonoids, Phenolic acids	Essential oils (phenolic compounds)	Antioxidants	Chang et al. (2019)
Sodium Bicarbonate, Citric acid	Inorganic salt, Organic acids	Carbon dioxide emitter	Tsironi et al. (2019)
Gum arabic, Galactomannans, Pectins	Plant extracts	Antimicrobial activity, Antioxidants	Ali and Said (2020)

1.2 Advantage of Nanotechnology in Food Packaging

Nanotechnology applications have opened new advances in the active food packaging industries. This extended till reach food safety and biosafety and innovation of active substances such as nanoparticles (https://www.ift.org/news-and-publications/food-technology-magazine/issues/2003/december/features/nanotechnology-a-new-frontier-in-food-science) (Fig. 7.1).

Many food packaging patents that utilized the applications of nanomaterials have been documented in different continents including the USA, Europe, and Asia, for which nano-clays and nano-silver were the most used ones. Recent studies have proved that the incorporation of allyl isothiocyanate and carbon nanotubes into the AP systems can reduce microbial activity and discoloration and regulate oxidation (Crisosto et al. 1993).

Nanoencapsulation has proved its competency in improving controlled release and provide accurate bioactive material delivery into targeted locations more efficiently than in bulky or microencapsulation techniques. This is due to the unique particle nanosize in the range of a nano diameter between 10 and 1000 nm that could be transported to various body organs in contrast to microencapsulation, in which the particles express a diameter of 3–800 μm (Hughes 2005).

Current efforts towards a "green environment" have put more focus on the fabrication of nanocomposite matrices and coatings through the utilization of biopolymers and bioactive components. The most recent example for the creation of nanocomposites is the use of polysaccharides synthetic matrices incorporated with inorganic materials. Fillers could be incorporated with biopolymer coatings to come up with bionanocomposite. For instance, nanoparticles can be used as fillers within

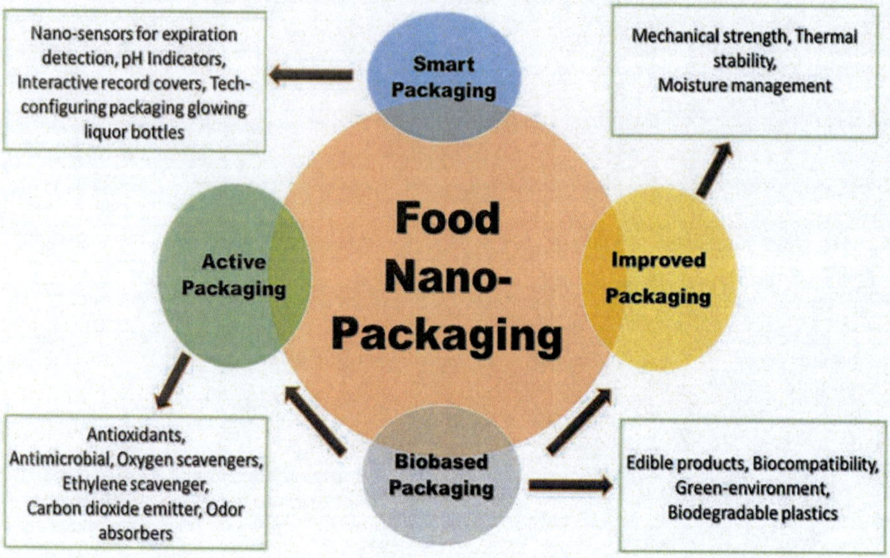

Fig. 7.1 Food nano-packaging categorize and its importance in food packaging industries

biopolymer matrices to enhance the plastic properties and biodegradability (Uysal Unalan et al. 2014). Halloysite nanotubes (HNTs) are one of the interesting developments in the sector of food AP. They can act as nanofillers since they are hollow tubular clay nanoparticles. They have many excellent advantages in delaying food spoilage and fruit aging by absorbing the naturally generated ethylene gas. Polyethylene (PE) can be treated with such HNTs to provide more ethylene gas capturers with lower permeability to water vapor than in pure PE (Tas et al. 2017).

1.3 Nanoencapsulation for Improved Food Packaging Properties

Recent investigations showed the effectiveness of HNT/polyethylene (HNT/PE) nanocomposite films in delaying ripening process of bananas due to the ethylene scavenging features. Results showed a. HNT/PE nanocomposite films proved their abilities in preserving food safety and enhancing food products' shelf life (Tas et al. 2017).

Moisture can reduce the quality of packed food and make it more liable to pathogenic microorganisms. Different moisture absorbers have been used in food packaging to reduce the moisture effect by combining desiccants substances like clay, zeolites, or silica, while others have tried bentonite and poly-acrylic acid sodium salts (Mahajan et al. 2008). To extend shelf life, ethylene scavengers have been used such as potassium permanganate, zeolite nanoparticles to absorb ethylene and prevent chlorophyll degradation. Essentially, carbon dioxide emitters are essential in inhibiting microbial invasion and expand the package shelf life. Conjugation of nanoparticles with antimicrobial properties like chitosan with gold, metal oxide, or silver nanoparticles has shown great versatility. Since they showed unparalleled antimicrobial effect when combined with other materials. For instance, silver nanoparticles have shown a typical attachment to the cell surface and penetration of microbial cells followed by DNA damage (Honarvar et al. 2016).

2 Encapsulation of Bioactive Molecules

Numerous studies have investigated bioactive encapsulation. Intensive studies have been made on food microbial spoilage and oxidation which affect consumer decisions in buying food. This usually happens to fruits including bananas, kiwi fruit, and tomatoes (Fernández-Pan et al. 2014). One study has used vacuum packaging to produce an oxygen-free environment in food packages (Dey and Neogi 2019). Others fabricated bionanocomposites packaging material using chitosan, zinc oxide nanoparticles, and carboxymethyl cellulose to prevent food spoilage by the effect of bacterial microorganisms (Winestrand et al. 2013). Anti-oxidants molecules are one of the crucial ones, especially in meat packaging. A study has utilized the antioxidant activity of rosemary extract and incorporated it into low-density polyethylene (LDPE) with the presence of α-tocopherol. Further, those materials were transformed into discs and came into contact with fresh beef. It has been reported over time that created a disc of rosemary were able to keep higher values of meat (redness) than control samples over a prolonged period (Moore et al. 2006) (Table 7.2).

3 Micro and Nano-encapsulation Techniques

The development of the micro and nano-encapsulated carriers that are needed for the entrapment of bioactive molecules into food packaging systems could be classified into two opposed approaches; (i) bottom-up, and (ii) top-down (Joye and McClements 2014). There is no optimum approach for encapsulation of all available bioactive compounds (Aguiar et al. 2016). The bottom-up one relies on small particles and elements association in a process that is controlled by the concentration of these monomers, pH of the media, a temperature that is needed for the self-assembly, and ionic strength. The bottom-up principle is found in several methods like electrospinning, spray drying, complex coacervation, and anti-solvent precipitation. This approach is characterized by its less energy consumption, and a good dominator on both shape and size of both micro and nanocarrier. Furthermore, the top-down principle is to focus on the downsizing of the large structure into a smaller one using external mechanical means. The top-down principle is found in techniques like emulsification, and extrusion processes. However, this approach shows

Table 7.2 Studies on food active packaging over the last 10 years

Function	Material	Application	Advantage	Reference
Oxygen scavengers	Nanoiron, silicon matrix	Sachets, films, bottles, containers	Limit the oxygen permeability & extend product shelf-life	Dey and Neogi (2019)
Carbon dioxide emitters	Latex polymer matrix with oxalate oxidase	CO_2 emitter bandages & dressings	Lower growth of pathogenic microorganisms & food spoilage	Winestrand et al. (2013)
Ethylene scavengers	Nanocomposite of polyethylene (PE) with nano-Ag, nano-TiO_2 & montmorillonite	Sachets, films	Prevent fruit ripening & control deterioration	Hu et al. (2011)
Odor absorbers	Essential oils	Films	Odor stabilization	Gutiérrez et al. (2009)
Moisture scavengers	Clay, Silica, & zeolites	Microwavable films, absorbent pads, Sachets	Management of moisture & humidity	Dey and Neogi (2019)
Antimicrobial activity	Silver & gold nanoparticles, & chitosan matrices	Bionanocomposite films	Prevent food spoilage	Youssef Ahmed et al. (2016)
Antioxidant releasers	Rosemary extract, & α-tocopherol incorporated into low-density polyethylene (LDPE)	Films	Delay food ripening	Moore et al. (2006)

less control over both the size and shape of micro and nanocarriers (Jia et al. 2016a). Different factors are affecting the selection of the optimum method for encapsulating the bioactive compounds for food packaging application such as particle size, solubility, and molecular weight of both bioactive compounds and encapsulating agents (Dias et al. 2017).

3.1 *Spray Drying*

One of the techniques that could be used for microencapsulating bioactive compounds is spray drying technique. It was established as a method of encapsulating different bioactive compounds such as; pumpkin seed oil (Ogrodowska et al. 2017), *Agaricus bisporus* extracts (Francisco et al. 2018), chia essential oil (Rodea-González et al. 2012), and *Lippia sidoides* oil (Abreu et al. 2012). It is characterized by its simplicity, rapid performance, as well as low-cost technique. The spray drying technique mode of action (Fig. 7.2) is based on introducing the bioactive compounds with the encapsulating material (solution, emulsion, or suspension in either

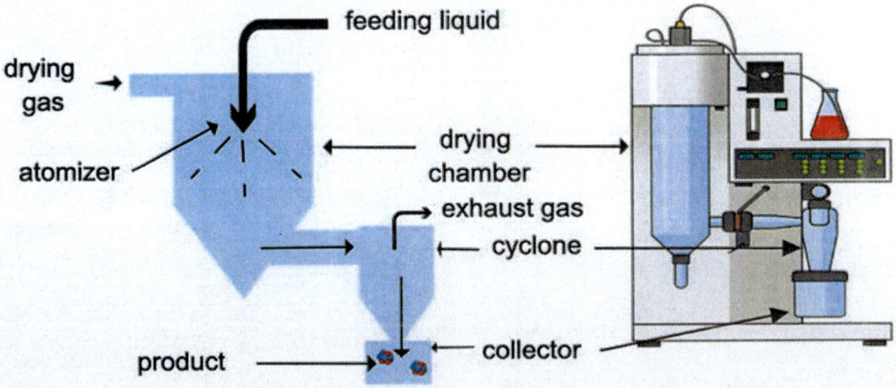

Fig. 7.2 Spray dryer components

aqueous or organic media), in form of fine mist or droplets of, in an atmosphere of hot air. The hot air is previously introduced through a tangential inlet generating as a centrifugation force inside the drying chamber. While spraying the liquid bioactive compounds in the hot atmosphere, the dehydration phase occurs, evaporating the exciting solvents forming a powder of bioactive compounds coated with the encapsulating agent. The drying process accomplished through the separation of the microencapsulated ingredients by the cyclone separator (Nesterenko et al. 2013a). The temperature inside the drying chamber could reach 130–180 °C. The main factor that infuences the encapsulation process is the solubility of the encapsulating material in the solvent. However, this technique suffers from the product low yield due to either adhesion of the dried powder on drying chamber wall or the degradability of the active compounds when they are facing this hot environment (Nesterenko et al. 2013a). The former disadvantage was controlled through generating ultrasound during the atomization phase that improves both encapsulation efficiency, product stability as well as the outcome yield. This is because the ultrasound waves produces less mechanical stress on the sprayed dried product (Tatar Turan et al. 2015). The latter disadvantage was minimized through using of the vacuum drying chamber, lowering the drying temperature to 40–60 °C (Islam et al. 2017).

3.2 *Freeze Drying*

Freeze drying is a common method for that have beed used for microencapsulation process. Basically, it relies on using vacuum for converting the bioactive compounds (especially the thermosensitive substances) from the liquid state into encapsulated powder state. Its mode of action is depending on three main stages; (i) freezing of the polymer-bioactive fluid at a temperature of –50 °C, followed by (ii) primary drying step where solvent start to sublime forming a wet powder, and finally (iii)

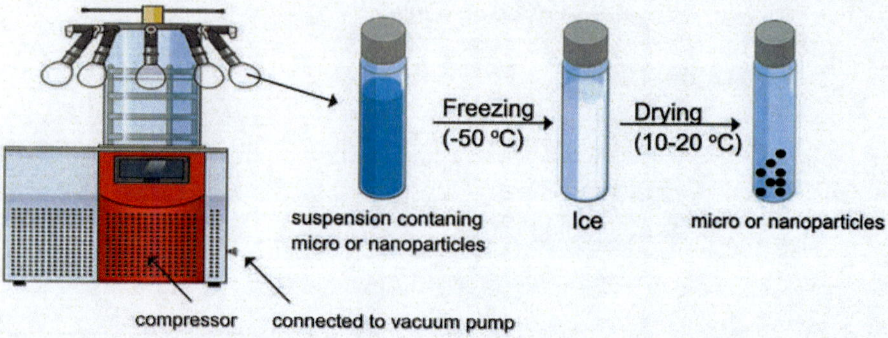

Fig. 7.3 Freeze dryer principle of operation

second drying step at which the remaining solvent on the surface of the powder starts to desorb. The second and third steps are achieved using a low temperature between 10 and 20 °C while applying vacuum till reaching the suitable moisture content (Fig. 7.3) (Abdul-Fattah et al. 2007). The main disadvantages of the freeze dryer is the long drying time that extends for days for a complete drying as well as its high cost (Fuciños et al. 2017a). However, there is another factor besides the drying time that can affect the encapsulation process which is the method of the preparation of the solution, suspension or emulsion that is needed to be freeze dried as it greatly affects the size and the morphology of the micro and nanosystems (Varshosaz et al. 2012). The freeze-drying techniques showed an efficiency in producing micro and nanosystems such as nano lipid carriers (Varshosaz et al. 2012), nanotubes (Fuciños et al. 2017a), and nanoparticles (Shetta et al. 2019; Attallah et al. 2020). Recently, spray freeze drying approach might be better alternative to the traditional freeze dryer by introducing an atomization means before starting the freezing phase (Ishwarya et al. 2015). Spray freeze drying decreases the drying time down to 8 h as well as increases the stability of the bioactive compounds (Hundre et al. 2015).

3.3 *Complex Coacervation*

One of the techniques that could be used for nano and microencapsulation is the complex coacervation approach. It is based on the ionic interaction between the core bioactive compounds and the coating material that are previously dissolved in immiscible solvents (Fig. 7.4) (Timilsena et al. 2017). After this interaction, the coating material starts to envelop the bioactive compounds after adjusting the pH of the media, concentration of both phases, and temperature of the solution. Finally, the resulted coacervate (complex) starts to solidify using thermal or chemical (cross-linking) means (Bakry et al. 2016). The coacervation process could be simple (by using one polymer), or complex (by using multiple polymeric materials) (Yuan et al. 2017a). There are many examples for encapsulating bioactive compounds

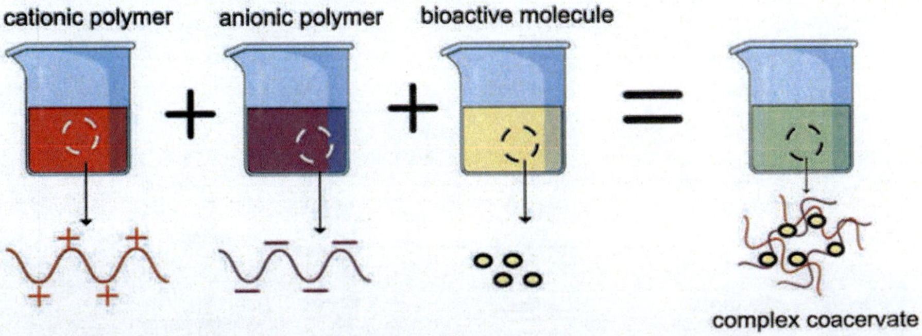

Fig. 7.4 The principle of complex coacervation

using complex coacervation such as; anthocyanin (Arroyo-Maya and McClements 2015), avocado extract (Calderón-Oliver et al. 2020), algal oil (Yuan et al. 2017a) and β-carotene (Jain et al. 2016a). The complex coacervation technique could modify the release of the bioactive ingredientes, and provides high encapsulation efficiency (Rutz et al. 2017). Although, the complex coacervation needs high preparation cost, it shows difficulty in controlling both shape and size of micro and nanosystems (Jia et al. 2016a).

3.4 *Emulsification*

The emulsification technique is widely used for encapsulating bioactive ingredients. The emulsion produced could be either simple emulsion such as oil in water (o/w) or water in oil (w/o), or multiple emulsions such (w/o/w) or (o/w/o). The emulsification technique is used mainly for encapsulation of both polar and nonpolar bioactive ingredients (Gumus et al. 2017). The micro-emulsification process could be performed via low and high energy approach. High energy emulsification could be achieved by homogenization, ultrasonication, and high shear mixer (Fig. 7.5). The high energy approach is based on the formation of coarse emulsion by high shearing followed by the application of high energy to produce micro and nanoemulsions that could be achieved by high pressure (Piorkowski and McClements 2014) and ultrasonic energy (Jafari et al. 2007). There are so many examples for microencapsulation of active molecules such as roasted coffee oil (Freiberger et al. 2015), and D-limonene (Jafari et al. 2007). However, this technique requires sophisticated equipment that uses high energy (Joye and McClements 2014). In contrast, low energy emulsification techniques show simple and economic approach.

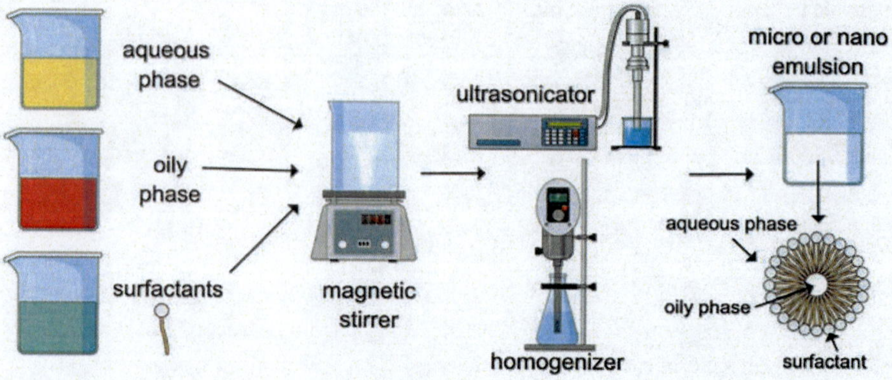

Fig. 7.5 The principle of Emulsification process

3.5 *Antisolvent Precipitation*

One of the main approaches for encapsulating bioactive compounds is the antisolvent precipitation method. It is a technique that is simple with low cost. The principle of this technique is relying on the dissolution of active molecules in binary solvent such as water and organic solvents. Then, a third solvent is added to act as antisolvent that decreases the solubility of the compounds leading to the formation of nanoparticles through the precipitation process (Fig. 7.6) (Jia et al. 2016a). One of the most common antisolvents used is supercritical carbon dioxide which is used for microencapsulation of active molecules such as curcumin molecules (Dias et al. 2017). Moreover, different bioactive molecules were encapsulated using the antisolvent precipitation such as epigallocatechin gallate (Donsì et al. 2017a), eugenol and thymol (Chen et al. 2015), and vitamins (David and Livney 2016). The antisolvent precipitation method could be improved through using of the ultrasound as it enhances the nucleation rates, and decreases agglomeration (Thorat and Dalvi 2012).

3.6 *Extrusion*

The passing of the mixture containing the bioactive compounds dispersed in the polymeric coating material across the nozzle, an area where gelling process occurs is called extrusion approach. It is a method that is used for the encapsulation of small as well as large molecules of both hydrophilic and hydrophobic nature (Fig. 7.7) (Han et al. 2018). Different materials could be encapsulated through this approach such as essential oil from marine resources (Bakry et al. 2016), gallic acid (Li et al. 2017), and riboflavin (He et al. 2015). However, the extrusion technique application is limited due to its inability to control the size. The reason behind that owing to the formation of a large and porous product during the extrusion and

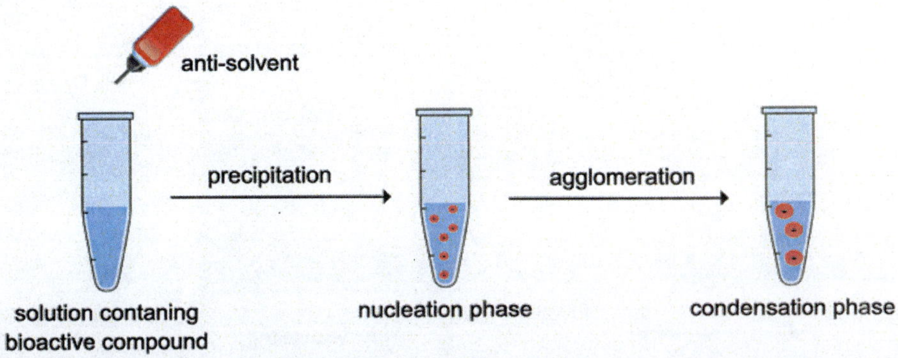

Fig. 7.6 The principle of antisolvent precipitation

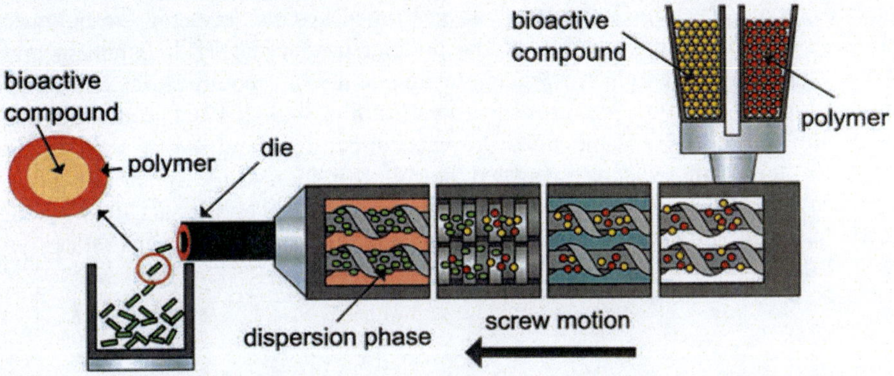

Fig. 7.7 Extrusion technique

gelling steps. Several modifications were applied on the extrusion system to improve its properties such as applying the multi nozzle-system, co-extrusion, and melt extrusion (Rodriguez et al. 2016).

3.7 *Electrospinning and Electro-Spraying*

Electrospinning and electro-spraying are commonly used methods for encapsulating bioactive molecules. They are characterized by the formation of the nanofibers or small droplets from a dispersion containing bioactive molecules and the polymeric material (Fig. 7.8 a, and b). The principle of both techniques depends on applying a voltage over a liquid ejected from a nozzle (Faridi Esfanjani and Jafari 2016). The main difference between both techniques is the physical form produced, in case of the electrospinning techniques, the outcome is nanofibers owing to the high polymer concentration and the stability of the product while in

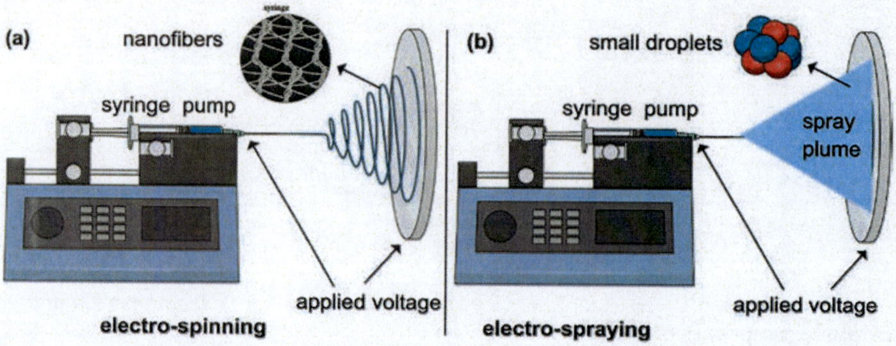

Fig. 7.8 Electro-spinning (**a**), Electro-spraying (**b**)

electrospraying, the small droplets will be formed because of using loe polymer concentration and the instability of the product produced (Faridi Esfanjani and Jafari 2016). Moreover, electrospraying technique doesn't require surfactant, while electrospinning technique requires surfactant (Tarhini et al. 2017). Different factors affect the nanofibers' and the small droplets' dimensions and shapes such as the applied voltage value, distance between the collector and syringe tip, the collection distance and the solution flow rate (Kegere et al. 2019; Wahbi et al. 2020). Both techniques are suitable for the production of micro and nanosystems (Elakkiya et al. 2014).

4 Various Forms of Micro and Nano-encapsulate Carriers Used in Food Packaging

Encapsulating technique is an inclusion approach of bioactive molecules or compounds into polymeric matrix or reservoir that is carried out to increase these compounds' stability. Encapsulation processes mainly depend on making the first droplet of the bioactive molecules as solid (as powder), liquid or gas. Then, these droplets get enclosed by the carrier material through different techniques that include but are not limited to coacervation, co-extruding, emulsifications, fluidized bed coating, spray-drying, spray-cooling, melt injection, etc. (Dubey et al. 2009; Parra 2010; Sanguansri and Augustin 2007; Raybaudi-Massilia and Mosqueda-Melgar 2012).

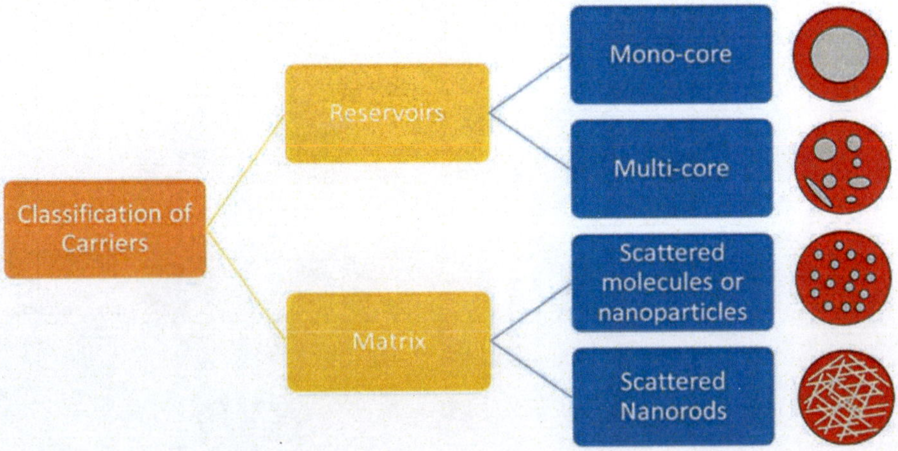

Fig. 7.9 Morphology of different encapsulated systems

4.1 *Reservoir and Matrix*

The encapsulating systems could possess different designs according to the nature and properties of the compounds incorporated as illustrated in Fig. 7.9. In addition, the design is controlled by the encapsulation method used to prepare the carriers either in the micro or nano scale. For instance, they can be classified into **(a) reservoirs** which can be either mono-core, where there is only a single chamber detected in the core of the capsule, or poly-core (multi-core), where there are more than one chamber inside the core that could be equal or different in shape and size, in addition to **(b) matrix** where the encapsulated compounds are scattered within the whole matrix of the carrier. In addition, the encapsulated compounds can be incorporated in their native form or as loaded nanoparticles or nanostructures (Dubey et al. 2009; Parra 2010).

4.2 *Emulsions*

4.2.1 Microemulsions

Microemulsions or traditional emulsions are colloidal systems that are made through mixing at least two immiscible liquids one of them is an aqueous phase while the another one is an oily phase (Brandelli et al. 2016). According to the type of dispersed material and dispersion medium, emulsions are classified into oil-in-water (O/W) emulsions and water-in-oil (W/O) emulsions as shown in (Fig. 7.10).

In O/W emulsions, oil is considered as the dispersed phase, while the continuous phase is the aqueous medium. On the contrary, the dispersed phase of the W/O

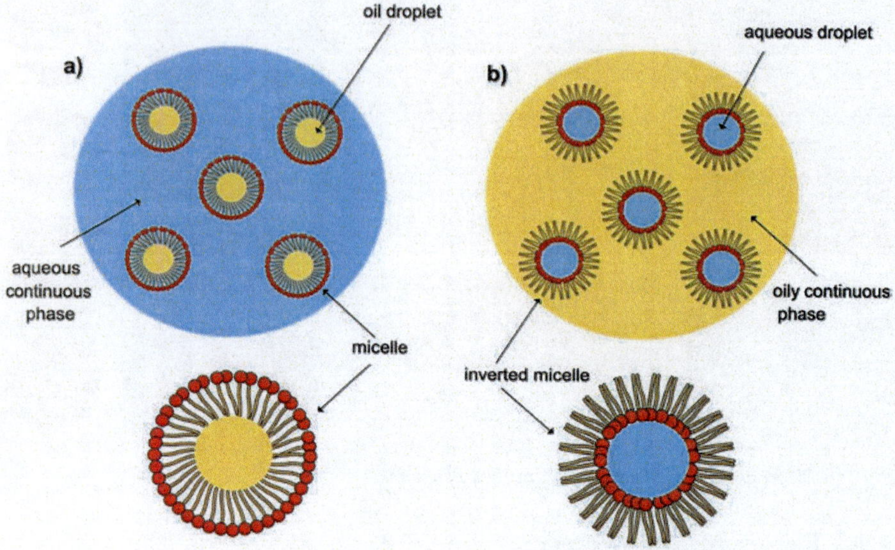

Fig. 7.10 (**a**) Oil-in-Water Emulsion (O/W), and (**b**) Water-in-Oil Emulsion (W/O)

emulsion contains the water droplets, while the oil constitutes the dispersion medium. Emulsions have to be stabilized through the addition of some amphiphilic macromolecules that act as emulsifying agents or stabilizers that act through increasing the steric hindrance, while minimizing the interfacial tension between the two immiscible media (Rayner et al. 2014). However, on long term duration, the dispersed droplets could coalesce again forming larger globules that could end up with complete separation of the two phases of microemulsions (Brandelli et al. 2016).

4.2.2 Nanoemulsions

Nanoemulsions are emulsified systems whose sizes range from 10 to 100 nm. They have the advantage of being more kinetically stable than microemulsions. Therefore, they require less surfactant concentrations as they are less susceptible to aggregation and coalescence. This returns back to the decrease in the attractive forces between the droplets after decreasing their size. Furthermore, stability of nanoemulsions was proven to be increased through addition of 5–10% alcohol as this amount was found to have an influence on decreasing the size of the droplets size owing to the interference with the interfacial tension among the droplets (Zeeb et al. 2014). Nanoemulsions can be prepared through different techniques either: (**a**) low energy techniques, or (**b**) high energy techniques (Gulotta et al. 2014; Ranjan et al. 2014). Table 7.3 summarizes some studies that involved the use of microemulsions and nanoemulsions as carriers incorporating bioactive molecules used in food applications.

Table 7.3 A summary of different incorporated compounds in microemulsions and nanoemulsions carriers

Microemulsion/nanoemulsion	Incorporated compounds	Example	References
Calcium caseinate stabilized flaxseed oil either with or without lecithin	Nutraceuticals	Vitamin D3 and ω-3	Mehmood (2015)
Sodium Alginate		Essential oils	Salvia-Trujillo et al. (2015)
Lipid mixtures such as glycerol monolaurate, glyceryl monostearate, and caprylic capric triglyceride	Antioxidants	Quercetin	Ni et al. (2015)
Ethyl acetate water			
Tween 80 with edible mustard oil	Nutraceuticals	Vitamin E acetate	Dasgupta et al. (2016)
Anhydrous milk fat		β-carotene	Zhang et al. (2013)

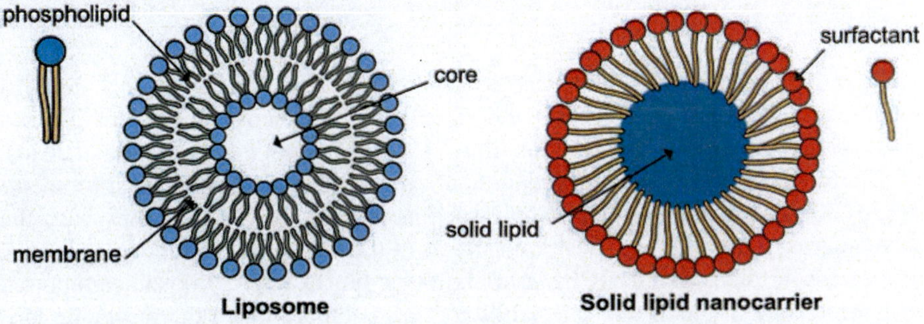

Fig. 7.11 Lipid-based carriers: (**a**) Liposomes (lipid vesicles) and (**b**) Solid lipid nanocarriers

4.3 *Lipid Based Carriers*

Lipid based carriers are fat-based encapsulating systems that are spherical structures capable of being dispersed in aqueous phase ranging from 0.1 to 100 um (Yalavarthi et al. 2014). Lipid based carriers can be mainly categorized into liposomes and solid lipid nanoparticles as shown in (Fig. 7.11).

4.3.1 Liposomes

Liposomes are spherical bilayer phospholipid colloidal systems. They are formed through dispersion of phospholipids in aqueous medium that self-assemble forming the vesicular structures. They are widely utilized as carriers for various bioactive molecules and nutraceuticals (Tan et al. 2013; Brandelli et al. 2017b; Thompson et al. 2006). Several approaches have been proven for their efficiency in preparation of liposomes. These approaches are classified into: (**a**) conventional techniques

Table 7.4 A summary of different compounds incorporated in liposomes carriers

Liposomes	Incorporated compounds	Example	Type of food	References
Phosphatidylcholine	Nutraceuticals	Folic acid, vitamin C, vitamin E, ω-3 and ω-6 fatty acids	Chocolate milk	Marsanasco et al. (2015)
	Bioactive	Bioactive peptides	Sea bream	Mosquera et al. (2014)
Soy phospholipid and whey protein isolate	Antimicrobial agent	Pediocin	–	de Mello et al. (2013)
	Antioxidants	Quercetin	–	Frenzel and Steffen-Heins (2015)
Milk phospholipids		Tea polyphenols	–	Gülseren et al. (2012)
Phospholipids	Antimicrobial agent	Nisin	Fresh fruit-cut	Taylor et al. (2005) and da Silva et al. (2010)

such as detergent removal, solvent evaporation & thin film hydration, and **(b)** emerging techniques such as modified electroformation, microfluidic devices and supercritical fluid procedures (Brandelli et al. 2016, 2017b; Patil and Jadhav 2014).

Liposomes bilayer phospholipid structure mimics the real structure of biomembranes. Therefore, they are commonly used as carriers. In addition, they have the advantage of masking the undesirable effects of the incorporated molecules as well as decreasing their toxicity (Chang et al. 2008). For example, liposomes reported high efficiency in increasing the stability of some incorporated thermo-labile bioactive compounds such as folic acid and many vitamins (Marsanasco et al. 2015). Some of the reported liposomal structure in food applications are summarized in Table 7.4.

4.3.2 Solid Lipid Nanoparticles

Solid lipid nanoparticles are spherical nanostructures composed of high food grade lipids capable of encapsulating hydrophobic compounds. They can be prepared by blending two lipid structures together differing in their melting points e.g. hydrogenated palm oil and cocoa butter. The pros of using solid lipid nanoparticles returns back to their capability of protecting the encapsulated compounds from chemical degradation in addition to their good physical stability. This is attributed to the strong steric repulsion among the particles generated by the non-ionic surfactants within the system (Brandelli et al. 2016; Qian et al. 2013).

4.4 Protein-Based Carriers

Proteins are natural biopolymers that could be extracted from either animal or plant sources. Proteins generally are insoluble in acidic medium while soluble in alkaline medium (Chen et al. 2006). Proteins are highly swellable materials and easily functionalized, therefore, they are considered as promising biomaterials for preparation of bioactive delivery systems for wide range of bioactive compounds such as fatty acids, fats, oils, and flavors (Jia et al. 2016b). Proteins extracted from animals include casein, collagen, gelatin, and whey proteins, while those obtained from plants include gliadin, soy and zein proteins (Tarhini et al. 2017). Compared to animal derived proteins, plant derived proteins have recently attracted many researches attention owing to their lower cost and minimized inflammatory properties (Nesterenko et al. 2013b). Many proteins have been reported as successful carrier systems for bioactive compounds.

4.4.1 Caseins

Caseins are the highest abundant proteins in milk. They are divided into α_{s1}-casein, α_{s2}- casein, β-casein, and κ-casein. They are characterized by possessing low solution viscosity, excellent emulsifying capabilities, in additional to their rich nutritional value. Hence, they are considered as promising materials for delivery systems preparation (Ho et al. 2017; Jain et al. 2016b).

Caseins are effective in encapsulating hydrophobic bioactive molecules such as β-carotene, vitamin D_3, essential oils and naringenin (Jarunglumlert et al. 2015; Moeiniafshari et al. 2015; Chen et al. 2014).

Furthermore, caseinates salts of calcium and sodium are also obtained through their solubilization in alkaline medium. They have the advantage of being highly water soluble, thus enhancing the dispersibility of highly hydrophobic bioactive compounds (Jarunglumlert et al. 2015).

In addition, caseins can form micellar structures as colloidal particles that are formed from caseinates solutions upon the addition of some ions such as calcium, phosphate or citrate (Shishir et al. 2018).

4.4.2 Cereal Proteins

Zein proteins, the most important cereal proteins, are classified into alpha, beta, gamma, and delta. Zein proteins are hydrophobic water insoluble proteins that could self-assemble forming various structures according to the solvents used. Therefore, zein proteins are considered as promising candidates for preparing carrier systems for bioactive compounds (Dai et al. 2017; Donsì et al. 2017b).

Wheat proteins are composed of gliadin and glutenin. Although, they have low water solubility, high allergenic properties, and high susceptibility to celiac disease, they have been reported to be incorporated within encapsulation systems either alone or with polysaccharides. Therefore, wheat gluten has been modified through acid heating deamination in order to minimize its cytotoxicity and enhance their entire performance. For instance, succinic acid deamidated wheat gluten microspheres has been prepared to encapsulate fish oil through double emulsion technique. The produced deamidated wheat gluten microspheres proved their effeciency in controlling release and maintaining the stability of fish oil (Sun et al. 2009; Liao et al. 2012; Andreani et al. 2009).

Potato protein has been recently utilized as biomaterials for carriers preparation owing to its good antioxidant activity, high emulsifying ability, low cost, and low non-allergenic properties. Potato protein based nano carriers have been used for the efficient encapsulation of hydrophobic bioactive molecules such as vitamin D. The prepared potato protein based carriers protected vitamin D and extended its shelf-life when tested under different storage conditions (David and Livney 2016; Waglay et al. 2014).

4.4.3 Gelatins

Gelatins are animal derived proteins that could be obtained through partial either acid or alkaline hydrolysis of collagen. The hydrolysis could undergo enzymatically as well. Gelatins are classified into type A and type B according to the method of preparation. For instance, Gelatin type A is obtained from the skin of bovine, porcine or fish via acid hydrolysis procedure. However, gelatin type B is extracted from the bones of bovine, fish, or porcine through alkaline hydrolysis procedure (Shishir et al. 2018; Patel et al. 2008).

Gelatins have been reported as good biomaterials for preparation of carrier systems owing to their high biocompatibility, biodegradability as well as water retention ability. In addition, they are characterized by being non-carcinogenic and non-immunogenic biopolymer (Shishir et al. 2018; Chen et al. 2017).

Gelatin has proven their effeciency as biomaterials for the preparation of carrier systems incorporating wide range of bioactive compounds e.g. antimicrobial agents, antioxidants as well as some nutraceuticals (Chen et al. 2017; Gómez-Mascaraque et al. 2017).

4.4.4 Soy Proteins

Soy proteins are the highest portions of legume proteins. Soy proteins are considered to be promising biomaterials owing to their good physicochemical properties in terms of emulsification, fat absorption, gel formation, water binding capability, and antioxidant ability. Soy proteins consist of glycinin (11S globulin) and conglycinin (7S globulin). Carriers made of soy protein have been used to encapsulate diffferent bioactive compounds such as algal oil, curcumin, tomato oleoresin, paprika oleoresin, and lycopene (Ho et al. 2017; Dai et al. 2017; Lia et al. 2015).

In addition, soy proteins have been used with polysaccharides for preparation of blended encapsulate systems that showed enhanced stability and anti-oxidative properties. For instance, a study showed that a carrier system composed of soy protein and chitosan together was a successful encapsulate of algal oil with enhanced encapsulation efficiency reaching around 97.36% as well as high oxidative stability compared to soy protein alone. In addition, another study showed that the use of soy protein and gum acacia together to encapsulate tomato oleoresin showed better emulsifying features, higher biocompatibility as well as encapsulation efficiency than single soy protein carrier systems. Furthermore, conjugated soy protein with some polysaccharides revealed good protection ability of lycopene from light, temperature and humidity during storage conditions. In addition, the system showed higher capability to control the lycopene release within gastric conditions (Shishir et al. 2018; Lia et al. 2015; Yuan et al. 2017b).

4.4.5 Whey Proteins

Whey proteins are characterized by having superior biological characteristics especially emulsifying and gelling properties. Therefore, they are typically used in hydrogels preparations as well as nanoparticle systems through their conjugation with several polysaccharides. For instance, β-lactoglobulin was utilized in different forms i.e. pre-heat treated, cross-linked or even untreated form to incorporate various bioactive molecules e.g. sour cherries anthocyanins. It was found that cross-linked protein form reached the highest encapsulation efficiency up to around 64.69%. While both the cross-linked and pre-treated forms showed higher protection effeciency for anthocyanins against the gastric juice in the stomach compared to the untreated form. Consequently, this allowed the anthocyanins be released into the gut (Oancea et al. 2017). On the other hand, α-lactalbumin nanotubes were used successfully to incorporate caffeine with encapsulation efficiency of almost 100%. This proves the high capability of α-lactalbumin nanotubes to encapsulate bioactive compounds (Fuciños et al. 2017b).

4.5 Polysaccharide Micro and Nanocarriers

Polysaccharides, long polymeric chains of monosaccharides linked together by glycosidic linkages, can be used as edible encapsulating polymeric matrix through either microencapsulation or nanoencapsulation approaches according to the final dimensions of the prepared formulation. Polysaccharides have been reported as promising encapsulating systems for incorporating both additive and bioactive compounds used in food packaging systems. Among the most commonly used encapsulated additive and bioactive materials used in food packaging systems are: **(a)** natural nutraceuticals e.g. minerals, probiotics & vitamins, **(b)** antimicrobial agents, **(c)** antioxidants, and **(d)** anti-softening agents (Raybaudi-Massilia and Mosqueda-Melgar 2012; Tapia et al. 2007; Rojas-Graü et al. 2009).

Encapsulation of the above-mentioned compounds have the advantage of: **(a)** preserving & protecting them against the surrounding environment, **(b)** prolonging their shelf life, and **(c)** enhancing their safety and sensory features (Dubey et al. 2009; Parra 2010; Raybaudi-Massilia and Mosqueda-Melgar 2012). The efficiency of these encapsulated systems has been reported for wide range of various food products including dairy products, fresh fruit & vegetables cuts, as well as meat and poultry products (Olivas and Barbosa-Cánovas 2005; Min and Ahn 2005; Waghmare 2020; Cerqueira et al. 2009). Many polysaccharides have been reported for their efficiency in encapsulating various additives and bioactive agents mandatory in food packaging.

4.5.1 Cellulose & Cellulose Derivatives

Cellulose, the first most abundant polymer in nature, as well as its derivatives including carboxymethyl cellulose (CMC), hydroxylpropyl methyl cellulose (HPMC), and methyl cellulose (MC) have been reported as efficient polysaccharides for encapsulating antimicrobial agents in food packaging to prevent the growth of microorganisms and prolong the shelf life of packaged food. For example, cellulose based systems incorporating various antimicrobial compounds e.g. nisin and Pediocin, an antimicrobial peptide produced by *Pediococcus sp.,* have shown good antimicrobial potency against *L. monocytogenes, L. innocua, and Salmonella* sp. This has been proven when tested on fresh fruits cuts such as strawberry and sliced meat products (Nguyen et al. 2008; Park et al. 2005).

In addition, CMC and HPMC based carriers of some antimicrobial agents such as sodium benzoate, sodium propionate, potassium sorbate, etc. have shown a high inhibitory effect of fungal and mold growth such as *Penicillum digitatum* and *Penicilllum italicum* when tested with fresh pistachio and oranges (Sayanjali et al. 2011; Valencia-Chamorro et al. 2009). Furthermore, HPMC based carrier systems incorporating some antioxidants that have shown good inhibitory effects on lipid oxidation of almond.

4.5.2 Chitosan

Another example for polysaccharides involved in encapsulation of antimicrobials in food packaging applications, is chitosan, which is obtained from chitin. Chitin is counted as the second most abundant natural polysaccharide after cellulose.

Chitosan has been reported for its ability to encapsulate different antimicrobial agents during food packaging e.g. acetic acid, sodium benzoate, sodium diacetate, sodium lactate, potassium sorbate, propionic acid, lauric acid, cinnamaldehyde, vanillin, lysozymes, nisin, etc. (Dutta et al. 2009; Ouattara et al. 2000; Duan et al. 2007; Ye et al. 2008). Chitosan based systems have shown high efficiency in protecting salmon, meat products and fresh fruit cuts against various microorganisms. These include bacteria, fungi and molds such as *Lactobacillus sakei, Serratia liquefaciens, L. monocytogenes, Pseudomonas Fluorescens, Saccharomyces cerevisiae, Escherichia coli, Cladosporium sp., Rhizopus sp*, among others (Ojagh et al. 2010; Fajardo et al. 2010; Siripatrawan and Noipha 2012).

4.5.3 Alginate

Alginates, anionic polysaccharide chains obtained from brown algae cell walls, have been used as encapsulating systems for some wide range of antimicrobial agents such as nisin, lysozymes, sodium diacetate, sodium lactate, enterocins A and B, etc., to protect different meat and poultry products against wide range of microorganisms such as *Brochothrix thermosphacta, Salmonella enterica, Staphylococcus aureus, Listeria monocytogenes* (Natrajan and Sheldon 2000; Millette et al. 2007; Datta et al. 2008; Neetoo et al. 2010; Marcos et al. 2008).

Furthermore, alginate based encapsulating systems have been reported as promising carriers for some essential oils as well as their active ingredients such as Spanish oregano, Chinese cinnamon, winter savory, lemongrass, oregano, vanillin, palmarose, eugenol, geraniol and citral. They showed high efficiency in protecting meat & poultry products as well as fresh fruit cuts in refrigerators against wide range of microorganisms such as *Salmonella enterica, Listeria innocua, Escherichia coli,* native flora, in addition to mesophilic and psychotropic bacteria, molds and yeasts (Oussalah et al. 2006, 2007; Rojas-Graü et al. 2007; Raybaudi-Massilia et al. 2008a; Olivas et al. 2007; Seol et al. 2009).

4.5.4 Starch

Starch, composed of polysaccharide chains consisting of glucose monomers linked together via α 1,4 linkages, based carriers systems incorporating antioxidants agents such as rosemary oleoresin, tocopherols, stearic acid, etc., reported inhibitory effect on lipid oxidation and moisture loss in food packaging (Wu et al. 2000, 2001; Hargens-Madsen et al. 1995). Tables 7.5 and 7.6 summarize the use of polysaccharides as carriers.

Table 7.5 A summary of different compounds incorporated in Chitosan carriers

Polysaccharide	Incorporated compounds	Example	Type of food	References
Chitosan	Antimicrobial agents	Lysozymes	Mozzarella cheese	Duan et al. (2010)
		Potassium sorbate	Whole strawberry	Park et al. (2005)
		Potassium sorbate, sodium lactate and diacetate	Roasted turkey, cold smoked salmon, and ham steaks	Ye et al. (2008) and Jiang et al. (2011a, b)
		Vanillin	Fresh-cut pineapple and melon	(Sangsuwan et al. 2008)
	Nutraceuticals	Omega 3 and vitamin E	Lingcod fillets	Raybaudi-Massilia and Mosqueda-Melgar (2012) and Duan et al. (2007)
		Calcium	Whole strawberry	Hernández-Muñoz et al. (2006)
		Calcium and vitamin E	Whole strawberry and red raspberry	Han et al. (2004)

Table 7.6 A summary of bioactives incorporated in alginate, cellulose & its derivatives, and starch carriers

Polysaccharide	Incorporated compounds	Example	Type of food	References
Alginate	Antimicrobial agents	Sodium lactate and diacetate	Roasted turkey slices, fillets, and cold-smoked salmon slices	Neetoo et al. (2010) and Jiang et al. (2011a)
		Malic acid, potassium sorbate, vanillin and essential oils of lemongrass, oregano, cinnamon, clove, cinnamaldehyde, eugenol, citral, eyc	Fresh-cut apple	Rojas-Graü et al. (2007), Raybaudi-Massilia et al. (2008a), and Olivas et al. (2007)
		Malic acid and essential oils of cinnamon, citral, eugenol, lemongrass, plamarose, geraniol	Fresh-cut melon	Raybaudi-Massilia et al. (2008b)

(continued)

Table 7.6 (continued)

Polysaccharide	Incorporated compounds	Example	Type of food	References
	Antioxidants	Glutathione and N-Acetyl-cysteine, Calcium Chloride	Fresh-cut apple	Tapia et al. (2007), Olivas et al. (2007), and Raybaudi-Massilia et al. (2008b)
		Glutathione and N-Acetyl-cysteine	Fresh-cut pears	Oms-Oliu et al. (2008)
		Tea polyphenols and vitamin C	Bream (fresh water fish)	Song et al. (2011)
	Anti-softening agents	Calcium chloride and calcium lactate	Fresh-cut apples, pears, melons and papayas	Tapia et al. (2007, 2008), Rojas-Graü et al. (2007), Raybaudi-Massilia et al. (2008a), Olivas et al. (2007), Oms-Oliu et al. (2008), and Tapia (2008)
	Nutraceuticals	Probiotics	Fresh-cut papayas	Tapia et al. (2007)
HPMC	Antimicrobial agents	Sodium benzoate, sodium propionate, and potassium sorbate	Whole strawberry and oranges	Park et al. (2005) and Valencia-Chamorro et al. (2009)
	Antioxidants	Ascorbic, citric and EO ginger	Toasted almond	Atarés et al. (2011)
CMC		Ascorbic acid and TBHQ	Fresh-cut apples and potatoes	Baldwin et al. (1996)
MC		Ascorbic acid	Fresh-cut apples	Raybaudi-Massilia and Mosqueda-Melgar (2012)
Starch	Antimicrobial agents	Sodium lactate and diacetate	Roasted Turkey	Jiang et al. (2011a)
		Green tea extract	Pork slices, fruit-based salad, and romaine hearts	Chiu and Lai (2010)

5 Technological Challenges, Food Integrity and Regulatory Manifestations in Encapsulation

5.1 Technological Challenges in Preparation of Micro and Nanoencapsuled Systems

Although micro- and nanoencapsulation techniques have been widely used to prepare successful carrier systems for bioactive compounds, not all of them have been applied on large scale for pharmaceutical and nutraceutical applications. This returns back to the necessity of more long term and extensive studies towards these techniques to assure their credibility to be used on large scale. The used techniques should be tested again in terms of their impact on the chemical and physical stabilities of both the encapsulating systems and encapsulated compounds. For instance, concerning the newly emerged techniques, only freeze drying and spray drying are the most extensively used techniques in the industrial production (de Souza et al. 2017; Đorđević et al. 2016).

Even the traditional techniques still need more investigations to overcome their drawbacks and maximize their benefits. For example, it is advisable to discover more biomaterials that could be used in the encapsulate development. Furthermore, studies should be conducted to investigate the limitation of each materials and study how these could be improved (Shishir et al. 2018).

For instance, it is highly challengeable to incorporate liposomes in food owing to the high semi-permeability while low physical stability of the membranes. For successful preparation of liposomal structures as carrier systems in food packaging, it is mandatory to understand deeply the stability and solubilization features of the encapsulated compounds as well as their effect on the structure and properties of bilayer membrane of lipids in terms of fluidity, molecule dynamics and micropolarity (Brandelli et al. 2016; Frenzel and Steffen-Heins 2015; Yoshimoto et al. 2007).

Another technological challenge in encapsulate system is the tailoring of the ratio between oily and aqueous phases during the preparation of emulsions. In addition, the selection of the appropriate type of surfactant or the emulsifying agent as well as the proper required concentrations are considered as challenging aspects (Brandelli et al. 2016; Rayner et al. 2014).

5.2 Safety Aspects

Nanotechnology applications have gained high popularity in encapsulation approach of bioactive compounds in food and pharmaceutical industry. Although they showed promising and interesting solutions to overcome the drawbacks of using the bare bioactive compounds, some environmental and safety concerns have been raised during the past few years. The raised fears are related to the effect of nano- sized delivery systems on both the environment and the human health. The point of view for many researchers is attributed to the nanoscale systems potential toxicity risk on both the human hygiene and the surrounding environment (Katouzian et al. 2017).

Therefore, safety information concerning the size of the nanoencapsulated systems should be furtherly assessed especially their long-term toxicity (Katouzian et al. 2017; He and Hwang 2016; Ezhilarasi et al. 2013).

For instance, the small size of nanoparticles allows them to cross the biological barriers easily leading to probable interference with some biological reaction resulting in possible toxicological effects. Furthermore, the high surface area and consequently the high reactivity of the small sized nanoparticles could lead to acute inflammation and irritation (Jafari et al. 2017; Zhang et al. 2007; Fröhlich and Roblegg 2012).

Finally, the chemicals such as surfactants, stabilizing agents, etc. could have some toxicological effects on cells. Therefore, it is advisable to use green and naturally derived and FDA approved biomaterials during the preparation procedures. Moreover, it has been proven through cell viability assessments that coating of the prepared nanoparticles leads to decreasing the adverse effects on the cells (Shishir et al. 2018; He and Hwang 2016; Cruz et al. 2015; El Badawy et al. 2011; Seeli and Prabaharan 2016).

5.3 Regulatory Aspects

Unfortunately, the regulatory aspect concerning the use of nanoparticle systems in food and drug industry has not been firmly legislated globally till now. Most of the countries do not have particular roles to estimate the risk of the encapsulated nanoproducts or even to limit or minimize their risk factors (de Souza et al. 2017; Katouzian et al. 2017).

However, only the European Union (EU) has started recently to initiate clear regulatory vision for the use of nano- materials in industry (de Souza et al. 2017). Furthermore, Regulation (EU) no. 10/2011 states that nanoparticles cold lead to different toxicological features. Therefore, they need to be assessed on a case-by-case basis. For instance, nanoparticles can be utilized only if they have been approved or mentioned in Annex I of Regulation (EU) no. 10/2011.

The US Food and Drug Administration (FDA) does not possess precise principle relating the nanomaterials that could be used in the food and drug industry. However, it has issued guidelines concerning the regulatory and safety manifestations in new food industry technological approaches entitled "Draft Guidance for Industry" (Duvall and Knight 2011). The guidance stated some responsibilities that should be considered by the industrial sectors which include: **(a)** monitoring the changes that could occur to the food materials in terms of appearance of impurities or alteration in physicochemical properties, **(b)** evaluating the food products safety after being modified, **(c)** submitting a regulatory assessment to US FDA, and **(d)** specifying a regulatory concern for the new food products consumption. USA FDA states that the current law protocols are proper for the assessment of nanomaterials safety (Shishir et al. 2018; Katouzian et al. 2017).

6 Future Trends

Active food packaging through nano and micro-encapsulation is a method that has been gradually integrated into food industries to improve the quality and safety of food. Numerous gaps are still existing and hinder the development of this widely growing field. It needs to be easily accessible by industry and consumers. Safety and regulatory matters are being improved by the national organizations to establish convenient management for the progression of active food packaging.

The great challenge is to come with a product that is safe, high in quality, with extended shelf life, and able to prevent active microorganisms, moisture, gases, and mechanical forces. Thus, the process of active packaging and intelligent packaging has progressed to overcome those challenges for safer and healthier food.

Food pathogenic microorganisms present one of the serious problems that shortened the product shelf life and lead to food spoilage, and so reducing its quality. Multiple methods have come with promising ways to incorporate bioactive materials (e.g., essential oils) that have antimicrobial activity. Nevertheless, an urgent need to develop new substances with higher features is essential to get the most use of that active materials in the food industry. In this regard, the role of nanotechnology is an attractive system to improve and fulfill the different related issues.

The utilization of antimicrobial materials is remarkable since it allows the control of the pathogenic microorganisms by delaying or terminating their actions. This method can guarantee microbiologically safe food with no chemical additives and prolonged shelf-life. Multiple plant extracts showed magnificent antimicrobial behavior without negatively affecting the organoleptic properties of the food and so-called as Generally Recognized As Safe" (GRAS).

Nanotechnology has shown outstanding progression in numerous sectors including food technology. This extended to involve different aspects of active food packaging (e.g. controlled release of the delivery systems, nano-sensors for food safety, and nanoencapsulation or emulsion for low stability bioactive. Although nanotechnology strategy is promising in many sectors, food is not progressed as like as other fields (e.g., cosmetics). Extensive research studies are substantially needed to evaluate the efficiency and biosafety of such novel packaging in food through regulator bodies (e.g., the Food and Drug Administration). Scientists still required a research's increase and knowledge in the field of active food packaging by investigating the control of the release kinetics and the development of products that economically feasible.

7 Conclusion

The encapsulation of the bioactive compounds helps in preserving their activities, enhancing the stability, and improving the release profile. Different materials could be used for the microencapsulataion of the bioactive ingredients such as polysaccharides, proteins, and lipids. The successful encapsulation in either micro or nano-forms relies on different properties of both encapsulating agents and bioactive compounds. Different encapsulating techniques such as spray drying, molecular inclusion, freeze-drying, and emulsification are used to prepare both micro and nano-systems as well. Nanofibers are considered one of the main platforms that could be used to to encapsulate the bioactive ingredients in food packaging. Future studies are needed to solve any problems regarding micro and nano encapsulation techniques to enhance their efficiency as well as scalability.

References

A concise guide to active agents for active food packaging | Elsevier Enhanced Reader [Internet]. [cited 2020 Nov 30]. Available from: https://reader.elsevier.com/reader/sd/pii/S0924224418302760?token=022AB9E1FF4BBDFD1F36E75608D69A35F711C9EAD3A6DAD330D36E5DB5D6887A3E7612D769D095AE946C531685EFDA1C

Abdul-Fattah AM, Kalonia DS, Pikal MJ. The challenge of drying method selection for protein pharmaceuticals: product quality implications. J Pharm Sci. 2007;96(8):1886–916.

Abreu FO, Oliveira EF, Paula HC, de Paula RC. Chitosan/cashew gum nanogels for essential oil encapsulation. Carbohydr Polym. 2012;89(4):1277–82.

Aguiar J, Estevinho BN, Santos L. Microencapsulation of natural antioxidants for food application – the specific case of coffee antioxidants – a review. Trends Food Sci Technol. 2016;58:21–39.

Ali MR, Said RME. Assessment of the potential of Arabic gum as an antimicrobial and antioxidant agent in developing vegan "egg-free" mayonnaise. J Food Saf. 2020;40(2):e12771.

Amenta V, Aschberger K, Arena M, Bouwmeester H, Botelho Moniz F, Brandhoff P, et al. Regulatory aspects of nanotechnology in the agri/feed/food sector in EU and non-EU countries. Regul Toxicol Pharmacol. 2015;73(1):463–76.

Andreani L, Cercená R, Ramos BGZ, Soldi V. Development and characterization of wheat gluten microspheres for use in a controlled release system. Mater Sci Eng C. 2009;29(2):524–31.

Arroyo-Maya IJ, McClements DJ. Biopolymer nanoparticles as potential delivery systems for anthocyanins: fabrication and properties. Food Res Int. 2015;69:1–8.

Atarés L, Pérez-Masiá R, Chiralt A. The role of some antioxidants in the HPMC film properties and lipid protection in coated toasted almonds. J Food Eng. 2011;104(4):649–56.

Attallah OA, Shetta A, Elshishiny F, Mamdouh W. Essential oil loaded pectin/chitosan nanoparticles preparation and optimization *via* Box–Behnken design against MCF-7 breast cancer cell lines. RSC Adv. 2020;10(15):8703–8.

Bakry AM, Abbas S, Ali B, Majeed H, Abouelwafa MY, Mousa A, et al. Microencapsulation of oils: a comprehensive review of benefits, techniques, and applications. Compr Rev Food Sci Food Saf. 2016;15(1):143–82.

Baldwin EA, Nisperos MO, Chen X, Hagenmaier RD. Improving storage life of cut apple and potato with edible coating. Postharvest Biol Technol. 1996;9(2):151–63.

Ben-Yehoshua. S. Individual seal-packaging of fruit and vegetables in plastic film-a new postharvest technique. HortScience. 1985;20(1):32–7.

Bhat M, Sharma A, Rao NN. Carrageenan-based edible biodegradable food packaging: a review. J Food Sci Nutr. 2020;(5):69–75.

Brandelli A, Brum LFW, dos Santos JHZ. Nanobiotechnology methods to incorporate bioactive compounds in food packaging. 2016. p. 27–58.

Brandelli A, Ritter AC, Veras FF. Antimicrobial activities of metal nanoparticles. In: Mahendra R, Ranjita S, editors. Metal nanoparticles in pharma [Internet]. Cham: Springer; 2017a. p. 337–63. [cited 2020 Dec 1]. Available from: https://doi.org/10.1007/978-3-319-63790-7_15.

Brandelli A, Brum LFW, dos Santos JHZ. Nanostructured bioactive compounds for ecological food packaging, Environmental Chemistry Letters, vol. 15. Springer; 2017b. p. 193–204.

Calderón-Oliver M, Escalona-Buendía HB, Ponce-Alquicira E. Effect of the addition of microcapsules with avocado peel extract and nisin on the quality of ground beef. Food Sci Nutr. 2020;8(3):1325–34.

Cerqueira MA, Lima ÁM, Souza BWS, Teixeira JA, Moreira RA, Vicente AA. Functional polysaccharides as edible coatings for cheese. J Agric Food Chem. 2009;57(4):1456–62.

Chang CC, Liu DZ, Lin SY, Liang HJ, Hou WC, Huang WJ, et al. Liposome encapsulation reduces cantharidin toxicity. Food Chem Toxicol. 2008;46(9):3116–21.

Chang SK, Alasalvar C, Shahidi F. Superfruits: phytochemicals, antioxidant efficacies, and health effects – a comprehensive review. Crit Rev Food Sci Nutr. 2019;59(10):1580–604.

Chen L, Remondetto GE, Subirade M. Food protein-based materials as nutraceutical delivery systems, Trends in Food Science and Technology, vol. 17. Elsevier; 2006. p. 272–83.

Chen H, Zhang Y, Zhong Q. Physical and antimicrobial properties of spray-dried zein-casein nanocapsules with co-encapsulated eugenol and thymol. J Food Eng. 2014;144:93–102.

Chen H, Zhang Y, Zhong Q. Physical and antimicrobial properties of spray-dried zein–casein nanocapsules with co-encapsulated eugenol and thymol. J Food Eng. 2015;144:93–102.

Chen H, Yang M, Shan Z, Mansouri S, May BK, Chen X, et al. On spray drying of oxidized corn starch cross-linked gelatin microcapsules for drug release. Mater Sci Eng C. 2017;74:493–500.

Chiu PE, Lai LS. Antimicrobial activities of tapioca starch/decolorized hsian-tsao leaf gum coatings containing green tea extracts in fruit-based salads, romaine hearts and pork slices. Int J Food Microbiol. 2010;139(1–2):23–30.

Crisosto CH, Garner D, Doyle J, Day KR. Relationship between fruit respiration, bruising susceptibility, and temperature in sweet cherries. HortScience. 1993;28(2):132–5.

Cruz Z, García-Estrada C, Olabarrieta I, Rainieri S. Lipid nanoparticles: delivery system for bioactive food compounds. In: Microencapsulation and microspheres for food applications. Elsevier Science Ltd.; 2015. p. 313–31.

da Silva MP, Daroit DJ, Brandelli A. Food applications of liposome-encapsulated antimicrobial peptides, Trends in Food Science and Technology, vol. 21. Elsevier; 2010. p. 284–92.

Dai L, Sun C, Li R, Mao L, Liu F, Gao Y. Structural characterization, formation mechanism and stability of curcumin in zein-lecithin composite nanoparticles fabricated by antisolvent coprecipitation. Food Chem. 2017;237:1163–71.

Dasgupta N, Ranjan S, Mundra S, Ramalingam C, Kumar A. Fabrication of food grade vitamin E nanoemulsion by low energy approach, characterization and its application. Int J Food Prop. 2016;19(3):700–8.

Datta S, Janes ME, Xue QG, Losso J, La Peyre JF. Control of Listeria monocytogenes and Salmonella anatum on the surface of smoked salmon coated with calcium alginate coating containing oyster lysozyme and nisin. J Food Sci. 2008;73(2):M67–71.

David S, Livney YD. Potato protein based nanovehicles for health promoting hydrophobic bioactives in clear beverages. Food Hydrocoll. 2016;57:229–35.

de Mello MB, da Silva MP, Brandelli A, da Silveira NP, Jantzen MM, de Souza da Motta A. Characterization and Antilisterial effect of phosphatidylcholine Nanovesicles containing the antimicrobial peptide Pediocin. Probiotics Antimicrob Proteins. 2013;5(1):43–50.

de Souza SL, Madalena DA, Pinheiro AC, Teixeira JA, Vicente AA, Ramos ÓL. Micro- and nano bio-based delivery systems for food applications: in vitro behavior, Advances in Colloid and Interface Science, vol. 243. Elsevier B.V; 2017. p. 23–45.

Devi N, Sarmah M, Khatun B, Maji TK. Encapsulation of active ingredients in polysaccharide–protein complex coacervates, Advances in Colloid and Interface Science, vol. 239. Elsevier B.V; 2017. p. 136–45.

Dey A, Neogi S. Oxygen scavengers for food packaging applications: a review. Trends Food Sci Technol. 2019;90:26–34.

Dias DR, Botrel DA, Fernandes RVDB, Borges SV. Encapsulation as a tool for bioprocessing of functional foods. Curr Opin Food Sci. 2017;13:31–7.

Donsì F, Voudouris P, Veen SJ, Velikov KP. Zein-based colloidal particles for encapsulation and delivery of epigallocatechin gallate. Food Hydrocoll. 2017a;63:508–17.

Donsì F, Voudouris P, Veen SJ, Velikov KP. Zein-based colloidal particles for encapsulation and delivery of epigallocatechin gallate. Food Hydrocoll. 2017b;63:508–17.

Đorđević V, Paraskevopoulou A, Mantzouridou F, Lalou S, Pantić M, Bugarski B, et al. Encapsulation technologies for food industry, Food Engineering Series. Springer; 2016. p. 329–82.

Duan J, Park SI, Daeschel MA, Zhao Y. Antimicrobial chitosan-lysozyme (CL) films and coatings for enhancing microbial safety of mozzarella cheese. J Food Sci. 2007;72(9):M355–62.

Duan J, Cherian G, Zhao Y. Quality enhancement in fresh and frozen lingcod (Ophiodon elongates) fillets by employment of fish oil incorporated chitosan coatings. Food Chem. 2010;119(2):524–32.

Dubey R, Shami TC, Rao B. Microencapsulation technology and applications. Def Sci J. 2009;59(1):82–95.

Dutta PK, Tripathi S, Mehrotra GK, Dutta J. Perspectives for chitosan based antimicrobial films in food applications. Food Chem. 2009;114(4):1173–82.

Duvall MN, Knight K. FDA regulation of nanotechnology. Food Drug Agency FDA USA; 2011.

El Badawy AM, Silva RG, Morris B, Scheckel KG, Suidan MT, Tolaymat TM. Surface charge-dependent toxicity of silver nanoparticles. Environ Sci Technol. 2011;45(1):283–7.

Elakkiya T, Malarvizhi G, Rajiv S, Natarajan TS. Curcumin loaded electrospun Bombyx mori silk nanofibers for drug delivery. Polym Int. 2014;63(1):100–5.

Ezhilarasi PN, Karthik P, Chhanwal N, Anandharamakrishnan C. Nanoencapsulation techniques for food bioactive components: a review, Food and Bioprocess Technology, vol. 6. Springer; 2013. p. 628–47.

Fajardo P, Martins JT, Fuciños C, Pastrana L, Teixeira JA, Vicente AA. Evaluation of a chitosan-based edible film as carrier of natamycin to improve the storability of Saloio cheese. J Food Eng. 2010;101(4):349–56.

Faridi Esfanjani A, Jafari SM. Biopolymer nano-particles and natural nano-carriers for nano-encapsulation of phenolic compounds. Colloids Surf B Biointerfaces. 2016;146:532–43.

Fernández-Pan I, Carrión-Granda X, Maté JI. Antimicrobial efficiency of edible coatings on the preservation of chicken breast fillets. Food Control. 2014;36(1):69–75.

Francisco CRL, Heleno SA, Fernandes IPM, Barreira JCM, Calhelha RC, Barros L, et al. Functionalization of yogurts with Agaricus bisporus extracts encapsulated in spray-dried maltodextrin crosslinked with citric acid. Food Chem. 2018;245:845–53.

Freiberger EB, Kaufmann KC, Bona E, Hermes de Araújo PH, Sayer C, Leimann FV, et al. Encapsulation of roasted coffee oil in biocompatible nanoparticles. LWT Food Sci Technol. 2015;64(1):381–9.

Frenzel M, Steffen-Heins A. Impact of quercetin and fish oil encapsulation on bilayer membrane and oxidation stability of liposomes. Food Chem. 2015;185:48–57.

Fröhlich E, Roblegg E. Models for oral uptake of nanoparticles in consumer products. Vol. 291, toxicology. Elsevier; 2012. p. 10–7.

Fuciños C, Míguez M, Fuciños P, Pastrana LM, Rúa ML, Vicente AA. Creating functional nano-structures: encapsulation of caffeine into α-lactalbumin nanotubes. Innov Food Sci Emerg Technol. 2017a;40:10–7.

Fuciños C, Míguez M, Fuciños P, Pastrana LM, Rúa ML, Vicente AA. Creating functional nano-structures: encapsulation of caffeine into α-lactalbumin nanotubes. Innov Food Sci Emerg Technol. 2017b;40:10–7.

Gaikwad KK, Singh S, Negi YS. Ethylene scavengers for active packaging of fresh food produce. Environ Chem Lett. 2020;18(2):269–84.

Gómez-Mascaraque LG, Casagrande Sipoli C, de La Torre LG, López-Rubio A. Microencapsulation structures based on protein-coated liposomes obtained through electrospraying for the stabilization and improved bioaccessibility of curcumin. Food Chem. 2017;233:343–50.

Gulotta A, Saberi AH, Nicoli MC, McClements DJ. Nanoemulsion-based delivery systems for polyunsaturated (ω-3) oils: formation using a spontaneous emulsification method. J Agric Food Chem. 2014;62(7):1720–5.

Gülseren I, Guri A, Corredig M. Encapsulation of tea polyphenols in Nanoliposomes prepared with Milk phospholipids and their effect on the viability of HT-29 human carcinoma cells. Food Dig. 2012;3(1–3):36–45.

Gumus CE, Decker EA, McClements DJ. Gastrointestinal fate of emulsion-based ω-3 oil delivery systems stabilized by plant proteins: lentil, pea, and faba bean proteins. J Food Eng. 2017;207:90–8.

Gutiérrez L, Escudero A, Batlle R, Nerín C. Effect of mixed antimicrobial agents and flavors in active packaging films. J Agric Food Chem. 2009;57(18):8564–71.

Han C, Zhao Y, Leonard SW, Traber MG. Edible coatings to improve storability and enhance nutritional value of fresh and frozen strawberries (Fragaria x ananassa) and raspberries (Rubus ideaus). Postharvest Biol Technol. 2004;33(1):67–78.

Han J-W, Ruiz-Garcia L, Qian J-P, Yang X-T. Food packaging: a comprehensive review and future trends. Compr Rev Food Sci Food Saf. 2018;17(4):860–77.

Hargens-Madsen M, Schnepf M, Hamouz F, Weller C, Roy S. Use of edible films and tocopherols in the control of warmed over flavor. J Am Diet Assoc. 1995;95(9 Suppl):A41.

He X, Hwang HM. Nanotechnology in food science: functionality, applicability, and safety assessment. J Food Drug Anal. Elsevier Taiwan LLC. 2016;24:671–81.

He Z, Zhang X, Qi W, Huang R, Su R. Alginate-casein microspheres as bioactive vehicles for nutrients. Trans Tianjin Univ. 2015;21(5):383–91.

Hernández-Muñoz P, Almenar E, Ocio MJ, Gavara R. Effect of calcium dips and chitosan coatings on postharvest life of strawberries (Fragaria x ananassa). Postharvest Biol Technol. 2006;39(3):247–53.

Ho KKHY, Schroën K, San Martín-González MF, Berton-Carabin CC. Physicochemical stability of lycopene-loaded emulsions stabilized by plant or dairy proteins. Food Struct. 2017;12:34–42.

Honarvar Z, Hadian Z, Mashayekh M. Nanocomposites in food packaging applications and their risk assessment for health. Electron Physician. 2016;8(6):2531–8.

Hu Q, Fang Y, Yang Y, Ma N, Zhao L. Effect of nanocomposite-based packaging on postharvest quality of ethylene-treated kiwifruit (Actinidia deliciosa) during cold storage. Food Res Int. 2011;44(6):1589–96.

Hughes GA. Nanostructure-mediated drug delivery. Nanomed Nanotechnol Biol Med. 2005;1(1):22–30.

Hundre SY, Karthik P, Anandharamakrishnan C. Effect of whey protein isolate and β-cyclodextrin wall systems on stability of microencapsulated vanillin by spray–freeze drying method. Food Chem. 2015;174:16–24.

Ishwarya SP, Anandharamakrishnan C, Stapley AGF. Spray-freeze-drying: a novel process for the drying of foods and bioproducts. Trends Food Sci Technol. 2015;41(2):161–81.

Islam MZ, Kitamura Y, Kokawa M, Monalisa K, Tsai F-H, Miyamura S. Effects of micro wet milling and vacuum spray drying on the physicochemical and antioxidant properties of orange (Citrus unshiu) juice with pulp powder. Food Bioprod Process. 2017;101:132–44.

Jafari SM, He Y, Bhandari B. Production of sub-micron emulsions by ultrasound and microfluidization techniques. J Food Eng. 2007;82(4):478–88.

Jafari SM, Esfanjani AF, Katouzian I, Assadpour E. Release, characterization, and safety of Nanoencapsulated food ingredients. In: Nanoencapsulation of food bioactive ingredients. Elsevier; 2017. p. 401–53.

Jain A, Thakur D, Ghoshal G, Katare OP, Shivhare US. Characterization of microcapsulated β-carotene formed by complex coacervation using casein and gum tragacanth. Int J Biol Macromol. 2016a;87:101–13.

Jain A, Thakur D, Ghoshal G, Katare OP, Shivhare US. Characterization of microcapsulated β-carotene formed by complex coacervation using casein and gum tragacanth. Int J Biol Macromol. 2016b;87:101–13.

Jarunglumlert T, Nakagawa K, Adachi S. Influence of aggregate structure of casein on the encapsulation efficiency of β-carotene entrapped via hydrophobic interaction. Food Struct. 2015;5:42–50.
Jia Z, Dumont M-J, Orsat V. Encapsulation of phenolic compounds present in plants using protein matrices. Food Biosci. 2016a;15:87–104.
Jia Z, Dumont MJ, Orsat V. Encapsulation of phenolic compounds present in plants using protein matrices, Food Bioscience, vol. 15. Elsevier Ltd.; 2016b. p. 87–104.
Jiang Z, Neetoo H, Chen H. Efficacy of freezing, frozen storage and edible antimicrobial coatings used in combination for control of listeria monocytogenes on roasted Turkey stored at chiller temperatures. Food Microbiol. 2011a;28(7):1394–401.
Jiang Z, Neetoo H, Chen H. Control of listeria monocytogenes on cold-smoked Salmon using chitosan-based antimicrobial coatings and films. J Food Sci. 2011b;76(1):M22–6.
Jouki M, Yazdi FT, Mortazavi SA, Koocheki A. Quince seed mucilage films incorporated with oregano essential oil: physical, thermal, barrier, antioxidant and antibacterial properties. Food Hydrocoll. 2014;36:9–19.
Joye IJ, McClements DJ. Biopolymer-based nanoparticles and microparticles: fabrication, characterization, and application. Curr Opin Colloid Interface Sci. 2014;19(5):417–27.
Katouzian I, Faridi Esfanjani A, Jafari SM, Akhavan S. Formulation and application of a new generation of lipid nano-carriers for the food bioactive ingredients, Trends in Food Science and Technology, vol. 68. Elsevier Ltd.; 2017. p. 14–25.
Kegere J, Ouf A, Siam R, Mamdouh W. Fabrication of poly(vinyl alcohol)/chitosan/Bidens pilosa composite electrospun nanofibers with enhanced antibacterial activities. ACS Omega. 2019;4(5):8778–85.
Lazić V, Vivod V, Peršin Z, Stoiljković M, Ratnayake IS, Ahrenkiel PS, et al. Dextran-coated silver nanoparticles for improved barrier and controlled antimicrobial properties of nanocellulose films used in food packaging. Food Packag Shelf Life. 2020;26:100575.
Li M, Ioannidis N, Gogos C, Bilgili E. A comparative assessment of nanocomposites vs. amorphous solid dispersions prepared via nanoextrusion for drug dissolution enhancement. Eur J Pharm Biopharm. 2017;119:68–80.
Lia C, Wang J, Shi J, Huang X, Peng Q, Xueb F. Encapsulation of tomato oleoresin using soy protein isolate-gum aracia conjugates as emulsifier and coating materials. Food Hydrocoll. 2015;45:301–8.
Liao L, Luo Y, Zhao M, Wang Q. Preparation and characterization of succinic acid deamidated wheat gluten microspheres for encapsulation of fish oil. Colloids Surf B Biointerfaces. 2012;92:305–14.
Mahajan PV, Rodrigues FAS, Motel A, Leonhard A. Development of a moisture absorber for packaging of fresh mushrooms (Agaricus bisporous). Postharvest Biol Technol. 2008;48(3):408–14.
Marcos B, Aymerich T, Monfort JM, Garriga M. High-pressure processing and antimicrobial biodegradable packaging to control Listeria monocytogenes during storage of cooked ham. Food Microbiol. 2008;25(1):177–82.
Marsanasco M, Márquez AL, Wagner JR, Chiaramoni NS, Del V, Alonso S. Bioactive compounds as functional food ingredients: characterization in model system and sensory evaluation in chocolate milk. J Food Eng. 2015;166:55–63.
Mehmood T. Optimization of the canola oil based vitamin E nanoemulsions stabilized by food grade mixed surfactants using response surface methodology. Food Chem. 2015;183:1–7.
Millette M, Le Tien C, Smoragiewicz W, Lacroix M. Inhibition of Staphylococcus aureus on beef by nisin-containing modified alginate films and beads. Food Control. 2007;18(7):878–84.
Min B, Ahn DU. Mechanism of lipid peroxidation in meat and meat products – a review. Food Sci Biotechnol. 2005;14(1):152–63.
Moeiniafshari AA, Zarrabi A, Bordbar AK. Exploring the interaction of naringenin with bovine beta-casein nanoparticles using spectroscopy. Food Hydrocoll. 2015;51:1–6.
Moore ME, Han I, Acton J, Ogale A, Barmore C, Dawson P. Effects of antioxidants in polyethylene film on fresh beef color. J Food Sci. 2006;68:99–104.

Mosquera M, Giménez B, Da Silva IM, Boelter JF, Montero P, Gómez-Guillén MC, et al. Nanoencapsulation of an active peptidic fraction from sea bream scales collagen. Food Chem. 2014;156:144–50.

Nanotechnology: a New Frontier in Food Science [Internet]. [cited 2020 Nov 27]. Available from: https://www.ift.org/news-and-publications/food-technology-magazine/issues/2003/december/features/nanotechnology-a-new-frontier-in-food-science

Natrajan N, Sheldon BW. Inhibition of Salmonella on poultry skin using protein- and polysaccharide-based films containing a nisin formulation. J Food Prot. 2000;63(9):1268–72.

Neetoo H, Ye M, Chen H. Bioactive alginate coatings to control Listeria monocytogenes on cold-smoked salmon slices and fillets. Int J Food Microbiol. 2010;136(3):326–31.

Nesterenko A, Alric I, Silvestre F, Durrieu V. Vegetable proteins in microencapsulation: a review of recent interventions and their effectiveness. Ind Crop Prod. 2013a;42:469–79.

Nesterenko A, Alric I, Silvestre F, Durrieu V. Vegetable proteins in microencapsulation: a review of recent interventions and their effectiveness, Industrial Crops and Products, vol. 42. Elsevier; 2013b. p. 469–79.

Nguyen VT, Gidley MJ, Dykes GA. Potential of a nisin-containing bacterial cellulose film to inhibit Listeria monocytogenes on processed meats. Food Microbiol. 2008;25(3):471–8.

Ni S, Sun R, Zhao G, Xia Q. Quercetin loaded nanostructured lipid carrier for food fortification: preparation, characterization and *in vitro* study. J Food Process Eng. 2015;38(1):93–106.

Oancea AM, Aprodu I, Ghinea IO, Barbu V, Ioniţă E, Bahrim G, et al. A bottom-up approach for encapsulation of sour cherries anthocyanins by using β-lactoglobulin as matrices. J Food Eng. 2017;210:83–90.

Ogrodowska D, Tańska M, Brandt W. The influence of drying process conditions on the physical properties, bioactive compounds and stability of encapsulated pumpkin seed oil. Food Bioprocess Technol. 2017;10(7):1265–80.

Ojagh SM, Rezaei M, Razavi SH, Hosseini SMH. Effect of chitosan coatings enriched with cinnamon oil on the quality of refrigerated rainbow trout. Food Chem. 2010;120(1):193–8.

Olivas GI, Barbosa-Cánovas GV. Edible coatings for fresh-cut fruits. Crit Rev Food Sci Nutr. 2005;45:657–70.

Olivas GI, Mattinson DS, Barbosa-Cánovas GV. Alginate coatings for preservation of minimally processed "Gala" apples. Postharvest Biol Technol. 2007;45(1):89–96.

Oms-Oliu G, Soliva-Fortuny R, Martín-Belloso O. Edible coatings with antibrowning agents to maintain sensory quality and antioxidant properties of fresh-cut pears. Postharvest Biol Technol. 2008;50(1):87–94.

Ouattara B, Simard RE, Piette G, Bégin A, Holley RA. Inhibition of surface spoilage bacteria in processed meats by application of antimicrobial films prepared with chitosan. Int J Food Microbiol. 2000;62(1–2):139–48.

Oussalah M, Caillet S, Salmiéri S, Saucier L, Lacroix M. Antimicrobial effects of alginate-based film containing essential oils for the preservation of whole beef muscle. J Food Prot. 2006;69(10):2364–9.

Oussalah M, Caillet S, Salmiéri S, Saucier L, Lacroix M. Antimicrobial effects of alginate-based films containing essential oils on Listeria monocytogenes and Salmonella typhimurium present in bologna and ham. J Food Prot. 2007;70(4):901–8.

Pant A, Sängerlaub S, Müller K. Gallic acid as an oxygen scavenger in bio-based multilayer packaging films. Materials. 2017;10(5):489.

Park SI, Stan SD, Daeschel MA, Zhao Y. Antifungal coatings on fresh strawberries (Fragaria × ananassa) to control mold growth during cold storage. J Food Sci. 2005;70(4):202–7.

Parra RR. Microencapsulación de Alimentos. Rev Fac Nac Agron. 2010;63(2):5669–84.

Patel ZS, Yamamoto M, Ueda H, Tabata Y, Mikos AG. Biodegradable gelatin microparticles as delivery systems for the controlled release of bone morphogenetic protein-2. Acta Biomater. 2008;4(5):1126–38.

Patil YP, Jadhav S. Novel methods for liposome preparation, Chemistry and Physics of Lipids, vol. 177. Elsevier Ireland Ltd.; 2014. p. 8–18.

Piorkowski DT, McClements DJ. Beverage emulsions: recent developments in formulation, production, and applications. Food Hydrocoll. 2014;42:5–41.
Qian C, Decker EA, Xiao H, McClements DJ. Impact of lipid nanoparticle physical state on particle aggregation and β-carotene degradation: potential limitations of solid lipid nanoparticles. Food Res Int. 2013;52(1):342–9.
Ranjan S, Dasgupta N, Chakraborty AR, Melvin Samuel S, Ramalingam C, Shanker R, et al. Nanoscience and nanotechnologies in food industries: opportunities and research trends. J Nanoparticle Res. Kluwer Academic Publishers. 2014;16:1–23.
Raybaudi-Massilia R, Mosqueda-Melgar H. Polysaccharides as carriers and protectors of additives and bioactive compounds in foods. In: The complex world of polysaccharides. InTech; 2012. p. 27.
Raybaudi-Massilia RM, Rojas-Graü MA, Mosqueda-Melgar J, Martín-Belloso O. Comparative study on essential oils incorporated into an alginate-based edible coating to assure the safety and quality of fresh-cut Fuji apples. J Food Prot. 2008a;71(6):1150–61.
Raybaudi-Massilia RM, Mosqueda-Melgar J, Martín-Belloso O. Edible alginate-based coating as carrier of antimicrobials to improve shelf-life and safety of fresh-cut melon. Int J Food Microbiol. 2008b;121(3):313–27.
Rayner M, Marku D, Eriksson M, Sjöö M, Dejmek P, Wahlgren M. Biomass-based particles for the formulation of Pickering type emulsions in food and topical applications. Colloids Surf Physicochem Eng Asp. 2014;458(1):48–62.
Rodea-González DA, Cruz-Olivares J, Román-Guerrero A, Rodríguez-Huezo ME, Vernon-Carter EJ, Pérez-Alonso C. Spray-dried encapsulation of chia essential oil (Salvia hispanica L.) in whey protein concentrate-polysaccharide matrices. J Food Eng. 2012;111(1):102–9.
Rodriguez J, Martín MJ, Ruiz MA, Clares B. Current encapsulation strategies for bioactive oils: from alimentary to pharmaceutical perspectives. Food Res Int. 2016;83:41–59.
Rojas-Graü MA, Raybaudi-Massilia RM, Soliva-Fortuny RC, Avena-Bustillos RJ, McHugh TH, Martín-Belloso O. Apple puree-alginate edible coating as carrier of antimicrobial agents to prolong shelf-life of fresh-cut apples. Postharvest Biol Technol. 2007;45(2):254–64.
Rojas-Graü MA, Soliva-Fortuny R, Martín-Belloso O. Edible coatings to incorporate active ingredients to fresh-cut fruits: a review. Trends Food Sci Technol. 2009;20(10):438–47.
Rutz JK, Borges CD, Zambiazi RC, Crizel-Cardozo MM, Kuck LS, Noreña CPZ. Microencapsulation of palm oil by complex coacervation for application in food systems. Food Chem. 2017;220:59–66.
Salvia-Trujillo L, Rojas-Graü A, Soliva-Fortuny R, Martín-Belloso O. Physicochemical characterization and antimicrobial activity of food-grade emulsions and nanoemulsions incorporating essential oils. Food Hydrocoll. 2015;43:547–56.
Sangsuwan J, Rattanapanone N, Rachtanapun P. Effect of chitosan/methyl cellulose films on microbial and quality characteristics of fresh-cut cantaloupe and pineapple. Postharvest Biol Technol. 2008;49(3):403–10.
Sanguansri L, Augustin M-A. Microencapsulation and delivery of Omega-3 fatty acids. In: Functional food ingredients and nutraceuticals processing technologies. Taylor & Francis; 2007. p. 297–327.
Sayanjali S, Ghanbarzadeh B, Ghiassifar S. Evaluation of antimicrobial and physical properties of edible film based on carboxymethyl cellulose containing potassium sorbate on some mycotoxigenic Aspergillus species in fresh pistachios. LWT Food Sci Technol. 2011;44(4):1133–8.
Seeli DS, Prabaharan M. Guar gum succinate as a carrier for colon-specific drug delivery. Int J Biol Macromol. 2016;84:10–5.
Seol KH, Lim DG, Jang A, Jo C, Lee M. Antimicrobial effect of κ-carrageenan-based edible film containing ovotransferrin in fresh chicken breast stored at 5 °C. Meat Sci. 2009;83(3):479–83.
Shetta A, Kegere J, Mamdouh W. Comparative study of encapsulated peppermint and green tea essential oils in chitosan nanoparticles: encapsulation, thermal stability, in-vitro release, antioxidant and antibacterial activities. Int J Biol Macromol. 2019;126:731–42.

Shishir MRI, Xie L, Sun C, Zheng X, Chen W. Advances in micro and nano-encapsulation of bioactive compounds using biopolymer and lipid-based transporters. Trends Food Sci Technol. 2018;78:34–60.

Siracusa V, Blanco I. Bio-polyethylene (bio-PE), bio-polypropylene (bio-PP) and bio-poly(ethylene terephthalate) (bio-PET): recent developments in bio-based polymers analogous to petroleum-derived ones for packaging and engineering applications. Polymers. 2020;12(8):1641.

Siripatrawan U, Noipha S. Active film from chitosan incorporating green tea extract for shelf life extension of pork sausages. Food Hydrocoll. 2012;27(1):102–8.

Song Y, Liu L, Shen H, You J, Luo Y. Effect of sodium alginate-based edible coating containing different anti-oxidants on quality and shelf life of refrigerated bream (Megalobrama amblycephala). Food Control. 2011;22(3–4):608–15.

Sun S, Song Y, Zheng Q. Rheological behavior of heat-induced wheat gliadin gel. Food Hydrocoll. 2009;23(3):1054–6.

Suppakul P, Miltz J, Sonneveld K, Bigger SW. Active packaging technologies with an emphasis on antimicrobial packaging and its applications. J Food Sci. 2003;68(2):408–20.

Tan C, Xia S, Xue J, Xie J, Feng B, Zhang X. Liposomes as vehicles for lutein: preparation, stability, liposomal membrane dynamics, and structure. J Agric Food Chem. 2013;61(34):8175–84.

Tapia M. Using polysaccharide-based edible coatings to maintain quality of fresh-cut Fuji apples. LWT Food Sci Technol. 2008;41:139–47.

Tapia MS, Rojas-Graü MA, Rodríguez FJ, Ramírez J, Carmona A, Martin-Belloso O. Alginate- and gellan-based edible films for probiotic coatings on fresh-cut fruits. J Food Sci. 2007;72(4):E190–6.

Tapia MS, Rojas-Graü MA, Carmona A, Rodríguez FJ, Soliva-Fortuny R, Martin-Belloso O. Use of alginate- and gellan-based coatings for improving barrier, texture and nutritional properties of fresh-cut papaya. Food Hydrocoll. 2008;22(8):1493–503.

Tarhini M, Greige-Gerges H, Elaissari A. Protein-based nanoparticles: from preparation to encapsulation of active molecules. Int J Pharm Elsevier BV. 2017;522:172–97.

Tas CE, Hendessi S, Baysal M, Unal S, Cebeci FC, Menceloglu YZ, et al. Halloysite nanotubes/polyethylene nanocomposites for active food packaging materials with ethylene scavenging and gas barrier properties. Food Bioprocess Technol. 2017;10(4):789–98.

Tatar Turan F, Cengiz A, Kahyaoglu T. Evaluation of ultrasonic nozzle with spray-drying as a novel method for the microencapsulation of blueberry's bioactive compounds. Innov Food Sci Emerg Technol. 2015;32:136–45.

Taylor TM, Davidson PM, Bruce BD, Weiss J. Ultrasonic spectroscopy and differential scanning calorimetry of liposomal-encapsulated nisin. J Agric Food Chem. 2005;53(22):8722–8.

Thompson AK, Hindmarsh JP, Haisman D, Rades T, Singh H. Comparison of the structure and properties of liposomes prepared from milk fat globule membrane and soy phospholipids. J Agric Food Chem. 2006;54(10):3704–11.

Thorat AA, Dalvi SV. Liquid antisolvent precipitation and stabilization of nanoparticles of poorly water soluble drugs in aqueous suspensions: recent developments and future perspective. Chem Eng J. 2012;181–182:1–34.

Timilsena YP, Wang B, Adhikari R, Adhikari B. Advances in microencapsulation of polyunsaturated fatty acids (PUFAs)-rich plant oils using complex coacervation: a review. Food Hydrocoll. 2017;69:369–81.

Tsironi T, Ntzimani A, Gogou E, Tsevdou M, Semenoglou I, Dermesonlouoglou E, et al. Modeling the effect of active modified atmosphere packaging on the microbial stability and shelf life of Gutted Sea Bass. Appl Sci. 2019;9(23):5019.

Uysal Unalan I, Cerri G, Marcuzzo E, Cozzolino CA, Farris S. Nanocomposite films and coatings using inorganic nanobuilding blocks (NBB): current applications and future opportunities in the food packaging sector. RSC Adv. 2014;4(56):29393–428.

Valencia-Chamorro SA, Pérez-Gago MB, del Río MÁ, Palou L. Effect of antifungal hydroxypropyl methylcellulose (HPMC)-lipid edible composite coatings on postharvest decay development and quality attributes of cold-stored "Valencia" oranges. Postharvest Biol Technol. 2009;54(2):72–9.

Varshosaz J, Eskandari S, Tabbakhian M. Freeze-drying of nanostructure lipid carriers by different carbohydrate polymers used as cryoprotectants. Carbohydr Polym. 2012;88(4):1157–63.

Waghmare RB. Encapsulation techniques for delivery of bioactive compounds in Milk and dairy products – a review. J Dairy Res Technol. 2020;3(1):1–9.

Waglay A, Karboune S, Alli I. Potato protein isolates: recovery and characterization of their properties. Food Chem. 2014;142:373–82.

Wahbi W, Siam R, Kegere J, El-Mehalmey WA, Mamdouh W. Novel inulin electrospun composite nanofibers: prebiotic and antibacterial activities. ACS Omega. 2020;5(6):3006–15.

Winestrand S, Johansson K, Järnström L, Jönsson LJ. Co-immobilization of oxalate oxidase and catalase in films for scavenging of oxygen or oxalic acid. Biochem Eng J. 2013;72:96–101.

Wu Y, Rhim JW, Weller CL, Hamouz F, Cuppett S, Schnepf M. Moisture loss and lipid oxidation for precooked beef patties stored in edible coatings and films. J Food Sci. 2000;65(2):300–4.

Wu Y, Weller CL, Hamouz F, Cuppett S, Schnepf M. Moisture loss and lipid oxidation for precooked ground-beef patties packaged in edible starch-alginate-based composite films. J Food Sci. 2001;66(3):486–93.

Yalavarthi P, Dudala T, Mudumala N, Pasupati V, Thanniru J, Vadlamudi H, et al. A perspective overview on lipospheres as lipid carrier systems. Int J Pharm Investig. 2014;4(4):149.

Ye M, Neetoo H, Chen H. Control of Listeria monocytogenes on ham steaks by antimicrobials incorporated into chitosan-coated plastic films. Food Microbiol. 2008;25(2):260–8.

Yoshimoto M, Sakamoto H, Yoshimoto N, Kuboi R, Nakao K. Stabilization of quaternary structure and activity of bovine liver catalase through encapsulation in liposomes. Enzym Microb Technol. 2007;41(6–7):849–58.

Youssef Ahmed M, EL-Sayed SM, EL-Sayed HS, Salama HH, Dufresne A. Enhancement of Egyptian soft white cheese shelf life using a novel chitosan/carboxymethyl cellulose/zinc oxide bionanocomposite film. Carbohydr Polym. 2016;151:9–19.

Yuan Y, Kong Z-Y, Sun Y-E, Zeng Q-Z, Yang X-Q. Complex coacervation of soy protein with chitosan: constructing antioxidant microcapsule for algal oil delivery. LWT Food Sci Technol. 2017a;75:171–9.

Yuan Y, Kong ZY, Sun YE, Zeng QZ, Yang XQ. Complex coacervation of soy protein with chitosan: constructing antioxidant microcapsule for algal oil delivery. LWT Food Sci Technol. 2017b;75:171–9.

Zeeb B, Herz E, McClements DJ, Weiss J. Impact of alcohols on the formation and stability of protein-stabilized nanoemulsions. J Colloid Interface Sci. 2014;433:196–203.

Zhang X, Sun H, Zhang Z, Niu Q, Chen Y, Crittenden JC. Enhanced bioaccumulation of cadmium in carp in the presence of titanium dioxide nanoparticles. Chemosphere. 2007;67(1):160–6.

Zhang L, Hayes DG, Chen G, Zhong Q. Transparent dispersions of milk-fat-based nanostructured lipid carriers for delivery of β-carotene. J Agric Food Chem. 2013;61(39):9435–43.

Chapter 8
Effect of the Presence of Bioactive Compounds Embedded into Active Packaging Films on Their Mechanical and Barrier Properties

M. Stanzione, R. Zullo, G. G. Buonocore, and M. Lavorgna

Abstract Active natural additives can be loaded into bio-active packaging films in order to impart new functionalities and to broaden their potential application fields. However, they modify the film structure and properties, thus in this chapter, the authors aim to review the influence of the incorporation of natural additives and (bio)active compounds on the properties of active films potentially used as packaging materials able to prolong the shelf-life of packaged foodstuff. The chapter is focused on the effect of several active substances on mechanical behavior and barrier properties of active packaging materials, namely on their tensile stress, elastic modulus and elongation at break as well as oxygen, water vapor and UV light barrier properties. Several categories of (bio)active compounds are reviewed and particular attention is paid to active packaging films based on both polyolefin and biodegradable/bio-based polymers flexible films obtained by means of both solvent-casting and melt-mixing/extrusion techniques.

Keywords Bioactive compounds · Flexible films · Mechanical behavior · Barrier properties · Active packaging · Releasing systems

M. Stanzione · R. Zullo · G. G. Buonocore (✉) · M. Lavorgna
Institute for Polymers, Composites and Biomaterials, National Research Council of Italy – P.le E. Fermi, Naples, Italy
e-mail: giovannagiuliana.buonocore@cnr.it

S. M. Jafari, A. S. Silva (eds.), *Releasing Systems in Active Food Packaging*, Food Bioactive Ingredients, https://doi.org/10.1007/978-3-030-90299-5_8

1 Introduction

Packaging materials have to preserve the quality and safety of the content during storage and transportation. In the field of food packaging, the protection against external biological, physical, and chemical factors is of utmost importance to preserve the food quality, to prolong the food shelf life and to insure its safety by slowing down the microbial spoilage and chemical reactions leading to food deterioration. Moreover, it must provide convenience for the consumer and act as a communicating interface between the producer and consumer. Proper packaging material must be selected and/or developed taking into account the required functions as well as the price, the weight, the esthetic and processing issues (Buonocore et al. 2014).

Active packaging systems have been already described in detail in the first chapter of the present book. In particular, the formulation of bio-active films generally requires the incorporation of a minimal amount of natural additives to impart new functionalities to the resulting material such as antioxidant and antimicrobial activities, but also to modify its final and functional properties such as mechanical properties, water vapor or oxygen barrier properties and UV-barrier. Indeed, the chemical or physical interaction between polymeric matrix and natural additives or bioactive compound can affect the film structure and modify their characteristics.

Generally speaking, food packaging materials can be categorized as rigid, flexible, or semi-flexible systems. Flexible packaging films produced by cast or blown extrusion are the most common and used ones. Generally, they have a multilayer structure obtained by coextrusion, by adhesive lamination or by extrusion lamination. Packaging films, generally printed for marketing purposes and for providing information to the consumers, are purchased as reels and then used to prepare bags and pouches using confectionary machines to form, fill and seal the packaging container.

As for mechanical properties, it is worth noting that food packaging materials are primarily developed to preserve the quality and safety of the food but also to provide adequate conditions of transport and storage in order to reach the consumer without damages and to avoid losses in the production chain. The quality features such as barrier properties, volatiles and liquid adsorption, thermal stability and antimicrobial or antioxidant activities are not the only characteristic to evaluate for promoting an active food packaging material. In fact, mechanical performance of food packaging are fundamental issues to be taken into account when a novel material is designed for later industrial implementation. Indeed, suitable mechanical performances must be achieved and guaranteed in various conditions of external and internal temperature and humidity.

Mechanical protection of food packaging system toward packaged foodstuff must take into account the exposure to environmental factors but also to the intentional modification of the product (i.e. tampering). Indeed, the protection and preservation of the integrity of the contained food are fundamental issues but, in addition, it is requested that both primary and secondary food packaging materials must be suitable for printing information or legislation prescription and capable to sustain internal load weight and to resist to external load or forces, due to handling, storage and operational shipment.

Packaging materials come in different shapes with various functions associated to their properties and also to their specific industrial food application. To maximize the performances and the design of a packaging material, it is essential to reach a balance between its shape and its function, as the feasibility to adopt the same material for different food packaging industry stands as a fundamental goal for both economic and technological opportunities and constraints.

As already discussed, flexible films represent the most used and developed product of the whole packaging manufacturing industry. Appropriate mechanical and thermal resistance have a significant impact on the selection and on the efficiency of various groups of materials, products and packaging solutions. The mechanical properties of packaging films are strictly related to the behavior of the selected materials against forces applied on their surface due to operational causes. In the food and food packaging market, storage and distribution operations are finalized to avoid or significantly reduce losses and to assure adequate protection of the food from physical damages and leakage for long time. For these reasons, evaluation of mechanical parameters, in terms of film flexibility and resistance, is extremely important in this field.

As for barrier properties, it is worth noting that, due to their permeability to low molecular weight compounds, plastic packages are able to control the exchange of molecules and compounds between the products and the environment. In order to preserve the packaged foodstuff, polymeric packaging materials should provide adequate 'barrier' to gases (i.e. oxygen, carbon dioxide, nitrogen), vapors (i.e. water vapors) and aromas. The needed barrier properties strongly depend on the type of food and on the desired shelf life: oxygen barrier is needed to avoid or limit degradative oxidation of fats, carbon dioxide barrier for carbonated beverages, and a proper ratio of several gas permeability is needed to keep the desired 'modified atmosphere' in the package headspace. The protection from external oxygen and water molecules is the most common requirement in food packaging. In fact, water and oxygen molecules may induce many degradation reactions in food, such as fat and oil rancidity, spoilage microorganism growth, Maillard reaction and enzymatic browning. For this reason, one of the main characteristic to be exhibited by a flexible food packaging film is its barrier property toward low molecular weight compounds with the aim to inhibit molecules permeation into the polymer and to minimize inward/outward diffusion as much as possible. High barrier property is the main requirement to be exhibited by a polymeric packaging material since the increase of this feature allows the shelf-life extension of packaged food (Lagaron et al. 2004).

As for electromagnetic properties the interaction of the packaging materials with electromagnetic radiation is a very important feature, namely for the protection against the light, for the interaction with microwaves, and for esthetic issues. The parameters which are usually taken into account are the transmittance spectrum in ultraviolet (UV), Vis, and IR, the transparency which measures the transmittance of a radiation through the material, the haze which measures the opacity of the material and gloss which measures the appearance of the film surface.

In this chapter, the influence of the presence of natural bio-active compounds on these functional properties of the resulting bioactive packaging films will be discussed in detail. Natural active compounds are a safely alternative to the use of chemical additives in the food industry. These compounds can be essential oils and/or can be extracted from plants materials such as leaves, flowers, buds and barks, they exhibit very good antioxidant (AO) and antimicrobial (AM) properties and can be used to extend the shelf life of the food. Essential oils are generally in the form of liquids containing aromatic and volatile compounds, and plant extracts also show very good AO and AM properties due to their bioactive phenolic and flavonoid compounds. These bioactive compounds can be used to preserve the food or by adding them directly into the food formulation or by introducing and embedding them in the active packaging films and to allow their controlled release toward the food in order to exert their action all over the storage period. Besides these essential oils and plant extracts, in this chapter the attention will be also focused on the use of some active natural acids, active lipids and active polysaccharides in form of particles and nanoparticles.

The incorporation of bioactive species in the films can modify the polymeric structure of the matrices and, consequently, their functionality. Indeed, hydrophobic compounds can interact with hydrophilic polymer chain and act as plasticizers or crosslinkers for the polymer matrix. The effect of the presence of natural active compounds on mechanical and barrier properties of the resulting film is rather different and depends on several variables such as the type of active compound and of the polymeric matrix, polymer chain mobility and interactions between the active components and the polymer, leading sometimes to opposite effects. For example, when the matrix is an hydrophilic polymer, the inclusion of essential oils may modify its nature by increasing its hydrophobic behavior, thus causing an enhancement of water vapor barrier properties, however it may also lead to the presence of polymer discontinuity, favoring the diffusion of water molecules and thus increasing water permeability.

Given the above scenario, in the following paragraphs the attention will be focused on the description of the effect of the presence of bioactive compounds on the mechanical, barrier and electromagnetic properties of the resulting active packaging films. As for mechanical properties, the influence of type and concentration of active substances on parameters such tensile strength, Young's modulus and elongation at break will be thoroughly discussed. As for barrier properties, particular attention will be devoted to the improvement or worsening of water and oxygen permeability, which are very important factors to be optimized in order to slow down those detrimental mechanisms responsible for the unacceptability of packaged foodstuff. As for electromagnetic properties, UV barrier properties of bioactive packaging films will be analyzed, since the quantification of the protection degree provided by the packaging material against the photo-oxidation activated by UV radiation is particularly interesting. Mechanisms such as crosslink and/or plasticizing effect induced by the presence of the bioactive compound will be described in the following paragraphs.

2 Mechanical Properties of Active Packaging Flexible Films Containing Various Bioactives

In order to provide a significant improvement in the development of new products within food packaging industry and technology, the main mechanical performances of food packaging films such as tensile property, sealing strength, tear property, impact resistance and puncture performance should be thoroughly considered and carefully modulated.

In general, requirements for good mechanical properties of packaging are good tensile strength and adequate elongation and flexibility to promote a good protection against external agents (Robertson 2013). Tensile test represents the most common experimental testing configuration used to investigate mechanical properties of food packaging film, and the main parameters gained from the analysis of tensile data and needed to validate their potential application within the food packaging industry are: tensile strength (TS), elongation at break (EB) and Young's modulus (YM). For sake of completeness, resistance to preformation or drilling should be also mentioned, but most of the current literature works do not consider this parameter being more an *industrial quality* rather than a constitutive material property. Tensile strength measures the force required to break the tested material under the action of uniaxial external force or internal pressure normalized by the film cross-sectional area; whereas elongation at break is commonly defined as the capacity to increase the film length under a loading force on its surface and it is measured as a percentage of the initial sample extension. Mechanical behavior of food packaging film can be affected by several variables such as material composition and formulation, manufacturing techniques and their setting parameters, type of embedded additives and active compounds as well as their synergy within the polymer matrix. Moreover, environmental factors, such as temperature, humidity, total and partial pressure must be carefully taken into account.

The mechanical behaviour variation of polymeric films caused by the dispersion of single or multiple fillers into the polymer matrix can be associated to various and sometimes concomitant occurring mechanisms. Generally, the influence of the filler depends not only on its specific amount and characteristic, responsible for chemical bonding, physical entanglements or free volume location into the polymer network, but also on its level of dispersion, uniformity of distribution, processing parameters and other external issues.

Among the different mechanisms which could be analysed to explain the modification of mechanical properties of bioactive packaging films, crosslinking (in terms of both covalent bonding or hydrogen-bonding), plasticizing effect and a series of unpredictable events such as dispersion, thermal degradation and filler aggregations (herein referred to as other mechanisms) appear to be the most relevant ones.

As for the possible crosslinking effect induced by the presence of additives and (bio)active compounds, two different effects must be taken into account. The first one is related to the hydrogen bond formation capacity (hereafter reported as *H-bonding mechanism*) that can significantly change the polymer-filler interfacial interaction. The presence of H-containing groups could promote the formation of hydrogen bonds which improves the stress transfer between the matrix and the fillers by increasing the surface adhesion and therefore improving the tensile strength of the final product. It has been found that the forming hydrogen bonds could result, in some cases, in different crystalline structures leading to a higher rigidity of the final structure. Biopolymers such as proteins show an extensive capacity to develop intramolecular hydrogen bonding as secondary links among the polymer networks, thus the incorporation of bioactive compounds could directly affect the mechanical performance of the resulting bioactive film (Berthet et al. 2015). The second one is related to the availability of different typologies of fillers either for chemical, geometrical, structural or morphological features which paves the way to the interesting strategy of linking the filler to the neat hosting matrix by means of strong chemical interaction. The crosslinking effects, mostly related to the chemistry of the filler, will indeed enhance the original matrix resistance improving tensile strength and, generally, reducing the elongation at break.

The presence of filler can also induce, intentionally or not, a plasticizing effect (hereafter referred as plasticizing mechanism) thus reducing the fragility of the resulting films conferring a more flexible behaviour. The plasticizing effect can be regarded as the capability of modifying the three dimensional organization of the polymeric matrix leading to a decrease of the intermolecular forces and also to an increase of the available free volume and chain mobility. The level of plasticization is related to the filler size, shape, molecular weight and compatibility with the matrix. The occurring mechanism may improve elongation at break and flexibility, ultimately lowering the elastic modulus. The use of food-grade plasticizers, such as sorbitol, glycerol, mannitol, sugar and polyethylene glycol, generally lead to an overall decrease of the mechanical resistance of the film and to an enhancement of its flexibility. Water is also an effective plasticiser in many polymer matrices (Pérez et al. 2016; Castro-Rosas et al. 2016).

Moreover, bioactive composites can be also regarded as immiscible phases formed by continuous polymer network loaded with dispersed fillers and characterized by a new structure with specific properties different from the pristine polymer matrix. The modification of the final functional properties, induced by the filler loading, is inherently associated to the processing steps and parameters either in term of temperature and mechanical stress. In general, the thermal degradation level of the loaded filler, its non-uniform dispersion or rather its poor disaggregation can be considered final properties' key issues of the packaging composites. All these variables may lead to detrimental effects limiting the expected efficiency of the filler. All these undesired and sometimes unpredictable circumstances will be herein taken into account by the authors under the term *other mechanism* (Alvim et al. 2016; Collazo-Bigliardi et al. 2018).

In the following sections, the effect of some bioactive fillers on the final mechanical response of the resulting bioactive packaging films will be reported and analysed, paying attention to classify the observed effect on tensile strength, elongation at break and Young's modulus with respect to the aforementioned mechanisms.

2.1 Mechanical Properties of Flexible Bioactive Films Obtained by Solvent-Casting Technique

Among the different techniques to manufacture an active packaging film, the solvent casting process consists of three different phases: solubilization, casting and drying. Active compounds, as well as additives, crosslinking agents or nanoparticles, are firstly dispersed or solubilized either at isothermal condition (room temperature) or by heating in a suitable solvent to achieve an intermediate product called *film forming solution*. Then, the obtained mixture is poured onto a glass/aluminum plate to achieve a thin filmed product which is finally dried to reach a complete solvent evaporation before demolding of the active films. Solution cast films can be potentially used as packaging materials themselves and can also be applied directly to various food products, ensuring safety and enhancing the shelf life. It is worth to mention that this technique presents specific technological advantages but also some drawbacks which concur to limit its widespread use and production at industrial level. In fact, slight change in the films' formulation can significantly affect the physical properties of the final composite films and also solution casting technique is not yet fully employed at industrial scale, being sometimes uneconomical and time-consuming (Zhang et al. 2019). For this reason, this kind of films can be more easily used as coatings on packaging substrates.

Essential oils represent interesting active agents, which exhibit desirable effects in the active food packaging development, even at low concentrations. Bonilla et al. (2013) have highlighted that the addition of *essential oils* (i.e. *basil and thyme*) or other antioxidant substances, such as *α-tocopherol* or *citric acid*, leads to a functional properties' enhancement of wheat starch/chitosan based films which can potentially be used to prolong the shelf life of high-fat-content foods. In details, the basil essential oil or α-tocopherol based films showed a slight increase, in terms of elastic modulus and tensile strength, with a very negligible effect on elongation at break. The citric acid loading and consequently the forming cross-links with starch increased the elastic modulus but at the same time, reduced the film resistance at break and its extensibility. On the contrary, the incorporation of *garlic, cinnamon and clove oil* into PLA matrix shows a different trend in the mechanical properties (Ahmed et al. 2016). In their work, composite films based on polylactide, polyethylene glycol and cinnamon oil were developed and investigated. In details, when the concentration of cinnamon embedded in the polymer increased, a decrease of tensile strength (46.4%) and modulus (67.8%) and an increase of elongation at break (367.3%), compared to the pristine PLA film, have been observed. In this case, cinnamon essential oils act as a plasticizer agent reducing mechanical properties due to the polymer chain mobility increase. The cinnamaldehyde plasticizing effect is likely related to its low molecular weight which induces a slipping phase within the hosting matrix.

Many researchers have reported that essential oils act as plasticizers increasing the chain mobility and thus the extensibility and flexibility of the films along with a tendency to reduce the cohesive forces of the polymer macromolecules. The EOs plasticization effect within polymer backbone has been also observed in the preparation of antimicrobial and antioxidant bioactive films based on poly(lactic acid)/poly(trimethylene carbonate) loaded with different *oregano essential oil* (OEO) amounts (Liu et al. 2016). In addition, in this case, the OEO plasticization effect was likely due to the interactions of the additive and the polymer macromolecular network. The addition of oregano essential oils hindered polymer chain-to-chain interactions causing a reduction in the tensile strength of the films within the range from 3.6% to 18.7%. The same effect was confirmed by Mahcene et al. (2020), according to which the incorporation of essential oils extracted by Algerian medicinal and aromatic plants in sodium alginate-based films tends to reduce the mechanical strength of the active films. In details, the presence of plasticizers weakened the intermolecular forces in the macromolecule chains by increasing the available free volume.

Plant and/or spice extracts represent a different typology of antimicrobial compounds to be potentially used to develop novel active food packaging. Their influence on mechanical performance is related to the presence of the phenolic compounds, characteristic of these extracts. Kanatt et al. (2012) have demonstrated their effect on composite active films based on chitosan and polyvinyl alcohol (PVA) loaded with *mint extract* (ME)/*pomegranate peel extract* (PE). At the highest concentration of both ME or PE and PVA, tensile strength is positively affected, increasing its value of about 37.0% and 52.1% respectively for ME and PE. The strengthening effect was in particular related to the interaction between OH groups of natural extracts and chitosan matrix by means of the formation of new hydrogen bonds. In this work, the authors have also investigated the puncture resistance of the manufactured composites, which represents a fundamental parameter in the design of bagging or packaging products. Obtained results revealed that puncture strength of composite chitosan-PVA films was negligibly affected by the presence of ME/PE extracts into the matrix, thus avoiding damages by penetration/indentation, which could lead to a complete loss of package integrity. Also, the addition of *green tea extract (GTE)* into chitosan-based films showed improvement in tensile strength, likely due to an intermolecular interaction among chitosan polymer chains and phenolic compounds contained in GTE filler (Siripatrawan and Harte 2010).

Recently, Aparicio-Fernández et al. (2018) investigated the effect of prickly pear (*Opuntia ficus-indica L. cv. San Martín*) peel powder and its aqueous extract incorporated in carboxymethyl cellulose (CMC) films on the mechanical performances of the resulting material. They highlighted that the addition of peel powder in the film formulation led to an increase of tensile strength (41.8%). This can be likely due to the presence of fiber molecules in peel, which induce a reinforcing action within the hosting polymer matrix by the formation of inter-molecular hydrogen bonds. Due to this reinforcement effect, the obtained composite films resulted stiffer, showing lower values of elongation at break and thus lower flexibility. On the contrary, the presence of the aqueous extract did not alter the mechanical properties of the composite films, probably because, despite the formation of hydrogen bonds, the soluble compounds in the extract did not interrupt the CMC polymer structure. Similar results have been recently observed by Cejudo-Bastante et al. (2020) in the preparation of composite on ethylene vinyl alcohol (EVOH) based films, loaded with *betalain-rich beetroot.* In this research, the mechanical performances were not significantly influenced by the presence of betalain-rich beetroot extract and in particular, by its soluble compounds which did not intercalate the EVOH matrix despite the hydrogen bond interactions.

Guava leaf water extract (WE) *and ethanolic extract* (EE) have been recently used to produce antioxidant and antibacterial sodium alginate-based films by Luo et al. (2019). They proved that the mechanical strength and flexibility of the composite films were strictly dependent on the extracts concentration. At low extract contents (i.e. 10%), tensile strength of the composite films was enhanced of about 878.4% and 1445.0%, respectively for WE and EE, and this effect could be explained since the structure of the composite films became more compact due to sodium alginate/guava leaf extract interfacial hydrogen interactions. However, at the highest concentration value (20%), the improvement of tensile strength was drastically reduced compared to the values obtained at lower concentrations. The improvement was reduced at about 431.5% and 667.6%, respectively for WE and EE, probably due to a non-homogeneous dispersion of the extracts within polymer matrix. In order to analyze the effect of filler within the considered matrix, a dimensionless parameter defined as the ratio between the property variation and the percentage of filler content, has been calculated. These values represent the filler efficiency to induce a property variation. Figure 8.1 reports the filler efficiency, respectively for tensile strength and elongation at break, calculated at various percentages of guava leaf water and ethanolic extract incorporated into the sodium alginate matrix. The bi-linear trend of the efficiency curve is likely related to the non-homogeneous dispersion of the filler within the matrix.

Vidal et al. (2020) developed carboxymethyl cellulose-based bioactive films filled with *green coffee oil* (GCO), *cake* (CE) and *sediment extracts* (SE) and explored the synergistic contribution of these three different typologies of additives on the mechanical properties of the resulting material. The tensile strength and elastic modulus were inversely proportional to the amount of both SE and CE, this latter mainly constituted by linoleic and palmitic acid. In fact, within the concentration range from 20% to 40%, the percentage variations of TS and YM with respect to pristine matrix were respectively about −79.3% and −93.1% (TS) and −82.2% and −94.1% (YM) for SE. Similar values were recorded in the case of GCO and CE, showing, respectively, values of −82.7% and −94.8% for TS and −83.6% and −95.7% for YM. A completely different effect was induced by the active

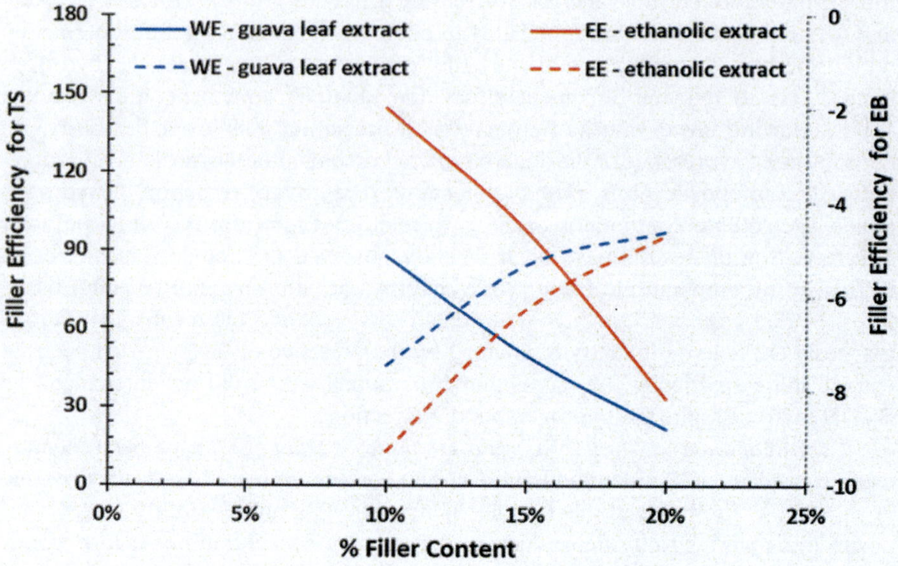

Fig. 8.1 Active compounds efficiency for TS (solid curves) and EB (dash curves) vs AC content into the matrix: guava leaf water extract (WE) and ethanolic extract (EE) embedded into sodium alginate matrix. (Data adapted by Luo et al. 2019)

compounds on the elongation at break values. In fact, in term of percentage variations respectively for the 20% and 40% concentration values, the elongation at break significantly improved of about 246.4% and 232.1% in the case of SE and of about 407.1% and 457.1% with GCO and CE.

The influence of *blackberry pulp* incorporated into arrowroot starch on mechanical properties of the resulting bioactive films was recently evaluated by Nogueira et al. (2019). Tensile strength was reduced of about 67.4%, 84.8%, and 84.1% respectively at 20%, 30% and 40% concentration values of blackberry pulp. Whereas the elongation at break was increased significantly of about 15.1%, 293.9% and 318.2% for analogous concentrations, inducing more flexibility and lower rigidity to the composite films. Blackberry is a source of bioactive compounds, vitamins, essential minerals and fibers, which contain proteins, lipids and sugars; when these substances were added in arrowroot starch films, due to the occurring of discontinuities in the polymer macromolecules, the matrix became more permeable to water and thus more flexible and less rigid.

Some studies have reported on the plasticizing effect of sugars contained in fruit pulps added into a biopolymer film based on pectin (Azeredo et al. 2016). It has been found that, as the amount of pomegranate juice/water volume ratio embedded in pectin films increased, tensile strength and elastic modulus of the films decreased, leading to a higher extension of the system before its break. In particular, the presence of pomegranate acts as a plasticizing agent modifying the interactions of the macromolecules, increasing the inherent motility of the chains and softening the resulting bioactive film which shows high extensibility and flexibility (+318.2%).

Organic acids such as *acetic acid, lactic acid, sorbic acid, citric acid, and their salts* are commonly used as antimicrobial agents in bioactive film packaging, as the US Food and Drug Administration has recognized them safe (GRAS) food preservatives. Many researchers have studied the effect of their incorporation into polymeric films on the physical properties and mechanical behavior of the resulting bioactive films. *Sorbic acid* (SA) *and benzoic acid* (BA) have been embedded into Argentine anchovy protein isolate (API) films by Da Rocha et al. (2018). They found that the organic acid concentration significantly influenced the tensile strength and elongation at break of API based composites. In fact, both the organic acids act as plasticizers within the hosting matrix lowering the tensile strength with an improvement of extension capacity at break. This EB improvement is likely due to different reinforcement behavior, characterized by two different increasing trends in term of efficiency. In the case of sorbic acid addition, the path is directly proportional to its amount, whereas in the case of benzoic acid the effect seems to reach a saturation level as the percentage content increases (Fig. 8.2). This latter behavior is reasonably associated to the linear structure of organic acids chain which interpenetrate the protein chains, with a consequent enhancement of the protein network mobility.

Reddy and Yang (2010) have showed that starch matrix is crosslinked by the addition of citric acid, improving both mechanical performance and dissolution level in water and formic acid. At specific citric acid content (5%), starch molecular weight significantly increased inducing stronger intermolecular interactions among polymer chains. From the mechanical point of view, at citric acid content less than

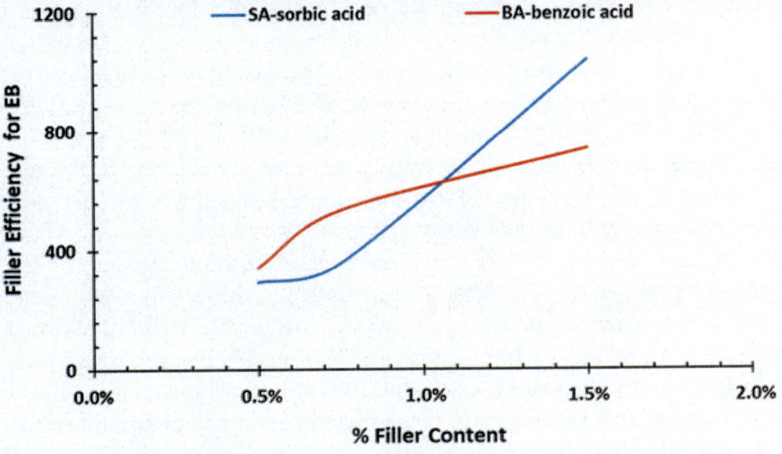

Fig. 8.2 Active compounds efficiency for EB vs AC content into the matrix: sorbic acid (SA) and benzoic acid (BA) embedded into Argentine anchovy protein isolate. (Data adapted by Da Rocha et al. 2018)

5%, the experiments revealed an improvement of the tensile strength, whereas at higher amount, this property drastically decreased.

Both *citric acid and chitosan* synergy on the functional properties of fish gelatin-based films was studied by Uranga et al. (2019), showing a very interesting behavior. The addition of citric acid, in particular its residual free at 10% and 20% of concentration, plasticized the composite film significantly lowering the tensile strength but increasing the elongation at break likely due to the filler interactions which take place within the polymer matrix favoring the chain mobility. The synergistic effect of chitosan addition brought about an embrittlement of the resulting films in the case of 10% w/w of citric acid, showing a further reduction of both tensile strength and elongation at break; whereas in the case of biocomposites containing 20% w/w of citric acid, the effect of chitosan leads to an overall enhancement of the system resilience showing a higher value of absorbed energy before failure (i.e. higher tensile strength and elongation at break). This different trend can be associated to the hydrogen bonding formation between gelatin carboxyl, amino and hydroxyl groups and chitosan amino and hydroxyl groups. The forming H-bond density became relevant only at highest concentration. The obtained experimental data clearly showed the mentioned switching effects as a function of the CA content.

An interesting research, based on the development of a modified poly glutamic acid (PG) films by ε-Poly-L-lysine (PL), was carried out by Karimi et al. (2020). ε-Poly-L-lysine is an antimicrobial peptide cationic molecule with potential surface activity due to the positive charge of the amino groups in water. This additive has a wide antimicrobial efficiency on different stains such as yeast, fungi, Gram-positive and Gram-negative bacteria. The PL addition in PG based film significantly improved the elongation at break, which increased of about 129.1%, 214.7%, and 314.2% with the PL content increasing, respectively at 2%, 4% and 6%. The observed plasticizing effect associated to the addition of PL was related to its presence in the PG backbone increased the negative charge thus leading to a higher electrostatic repulsion among the polymer chains, inducing a decrease of hydrogen bonds and a higher mobility of polymer chains. Increasing PL content up to 2%, the tensile strength significantly increase of about 54.5%, whereas for higher concentration (i.e. up to 4%) the improvement was attenuated (~19.2%), whereas a further increase of the concentration (up to 6%) determined a reduction of the measured value of about 12.1% since, as the PL content increased, a lower number of free OH groups were available to create hydrogen bonds.

Recently, *lipid-based polymers* have been largely used to develop novel active food packaging materials. The utmost potentiality of these polymers is inherently related to their biodegradable, biocompostable, sustainable and non-toxic features, although the high sensitivity to moisture and low mechanical resistance could represent their main disadvantages. Gao et al. (2019) have evaluated the effect of different *tea polyphenols* loadings into cross-linked pectin and chitosan matrix on the physical and functional properties of these novel polysaccharide films (CPTF). It was assessed that the addition of tea polyphenols improved the tensile strength and elongation at break of the resulting films. Tea polyphenols are constituted by a large numbers of hydrophilic groups (such as phenol hydroxyl), which can trap unbounded

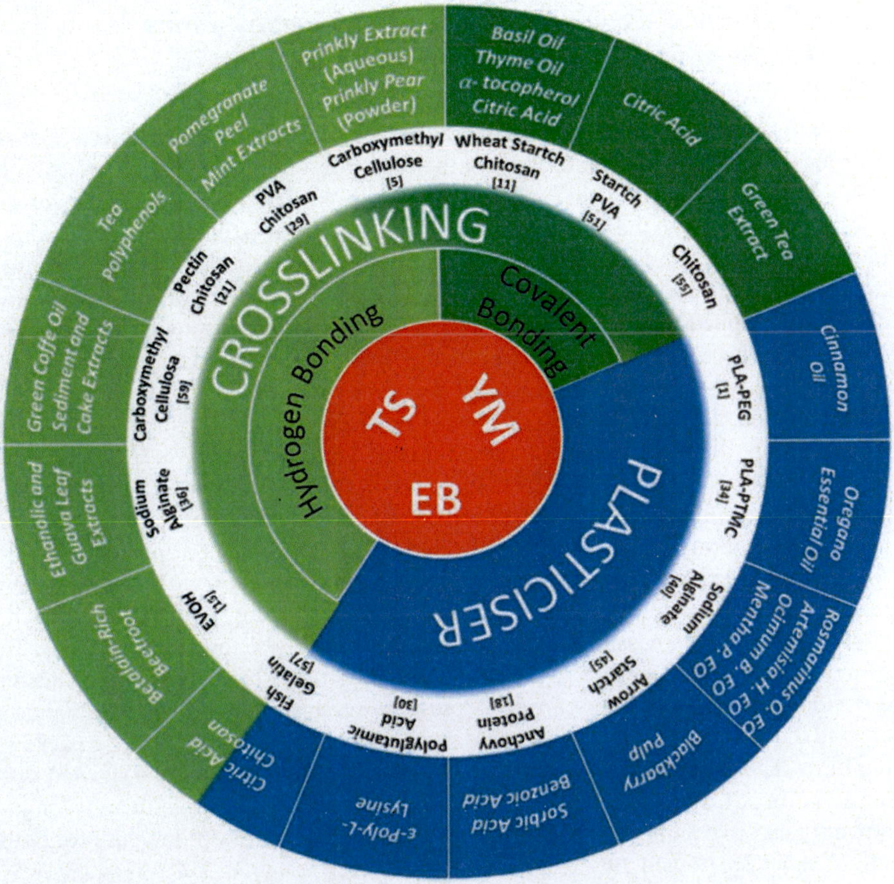

Fig. 8.3 Schematic representation of the various bioactive packaging films obtained by solvent-casting technique reviewed in this paragraph. They are catalogued according to the different interaction mechanisms among the bioactive compounds and the polymeric matrices, which can lead to the modification of the mechanical properties of the resulting biocomposites

water molecules and promote the water penetration allowing the formation of many hydrogen bonds with polymeric substrates (including chitosan and pectin). This weak bond architecture may reasonable trigger the formation of a more complex network liable for the physical property improvement.

The various bioactive packaging films obtained by solvent-casting technique reviewed in this paragraph are schematically reported in Fig. 8.3, cataloging them according to the different interaction mechanisms among the bioactive compounds and the polymeric matrices.

2.2 *Mechanical Properties of Flexible Bioactive Films Obtained by Melt-Mixing/Extrusion Techniques*

Melt extrusion technique is mainly used to process active-packaging films at industrial level and it is often preferred to other solution casting methods due to both faster processing time and lower energy consumption. This process mainly involves two subsequent steps: the preparation of different active masterbatches by blending raw materials and suitable active compounds and additives under specific processing conditions and the extrusion of the pellets to flat sheets by means of twin-screw extruder, equipped with a flat die, or to blow films by using a blown film extruder with an annular die. It has been found that active films obtained by extrusion processes generally show good mechanical behavior and thermal stability (Martinez-Camacho et al. 2013). Nevertheless, to develop active food packaging by extrusion, it is very important to take into account that, due to the sensitivity to high temperatures of the embedded natural substances and/or organic compounds, the integrity and the bioactive efficiency preservation through the optimal setting of processes parameters in terms of temperature, time and heating rate must be ensured.

The incorporation of natural antimicrobial additives, such as *citral* (CI) and *trans-cinnamaldehyde* (TC) into packaging films by using melt extrusion techniques is not always possible due to the active compounds' thermal sensitivity. To overcome this limit, Chen et al. (2019) have encapsulated bioactive components in β-cyclodextrin (β-CD), as a carrier to preserve their antimicrobials effectiveness beyond the film extrusion stage, in the development of EVOH-based active packaging films. The experimental tests on the manufactured composite films showed that the addition of CI and β-CD-CI acted differently on the tensile strength and elongation at break. In details, the analyzed values revealed a significant increase for both TS, of about 38.7% and 29% respectively, and for EB, of about 122.9% and 30.8%. A completely different effect is recorded in the case of active films containing neat and encapsulated *trans-cinnamaldehyde (*EVOH-TC and EVOH- β-CD-TC). Indeed, in both cases the value of TS was negligibly modified, whereas the elongation at break sensibly decreased in the EVOH-TC films due to an embrittlement of the final composite and a significantly increased (84.6%) in the EVOH-β-CD-TC films due to a strong plasticization effect.

Biodegradable active films based on polylactic acid (PLA) and poly(3-hydroxybutyrate-4-hydroxybutyrate) (P3,4HB) loaded with *ginger (Zingiber officinale Rosc.)* and *angelica (Angelica acutiloba – Sieb.et Zucc. – Kitag) essential oil* were developed and investigated by Jiang et al. (2020). The resulting composite films showed an increase in terms of elongation at break, improving their ductility and brittleness. The two essential oils probably acted as plasticizing agents, by enhancing the polymer chain mobility.

Gavril et al. (2019) have developed an active polylactic acid (PLA) films containing powdered dry *leaves of sage and lemon balm.* Results of tensile tests indicated that the mechanical performances of PLA matrix were not substantially affected by the addition of these two natural active compounds. In fact, the incorporation of lemon balm brought no significant changes in tensile strength with respect to pristine PLA films, whereas in presence of sage leaves it significantly decreased, likely due to its uncomplete dispersion within the polymer matrix, thus forming high number of agglomerates. The resulting composite films did not show any difference also in terms of elongation at break.

Rosemary extract was used to develop bioactive nanocomposite films of starch-glycerol-based films by extrusion followed by thermo-compression by Estevez-Areco et al. (2019). At low concentration loading of spherical rosemary nanoparticles, starch-glycerol based composite showed a similar Young's modulus and higher stress and strain at break compared to pristine film. Indeed, the rosemary nanofiller was homogeneously dispersed into the hosting polymer, avoiding cracks propagation without lowering the Young's modulus. As the extract amount, increased, composite films showed a significant improvement of the tensile toughness values, but also a decrease of Young's modulus and stress at break values. These results can be related to two concurrent effects: the reduced reinforcement efficiency of rosemary nanoparticles due to the formation of agglomerates and the plasticizing effect due to the low molecular weight of rosemary, which can interact with the polymer chains increasing their mobility. The trend of the reported date for TS and EB clearly shows that above a specific percentage of the rosemary filler (~3%), the induced reinforcement effect is attenuated and inverted. Likely, due to the worse level of dispersion, at filler content higher than 3% the composite is less resistant and the maximum extension before failure is drastically reduced.

Recently, Menzel et al. (2020) have valorized an extract from rice straw antioxidant behavior to develop potential biodegradable packaging materials incorporating it into potato starch films by melt blending and compression molding techniques. As plasticizing agent, they used glycerol, which reduced the inter-chain hydrogen bonding associated to an increase of the starch polymers space and induced a gliding of starch chains and consequently the material elongation. The antioxidant *rice straw extract* addition to the starch-glycerol system increased both tensile strength and Young's modulus, but it detrimentally affected the extensibility of the final composite films. This latter finding became more relevant if the antioxidant extract partially replaced glycerol in the starch-glycerol films. Reasonably, due to cross-linking between the phenolic compounds and starch, the intermolecular movements were hindered, thus making the films more rigid.

Moura et al. (2018) developed active low density polyethylene-based films containing a natural commercial *carotenoid (Vegex – NC 3c WSP mct) and yerba mate (Ilex paraguariensis) extracts* (CE and YME). The addition of both additives in the film formulations induced a reduction in tensile strength, which became more pronounced at high concentration of yerba mate extract (~26,7% at YME concentration of 1.76%). Generally, among flavonoids, carotenoids, vitamins and derivatives, the extracts are substances rich in phenolic compounds, constituted by one or more aromatic rings with hydroxyl substituents that cannot interact with the polymer matrix. Probably the stress reduction can be attributed to the aromatic incompatibility with LDPE polymer at high concentration. Most of the developed formulations containing the two essential oils show a reduction in the elongation at break with respect to pristine LDPE. Only two formulations, characterized by 0.88% YME and 0.88% CE respectively, revealed an increase of this property, leading to a plasticizing effect. Probably the synergistic action of both additives in the other formulations reduced their plasticizing action. The Young's modulus values also decreased compared with the control for most of the formulations, indicating that the polymer-additive interaction negatively affected this parameter.

Mechanical properties of polycaprolactone (PCL)-based composite films embedding grapefruit seed extract (GSE) different concentrations were investigated by Lyu et al. (2019). At high GSE amount, the tensile strength significantly decreased, whereas the elongation at break gradually increased. This phenomenon can be related to the presence of glycerol in GSE, which acts as plasticizing agent, enhancing the polymer chain mobility and consequently the flexibility of the resulting final composite films. The trend of the calculated filler efficiency is reported in Fig. 8.4, it can be clearly observed that the plasticizing mechanism is reduced with increasing filler content until it reached the plateau level for percentage above 3%.

The effect of organic acid content, such as citric acid (CA), added to fish gelatin films on optical, mechanical and barrier properties was studied by Uranga et al. (2016). They showed that mechanical performances of fish gelatin films improved with the addition of CA. In details, the tensile strength and elongation at break significantly increased when the percentage of citric acid reached 30 wt.%. This result could be explained considering that the reaction between citric acid and fish gelatin induced stronger interaction than those existing among the gelatin polymer chains. When the amount of filler became higher than 20%, some citric acid remained

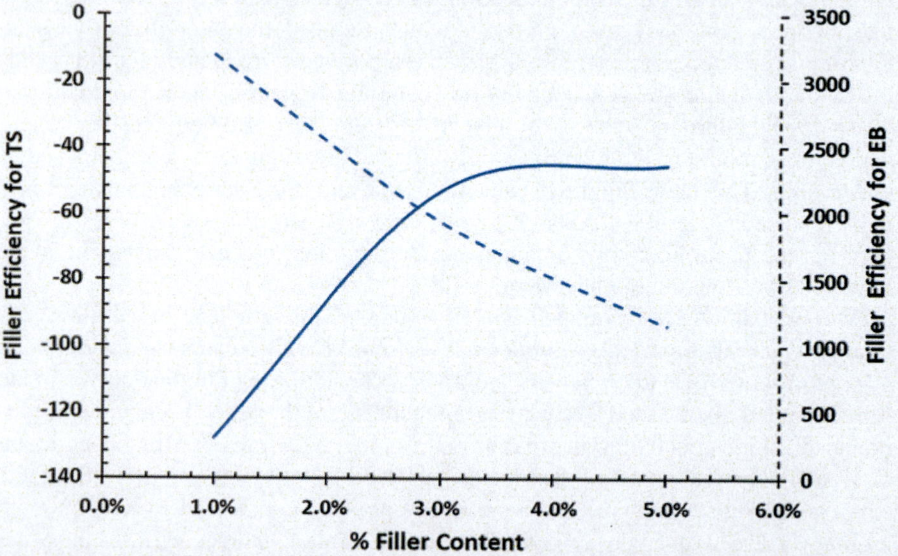

Fig. 8.4 Active compounds efficiency for TS (solid curve) and EB (dash curve) vs AC content into the matrix: grapefruit seed extract embedded into polycaprolactone. (Data adapted by Lyu et al. 2019)

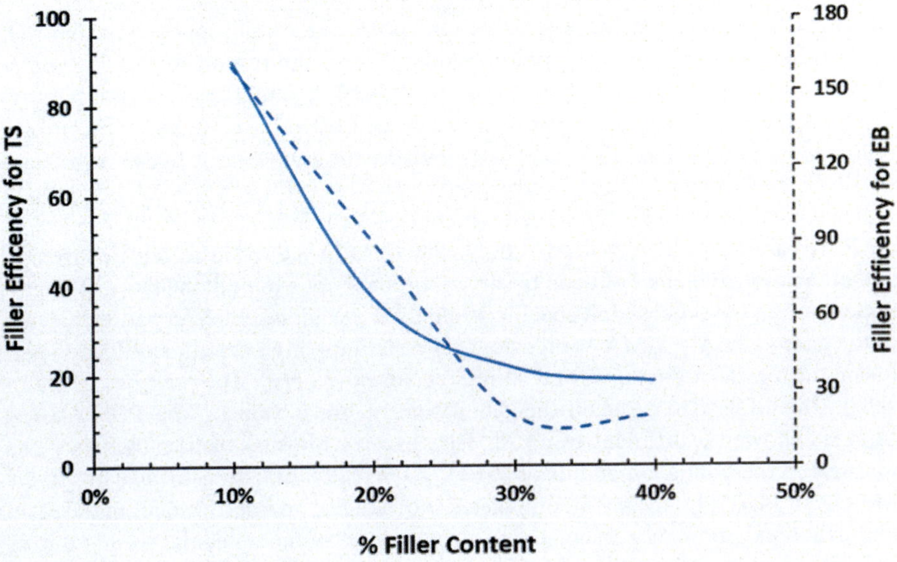

Fig. 8.5 Active compounds efficiency for TS (solid curve) and EB (dash curve) vs AC content into the matrix: citric acid embedded into fish gelatin films. (Data adapted by Uranga et al. 2016)

unreacted. Analyzing the calculated efficiency of the filler (Fig. 8.5), it can be highlighted that the reaction between citric acid and fish gelatin induced a significant variation of both TS and EB at low loading percentage. At higher filler content, it was attenuated, particularly, at values above 30%, the citric acid was unreacted, and thus the efficiency was levelled to a plateau value.

In the following the use of chitin and/or chitosan as bioactive filler in various matrices has been described. Salaberria et al. (2017) have investigated the compatibility of *chitin nanocrystals* with PLA matrix to produce potential active food packaging with antifungal activity. The incorporation of modified chitin nanocrystals into PLA matrices led to a reduction in terms of Young's modulus, tensile strength and elongation at break (~ – 33% for the sample loaded with acetic anhydride at 1%); this effect could be associated to the C_2CHNC and C_{12}CHNC aggregation.

Bie et al. (2013) have investigated polylactic acid (PLA)/starch/chitosan antimicrobial packaging materials, in which, *chitosan* was the antimicrobial bioactive agent and starch was the additive aimed to improve the hydrophilicity of the matrix. Pristine PLA/starch/chitosan film showed a decrease in the tensile strength and an increase in the elongation at break. When the compatibilizers, methylenediphenyl diisocyanate (MDI) and the maleic anhydride (MA), were added to the blend formulations, the tensile properties of PLA/starch blends were maintained similar to the pure PLA values. It was found that tensile strength of the final blends was partially reduced due to chitosan release. The detrimental effect on the mechanical properties was likely associated to the migration of the starch/chitosan aggregates, which, in presence of water, diffused from the PLA matrix thus leaving a porous structure, which resulted progressively more brittle as the chitosan release occurred. As already discussed, chitosan hydrophilicity is the main responsible of incompatibility with most of bio-based polymers, such as poly(lactic acid) (PLA) or polyhydroxyalkanoates (PHA). Recently Vernaez et al. (2019) have studied the *chitosan* (CS) compatibility with PLA and poly(3-hydroxybutyrate-co-3-hydroxyvalerate) (PHBV), functionalizing the thermoplastic polyester by reactive blending with maleic anhydride (MAH) and glycidyl methacrylate (GMA) in an internal mixer. In all PLA-based and PLA-g-MAH composite films, a reduction in tensile strength and elongation at break has been observed, probably due to both the thermal degradation during the process and specific strength of the added chitosan particles. Only in the case of PLA-g-GMA composite film containing high amounts of loaded chitosan, the tensile strength showed a high increase (~39.5%). The presence of chitosan in the PLA also reduced the elongation at break values. The PHBV-based samples showed a different trend. In fact, as the chitosan loading increased, an increase in the Young's modulus of PHBV-g-GMA/CS films was clearly observed. Moreover, PHBV-g-GMA/CS composites showed also an improvement in the elongation at break, probably associated to a small decrease in molecular weight during the processing and to a negligible change in crystallinity of the produced composites. It has been highlighted that the functionalization of the polymer chains by glycidyl methacrylate led to an increase of the compatibility of PHBV with chitosan, thus improving the overall mechanical performances of the resulting composites.

The various bioactive packaging films obtained by melt mixing/extrusion technique reviewed in this paragraph are schematically reported in Fig. 8.6, cataloging them according to the different interaction mechanisms among the bioactive compounds and the polymeric matrices.

Since it is necessary to take into account all the variables involved in the behavior of the different investigated systems, it is not possible to define a specific trend in terms of modification of mechanical properties, and mechanical parameters such as TS, EB and YM for the several papers reviewed in these paragraphs. However, for the sake of comparison, in Fig. 8.7 a representation of some of the collected data are reported in terms of TS and EB and compared with those ones of LPDE, PLA, EVOH and PET that can be considered among the mostly widely used and commercially available polymers employed in the food packaging industry. It can be highlighted that bioactive films based i.e. on proteins, alginates and CMC and obtained by solvent casting technique exhibit, mainly for TS, values of many tenths of percentages lower than the considered benchmarks (i.e. PLA, PET and EVOH), whatever bioactive filler is considered. Active films obtained by melt mixing/extrusion technique exhibit TS values closer to those ones of the most industrially adopted systems. It is worth noting that mechanical properties of biopolymers obtained by solvent casting technique are still quite far from reaching values suitable for their use as self-standing materials. For this reason, they are generally more suitably used as coatings on traditional/commercial polymeric substrates.

Fig. 8.6 Schematic representation of the various bioactive packaging films obtained by melt mixing/extrusion technique reviewed in this paragraph. They are catalogued according to the different interaction mechanisms among the bioactive compounds and the polymeric matrices, which are involved in the modification of mechanical properties of the resulting biocomposites

3 Barrier Properties of Packaging Flexible Films Containing Various Bioactive Compounds

As already described in the Introduction Section, barrier properties toward low molecular weight compounds, particularly water and oxygen molecules, and toward UV-light are of utmost importance in the study and development of food packaging materials. In the following paragraphs the influence of bioactive compounds on these two important properties will be carefully described and analyzed.

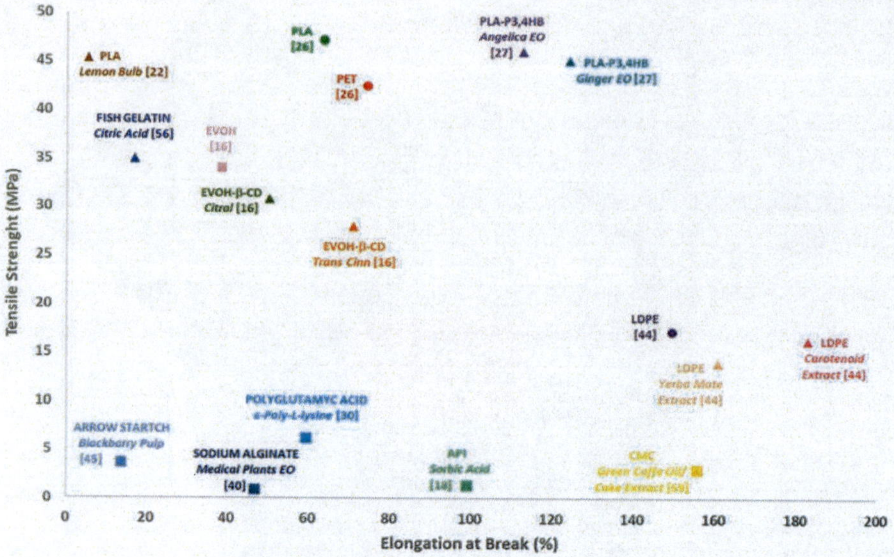

Fig. 8.7 Representation of TS vs EB data from some of the papers reviewed in these paragraphs, compared with those ones of LPDE, PLA, EVOH and PET that can be considered among the most widely used and commercially available polymers employed in the food packaging industry (● round symbols have been used for benchmarks, ■ rectangular symbols for films obtained by solvent-casting technique ▲triangular symbols for films obtained by melt mixing/extrusion)

3.1 Water and Oxygen Barrier Properties of Flexible Bioactive Films

Food packaging materials generally requires barrier properties that must be adequate to the specific features of the packaged food product and to their final use. In food packaging applications, oxygen and water mass transport are very important factors to be minimized in order to not alter the quality and improve the preservation of packaged food products (Brody et al. 2008).

The transfer of vapor water molecules is influenced by the ratio between the hydrophilic and hydrophobic compounds in the film and, in order to prevent the dehydration of the food, water vapor permeability must be as low as possible (Ma et al. 2008). A potential approach to reduce water vapor permeability is to use hydrophobic species in the film or multilayer structures in which laminates exhibit different barrier properties. In general, water vapor permeability of traditional polymer and biopolymer film is markedly influenced by many parameters such as hydrophilicity of the plasticizer, type, structure, quantity, particle size of additives as well as the external humidity, internal water activity of the food and thickness of the film (Gutiérrez et al. 2015). Indeed, the additive typology and its particles size can create different interactions and/or a good dispersion with/within the polymer matrix and consequently modify the properties of the film. Internal water activity, which depends on the moisture content of food, and external humidity strongly influence water permeability of the films due to the strong interactions which can occur among water molecules, other chemical species and polymer chain.

Also, oxygen permeability is strongly related to the environmental relative humidity and the moisture content in foodstuff. In fact, water molecules have a plasticizer effect for the polymeric film, by increasing the chain mobility, thus influencing the solubility and the oxygen molecules diffusion. The reduction of oxygen permeability is generally achieved with the use of suitable additives which modify the polymer structure and its free volume, the use of high barrier thin coatings on the polymeric substrates as well as the use of suitable (nano)particles which slow down the diffusion process of oxygen molecules.

Generally speaking, additives, which act as plasticizers, can interact with the polymer network in different ways, increasing chain mobility or creating discontinuities in the polymer chain. In the first case the polymer chain can relax and shift its structure, allowing the access into the formed void spaces. In the other case, the polymer matrix can reorganize in a denser packed structure, reducing the molecular transport. Moreover, diffusion of low molecular weight compounds occurs through the polymeric film via the amorphous region, thus the crystallinity degree of the polymer matrix strongly influences the mass transport of the gas and vapor through the film.

The diffusion of small molecules in polymers depends on the available space in the polymer matrix (free volume and its distribution), on the flexibility of the polymer chains and their side groups and on the interaction forces between the polymer and the diffusing molecules. The concept of free volume is based on the unoccupied space between the end polymer chains, it depends on the number of chain ends and hence on the degree of polymerization. The adsorption and diffusion of molecules in polymers greatly depend on the available free volume, particularly, all the effects that induce an increase of free volume enhance the molecular motion into the polymeric system.

It is well established that the introduction of specific small molecules can act as plasticizer for polymers. The effect of plasticization is to increase the free volume of the system (Fig. 8.8a). Consequently, the polymer chains can move more easily increasing their flexibility and, at the same time, the low molecular weight compounds can diffuse more rapidly through the polymer chains (Machin and Rogers 1972).

On the contrary, polymer crosslinking reduces the polymer free volume (Fig. 8.8b). As the crosslinking density increase, the free volume available in the structure continuously decrease, thus reducing the diffusion rate low molecular weight chemical species (Krongauz 2010).

Generally speaking, the structure of the polymer, the presence of additives, the solvent molecules, and the chemical-physical interaction between polymer end chains and small molecules influence the diffusion process and thus the water and oxygen permeability of the composite films. Bioactive additives can act as plasticizers or crosslinkers depending on the polymer matrices, and hence can modify their functionality and barrier properties. In the following paragraphs a detailed analysis

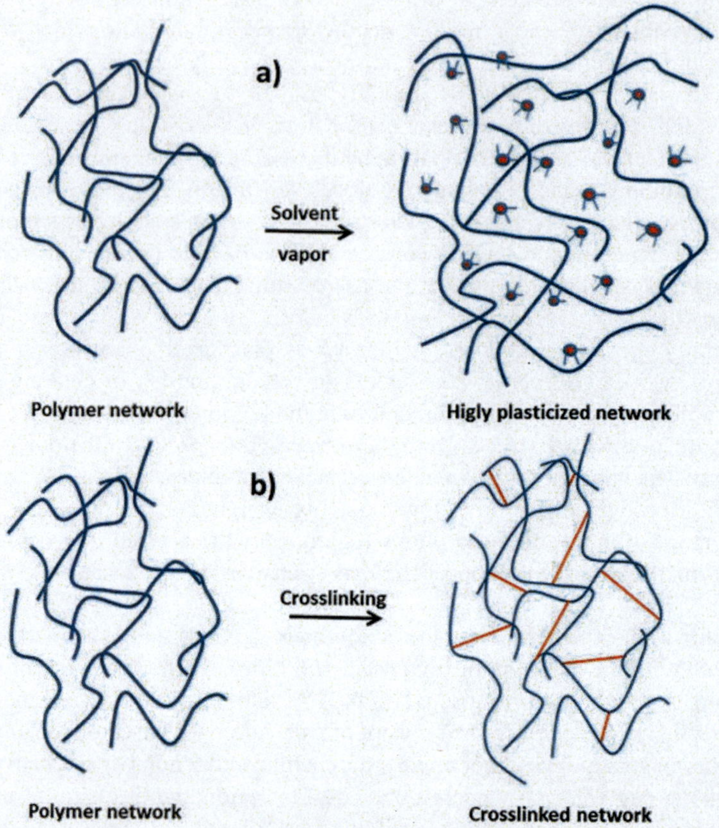

Fig. 8.8 Effect of plasticization (**a**) and of crosslinking (**b**) on polymer chains

of the effect of plant extracts, essential oils and other bioactive compounds on the water vapor and oxygen permeability of the resulting bioactive packaging films will be reported.

Plant Extract/Essential Oil

In the recent literature, numerous papers deal with the study of the effect of various essential oils and plant extracts on the water and oxygen barrier properties of the resulting bioactive films. In some cases, water vapor and oxygen permeability are not influenced by the presence of these active compounds, whereas in many others, barrier properties are significantly modified. In the following a detailed description of the effects and possible interaction mechanisms of natural additives with polymeric matrices are described and discussed.

As far as thymol essential oil is concerned, different results were found according to the polymer matrix in which it was embedded. Kavoosi et al. (2013) proved that the addition of thymol in a gelatin film containing also glycerol and glutaraldehyde (GA) as crosslinkers, led to a reduction of water barrier of the resulting film.

The cross-linking of the matrix due to the presence of GA reduced the penetration of the water vapor molecules across the film and brought a significant decrease of the water vapor permeability; however, the addition of thymol, especially at high concentration, likely hindered the polymer chain-chain interaction resulting in a reduction of crosslinking and consequently leading to an increase of water vapor permeability. On the contrary, when thymol essential oil was added to poly(lactic acid)/poly(trimethylene carbonate) (PLA/PTMC) matrix (Wu et al. 2014), water vapor barrier increased significantly, especially at high oil content. In particular, as the thymol concentration increases from 0 to 12%, the water vapor permeability of the loaded film decreased of about 45%, thus potentially increasing the conservation of packaged foodstuff also due to a significant increase of antimicrobial activity of the film.

The incorporation of several essential oils of *R. officinalis* L., *A. herba alba* Asso, *O. basilicum* L. *and M. pulegium* L. in sodium alginate films showed an increase of barrier properties of the composite bioactive films (Mahcene et al. 2020). The essential oils were uniformly dispersed in the polymer matrix, thus affecting the hydrophilic characteristic of the film and increasing the solubility of the oxygen molecules in the non-polar oily phase. Water and oxygen permeability of the films with different essential oils did not show significant differences during the same days of storage, but an increase of permeability of all the samples (control and active films) has been observed during the analysis period. However, the resulting films were effective in protecting the model food, since due to the presence of sodium alginate film, the sunflower oil showed a reduction of peroxide value of about 45%. The observed decrease is the result of the interaction between alginate matrix and oxygen of bioactive substance.

As for cinnamon essential oil, it has been shown that the use of *cinnamon bark oil* and *sea squirt shell protein* into polylactic acid film showed a reduction of water vapor permeability of about 50% with respect to the film without active compound (Beak et al. 2018). This is due to the hydrophobicity of cinnamon bark oil that increased the hydrophobic character of the polymer and improved its barrier properties, thus showing promising features as bioactive packaging film. A different effect has been highlighted by Perdones et al. (2014) who studied the effect of *cinnamon leaf essential oil* and *oleic acid* on barrier properties of film incorporated with chitosan. The presence of oleic acid reduced water vapor permeability of chitosan composite film while the addition of cinnamon essential oil enhances water vapor permeability. In fact, cinnamon oil acted as plasticizer resulting in an increase of polymer chain mobility and thus in an increase of the motion of water molecules into the film. Only the synergic addition of both cinnamon leaf essential oil and oleic acid led to a decrease of water vapor permeability because oleic acid avoids the cinnamon oil migration from the film during drying.

Active bio-based nanocomposite films were developed by Hosseini et al. (2016) by adding *oregano essential oil* and *chitosan nanoparticles* into fish gelatin. The introduction of low amount of oregano oil (0.8% w/v) in the film reduced of about 35% the water vapor permeability. Pelissari et al. (2009) founded comparable results when low amount of oregano oil were embedded in starch-chitosan films (WVP for the maximum concentration of oregano oil (1.0%) was 0.6210–10 g/Pa m s). However, they also found that as the oregano oil amount increases, the water vapor permeability of bio-nanocomposite increased. Indeed, the resulting water absorption capacity of the polymer network was significantly influenced by the different interaction between oil constituents and some protein tails of the matrix which, in turn, were significantly depending on the oil concentration. Liu et al. (2016) also reported on the addiction of *oregano essential oil* into Poly(Lactic Acid)/Poly(Trimethylene Carbonate) films. They found that water vapor permeability increased because, due to the presence of the EO, the average pore size of the films increased and the crystallinity of the polymer matrix significantly decreased, thus affecting its mass transport properties.

Recently Vidal et al. (2020) embedded *green coffee oil* into carboxymethyl cellulose (CMC)-based film. CMC has a hydrophilic character due to the presence in the structure of free -OH and -CH groups that can interact with water molecules. Green coffee oil at 0.2% was added in the film formulation since compounds with polar groups can likely form hydrogen bonds with the free groups of -OH and -CH of CMC and make the polymer structure less accessible to water molecules. Despite this feature, the presence of green coffee oil was able to decrease water permeability of only about 15% whereas a significant improvement of elongation at break has been observed, as previously reported.

As for the addition of extracts by natural sources, the introduction of phenolic compounds derived by *rice straw source* in starch-based film has been recently investigated by Menzel et al. (2020). An increase of barrier properties of the resulting bioactive film has been highlighted, in particular, oxygen barrier improved without compromising water barrier properties. Even if glycerol and water acted as plasticizers in the film, the addition of the antioxidant extract led to a reduction of oxygen diffusion with respect to control film of starch and glycerol, due to the molecular interaction between phenolic compounds and starch hydroxyl groups, thus showing a reduction of oxygen permeability of about 70%.

When natural extracts of *aqueous mint* and *pomegranate* were embedded into polyvinyl alcohol and chitosan matrix, the addition of the antioxidant compounds did not change the barrier properties of the resulting bioactive film (Kanatt et al. 2012). Water vapor transmission rate decreased significantly as the amount of chitosan increased, being dependent on the ratio of the hydrophilic/hydrophobic components of the compound. As for the oxygen barrier properties, chitosan formed an intermolecular hydrogen bond with PVA; the polymer chain mobility was restricted and the oxygen motion through the polymeric film was hindered.

As for extracts from leaves, recently biodegradable films were produced by embedding different concentration of *olive leaf extract* into biopolymeric matrix of carrageen (Silveira da Rosa et al. 2020) and *guava leaf extract* into sodium alginate (Luo et al. 2019), each of them leading to different effects. The presence of polar compounds of olive leaf extract increased the hydrophilicity of the carrageen-based film, thus enhancing the water vapor permeability. In the case of guava leaf extract, the water vapor permeability of sodium alginate-based film decreased because morphological analysis showed that the network formed by the addition of polyphenols made the polymer structure denser and packed, thus being less accessible to the diffusion of water molecules.

Interesting results in terms of improvement of water barrier properties have been obtained by Wu et al. (2019) who added *tea polyphenols* (TP) into pomelo peel flour to realize bioactive packaging films. They showed that low concentration values of TP led to an enhancement of water barrier properties, due to the interfacial interaction between the polyphenols and the biopolymer leading to the formation of a more packed structure which hinders water molecules diffusion. On the other hand, as the loading content increases, TP formed agglomerates and voids in the polymer matrix, thus resulting in an increase in water vapor permeability. Wang et al. (2013) found similar results, highlighting a significant enhancement of water vapor barrier properties (about 50%) incorporating TP in chitosan based film. In fact, the decrease of the water permeability of the bioactive chitosan/TP films is due to the fact that the presence of TP lead to a denser structure of the chitosan chain in the film, thus reducing the interstitial space in chitosan matrix and making more difficult the diffusion of water molecules through the films.

As for the introduction of natural colorants extracted from fruit or vegetables, the incorporation of natural red color of *betacyanin (NRC)* into hydroxypropyl methylcellulose affected the barrier properties of the resulting film (Akhtar et al. 2013). The presence of phenolic group of NRC increased polymer chain mobility, the polymer network became less dense and water vapor diffusion through the film was enhanced. SEM results clearly showed that the porosity and the smoothness of the film are diminished with the incorporation of the colorant, thus leading to an increase of the hydrophilic nature of the film and to an increase in water vapor permeability. On the contrary, morphological analysis highlighted the presence of the NRC in the pores resulting in a denser structure, less permeable to oxygen molecules. Indeed, film containing 4%w/w of NRC exhibited oxygen permeability values significantly reduced, of about 80%, with respect to pristine film, thus being potentially capable to protect the food product against oxidative deterioration.

Acids, Alpha-Tocopherol and Bioactive Nanoparticles

Organic acids are largely used as antimicrobial agents in food industry and can be potentially incorporated into active packaging film. The most common organic acids used in film packaging are sorbic acid, citric acid, lactic acid, acetic acid and their salt. It has been shown that their presence generally increases water and oxygen permeability of the resulting bioactive films, regardless the type of polymeric matrix. López-de-Dicastillo et al. (2012) report on the incorporation of *ferulic acid* and *ascorbic acid* together with *quercetin* and *green tea* into ethylene vinyl alcohol (EVOH) films. Barrier properties of the hydrophilic EVOH depends strongly on the relative humidity (RH) of external environment. In fact, water molecules act as plasticizer enhancing the flexibility of the polymer chains and thus reducing barrier properties. For this reason, different behaviors have been observed, depending on the values of external RH. At 35% of relative humidity, the addition of ferulic acid, quercetin or green tea led to an increase of water permeability of the EVOH film, whereas the addition of ascorbic acid led to a reduction of WP due to the interaction between hydroxyl groups of the polymer and the acid. At 90% of relative humidity the results are different: the incorporation of quercetin or green tea increased water barrier properties due to an increase of hydrophobicity of the composite due to a low affinity between additives and water. Differently, the ascorbic acid solubilized in the polymer and acted as plasticizer resulting in an increase of water permeability.

When *ellagic acid* was embedded by solvent casting technique into chitosan matrix (Vilela et al. 2017), it has been shown that its addition increased the water vapor permeability of about 30%. However, this increase was counterbalanced by a very good antioxidant activity of the resulting bioactive films which, in turn, were able to increase the shelf life of the packed product. Also the synergistic use of *sorbic acid* (SA) and *benzoic acid* (BA) embedded into an antimicrobial film obtained from protein isolate from Argentine anchovy (Da Rocha et al. 2018) showed an increase in water permeability (WVP) of the film which increased of about 30% and 45% with the addition of SA and BA, respectively, due to the effect of both acids to open the polymer chain thus increasing the water mobility.

Alpha-tocopherol is a lipophilic antioxidant largely used in food packaging to preserve the food by degradative oxidation phenomena. Jongjareonrak et al. (2008) investigated the antioxidant activities and properties of gelatin film loaded with α-tocopherol and butylated-hydroxy-toluene (BHT). It was found that the water vapor permeability of the film was reduced because the additives increased the hydrophobicity of the polymeric composite. BHT and α-tocopherol are not soluble in water but can be well dispersed in gelatin film. Aliphatic chain of α-tocopherol is longer than BHT, thus the smaller molecules of BHT can be better and more regularly dispersed in the polymer than α-tocopherol. For this reason, the composites containing BHT exhibited more effective water barrier properties than those ones containing α-tocopherol. As the time of storage increased, water vapor permeability continuously decreased due to a renaturation of gelatin molecules as well as to a more pronounced cross-linking effect between gelatin backbone or between antioxidants and gelatin during extended storage, resulting in a denser and less permeable network structure.

The synergistic incorporation of *α-tocopherol*, *citric acid*, *basil essential oil* and *thyme essential oil* into a film of wheat starch and chitosan with glycerol as plasticizer (Bonilla et al. 2013) has showed an increase of water vapor barrier properties. Especially, the presence of α-tocopherol increased the pathway tortuosity caused by the low miscibility between the antioxidant and the polymers resulting in a reduction of water vapor permeability. Citric acid acted as crosslinker of the polymer chain inhibiting the diffusion of the water molecules through the polymeric matrix, whereas basil and thyme essential oils did not affect significantly water vapor permeability. On the contrary, oxygen barrier properties were significantly enhanced (of nearly 50%) as compared to neat films, likely due to the blocking effect of the antioxidant. In general, when liquid lipids are embedded into hydrocolloid films, oxygen permeability increases due to an increase of oxygen solubility in the polymeric matrices (Fabra et al. 2016). However, when the lipid compound exhibits also antioxidant activity, a chemical oxygen blocking mechanism occurs thus leading to an enhancement of oxygen barrier properties. Alpha-tocopherol has been often used as a model antioxidant and it was found that its antioxidant activity was preserved during encapsulation process in hydrocolloid matrices (whey protein isolate, zein and soy protein isolate) and the resulting active coating led to an increase of barrier properties of the packaging film. *Bioactive micro/nanoparticles* in composite packaging films act as physical barrier in the polymer matrix. The structure becomes more complex thus, gas and vapor molecules diffuse more difficultly through the polymeric film, and this generally results in an improvement of gas and vapor barrier properties. Recently Yang et al. (2020) developed a composite film with improved barrier properties by embedding into grafted PLA star-like-*lignin microparticles* (NPG), the second most plentiful natural polyphenol, which have drawn much attention due their antioxidant, UV protection, antimicrobial and strengthen effects. Indeed, the addition of NPC significantly enhanced PLA crystallization, moreover the good compatibility between epoxy group of NPC and carboxyl and hydroxyl group of (LMP-g-PLA) acted as physical barrier in PLA matrix, resulting in a more tortuous polymer pathway. Hence, the transport of water vapor molecules through the film result more difficult, thus leading to a water barrier improvement of about 50%.

Arkoun et al. (2018) investigated the oxygen and water barrier properties of the bioactive film obtained by direct electrospinning the *chitosan/PEO nanofibers* (CNFs) on top of a multilayer film used to package meat. They found that oxygen permeability remained almost unaltered whereas water vapor permeability decreased of about 20% with respect to pristine film. In this case, the improved barrier properties, the presence of the hydrophilic chitosan which promoted the absorption of water molecules of the food acting as "moisture absorber" as well as its antimicrobial activity led to the reduction of bacterial growth and to the preservation of the quality and freshness of meat, along with shelf life extension to 1 week, without altering the organoleptic quality of the product. The addition of microfibrillated cellulose (MFC) and *clove essential oil* in active nanocomposite protein films based on soy protein isolate has been investigated by Ortiz et al. (2018). They highlighted an increase of oxygen permeability and a significant decrease of water vapor permeability. The addition of MFC showed an improvement in barrier properties caused by the good interaction between MFC and soybean proteins which likely causes a network of fiber inside soybean matrix. The incorporation of clove essential oil in protein and nanocomposite film further decreases WVP, achieving a reduction of about 40% due to a better dispersion of clove essential oil in synergy with MFC. In Table 8.1 is reported a summary of the influence of various bioactive compounds embedded into different polymer matrices on the water and oxygen barrier properties of the bioactive films reviewed in this paragraph. They are catalogued according to the different interaction mechanisms among the bioactive substances and the polymers which are involved in the modification of barrier properties of the resulting bioactive packaging films.

Table 8.1 Summary of the influence of various bioactive compounds embedded into different polymer matrices on the water permeability (WP) and oxygen permeability (OP) of the bioactive films reviewed in this paragraph, catalogued according to the different interaction mechanisms among the bioactive substances and the polymers

Active compound (AC)	Polymer matrix	Effect of AC on *WP*	Effect of AC on *OP*	Mechanism	References
Several essential oils *R. officinalis* L., *A. herba alba* Asso, *O.* basilicum L. And *M. pulegium* L.	Sodium alginate	↓	↓	Plasticizing	[40]
Cinnamon bark oil	PLA	↑		Plasticizing	[8]
Oregano essential oil	PLA/PTMC	↑		Plasticizing	[34]
Betacyanina	HPMC	↑	↓	Plasticizing	[3]
Ascorbic acid	EVOH	↓ (RH = 35%) ↑ (RH = 90%)	↓ (RH = 35%) ↑ (RH = 90%)	Plasticizing	[35]
Thymol/ glutaraldehyde	Gelatin	↑		Covalent bonding	[31]
Rice straw source	Starch	↑	↑	Covalent bonding	[42]
Tea polyphenol	Pomelo peel flour	↑ (c < 10%)	↑ (c < 10%)	Covalent bonding	[63]
α-Tocopherol and BHT	Skin gelatin	↓		Covalent bonding	[28]
α-Tocopherol	Starch and chitosan	↓	↓↓	Covalent bonding	[11]
Oregano essential oil	Starch and chitosan	↓		Hydrogen bonding	[47]
Green coffee oil	CMC	Not altered		Hydrogen bonding	[60]
Mint and pomegranate extract	Chitosan and PVA	Not altered	Not altered	Hydrogen bonding	[29]
Guava leaves extract	Sodium alginate	↓		Hydrogen bonding	[36]
Tea polyphenol extract	Chitosan	↓		Hydrogen bonding	[62]
Ellagic acid	Chitosan	↑		Other	[61]
Ag-cu NPs + cinnamon oil	LDPE	↑	↑	Other	[2]
Chitosan nanofibers	Chitosan	↑	Not altered	Other	[6]
Thymol	PLA/PTMC	↑		Other	[64]
Olive leaves extract	Carrageen based film	↑		Other	[54]
MF cellulose and clove EO	Soybean protein	↓	↓	Other	[46]
Oregano essential oil	Fish gelatin/ chitosan	↓		Other	[25]
Star-like-lignin MPs	PLA	↓		Other	[65]

3.2 UV Barrier Properties of Flexible Bioactive Films

Optical properties of packaging materials are very important features not only from an aesthetic point of view but also for the quality preservation of the food. When a new packaging material is developed, two opposite effects must be taken into account: the barrier performances against UV radiations and the use of transparent films due to the demand of industries and consumers to "see through" the package in order to see the product before purchasing it. For this latter reason, the use of transparent packaging materials within the entire food sector increased significantly. However, the risk of light-induced oxidation in foods packaged in transparent materials is relevant. Indeed, several degradation reactions of both packaging polymeric matrices and packed foodstuff are due to exposure to ultraviolet radiation which, even at small amounts, can induce the breakage of chemical bond and the formation of highly reactive radical species, leading to photolytic autoxidation. Generally speaking, when the UV radiation transmission decreases, the packed food is protected against UV light. Indeed, the production of films with a very good "UV-blocking" capability coupled with the proper transparency would be highly desirable. On the basis of these considerations, the study of UV barrier properties of packaging films is very important and, thus, also the evaluation of the influence of active and bioactive compounds on the UV-visible transmission spectra of the prepared composite films.

Ahmed et al. (2018) reported on the development of compression molded LLDPE films loaded with *cinnamon essential oil* (CEO) and *bimetallic (Ag-Cu) nanoparticles* (NPs) for the packaging of chicken meat. It has been shown that the use of high amount of CEO and NPs turned the nanocomposite film to a more red and yellow color with respect to the transparent LLDPE film, due to the characteristic surface plasmon resonance of Ag-Cu NPs and to the effect of phenolic components of the CEO. The presence of bimetallic nanoparticles led to a significant decrease of the transmittance, likely due to the absorbance of UV light and to the light scattering due to the distribution of NPs in the film matrix.

Hamdi et al. (2019) developed a bioactive film by embedding *carotenoproteins extract* (CPE) into the blue crab chitosan-based films. In the UV range of 200–280 nm, control film exhibited poor barrier property whereas films containing CPE demonstrated excellent UV barrier properties. In the range of 400–800 nm, the incorporation of CPE led to a decrease of visible light transmission absorbance values, at all wavelengths. This is likely due to the light scattering related to the embedment of both CPE and glycerol in the inter-molecular spaces of the chitosan matrix which hinder light transmittance. The opacity values of the bioactive films were significantly higher (about 32%) with respect to the control film, thus being less transparent. This occurred because the carotenoproteins were well solubilized in the matrix, thus leading to a homogeneous matrix as also confirmed by the morphological analysis. Moreover, the presence of the hydrophobic CPE improved films opacity by breaking the continuous ordered structure of chitosan. Particularly, CPE-loaded films became darker as the concentration of CPE increased, thereby light could be scattered, refracted or blocked and, thus, they could be advantageously used to coat or package food sensitive to light.

Recently Moreirinha et al. (2009) embedded *ferulic acid* (FA) and *feruloylated arabinoxylo-oligosaccharides* (FAXOS) into arabinoxylans (AX) suspensions containing different amounts of nanofibrillated cellulose (NFC) to obtain films with antioxidant and antimicrobial properties.

It has been proven that the UV-blocking properties of the control AX-NFC film improved by the addition of both the active compounds FA and FAXOS, thus showing UV absorbing properties in the UVC, UVB and UVA regions. These UV-blocking properties of both AX/NFC/FA and AX/NFC/FAXOS nanocomposite films can contribute to prevent food photo oxidation and, thus, to increase the shelf-life of light-sensitive foodstuffs by acting as UV absorbers.

Pereira et al. (2016) studied the incorporation of *lycopene*, as active compound, in whey protein concentrate (WPC) film containing montmorillonite as reinforcing filler. Results showed that the increase of lycopene content in WPC films led to a decrease in UV and visible light transmission through films. In UV light range (200 to 400 nm), very low values of transmission were noted for all films, especially for those containing lycopene. It is worth noting that, beyond lycopene, the UVC blocking ability can also be attributed to the UV-absorption ability of the aromatic amino acid residues in the protein structure. Moreover, the synergy between lycopene and MMT contributed significantly to reduce film transparency. Control films (WPC and WPC/MMT) exhibited the highest values which gradually decreased as the amount of lycopene increased, due to the coloring ability provided this latter.

Tea polyphenols (TP) have been embedded as bioactive compounds, into pomelo peel flour (PPF) composite film to prepare active food packaging films (Wu et al. 2019). However, due to the good dispersion of TP in the FFP substrate, its presence did not significantly affect the light transmittance of PPF films (63.27%), which was only slightly decreased (to ~58.1%) after addition of 5–20% TP.

Recently Vidal et al. (2020) developed and characterized bioactive films based on CMC incorporated with *green coffee oil* (GCO) and *extracts from green coffee residue*, cake and sediments. The light transmission values of the resulting films increased with the increase of wavelength, showing a maximum at 800 nm. The presence of GCO and extracts led to very low values of transmittance compared to the control film. The film containing sediment extract at 40% showed a 90% reduction of light transmission values with respect to the control films. The very high barrier effect to UV–Vis radiation of the active films can be attributed to polyphenols contained in green coffee cake (CE) and sediments (SE), especially chlorogenic acid, as well as to the synergistic effect of the phenolic compounds and the components of the lipid fraction. Films containing SE and CE were opaquer than control films; in particular, those ones loaded with 40% of SE showed a value of about 10 times higher than the control film, thus being less transparent and consequently more efficient as light barrier.

4 Conclusions

The effect induced by the presence of various bioactive compounds on the mechanical and barrier properties of active packaging films obtained by means of different manufacturing techniques, namely solvent-casting and melt-mixing/extrusion, represents the main focus of this chapter. It has been highlighted that the influence of the filler depends on its specific amount and characteristics, responsible for chemical bonding, physical entanglements or free volume into the polymer network. Nevertheless, also the manufacturing processing, inducing the dispersion and homogenization of the filler within the matrix, can contribute to affect differently the final performances of the active packaging film. Taking into account the above features, various occurring and likely concomitant mechanisms associated to the effect of the specific or multiple fillers were introduced and analysed, namely: (i) *crosslinking*, in terms of both covalent bonding or hydrogen-bonding, (ii) *plasticizing* and (iii) unpredictable events such as worse dispersion, thermal degradation and filler aggregations (herein referred to as *other mechanisms*).

The crosslinking can be related to the hydrogen bond formation tendency which can significantly change the polymer-filler interfacial interaction and also to the bonding of the filler to the neat hosting matrix by means of strong chemical (i.e. covalent) interactions. In the case of mechanical performance, the presence of H-containing groups in the active compound and in the hosting matrix can promote the formation of hydrogen bonds improving the stress transfer at interface matrix/filler by increasing adhesion which leads to a tensile strength enhancement of the final product. When the chemistry of the bioactive filler is capable of forming covalent bonds, an improvement of the pristine matrix mechanical resistance is revealed with a reduction of the elongation at break, generally associated to an overall increase of tensile strength. As for the barrier properties, the H-bonding or covalent crosslinking reduces the polymer free volume and hence the rate of diffusion of low molecular weight compounds, thus enhancing the barrier properties of the resulting active film.

On the contrary, when the presence of the bioactive filler induces a plasticization effect, the three dimensional network of the polymeric matrix is modified, weakening the intermolecular forces with a resulting increase of the available free volume and chain mobility. In this case, mechanical performance is generally modified toward a more flexible behaviour of the final films characterized by a higher elongation at break and a lower elastic modulus. Due to the increase of the free volume, the polymer backbone can move more easily and the low molecular weight compounds can diffuse more rapidly through the chains, thus generally worsening the barrier properties.

Moreover, it must be taken into account that the presence of bioactive additives generates new structures which exhibit different properties with respect to the original polymer matrix. The thermal degradation level of the loaded filler likely due to the processing condition, its non-uniform dispersion or its poor disaggregation can be considered as very important parameters in the final properties of the packaging composites. All these variables, sometimes undesired and unpredictable, may lead to detrimental effects limiting the expected efficiency of the filler in improving the final performances of the bioactive films.

The review work presented in this chapter has highlighted that it is not possible to define a unique behaviour of a specific class of bioactive compounds and/or additives on the mechanical and barrier properties of the resulting active composites. The final performances strongly depend on the material formulation, particularly on the specific polymer/additive interaction and on the active compound concentration, on its dispersion level and distribution uniformity, on the employed manufacturing techniques and their setting parameters, as well as on environmental factors, such as temperature, humidity and partial pressure.

Moreover, it has been highlighted that the UV barrier properties are very important features of packaging material for the protection of food against the photo-oxidation activated by UV radiation. Generally, the addition of bioactive compounds, led to a decrease in light transmission of films for both UV and visible light regions, thus improving their UV barrier. This can be attributed in some cases to the addition of active compounds containing phenolic compounds which exhibit a coloring effect and in other cases to active substances containing aromatic amino acid residues which exhibit UV-absorption ability and, thus an UV blocking effect which can retard lipid oxidation in packaged food. In conclusion, research is still needed on finding new methods and formulations for the development of bioactive packaging film and coatings. Indeed, with almost infinite combinations of polymers, active compounds, suitable additives and manufacturing methods, many solutions can be developed, each adapted to specific food products and consumer needs.

References

Ahmed J, Mulla M, Arfat YA. Thermo-mechanical, structural characterization and antibacterial performance of solvent casted polylactide/cinnamon oil composite films. Food Control. 2016;69:196–204.

Ahmed J, Mulla M, Arfat YA, Bher A, Jacob H, Auras R. Compression molded LLDPE films loaded with bimetallic (ag-cu) nanoparticles and cinnamon essential oil for chicken meat packaging applications. LWT. Food Sci Technol. 2018;93:329–38.

Akhtar MJ, Jacquot M, Jamshidian M, Imran M, Arab-Tehrany E, Desobry S. Fabrication and physicochemical characterization of HPMC films with commercial plant extract: influence of light and film composition. Food Hydrocoll. 2013;31:420–7.

Alvim ID, Stein MA, Koury IP, Dantas FBH, Cruz CLCV. Comparison between the spray drying and spray chilling microparticles contain ascorbic acid in a baked product application. LWT. Food Sci Technol. 2016;65:689–94.

Aparicio-Fernández X, Vega-Ahuatzin A, Ochoa-Velasco CE, Cid-Pérez S, Hernández-Carranza P, Ávila-Sosa R. Physical and antioxidant characterization of edible films added with red prickly pear (Opuntia ficus-indica L.) cv. San Martín Peel and/or its aqueous extracts. Food Bioprocess Technol. 2018;11:368–79.

Arkoun M, Daigle F, Holley RA, Heuzey MC, Ajji A. Chitosan-based nanofibers as bioactive meat packaging materials. Pack Technol Sci. 2018:1–11.

Azeredo HMC, Morrugares-Carmona R, Wellner N, Cross K, Bajka B, Waldron KW. Development of pectin films with pomegranate juice and citric acid. Food Chem. 2016;198:101–6.
Beak S, Kim H, Song KB. Sea squirt Shell protein and Polylactic acid laminated films containing cinnamon bark essential oil. J Food Sci. 2018;83(7):1896–903.
Berthet MA, Angellier-Coussy H, Chea V, Guillard V. Sustainable food packaging: valorising wheat straw fibres for tuning PHBV-based composites properties. Compos Part A Applied Sci Manufacturing. 2015;72:139–47.
Bie P, Liu P, Yu L, Li X. The properties of antimicrobial films derived from poly(lactic acid)/starch/chitosan blended matrix. Carbohydr Polym. 2013;98:959–66.
Bonilla J, Talón E, Atarés L, Vargas M, Chiralt A. Effect of the incorporation of antioxidants on physicochemical and antioxidant properties of wheat starch–chitosan films. J Food Eng. 2013;118:271–8.
Brody AL, Bugusu B, Jung HH, Sand CK, Mchugh TH. Innovative food packaging solutions. J Food Sci. 2008;73(8):R107–16.
Buonocore G, Sico G, Mensitieri G. Safety of food and beverages: packaging material and auxiliary items in encyclopedia of food safety, vol. 3. Waltham: Academic Press; 2014. p. 384–96.
Castro-Rosas J, Cruz-Galvez AM, Gomez-Aldapa CA, Falfan-Cortes RN, Guzman-Ortiz FA, Rodríguez-Marín ML. Biopolymer films and the effects of added lipids, nanoparticles and antimicrobials on their mechanical and barrier properties: a review. Int J Food Sci Technol. 2016;51:1967–78.
Cejudo-Bastante MJ, Cejudo-Bastante C, Cran MJ, Heredia FJ, Bigger SW. Optical, structural, mechanical and thermal characterization of antioxidant ethylene vinyl alcohol copolymer films containing betalain-rich beetroot. Food Packag Shelf Life. 2020;24:100502.
Chen H, Li L, Ma Y, Mcdonald TP, Wang Y. Development of active packaging film containing bioactive components encapsulated in β-cyclodextrin and its application. Food Hydrocoll. 2019;90:360–6.
Collazo-Bigliardi S, Ortega-Toro R, Chiralt A. Chapter 4 Properties of micro- and nano-reinforced biopolymers for food applications from Tomy J. Gutiérrez in Polymers for Food Book springer Editor. 2018.
Da Rocha M, Prietto L, Souza MM, Furlong EB, Prentice C. Effect of organic acids on physical-mechanical and antifungicidal properties of anchovy protein films. J Aquatic Food Product Technol. 2018;27(3):316–26.
Estevez-Areco S, Guz L, Famá L, Candal R, Goyanes S. Bioactive starch nanocomposite films with antioxidant activity and enhanced mechanical properties obtained by extrusion followed by thermo-compression. Food Hydrocoll. 2019;96:518–28.
Fabra MJ, Lopez-Rubio A, Lagaron JM. Use of the electrohydrodynamic process to develop active/bioactive bilayer films for food packaging applications. Food Hydrocoll. 2016;55:11–8.
Gao HX, He Z, Sun Q, He Q, Zeng WC. A functional polysaccharide film forming by pectin, chitosan, and tea polyphenols. Carbohydr Polym. 2019;215:1–7.
Gavril GL, Wrona M, Bertella A, Świeca M, Râpă M, Salafranca J, Nerín C. Influence of medicinal and aromatic plants into risk assessment of a new bioactive packaging based on Polylactic Acid (PLA). Food Chem Toxicol. 2019;132:110662.
Gutiérrez TJ, Tapia MS, Pérez E, Famá L. Structural and mechanical properties of edible films made from native and modified cush-cush yam and cassava starch. Food Hydrocoll. 2015;45:211–7.
Hamdi M, Nasri R, Li S, Nasri M. Bioactive composite films with chitosan and carotenoproteins extract from blue crab shells: biological potential and structural, thermal, and mechanical characterization. Food Hydrocoll. 2019;89:802–12.
Hosseini SF, Rezaei M, Zandi M, Farahmandghavi F. Development of bioactive fish gelatin/chitosan nanoparticles composite films with antimicrobial properties. Food Chem. 2016;194:1266–74.
http://ww.matweb.com/search/datasheet.
Jiang J, Gong L, Dong Q, Kang Y, Osako K, Li L. Characterization of PLA-P3,4HB active film incorporated with essential oil: application in peach preservation. Food Chem. 2020;313:126134.

Jongjareonrak A, Benjakul S, Visessanguan W, Tanaka M. Antioxidative activity and properties of fish skin gelatin films incorporated with BHT and a-tocopherol. Food Hydrocoll. 2008;22:449–58.

Kanatt SR, Rao MS, Chawla SP, Sharma A. Active chitosan-polyvinyl alcohol films with natural extracts. Food Hydrocoll. 2012;29:290–7.

Karimi M, Yazdi FT, Mortazavi SA, Shabibi-Ghahfarrokhi IS, Chamani J. Development of active antimicrobial poly (l-glutamic) acid-poly (l-lysine) packaging material to protect probiotic bacterium. Polym Test. 2020;83:106338.

Kavoosi G, Dadfar SMM, Purfard AM. Mechanical, physical, antioxidant, and antimicrobial properties of gelatin films incorporated with thymol for potential use as Nano wound dressing. J Food Sci. 2013;78(2):E244–50.

Krongauz VV. Diffusion in polymers dependence on crosslink density Eyring approach to mechanism. J Therm Anal Calorim. 2010;102:435–45.

Lagaron JM, Catala R, Gavara R. Structural characteristics defining high barrier properties in polymeric materials. Mater Sci Technol. 2004;20(1):1–7.

Liu D, Li H, Jiang L, Chuan Y, Yuan M, Chen H. Characterization of active packaging films made from poly(lactic acid)/poly(trimethylene carbonate) incorporated with oregano essential oil. Molecules. 2016;21:695.

López-de-Dicastillo C, Gómez-Estaca J, Catalá R, Gavara R, Hernández-Muñoz P. Active antioxidant packaging films: development and effect on lipid stability of brined sardines. Food Chem. 2012;131:1376–84.

Luo Y, Liu H, Yang S, Zeng J, Wu Z. Sodium alginate-based green packaging films functionalized by guava leaf extracts and their bioactivities. Materials. 2019;12:2923.

Lyu JS, Lee J-S, Han J. Development of a biodegradable polycaprolactone film incorporated with an antimicrobial agent via an extrusion process. Sci Rep. 2019;9:20236.

Ma X, Chang P, Yu J. Properties of biodegradable thermoplastic pea starch/carboxymethyl cellulose and pea starch/microcrystalline cellulose composites. Carbohydr Polym. 2008;72(3):369–75.

Machin D, Rogers CE. Free volume theories for penetrant diffusion in polymers. Die Mdkromolekulare Chemie. 1972;155:269–81.

Mahcene Z, Khelil A, Hasni S, Akman PK, Bozkurt F, Birech K, Goudjil MB, Tornuk F. Development and characterization of sodium alginate based active edible films incorporated with essential oils of some medicinal plants. Int J Biol Macromol. 2020;145:124–32.

Martinez-Camacho AP, Cortez-Rocha MO, Graciano-Verdugo AZ, Rodriguez-Félix F, Castillo-Ortega MM, Burgos-Hernández A, Ezquerra-Brauera JM, Plascencia-Jatomea M. Extruded films of blended chitosan, low density polyethylene and ethylene acrylic acid. Carbohydr Polym. 2013;91(2):666–74.

Menzel C, González-Martínez C, Vilaplana F, Diretto G, Chiralt A. Incorporation of natural antioxidants from rice straw into renewable starch films. Int J Biol Macromol. 2020;146:976–86.

Moreirinha C, Vilela C, Silva NHCS, Pinto RJB, Almeida A, Rocha MAM, Coelho E, Coimbra MA, Silvestre AJD, Freire CSR. Freire antioxidant and antimicrobial films based on brewers spent grain arabinoxylans, nanocellulose and feruloylated compounds for active packaging. Food Hydrocoll. 2009; https://doi.org/10.1016/j.foodhyd.2020.105836.

Moura LE, de Souza CO, de Oliveira EAS, Paulo Lemos VF, Druzian JI. Bioactive efficacy of low-density polyethylene films with natural additives. J Appl Polym Sci. 2018; https://doi.org/10.1002/APP.46461.

Nogueira GF, Soares CT, Cavasini R, Fakhouri FM, de Oliveira RA. Bioactive films of arrowroot starch and blackberry pulp: physical, mechanical and barrier properties and stability to pH and sterilization. Food Chem. 2019;275:417–25.

Ortiz CM, Salgado PR, Dufresne A, Mauri AN. Microfibrillated cellulose addition improved the physicochemical and bioactive properties of biodegradable films based on soy protein and clove essential oil. Food Hydrocoll. 2018;79:416–27.

Pelissari FM, Grossmann MVE, Yamashita F, Pineda EAG. Antimicrobial, mechanical, and barrier properties of cassava starch–chitosan films incorporated with oregano essential oil. J Agric Food Chem. 2009;57(16):7499–504.

Perdones A, Vargas M, Atarés L, Chiralt A. Physical, antioxidant and antimicrobial properties of chitosan-cinnamon leaf oil films as affected by oleic acid. Food Hydrocoll. 2014;36:256–64.

Pereira RC, de Carneiro JDS, Borges SV, Assis OBG, Alvarenga GL. Preparation and characterization of nanocomposites from whey protein concentrate activated with lycopene. J Food Sci. 2016;81(3):E637.

Pérez LM, Piccirilli GN, Delorenzi NJ, Verdini RA. Effect of different combinations of glycerol and/or trehalose on physical and structural properties of whey protein concentrate- based edible films. Food Hydrocolloid. 2016;56:352–35.

Reddy N, Yang Y. Citric acid cross-linking of starch films. Food Chem. 2010;118:702–11.

Robertson GL. Food packaging – principles and practice. 3rd ed. Taylor & Francis; 2013.

Salaberria AM, Diaz RH, Andrés MA, Fernandes SCM, Labidi J. The antifungal activity of functionalized chitin nanocrystals in poly (lactid acid) films. Materials. 2017;10:546. https://doi.org/10.3390/ma10050546.

Silveira da Rosa G, Vanga SK, Gariepy Y, Raghavan V. Development of biodegradable films with improved antioxidant properties based on the addition of carrageenan containing olive leaf extract for food packaging applications. J Polym Environ. 2020;28:123–30.

Siripatrawan U, Harte BR. Physical properties and antioxidant activity of Journal of the Science of Food and Agriculture in active film from chitosan incorporated with green tea extract. Food Hydrocoll. 2010;24(8)

Uranga J, Leceta I, Etxabide A, Guerrero P, de la Caba K. Cross-linking of fish gelatins to develop sustainable films with enhanced properties. Eur Polym J. 2016;78:82–90.

Uranga J, Puertas AI, Etxabide A, Dueñas MT, Guerrero P, de la Caba K. Citric acid-incorporated fish gelatin/chitosan composite films. Food Hydrocoll. 2019;86:95–103.

Vernaez O, Neubert KJ, Kopitzky K, R. S. Compatibility of chitosan in polymer blends by chemical modification of bio-based polyesters. Polymers. 2019;11:1939.

Vidal OL, Tsukui A, Garrett R, Rocha-Leão MHM, Carvalho CWP, Freitas SP, de Rezende MC, Ferreira L, M. S. Production of bioactive films of carboxymethyl cellulose enriched with green coffee oil and its residues. Int J Biol Macromol. 2020;146:730–8.

Vilela C, Pinto RJB, Coelho J, Domingues MRM, Daina S, Sadocco P, Santos SAO, Freire CSR. Bioactive chitosan/ellagic acid films with UV-light protection for active food packaging. Food Hydrocoll. 2017;73:120–8.

Wang L, Dong Y, Men H, Tong J, Zhou J. Preparation and characterization of active films based on chitosan incorporated tea polyphenols. Food Hydrocoll. 2013;32:35–41.

Wu H, Lei Y, Zhu R, Zhao M, Lu J, Xiao D, Jiao C, Zhang Z, Shen G, Li S. Preparation and characterization of bioactive edible packaging films based on pomelo peel flours incorporating tea polyphenol. Food Hydrocoll. 2019;90:41–9.

Wu Y, Yuan MW, Yang JY, Qin YY, Yuan ML, Cao JX. Physical properties and antimicrobial activity of a poly(lactic acid)/poly(trimethylene carbonate) film incorporated with thymol. Adv Mater Res. 2014;884-885:481–4.

Yang W, Weng Y, Puglia D, Qi G, Dong W, Kenny JM, Ma P. Poly(lactic acid)/lignin films with enhanced toughness and anti-oxidation performance for active food packaging. Int J Biol Macromol. 2020;144:102–10.

Zhang Y, Liu BL, Wang LJ, Deng YH, Zhou SY, Feng JW. Preparation, structure and properties of acid aqueous solution plasticized thermoplastic chitosan. Polymers. 2019;11(5):818.

Chapter 9
Effectiveness and Release Studies of Bioactive Systems

Victor Gomes Lauriano de Souza, Carolina Rodrigues, João Ricardo Afonso Pires, Isabel Coelhoso, and Ana Luisa Fernando

Abstract The recent advances in food science and technology have made available alternative methods to preserve foodstuff others than those traditional processes. Active food packaging is one of those current developing technologies that contributes to enlarge the shelf-life of food products. This type of packaging can be distinguished from the materials traditionally used to pack food due to the intentional interaction between the contained food and the packaging material. Compounds with activity are inserted into the packaging material, to functionalize the material with properties (e.g. antimicrobial and antioxidant) capable to retard deterioration processes. In major cases, the pattern of action is based on the delivery of active substances toward the food, thus the importance to study the releasing process and its effectiveness. This chapter summarizes the latest data available on the releasing studies of bioactive compounds used in food packaging material, as well as their effectiveness, focusing especially on polymeric packaging material incorporated with natural compounds. Current available guidelines and regulations are also focused once they represent an important part of those studies.

V. G. L. de Souza
MEtRICs, Departamento de Ciências e Tecnologia da Biomassa, NOVA School of Science and Technology | FCT NOVA, Universidade Nova de Lisboa, Caparica, Portugal

International Iberian Nanotechnology Laboratory (INL), Braga, Portugal
e-mail: v.souza@campus.fct.unl.pt; victor.souza@inl.int

C. Rodrigues · J. R. A. Pires · A. L. Fernando (✉)
MEtRICs, Departamento de Ciências e Tecnologia da Biomassa, NOVA School of Science and Technology | FCT NOVA, Universidade Nova de Lisboa, Caparica, Portugal
e-mail: cpe.rodrigues@campus.fct.unl.pt; jr.pires@campus.fct.unl.pt; ala@fct.unl.pt

I. Coelhoso
LAQV-REQUIMTE, Departamento de Química, NOVA School of Science and Technology | FCT NOVA, Universidade NOVA de Lisboa, Caparica, Portugal
e-mail: imrc@fct.unl.pt

S. M. Jafari, A. S. Silva (eds.), *Releasing Systems in Active Food Packaging*, Food Bioactive Ingredients, https://doi.org/10.1007/978-3-030-90299-5_9

Keywords Migration · Active food packaging · Antimicrobial · Antioxidant · Diffusion coefficient · Effectiveness

1 Introduction

The Commission Regulation (EC) 450/2009 (European Commission 2009) defines active materials as the materials "that are intended to extend the shelf-life or to maintain or improve the condition of packaged food"; also, according to the same regulation releasing active materials are those "designed to deliberately incorporate components that would release substances into the packaged food or the environment surrounding the food". By extension, bioactive releasing systems can be, therefore, understood as those type of active releasing systems where the deliberately incorporated components are bioactive compounds, most of them with antimicrobial and antioxidant properties that will help to preserve and enlarge the shelf-life of packaged food.

Dealing with food contact material, such packaging must also follow the European Regulation on materials and articles intended to come into contact with food (Regulation (EC) No 1935/2004) (European Union (EU), Regulation (EC) 2004). Moreover, the elements in charge of the active role should also be assessed to assure that they are harmless and observe the requirements mentioned in Regulation (EC) No 1935/2004 (European Union (EU), Regulation (EC) 2004) and its amendments, and Regulation (EC) No 450/2009 (European Commission 2009), and must be considered as GRAS (generally recognized as safe), i.e. do not pose any risk to human consumption. European Food Safety Authority (EFSA) is the authority responsible to carry out the safety assessment of a substance or of a combination of substances which constitutes the components, in Europe, while in the United States of America is the FDA (Food and Drug Administration).

Such materials can be prepared with one or more layers from different types of materials, namely: cardboard, coatings, paper, plastics and varnishes. Currently, the quest for more natural and environment-friendly products by the consumers has stimulated the attention of the scientific community and food packaging industry on biodegradable and renewable polymers, and its usage and studies to substitute the fossil-based polymeric materials (Abdollahi et al. 2012; Tang et al. 2012; Souza et al. 2018a), is being highlighted as the future generation of plastic materials (Souza and Fernando 2016).

Bioactive substances can be understood as those compounds obtained from biological sources, that have active properties (antimicrobial and/or antioxidant activities). Currently, extracts from a variety of plants and fruits have been investigated as possible natural additives to preserve food products (Brewer 2011; Regnier et al. 2012; Khatabi et al. 2016; Souza et al. 2018b). And, the synthetic food additives possible harmful effects to human have lead the consumers to prefer natural preservatives over the traditional synthetic ones (Santana et al. 2013; Ribeiro-Santos et al. 2017).

Traditional culinary plants are a good source of such bioactive compounds. Examples of these plants are thyme (*Thymus vulgaris*), rosemary (*Rosmarinus officinalis*), sage (*Salvia officinalis*), oregano (*Origanum vulgare*), marjoram (*Origanum majorana*), basil (*Ocimum basilicum*), tea (*Camellia sinensis*), among others; also other common food ingredients, including some roots, rhizomes, bulbs, barks, flower buds, are highlighted due to their high bioactivity: ginger (*Zingiber officinale*), turmeric (*Curcuma longa*), clove (*Syzygium aromaticum*), cinnamon (*Cinnamomum* spp.), onion (*Allium cepa*), garlic (*Allium sativum*) are few examples of those types of food ingredients (Brewer 2011; Andrade et al. 2018; Ribeiro-Santos et al. 2015). These active compounds are generally obtained through extraction (e.g. hydrodistillation, infusions in water, hydro-alcoholic extraction), some of them ultrasound assisted (Souza et al. 2018b), or even combined with pulsed electric fields to enhance the extraction yield (Abbas Syed 2017). Essential oils, volatile fat-soluble substances, rich in bioactive compounds, can be obtained by steam distillation or merely by mechanical processes, namely, through pressing. Currently, modern methods have been developed to extract essential oils using a solvent-free microwave-distillation (Filly et al. 2014). Essential oils (EOs) are a combination of aromatic molecules that can be extracted from different aromatic plants and used either as flavor agents by the food industry, or in aromatherapy and in other medical applications (Burt 2004). The rich content in phenolic compounds and terpenes confers to EOs the ability to hinder the growth of food spoilage microorganisms, thus enhancing the shelf life of processed foods, besides the improvement in the flavor of food, which may represent an increment on the consumer's satisfaction (Ribeiro-Santos et al. 2017; Burt 2004; Gharenaghadeh et al. 2017). Besides those bioactive properties (antimicrobial and antioxidant), EOs are also highlighted due to their other biological properties such as anti-tumor, anti-diabetic, analgesic, and anti-inflammatory properties, demonstrating also benefits to human health (Ribeiro-Santos et al. 2017).

Thus, in this context, extracts, EO's or their active components can be directly incorporated to the food matrix as natural preservatives or into the packaging material to be released through the shelf-life time aiming the product preservation. The latter alternative is the most common and studied in recent decades, as its main benefit is that it can offer a continuous liberation of active components during storage (Gómez-Estaca et al. 2014; Prasad and Kochhar 2014), i.e. providing an extension on the protection capacity over the shelf-life. The development of packaging material incorporated with bioactive components, intended to preserve a wide range of foodstuff, are thus being the aim of different studies reported recently in the literature: in ready-to-eat meat (Quesada et al. 2016), fresh poultry meat (Souza et al. 2019a; Pires et al. 2018), in muscle foods (Ahmed et al. 2017; Dias et al. 2018); to control lipid oxidation in fatty foods (Gómez-Estaca et al. 2014; Tian et al. 2013); in dairy products (Conte et al. 2007; Haghighi-Manesh and Azizi 2017); in breads (Melini 2018), among others.

Since the action mechanism of this technology lies on the release of the active component into the food contained in the packaging material, it is decisive to understand the dynamic of this process over time (Souza et al. 2018b, 2019b). Thus, this chapter summarizes the latest data available on the releasing studies of bioactive compounds used in food packaging material, as well as their effectiveness, focusing especially on polymeric packaging material incorporated with natural compounds. The current guidelines and regulations available will also be focused on, as they represent an vital issue in obtaining standardized data and to bridge the gap between laboratory research and commercial exploration.

2 How to Measure the Effectiveness of Bioactive Releasing Systems?

2.1 Migration Tests and Current Guidelines and Regulations

The migration of compounds from the food contact material (FCM) to the food packaged, in most cases, is considered undesired, once many of the chemical additives used in the production of traditional packaging can damage the health of consumers if ingested in doses superior than the limits legislated (Souza et al. 2018a; Souza and Fernando 2016; Azeredo et al. 2012). Contrarily, regarding the active packaging, the migration of the bioactive compounds, responsible for the material's active properties, is desirable, and aims to protect the food packaged through, for example, their antimicrobial or antioxidant properties (de Soares and Hotchkiss 1998), or to improve the properties of the foodstuff, in terms of flavor, color, etc.

The study of the diffusion process of the bioactive substances from the packaging materials toward the food packaged is done by migration tests. This test can be either done *in vitro*, by using food simulants, or *in situ* (i.e. directly on the food product though wrapping and packaging the food with the plastic material). The method consists on the quantification of the diffused compounds using analytical techniques such as chromatography and spectroscopy. Ideally, this quantification should be done directly in the food, after being in contact with FCM, however this could have some technical limitations, once food are complex matrices, which require further methods to separate and then proceed with the compounds quantification (Souza et al. 2018a; Souza and Fernando 2016; Silva et al. 2007). Alternatively, *in vitro* methods are used based on the Regulation (EC) n°10/2011 and its amendments, which establishes the authorized food simulants and the assays' conditions (time and temperature) to be followed (European Commission 2011).

Food simulants have physicochemical properties similar to foodstuff and mimic their behavior (Souza et al. 2018a; Souza and Fernando 2016; Ans 2016). Thus, simulants allow simplifying the identification and quantification of migrants from packaging material (Souza et al. 2018a). Approved food simulants defined by the Regulation (EU) 10/2011 are described on Table 9.1. Food simulants A, B and C can

Table 9.1 Food simulants and related types of food (Regulation (EC) 10/2011)

Abbreviation	Simulant	Type of food
Food simulant A	Ethanol 10% (v/v)	Food with hydrophilic character despite those described for food simulants B and C.
Food simulant B	Acetic acid 3% (w/v)	Food with pH below 4.5
Food simulant C	Ethanol 20% (v/v)	alcoholic foods with an alcohol content of up to 20% and those foods which contain a relevant amount of organic ingredients that render the food more lipophilic
Food simulant D1	Ethanol 50% (v/v)	Alcoholic foods with an alcohol content of above 20% and for oil in water emulsions.
Food simulant D2	Vegetable oil ([a]) or Ethanol 95%	Foods which contain free fats at the surface.
Food simulant E	Poly(2,6-diphenyl-p-phenylene oxide), particle size 60–80 mesh, pore size 200 nm	Dry foods.

[a]Any vegetable oil with the following fatty acid composition

No of carbon atoms in fatty acid chain: No of unsaturation	6–12	14	16	18:0	18:1	18:2	18:3
Range of fatty acid composition expressed % (w/w) of methyl esters by Gas chromatography	<1	<1	1.5–2.0	<7	15–85	5–70	<1.5

extract hydrophilic substances and are applied to matrices that have a hydrophilic character. Food simulants D1 and D2 can extract lipophilic substances and are applied to foods that have a lipophilic character. Specific migration into dry foods is evaluated with food simulant E. A more detailed description of which simulant suits each food category are available in the Regulation (EC) 10/2011 (European Commission 2011).

The conditions regarding the time and temperature that the sample shall be placed in contact with the food simulant must be the worst of the foreseeable conditions according to Table 9.2 (European Commission 2011).

Besides the amount of compound diffused, one parameter of interest in the migration study is the determination of the diffusion coefficient (D) (Souza et al. 2018b). The diffusion coefficient is a parameter that measures how fast the migrant diffuses toward the food or the food simulant (Stoll et al. 2019), i.e., the bigger is the D value the faster is the diffusion process, which means that the packaging material has lower capacity to retain the active compound trapped within the polymeric chain and the time to release it is smaller (Souza et al. 2018c).

Table 9.2 Test conditions extracted from the Regulation (EC) 10/2011

Contact time in worst foreseeable use	Test time	Contact temperature in worst foreseeable use	Test temperature
t ≤ 5 min	5 min	T ≤ 5 °C	5 °C
5 min < t ≤ 0.5 h	0.5 h	5 °C < T ≤ 20 °C	20 °C
0.5 h < t ≤ 1 h	1 h	20 °C < T ≤ 40 °C	40 °C
1 h < t ≤ 2 h	2 h	40 °C < T ≤ 70 °C	70 °C
2 h < t ≤ 6 h	6 h	70 °C < T ≤ 100 °C	100 °C or reflux temperature
6 h < t ≤ 24 h	24 h	100 °C < T ≤ 121 °C	121 °C ([b])
1 day < t ≤ 3 days	3 days	121 °C < T ≤ 130 °C	130 °C ([b])
3 days < t ≤ 30 days	10 days	130 °C < T ≤ 150 °C	150 °C ([b])
Above 30 days	See specific conditions ([a])	150 °C < T < 175 °C	175 °C ([b])
		T >175 °C	Adjust the temperature to the real temperature at the interface with the food ([a])

[a]For contact times above 30 days at room temperature and below the specimen shall be tested in an accelerated test at elevated temperature for a maximum of 10 days at 60 °C.

[b]This temperature should be used only for food simulant D2 and E. For food simulants A, B, C, and D1 the test may be placed by a test temperature at 100 °C or at a reflux temperature during of four times the time selected according to the contact time in worst foreseeable use.

Migration of components (additive or contaminant) is often controlled by the molecular diffusion of the migrant through the polymeric packaging material. This process can be described by Fick's Second Law:

$$\frac{\partial C_P}{\partial t} = D\frac{\partial^2 C_P}{\partial x^2} \tag{9.1}$$

where C_P is the concentration of the migrant in the packaging at time t and position x. The amount of migrant transferred from the packaging film to the food from time zero to time t, $M_{F;t}$, can be obtained using the migrant at packaging film/food interface ($J\,|_{x=LP}$).

$$M_{F,t} = \int_0^t AJ_{L_P}\, dt \text{ where } J_{L_P} = -D\frac{\partial C_P}{\partial x} \; for\; x = L_P \tag{9.2}$$

For large migration time the equation 2 solution is:

$$\frac{M_{F,t}}{M_{F,L}} = \frac{2}{L_P}(Dt)^{0.5}\left\{\frac{1}{\pi^{0.5}} + 2\sum_{n=1}^{\infty}(-1)ierfc\left[\frac{nL_P}{(Dt)^{0.5}}\right]\right\} \tag{9.3}$$

where $M_{F,L}$ is the total migrant diffused from the film to the food during the whole migration time. For short migration times the function ierfc tends to zero and Eq. 9.3 can be simplified to:

$$\frac{M_{F,t}}{M_{F,L}} = \frac{2}{L_P}\left(\frac{Dt}{\pi}\right)^{0.5} \tag{9.4}$$

The assessment of D has been frequently done using the two following equations (Chungy et al. 2002):

$$\frac{M_{F,t}}{M_{P,0}} = \frac{2}{L_P}\left(\frac{Dt}{\pi}\right)^{0.5} \tag{9.5}$$

$$\frac{M_{F,t}}{M_{F,\infty}} = \frac{2}{L_P}\left(\frac{Dt}{\pi}\right)^{0.5} \tag{9.6}$$

Where $M_{F,t}$ is the quantity of migrant in the food (or food simulant) at time t, $M_{P,0}$ is the initial amount of migrant in the packaging film, $M_{F,\infty}$ is the amount of migrant in the food (or food simulant) at equilibrium, D ($cm^2.s^{-1}$) is the diffusion coefficient of migrant in the packaging film and L_P (cm) is the thickness of the packaging film.

The two models are the same and known to provide truthful estimates of diffusion coefficients for complete migration ($M_{P,0} = M_{F,\infty}$) and when partitioning and resistance to mass transfer are negligible. The two models are less precise in determining the diffusion coefficients for partitioned migration ($M_{P,0} > M_{F,\infty}$) (Chungy et al. 2002). The diffusion coefficient can be determined from a single datum point using Eqs. 9.1 or 9.2, however, the use of multiple data points and linear regression from the plot $M_{F,t}/M_{P,0}$ or $M_{F,t}/ M_{F,\infty}$ versus $t^{0.5}$ provides a more reliable estimative of D (Chungy et al. 2002; Sanches Silva et al. 2009).

The migration study is important and should be carried out extensively and case by case, once migration is influenced by many factors, namely: (i) the chemical and physical nature of the migrant (its polarity, shape, chemical composition, solubility); (ii) the type of the food packaged; (iii) the surface area of the packaging in contact with the food product; (iv) the exposure conditions (temperature and time of contact); (v) the type of packaging material - its morphology, molecular weight, distribution, crystallinity, density, and orientation that may interfere on sizes, shapes and distribution of microcavities, therefore on diffusion path (as intrinsically diffusion coefficients of the material interfere on the migration levels); and (vi) the interaction between the polymer and the active compounds incorporated – which can be influenced by the plasticizing agents (Sanches Silva et al. 2009; Jamshidian et al. 2012). Also, compilation of results allows the construction of mathematical models that enable the prediction of migration levels. Some examples on the studies are described in the following sections, separated according to the active function of the migrant incorporated into the polymeric matrices.

2.2 *Antimicrobial Systems*

Microbiological growth is the major reason of the food spoilage, which can occur on food surfaces via contamination introduced by food handlers and equipment (de Soares et al. 2009), but also due to environmental contamination. Antimicrobial active releasing packaging is able to transfer the antimicrobial activity from the FCM to the food packaged and preserve eit. It is an alternative method to preserve food by the controlled release of the antimicrobial agents during storage, which maintain the critical concentration required to inhibit microbial growth on food surfaces (Appendini and Hotchkiss 2002).

Chitosan is a natural biopolymer widely researched as antimicrobial food packaging due to its intrinsic antimicrobial properties (Souza et al. 2020a). Novel bionanocomposites of chitosan / montmorillonite incorporated with essential oils (EOs), rosemary essential oil (REO) or ginger essential oil (GEO) were developed and their *in vitro* activity was assessed (Souza et al. 2019b) (Table 9.3). In this study, the group investigated the double-side total migration of total phenolic compounds (TPC) in four food simulants (ethanol 95%, ethanol 50%, ethanol 10% and distilled water) at 40 °C for 10 days. The quantification of TPC was done by UV/Vis spectroscopy. The diffusion profile observed was an "exponential growth to a maximum" followed by a decrease in the TPC, which was associated the bioactive compounds degradation observed at the end of the migration test incubation time. The diffusion coefficient was obtained, and the trend observed was an increase in the D with the decrease of the simulant hydrophilicity (distilled water < ethanol 10% < ethanol 50% < ethanol 95%). This behavior was associated to the high affinity to the fatty food simulant (more hydrophobic media) of the phenolic compounds present in the EOs (Souza et al. 2019b).

Lysozyme (LYS) (Ozer et al. 2016) and natamycin (NAT) (De Oliveira et al. 2007), two natural antimicrobials, were used as active compounds in whey protein isolate (WPI) and cellulose polymeric films, respectively. The diffusion process of lysozyme was studied from double-sided (16 cm^2 specimen) total immersion migration in 40 mL of release medium (0.05 M Na-Acetate buffer at pH 4.5) at 10 °C under continuous agitation at 100 rpm, and the quantification was done by HPLC

Table 9.3 Releasing studies of bioactive compounds from active food packaging

Polymer	Active compound	Simulant	Food	Conditions	References
Polypropylene	Catechin and green tea extract	10% ethanol and 50% ethanol	N/A	Double-sided (80 × 0.4 × 0.17 mm specimen) total immersion (10 mL of food simulant) at 40 °C for 5 and 10 days of storage. Quantification of antioxidant compounds by HPLC.	del Castro-López et al. (2014)
Chitosan	Rosemary essential oil and ginger essential oil	Ethanol 50% (v/v), ethanol 10% (v/v) and distilled water	N/A	Double-sided, total immersion migration tests were performed at 40 ± 2 °C for 10 days. The concentration of the active compounds in the simulants was analyzed by UV-spectroscopy.	Souza et al. (2019b)
EVOH	Ascorbic acid, ferulic acid, quercetin, green tea extract	Distilled water (aqueous food simulant), 10% ethanol (v/v) (alcoholic food simulant), and 95% ethanol (v/v) (fatty food simulant).	N/A	Double-sided, total immersion migration tests were performed at 37 °C. The concentration of the antioxidant in the simulants was analyzed by UV-spectroscopy.	López-de-Dicastillo et al. (2012)
LPDE	Thymol essential oil	N/A	N/A	EO concentration as a function of time in the headspace was determined by the headspace analysis using GC-MS.	Efrati et al. (2014)

(continued)

Table 9.3 (continued)

Polymer	Active compound	Simulant	Food	Conditions	References
Chitosan	ZnO nanoparticle	N/A	Poultry meat	Samples wrapped with the films were refrigerated stored (5 ± 2 °C) for 11 days. Zinc ion release was determined by atomic absorption spectrometer.	Souza et al. (2020b)
EVOH	Citral (CI) and trans-cinnamaldehyde (TC)	Distilled water (neutral food), 3% acetic acid solution (v/v) (acid food), 10% ethanol solution (v/v) (alcoholic food), 95% ethanol solution (v/v) (fatty food)	N/A	Double-sided total immersion migration at 20 ± 1 °C for at least 30 days to achieve equilibrium. Quantification using HPLC.	Chen et al. (2019)
Cellulose polymeric base	Natamycin	N/A	Gorgonzola cheese	The cheese protected with the films were stored for 45 days for ripening. After this period, approximately 2 mm in depth of the rind was removed and mixed with extraction solvent. The natamycin was analyzed in a spectrophotometer.	De Oliveira et al. (2007)

(continued)

Table 9.3 (continued)

Polymer	Active compound	Simulant	Food	Conditions	References
Whey protein isolate (WPI)	Lysozyme	0.05 M Na-Acetate buffer at pH 4.5	N/A	Double-sided (16 cm^2 specimen) total immersion migration (40 mL release medium) at 10 °C with continuous agitation at 100 rpm. The amount of Lysozyme released to the buffer solution was quantified by HPLC analysis.	Ozer et al. (2016)
LDPE and double layer (LDPE and HDPE)	Astaxanthin	95% (v/v) ethanol	N/A	Double-sided (4.4 × 6 cm^2 specimen) total immersion at 4 temperatures (10, 23, 30 and 40 °C until equilibrium. Astaxanthin quantification was done by HPLC.	Colín-Chávez et al. (2013)
Polypropylene	Carvacrol and thymol	3% (v/v) acetic acid, 10% (v/v) ethanol, 95% (v/v) ethanol and isooctane	N/A	Double-sided (12 cm^2 specimen) total immersion migration (20 ml of simulant – volume ratio around 6 dm^2/L) 40 °C for 10 days, except for isooctane studies which were done at 20 °C and 50% relative humidity for 2. Quantification was done by HPLC or GC-MS.	Ramos et al. (2014)
EVOH	Green tea extract (GTE)	Distilled water, 3% (v/v) acetic acid, 10% ethanol, and 95% (v/v) ethanol and isooctane	N/A	Double-sided (80 cm^2 specimen) total immersion migration (100 ml of simulant – volume ratio around 6 dm^2/L) at 40 ± 1 °C for 30 days. GTE was quantified using HPLC.	Lopez de Dicastillo et al. (2011)

(continued)

Table 9.3 (continued)

Polymer	Active compound	Simulant	Food	Conditions	References
Polystyrene	Silver nanoparticles	Distilled water, 1%, 2% and 3% acetic acid (v/v)	N/A	Double-sided (1 × 1 cm specimen) total immersion migration (10 mL simulant) at 4 temperatures (10, 20, 40 and 70 °C) for 5 days. Silver concentration was quantified using ICP-AES.	Hannon et al. (2017)
Gelatin	Cinnamon essential oil (CEO) nanoliposomes	Corn oil	N/A	Double-sided (4 cm^2 specimen) total immersion migration (20 mL simulant) at 4 and 40 °C for 30 days. The release of CEO nanoliposomes was evaluated by HPLC method.	Wu et al. (2015)
Paper sheets	Propolis – embedded in the paper or sprayed on the surface of the paper.	N/A	Sliced cooked ham	Samples protected with active paper were stored at 4 °C up to 4 days and samplings were carried out after 0, 2 and 4 days. Amount of total phenolic compounds and antioxidant activity was done from samples after extraction with methanol. The volatile composition of the samples was analyzed by HS-SPME GC–MS.	Rizzolo et al. (2016)
Polylactic acid (PLA)	Ascorbyl palmitate and α-tocopherol, and synthetic phenolic antioxidants.	Ethanol 95%, 50% and 10% (v/v)	N/A	Double-sided total immersion migration tests were performed at 40 °C and 20 °C for 60 days. Quantification using HPLC.	Jamshidian et al. (2012)

(continued)

Table 9.3 (continued)

Polymer	Active compound	Simulant	Food	Conditions	References
LDPE	Lemon aroma	N/A	Biscuit	The release was not analytical quantified, only the sensory acceptability of the food was assessed. Consumers evaluated the samples using a nine-point hedonic scale. The sensory analysis was done at 10 and 30 days of storage.	Dias et al. (2013a)
LDPE	Orange aroma or orange essential oil	N/A	Sugar biscuit	The product acceptance was at a supermarket with 102 consumers. The participants indicated how much they liked or disliked the orange aroma and taste of the sugar biscuit evaluated.	Dias et al. (2013b)
Metallocene polypropylene (sachets)	2-nonanone	N/A	Wild strawberry	2-nonanone was quantified in the headspace of the cups containing the fruits using CG-FID on day 2 and 4 of refrigerated storage.	Almenar et al. (2009)

analysis (Ozer et al. 2016). In this study, it was investigated a novel strategy for controlled release system based on pH-responsive polyacrylic acid (PAA)/LYS, and different films were produced either incorporated with free LYS or with (PAA/LYS). The incorporation of LYS into the film in complexed form prolonged its release time from less than 24 h up to 500 h and D value was reduced from ~10^{-9} to ~10^{-13} cm^2/s. Films with 50%-free-LYS + 50%-PAA/LYS complex reduced by 5.7 log the bacterial population within 72 h, while films with 100%-free-LYS could not inhibit *Listeria innocua* after 24 h.

Regarding the cellulose films incorporated with NAT, the authors investigated the diffusion of NAT directly at gorgonzola cheese after the contact with the film for the 45 days of ripening (De Oliveira et al. 2007). Samples of the cheese rind (removed at approximately 2 mm depth) were used for the extraction of NAT with acetonitrile: phosphoric acid 1M (4:1), and the filtered sample was analyzed in a spectrophotometer at 317 nm. The NAT levels reported for the samples protected with the active packaging were lower than control samples (cheese produced with traditional procedure – with NAT), which is a positive result since they also achieved greater microbial control, guaranteeing a product with better quality and healthier once less additive will be ingested (De Oliveira et al. 2007).

Other substances with antimicrobial activity such as metal nanoparticles and metal oxide nanoparticles are also used to produce antimicrobial food packaging (Table 9.3). Once nanomaterial migration can occur, which may influence the food's safety and, subsequently, the health of the consumers, it is important to assess this diffusion process as well (King et al. 2018). For instance, zinc oxide nanoparticles (ZnO NP) synthesized through an eco-friendly route using apple peel wastes was incorporated into chitosan film and used to protect fresh poultry meat (Souza et al. 2020b). After wrapping the fresh poultry meat with the bionanocomposites, the authors refrigerated and stored the samples (5 ± 2 °C) and evaluated after 11 days of contact the amount of zinc ions present in the meat. According to EFSA (2016) zinc oxide does not migrate in nanoform, and therefore, the safety evaluation is done based on the migration of soluble ionic zinc. Thus, to measure the amount of ionic zinc present in the meat packaged with the bionanocomposites, the samples were mineralized by dry combustion (550 °C), and the concentration of zinc was assessed in the ash residue dissolved with nitric acid by atomic absorption spectrometry. Over the storage time, the amount of zinc increased in the samples wrapped with the active bionanocomposites, associated with the diffusion of ZnO NPs present in the films (Souza et al. 2020b). Although, the amount of migrants found was yet below the maximum level permitted, 25 mg/person per day (European Food Safety Authority (EFSA) 2016), the release of zinc exceeded the specific migration limit of 5 mg/kg food (COMMISSION REGULATION (EU) 2016).

2.3 Antioxidant Systems

After the growth of microorganisms, oxidation of fats is the second most relevant mechanism leading to the deterioration of food (Prasad and Kochhar 2014; Souza et al. 2018c; Abdollahi et al. 2014). The deteriorative aspect of this process is related to the reduction on the foodstuff shelf-life due to changes on its sensory characteristics (taste, odor, color and texture), besides the depletion of essential nutrients such as polyunsaturated fatty acids (PUFAs). Free radicals (e.g. oxo, hydroxyl, and superoxide) are the major initiators of oxidation, and the elimination of such radicals as soon as they are formed is mandatory to avoid / retard this deteriorative process. Antioxidant food packaging is composed of materials capable to avoid oxidation process in the food packaged through different mechanisms, such as, stabilizing free-radicals, scavenging oxygen, blocking UV-light and chelating metals (Tian et al. 2013).

Once the purpose of antioxidant active packaging is to avoid the oxidative process in the food packaged, the simulants used on the migration assays always include those that mimic fatty foods (target of the antioxidants) and generally other types of simulants are also assessed as can be seen on Table 9.3.

The migration of natural antioxidants (catechin or green tea extract) incorporated in polypropylene (PP) films was studied in two food simulants (Simulant A and D), with a double-sided total immersion at 40 °C for 5 and 10 days of storage (del Castro-López et al. 2014). The quantification of catechin and quercetin was carried out in a filtered aliquot of the simulant by means of HPLC-PDA. The method employed was able to simultaneously determine catechin and quercetin in spite of the difference in the amounts released, which indicates the good performance of the analytical method used, demonstrating its potential use to analyze the release of several active compounds with distinguish chemical nature. The antioxidants were incorporated in two levels (2% and 5%) and the amount released was correlated to those percentages. Moreover, the compounds were present in higher amounts in simulant D than in simulant A, which was attributed to the compounds studied greater solubility in ethanol, i.e. higher release toward the simulants with high ethanol content (del Castro-López et al. 2014).

Similar studies using synthetic polymer ethylene vinyl alcohol (EVOH) (Lopez de Dicastillo et al. 2011, 2012; Chen et al. 2019), PP (Ramos et al. 2014), and low density polyethylene (LDPE) (Colín-Chávez et al. 2013) incorporated with natural antioxidants were carried out using at least the alternative fatty food simulant, 95% ethanol. The most used technique to quantify the active compounds was chromatography (HPLC or GS-MS). The differences on the studies rely on the time and temperature used in the assays and the simulants that were also tested besides the fatty food one. In general, the type of releasing profile presented for all simulants and samples was the "exponential growth to a maximum", however it was observed a dependency on the food simulant type and antioxidant agent in the release extension and kinetics (López-de-Dicastillo et al. 2012).

Another approach to quantify the amount of released active substance is related to the packaging material incorporated with EOs. Essential oils can have both antimicrobial and antioxidant properties (Souza et al. 2019a). The technique is based on the volatile characteristic of the active compounds in EO, which enables to determine their concentration over time using headspace analysis, for example in auto sampler headspace GC-MS (Efrati et al. 2014). This mimics those systems where the food is not directly in contact with the packaging material, and the preservative action occurs due to the vapor phase of the active compound present inside the atmosphere of the packaging (Pola et al. 2016). The sample is kept in sealed vials at room temperature, prior to the extraction procedure that occurs holding the vials in the auto sampler conditioning oven for 2 min at 40 °C, followed by injection in CG-MS (Efrati et al. 2014).

The migration rate will depend on how the material is exposed. In an open container, where the film is exposed to air, the driving force will be higher, since equilibrium is never attained. This will result in fast migration, while, for films kept in a closed compartment, the migration rate will be reduced and limited by the equilibrium and partitioning (Efrati et al. 2014). To evaluate both situations, Efrati et al. (2014) performed tests to quantify the amount of EO in films in open and sealed containers after different periods of time using UV–visible spectroscopy analysis (Efrati et al. 2014). It was demonstrated that in open containers the amount of active substance released decreased over time, while in the sealed vials increased. In this work, different configurations of film (multilayer films, with different materials – density, polarity) was also studied to assess how the crystallinity would interfere on the diffusion process. As conclusion, the study showed that the film's ability to hold the EO is affected by the material's crystallinity degree. The higher is the crystallinity, the higher are the periods that the material can hold/trap the EO (Efrati et al. 2014).

The measurement of total diffused active compound can also be done directly from the food after the contact with the packaging material. Sliced ham was packaged with paper either embedded (API) or sprayed onto the external polyethene surface (APP) with propolis extract (Rizzolo et al. 2016). After 0, 2 and 4 days of refrigerated storage, the content in phenolic compounds and their antioxidant activity (DPPH assay) were determined in the methanolic extract of the sliced hams. The authors observed that when paper the was embedded in propolis the migration process was slower, as the TPC and DPPH results were similar to the control samples, while for the ham protected with the other packaging presented higher diffusion values. The migration of such compounds also influenced the volatile profile of the sliced ham, which interfered in the product's acceptance by the sensory panel. In this case, a slower migration was more positive, despite no significant differences were observed in comparison to control. For the time evaluated, API was suitable for the preservation, but further studies should be performed for a longer storage time, once propolis extract also has antimicrobial properties that can help in the extension of the product shelf-life (Rizzolo et al. 2016).

2.4 *Others: Flavors, Colorants and Food ingredients*

Besides the antimicrobial and antioxidant agents, other components namely food additives, can also be incorporated in order to functionalize the food packaging (de Soares et al. 2009). The addition of aromas and colorants are good examples of such active food packaging applications. The incorporation of aromas in the packaging can enhance the desirability of the food packaged, enhance the food flavor perception in the moment that the packaging is opened, or even add the aroma of fresh product itself. For instance, their slow release during its shelf life can offset the natural loss of taste or smell of products with long shelf lives (Prasad and Kochhar 2014; Almenar et al. 2009).

An example of aroma incorporation into packaging film was studied by Almenar et al. (2009). Wild strawberry fruits were packaged in PP/EVOH/PP cups and then the cups were sealed using lids previously impregnated with 2-nonanone emitter system. The cups containing the fruits were stored at 10 °C and 77% RH and their headspace composition were tested at days 0, 2 and 4. The sachets used were produced with metallocene polypropylene containing alumine F-1 80/100 mesh (0.1 g per sachet) impregnated with 0, 0.1, 0.5, 1, 2 and 3 mL of 2-nonanone. This is a volatile compound naturally present in fruit aroma that also have antimicrobial activity. The quantification of the volatile was determined in 100 mL headspace gas injected into a GC equipped with a flame ionization detector (FID) and compared with GC calibration standards of know 2-nonanone concentrations. The amount of active compound was positively correlated with the amount of 2-nonanone incorporated in the sachets and the storage time (Almenar et al. 2009).

Natural food colorants can also be incorporated into FCM to obtain packaging with intelligent properties. In this type of packaging the purpose is not to have the active compound released to the food packaged, despite its non-toxicity, but to be used as sensor to monitor the quality of the product and the packaging over time. Therefore, only its influence on the color of the material is studied. A recent example is found in Latos-Brozio and Masek (2020), who developed PLA and polyhydroxybutyrate (PHB) incorporated with carotene, chlorophyll, curcumin, and lutein. In the study, the different natural colorants changed colour under the influence of UV radiation, elevated temperature and weathering, indicating that those composites of polymers with pigments can be used to make intelligent packaging once lifetime of the packaging material can be signaled.

Nutrients can also be incorporated into edible coatings aiming to produce enriched food in such nutrients. Common beans were coated with edible biopolymers (cassava starch) added with the aminoacids, methionine and cysteine. The amino acid quantification was carried out in the beans after cooking and in the resulting cooking broth using HPLC with fluorescence detection. To analyze the amino acid a derivatization step was done prior to the injection in the chromatograph (Sousa et al. 2018). Treatments containing cassava starch in the coating presented higher concentration of added amino acids both in the bean soup as in cooked bean grains, showing the opportunities in the nutrition field.

Bioactive substances can be incorporated into food packaging for many reasons (as antimicrobial, antioxidant, etc). Due to the importance of the releasing process, migration assays are commonly done, either *in vitro* or *in situ* conditions, as briefly summarized in this section. More examples of studies with their effectiveness are present in the next section.

3 Release Studies of Bioactive Packaging Systems and Their Effectiveness

3.1 Antimicrobial Packaging

Edible films and coatings loaded with antimicrobial substances are appropriate for the preservation of perishable products. These bioactive packaging systems extend the shelf-life of these products by retarding microbial spoilage and reducing the growth of pathogenic microorganisms (Pires et al. 2018). The antimicrobial activity effectiveness that a certain bioactive compound induces in food is highly dependent on the carrier system implemented, which in turn relies on other factors, like nature and properties associated to the materials used in the bionanocomposite structure, both for the matrix and filler as well as the consequent interaction between them, the time and type of contact (direct or indirect), the food matrix nature to be tested, and the ratio between the contact surface area of the active packaging and the volume of the food (Mousavi Khaneghah et al. 2018). Indeed, for the microorganisms present in the food to be properly inhibited through the antimicrobial action of these biomolecules, it is essential that there is a direct or very close contact between the food surface and the active packaging. That said, if the antimicrobial substance is not volatile it must be in direct contact with the food to be able to diffuse and solubilize easily, or if it is a volatile compound it must be close enough so that it can volatilize and be later absorbed by the food (Mousavi Khaneghah et al. 2018; Souza et al. 2017).

Nanofillers are nanometer-sized components that are introduced into biopolymers to improve their barrier and mechanical properties. In addition, they can equally serve as an aid to regulate the release of active compounds. These nanostructures possess the ability to trap the substances in a way that does not allow a fast and excessive release to the food (Souza et al. 2020a). The *in vitro* migration tests performed by Souza et al. (2019a, b) found that the introduction of montmorillonite (MMT) as a nano reinforcement in chitosan films reduces the diffusion of phenolic compounds present in the rosemary and ginger essential oil that had been previously introduced in the film. According to the authors, this is due to the network formed between the MMT that occupies the interstitial spaces of the chitosan, due to the interactions between functional groups present in the active compounds with hydroxyl and amino groups from chitosan chain, consequently entrapping the phenolic compounds and thus reducing their diffusion toward the simulante. Complementary, *in situ* experiments in which the same biofilms were used to wrap fresh poultry meat, confirmed, as expected, a lower microbiological inhibition effect in the samples wrapped with chitosan+MMT bionanocomposites (Souza et al. 2018c, 2019a). A similar effect was observed by Yu et al. (2019), where the incorporation of cellulose nanofibers weakens the antimicrobial activity in specific pathogens of soy protein-based nanocomposite films by attenuating the liberation of pine needle extract and lactic acid from the film structure.

To increase the bionanocomposite functionality, the introduction of nanoreinforcements into the biopolymeric matrix can provide not only value-added characteristics in mechanical and barrier terms, but could also offer an antimicrobial activity, which represents an interesting solution. The introduction in the bioactive packaging systems of inorganic nanoparticles that can confer this activity like metals (Ag, Au) or metal oxides, (ZnO, TiO_2, or organic nanoparticles capable of nanoencapsulation of bioactive compounds is seen as a promising approach with a lot of potential to be exploited (Giaconia et al. 2020; Brandelli 2019). Nanoemulsions, nanohydrogels, nanofibers, nanospheres, nanoliposomes or nanosponges are some current nanostructures assembled to encapsulate and deliver bioactive substances (Rezaei et al. 2019; Niu et al. 2020; Khorasani et al. 2019).

The applicability of extracts and essential oils is often limited by their lipophilic character and insolubility in water. As a means of overcoming this limitation, Paudel et al. (2019) incorporated essential oil of cinnamon in nanoemulsions. Subsequently, samples of two different species of melon, previously artificially inoculated with strains of *Listeria monocytogenes* and *Salmonella* spp, were exposed in direct contact with the nanoemulsion. The study observed that compared to the control sample, the nanoemulsion with 0.5% cinnamon oil inhibited in 7.7 and 5.5 log reductions in *Listeria monocytogenes* and *Salmonella* spp., respectively, thus suggesting its viable use in future bioactive packaging applications. In another interesting work regarding the potential of nanoemulsions, the antimicrobial activity of lemon oil-based nanoemulsion and two different concentrations of lemon essential oil (100% and 10%) were tested and compared against several foodborne pathogens and fish spoilage bacteria. The authors concluded that the antimicrobial activity of the nanoemulsion was higher than the original essential oil, which highlight them as possible natural antimicrobial agents in the preservation of processed fish products against foodborne pathogen and spoilage bacteria (Yazgan et al. 2019). More recently, chitosan edible coatings loaded with nanoemulsion containing *Zataria Multiflora* Boiss and *Bunium persicum* Boiss essential oils were developed and tested as an antimicrobial wrapping system in turkey meat. *In vitro* evaluations exposed that the application of these coatings decreases the microbial spoilage in about 2–3 log CFU/g for almost all samples, enlarging the shelf-life of the turkey meat sored at 4 °C in about 9–12 days (Keykhosravy et al. 2020).

Hydrogels, polymeric structures in three dimensions capable of retaining large amounts of water, are also a functional option for loading bioactive compounds. Due to good gelling properties, associated with biodegradability, biocompatibility, and safety, some biopolymers have been pointed out for this purpose. Carrageenan is one of these biopolymers, and in Oun and Rhim (2017) work, carrageenan-based hydrogels and films reinforced with metal oxide nanoparticles, ZnO and CuO and using KCl as a cross-linker, were produced. This study focused on the analysis of the effects caused by the introduction of metallic nanoparticles in the matrix, at a physical, mechanical, and microbiological level. The antibacterial activity against foodborne pathogenic bacteria, *Escherichia coli*, and *Listeria monocytogenes* was found to be better in carrageenan-based hydrogel films with the metallic nanoparticles. Another antimicrobial cellulosic hydrogel was prepared from olive oil industrial residue and filled with silver nanoparticles (AgNPs). The study remarkably observed a highly efficient antimicrobial activity against *Staphylococcus aureus*, *Pseudomonas aeruginosa*, and *Candida albicans*, mostly promoted by the AgNPs (Dacrory et al. 2018). Although the present research in this field is still restricted, the released studies demonstrate promising results for groundbreaking and potential applications for the food industry, including their consolidation into bioactive packaging systems and their incorporation directly into food matrices as an antimicrobial carrier system.

3.2 Antioxidant Packaging

The use of antioxidant agents in bioactive packaging systems is gaining high visibility through the recent years. Antioxidant agents have been used essentially to prevent oxidative degradation, one of the major problems associated to the degradation of food products, mostly related to those rich in lipid content (highly sensitive to oxidation process) (Vilela et al. 2018; Aziz and Karboune 2018). Oxidative reactions cause a decrease on food nutritional value, lead to the production of off-flavors and odors, and can cause color change (Vilela et al. 2018; Arroyo et al. 2018; Sanches-Silva et al. 2014). Other food degradation concern linked to oxidative reactions are enzymatic browning reactions and the growth of aerobic microorganisms (Bonilla et al. 2012).

Antioxidants can be included in food packaging systems for two purposes: acting as scavengers of undesirable substances (e.g. oxygen, metal ions or radical oxidative species) or for the release of them to the food (Gómez-Estaca et al. 2014). The inclusion of antioxidants into packaging systems appears to have less limitations rather than their application directly on the food surface. Once added to the food, the antioxidants are consumed in the reaction and become ineffective, leading to a rapid increase of the degradation rate (Gómez-Estaca et al. 2014).

The potential health risk associated to the use of synthetic antioxidants (e.g. butylated hydroxytoluene (BHT) and butylated hydroxyanisole (BHA)) has recently aroused the interest of the scientific community and food industry in the search for natural antioxidants (Nieva-Echevarría et al. 2015). Also, the consumers preference for natural food additives is another point of interest for the industry in the use of natural antioxidants (Aziz and Karboune 2018). The main issue on the use of natural antioxidants is that they generally require a larger amount than the synthetic ones to obtain the same antioxidant effect (Tian et al. 2013; Vilela et al. 2018).

Natural antioxidants are found in several plants, namely herbs and spices, and fruits (Silva-Weiss et al. 2013). Carotenoids, alkaloids, tannins, anthocyanins, and other phenolic compounds, as example, can be extracted and incorporated in bioactive packaging systems. The use of EOs extracted from plants such as essential oil of thyme, oregano, rosemary, basil or sage are also commonly used due to their strong antioxidant and antimicrobial properties (Valdés et al. 2015).

In general, the antioxidant systems are incorporated into packaging materials in edible films and coatings, or in the form of sachets, pads or labels giving some information of the foodstuffs (Gómez-Estaca et al. 2014; Vilela et al. 2018). The use of antioxidants blended on the packaging materials allows the reduction of contamination and the maintenance of organoleptic and nutritional properties of the food by its release to the food (Kadzińska et al. 2019). The incorporation of antioxidants in edible films, not only allows the creation of an environment to protect against food degradation releasing active compounds but also fits on the consumers demand for natural and safe food products (Aziz and Karboune 2018). The antioxidant coatings can also be applied by covalent immobilization laying the functional groups on food packaging material not altering their sensorial properties (Vasile 2018).

Some examples of several studies that have already been performed to demonstrate the effectiveness of antioxidants in bioactive packaging systems through the past few years and are discussed below.

Moringa oleifera L. has a strong antioxidant activity due to its high content in flavonoids and other phenolic compounds. Its combination with ascorbic acid or both individually (in the proportion of 6 wt% each) in papaya edible films were evaluated for pear preservation (Rodríguez et al. 2020). The study demonstrated that the combination of the two antioxidants or the single use of ascorbic acid, presented a high antioxidant activity. The use of ascorbic acid in the edible films demonstrated an effect on the shelf-life maintenance of minimally processed pears (Rodríguez et al. 2020).

The antioxidant capacity of mango leaf extracts (1–5 wt%) and its use in chitosan films were studied (Rambabu et al. 2019). The increment of extract concentration showed and improvement of the antioxidant properties of the chitosan films. Although the improvement of antioxidant capacity of the films, the increase of mango leaf extracts increased the films thickness and led to a decrease in moisture content. The application of films with 5% of mango leaf extract incorporation on cashew preservation for 28 days, showed a 56% resistance to oxidation when compared with a commercial polyamide/polyethylene film.

Souza et al. (2018b) studied the antioxidant migration effects of the incorporation of five essential oils (ginger, rosemary, sage, tea tree and thyme) and six hydroalcoholic extracts (HAEs) (ginger, green tea, black tea, kenaf leaves, rosemary and sage plants) on chitosan films into a food simulant (Souza et al. 2018b). The incorporation of all HAEs showed no diffusion from chitosan films, which demonstrates a good incorporation and interaction between them. This led to a greater barrier against light, water and oxygen. The essential oils incorporated in chitosan films maintained its antioxidant activity and demonstrated an exponential diffusion growth. From all the EOs and HAEs studied, ginger, sage and rosemary essential oils demonstrated the most promising results. In another study with essential oils (ginger and rosemary essential oil) by Souza et al. (2019b), incorporated in bionanocomposites based on chitosan and montmorillonite, their antioxidant activity on different food simulants (distilled water, ethanol 50% and ethanol 10%) by *in vitro* assays was evaluated. It was demonstrated that the essential oils after the migration assays maintain its antioxidant capacity by the ability to scavenge DPPH radical. Also, the release process depends on the EO incorporated and the presence/absence of the montmorillonite, and it was verified that the films with ginger EO have a higher diffusion into the food simulant, in special for ethanol 50% (Souza et al. 2019b).

Biopolymer films produced from potato peel wastes, reinforced with bacterial cellulose and curcumin (1–5 wt%) demonstrated to have positive effects on the preservation of fresh pork from lipid oxidation (Xie et al. 2020). Curcumin increased the antioxidant properties of the films leading to a lower content in malondialdehyde in fresh pork meat, an indicator for lipid oxidation.

Moghadam et al. (2020) developed edible films based on mung bean protein with antioxidant activity from pomegranate peel. The films with pomegranate peel presented high phenolic content with antibacterial and antioxidant properties, demonstrating an option for further bioactive packaging systems (Moghadam et al. 2020).

The antioxidant potential, among other properties, of the incorporation of anthocyanin-rich purple-fleshed sweet potato extracts on chitosan films were evaluated by Yong et al. (2019). It was demonstrated that these films could scavenge free radicals, exhibiting antioxidant capacity linked with the increase of purple-fleshed meat sweet potato extract amount on the film, and suggesting the chitosan films containing 10 wt% of the extract as the best combination to be used as antioxidant and intelligent packaging films for fish and pork (Yong et al. 2019).

Another studied linked with the use of natural antioxidant for food conservation was developed by Sganzerla et al. (2020). These authors produced a composite of starch and citric pectin incorporated with *Acca sellowiana* (feijoa) peel flour (0, 0.4, 1,2,3 and, 4% (w/w)) to apply on the postharvest conservation of apples. The use of feijoa peel flour at the highest concentration (4% (w/w)) on the films showed positive effects on all the parameters tested and, in especial, demonstrated a good antioxidant activity coupled with antimicrobial activity allowing the packaging produced to preserve the quality of the apples up to 5 days (Sganzerla et al. 2020).

The release of substances with antioxidant and antibacterial activity present on nanocomposite films were studied by Ramos et al. (2020). Poly(lactic acid) (PLA) based films with thymol (6 and 8 wt%) and silver nanoparticles (1 wt%) were produced to achieve active films. The results from the combination of thymol at 8 wt%, silver nanoparticles (1 wt%) and PLA showed positive results on antioxidant and antibacterial activity with a promising use for food bioactive packaging systems (Ramos et al. 2020).

The use of antioxidants in bioactive releasing systems for food applications still show some constraints and more studies are needed to better understand how the antioxidant compounds are released and interact not only with the packaging system but also with the food packaged. Moreover, it is important to make prior tests to identify the optimized balanced amount of bioactive compounds to be included in the FCM, once, for example, the incorporation of greater amounts of antioxidants in the polymers can originate pro-oxidant effects (Forester and Lambert 2011). Nevertheless, bioactive packaging systems are a promising area for the development of new, active and natural packaging systems fitting the actual global consumers demand for more natural/sustainable food products and packaging.

3.3 Others: Flavours, Colorants, Other Food ingredients

Flavours, colorants and other food ingredients can also be used in releasing systems for food active packaging. Essential oils are categorized as flavorings, and these volatile compounds are used as natural antimicrobial agents, but also due to their strong flavor, most of them additionally provide a characteristic aroma to the food packaged (Lucera et al. 2016), which can pose a negative aspect in the use of these compounds to preserve food products due to the changes in the sensory perception of the food packaged.

In the study of Gutiérrez et al. (2009), some aromas (banana, strawberry and vanilla) were evaluated to mask the negative odor and taste of essential oils or essential oils' components commonly used in the food industry (cinnamaldehyde, carvacrol, oregano essential oil and thymol). The authors found antimicrobial properties in thymol, carvacrol, and cinnamaldehyde, however none of them could be used together with the banana aroma, once these combinations were rejected by the sensory panel. Vanilla aroma was suitable to be combined with all antimicrobials tested providing both the antimicrobial property and the accepted odor, while for strawberry aroma resulted in satisfactory / acceptable sensory profile only with thymol.

Active packaging with flavor/odor releasers increases the attractiveness of the food to the customer by the improvement of the fresh aroma of the product itself, and the enhancement of the flavor of food when the package is opened. However, the gradual release of such flavor/odors can offset the natural loss of taste or smell of products with a long shelf life, masking their natural spoilage reactions (Almenar et al. 2009). For this reason, flavor/odor releasers have been effectively banned in Europe and in the USA, but yet are used in Japan in many commercial products with many applications that cannot be neglected (Lucera et al. 2016).

A Brazilian research group developed such type of active packaging by incorporating lemon aroma and/or lemon essential oil (Dias et al. 2013a) or orange aroma and/or orange essential oil (Dias et al. 2013b) into LPDE film to pack sugar biscuit (Table 9.3). The influence of the packaging on the biscuit packaged was only monitored through sensory analysis. The consumers preferred the samples packaged with the films incorporated with both EO and aroma of the respectively fruit, demonstrating that after 10 or 30 days of contact the food gained the fruit aroma desired with high acceptance score ("like very much" – from the hedonic scale used).

Nutrients can also be incorporated into the packaging material, in this case, aiming the enrichment of food packaged with such components. Beans were enriched with the amino acids methionine and cysteine through the use of active packaging based on edible biopolymers (cassava starch) added with these nutrients (Sousa et al. 2018). Two polymers were studied as the coating material, cassava starch and carnauba wax. The coatings were prepared with 1 wt% cassava starch or 4 wt% carnauba wax, and they were tested individually or in association 1:1 w/w. Samples coated with treatments containing cassava starch showed higher concentration of added amino acids in both broth or grain after cooking than those containing carnauba wax. As conclusions the authors stated that the addition of sulphur amino acids in edible coatings for foods poor with this nutrient, such as common bean, could represent a feasible alternative to enhance the daily ingestion of these amino acids in diets to fulfill the amount recommended by Food and Agricultural Organization/World Health Organization.

4 Conclusions

Active food packaging is a novel technology to preserve food products mostly by the release of bioactive compounds. The development of this technology follows the current trend in the market to look for more natural products with a greener approach, which has pushed the research to biopolymers incorporated with natural compounds, aiming the substitution of the traditional petroleum-based polymers. Moreover, nanotechnology is commonly associated with this type of packaging aiming not only the reinforcement of the material but also to add active properties to the system. In such bioactive releasing systems, migration assay is mandatory to understand its behavior and effectiveness to preserve the product packaged. Most of the releasing active systems claim to be effective, but their effectiveness depends on many factors (type of food being package/simulant, type of polymer/biopolymer, type of bioactive compounds, addition of plasticizers, reinforcements, other additives, and other factors). In order to simplify the analysis, simulants are often used although there is an increasing number of studies that use real foods to test the effectiveness of these materials (active releasing systems). It has been shown in literature the advantages of these systems, yet more studies need to be done to fully understand them and to critically assess this novel technology.

Acknowledgements This research was funded by national funding by FCT, Foundation for Science and Technology, through the individual PhD research grants (SFRH/BD/144346/2019) of J.P and (2020.04441.BD) of C.R. This work was also supported by the Mechanical Engineering and Resource Sustainability Center - METRiCs which is financed by national funds from FCT/MCTES (UIDB/04077/2020 and UIDP/04077/2020) and by the Associate Laboratory for Green Chemistry - LAQV which is financed by national funds from FCT/MCTES (UIDB/50006/2020 and UIDP/50006/2020).

References

Abbas Syed Q. Pulsed electric field technology in food preservation: a review. J Nutr Heal Food Eng. 2017;6(6).

Abdollahi M, Rezaei M, Farzi G. A novel active bionanocomposite film incorporating rosemary essential oil and nanoclay into chitosan. J Food Eng. 2012;111(2):343–50.

Abdollahi M, Rezaei M, Farzi G. Influence of chitosan/clay functional bionanocomposite activated with rosemary essential oil on the shelf life of fresh silver carp. Int J Food Sci Technol. 2014;49(3):811–8.

Ahmed I, et al. A comprehensive review on the application of active packaging technologies to muscle foods. Food Control. 2017;82:163–78.

Almenar E, Catala R, Hernandez-Muñoz P, Gavara R. Optimization of an active package for wild strawberries based on the release of 2-nonanone. LWT Food Sci Technol. 2009;42(2):587–93.

Andrade MA, Ribeiro-Santos R, Costa Bonito MC, Saraiva M, Sanches-Silva A. Characterization of rosemary and thyme extracts for incorporation into a whey protein based film. LWT. 2018;92(January):497–508.

Ans F. Re-evaluation of titanium dioxide (E 171) as a food additive. EFSA J. 2016;14(9):e04545.

Appendini P, Hotchkiss JH. Review of antimicrobial food packaging. Innov Food Sci Emerg Technol. 2002;3(2):113–26.

Arroyo BJ, Santos AP, de Almeida de Melo E, Campos A, Lins L, Boyano-Orozco LC. Bioactive compounds and their potential use as ingredients for food and its application in food packaging. In: Bioactive compounds: health benefits and potential applications. Elsevier Inc; 2018. p. 143–56.

Azeredo HMC, Faria JAF, Brito ES. Embalagens e suas interações com os alimentos. In: Azeredo HMC, editor. Fundamentos de estabilidade de alimentos. 2nd ed. Fortaleza: Embrapa Agroindústria Tropical; 2012. p. 328.

Aziz M, Karboune S. Natural antimicrobial/antioxidant agents in meat and poultry products as well as fruits and vegetables: a review. Crit Rev Food Sci Nutr. 2018;58(3):486–511.

Bonilla J, Atarés L, Vargas M, Chiralt A. Edible films and coatings to prevent the detrimental effect of oxygen on food quality: possibilities and limitations. J Food Eng. 2012;110(2):208–13.

Brandelli A. The interaction of nanostructured antimicrobials with biological systems: cellular uptake, trafficking and potential toxicity. Food Sci Human Wellness. 2019;9(1):8–20.

Brewer MS. Natural antioxidants: sources, compounds, mechanisms of action, and potential applications. Compr Rev Food Sci Food Saf. 2011;10(4):221–47.

Burt S. Essential oils: their antibacterial properties and potential applications in foods – a review. Int J Food Microbiol. 2004;94(3):223–53.

Chen H, Li L, Ma Y, Mcdonald TP, Wang Y. Development of active packaging fi lm containing bioactive components encapsulated in β -cyclodextrin and its application. Food Hydrocoll. 2019;90(December 2018):360–6.

Chungy D, Papadakis SE, Yamy KL. Simple models for assessing migration from food-packaging ® lms. Food Addit Contam. 2002;19(6):611–7.

Colín-Chávez C, Soto-valdez H, Peralta E, Lizardi-mendoza J, Balandrán-quintana R. Diffusion of natural astaxanthin from polyethylene active packaging fi lms into a fatty food simulant. Food Res Int. 2013;54(1):873–80.

COMMISSION REGULATION (EU) 2016/1416 of 24 August 2016 amending and correcting Regulation (EU) No 10/2011 on plastic materials and articles intended to come into contact with food. Off J Eur Union. 2016;L 230:22–42.

Conte A, Scrocco C, Sinigaglia M, Del Nobile MA. Innovative active packaging systems to prolong the shelf life of Mozzarella cheese. J Dairy Sci. 2007;90(5):2126–31.

Dacrory S, Abou-yousef H, Abouzeid RE, Kamel S, Abdel-aziz MS, El-badry M. Antimicrobial cellulosic hydrogel from olive oil industrial residue. Int J Biol Macromol. 2018;117:179–88.

De Oliveira TM, De Fátima Ferreira Soares N, Pereira RM, De Freitas Fraga K. Development and evaluation of antimicrobial natamycin-incorporated film in gorgonzola cheese conservation. Packag Technol Sci. 2007;20(2):147–53.

de Soares NFF, Hotchkiss JH. Naringinase immobilization in packaging films for reducing naringin concentration in grapefruit juice. J Food Sci. 1998;63:61–5.

de Soares NFF, Pires ACS, Camilloto GP, Santiago-Silva P, Espitia PJP, Silva WA. Recent patents on active packaging for food application. Recent Pat Food Nutr Agric. 2009;1(2):171–8.

del Castro-López M, López-Vilariño JM, González-Rodríguez MV. Analytical determination of flavonoids aimed to analysis of natural samples and active packaging applications. Food Chem. 2014;150:119–27.

Dias MV, et al. Development of low-density polyethylene films with lemon aroma. LWT Food Sci Technol. 2013a;50(1):167–71.

Dias MV, et al. Application of flavoring active films of low-density polyethylene in sugar biscuits. J Food Process Eng. 2013b;36(5):656–62.

Dias MV, et al. Effect of active films incorporated with montmorillonite clay and α-tocopherol: potential of nanoparticle migration and reduction of lipid oxidation in salmon. Packag Technol Sci. 2018;32(1):39–47.

Efrati R, et al. The effect of polyethylene crystallinity and polarity on thermal stability and controlled release of essential oils in antimicrobial films. J Appl Polym Sci. 2014;131(11).

European Commission. Commission Regulation (EC) No. 450/2009 of 29 May 2009 on active and intelligent materials and articles intended to come into contact with food. Off J Eur Union. 2009;L 135(450):3–11.

European Commission. Regulation (EU) No 10/2011 of 14 January 2011 on plastic materials and articles intended to come into contact with food. Off J Eur Union. 2011;L 12:1–89.

European Food Safety Authority (EFSA). Safety assessment of the substance zinc oxide, nanoparticles, for use in food contact materials. EFSA J. 2016;14(3).

European Union (EU), Regulation (EC) No 1935/2004 of the European Parliament and of the Council of 27 October 2004 on materials and articles intended to come into contact with food and repealing Directives 80/590/EEC and 89/109/EEC. Off J Eur Union. 2004;L 338:4–17.

Filly A, Fernandez X, Minuti M, Visinoni F, Cravotto G, Chemat F. Solvent-free microwave extraction of essential oil from aromatic herbs: from laboratory to pilot and industrial scale. Food Chem. 2014;150:193–8.

Forester SC, Lambert JD. The role of antioxidant versus pro-oxidant effects of green tea polyphenols in cancer prevention. Mol Nutr Food Res. 2011;55(6):844–54.

Gharenaghadeh S, et al. Application of salvia multicaulis essential oil-containing nanoemulsion against food-borne pathogens. Food Biosci. 2017;19(July):128–33.

Giaconia MA, dos Ramos SP, Pereira CF, Lemes AC, De Rosso VV, Braga ARC. Overcoming restrictions of bioactive compounds biological effects in food using nanometer-sized structures. Food Hydrocoll. 2020;107:105939.

Gómez-Estaca J, López-de-Dicastillo C, Hernández-Muñoz P, Catalá R, Gavara R. Advances in antioxidant active food packaging. Trends Food Sci Technol. 2014;35(1):42–51.

Gutiérrez L, Escudero A, Batlle R, Nerín C. Effect of mixed antimicrobial agents and flavors in active packaging films. J Agric Food Chem. 2009;57(18):8564–71.

Haghighi-Manesh S, Azizi MH. Active packaging systems with emphasis on its applications in dairy products. J Food Process Eng. 2017;40(5):e12542.

Hannon JC, Kerry JP, Cruz-Romero M, Azlin-Hasim S, Morris M, Cummins E. Kinetic desorption models for the release of nanosilver from an experimental nanosilver coating on polystyrene food packaging. Innov Food Sci Emerg Technol. 2017;44(October 2016):149–58.

Jamshidian M, Tehrany EA, Desobry S. Antioxidants release from solvent-cast PLA film: investigation of PLA antioxidant-active packaging. Food Bioprocess Technol. 2012;6(6):1450–63.

Kadzińska J, Janowicz M, Kalisz S, Bryś J, Lenart A. An overview of fruit and vegetable edible packaging materials. Packag Technol Sci. 2019;32(10):483–95.

Keykhosravy K, Khanzadi S, Hashemi M, Azizzadeh M. Chitosan-loaded nanoemulsion containing Zataria Multiflora Boiss and Bunium persicum Boiss essential oils as edible coatings: Its impact on microbial quality of turkey meat and fate of inoculated pathogens. Int J Biol Macromol. 2020;150:904–13.

Khatabi O, Hanine H, Elothmani D, Hasib A. Extraction and determination of polyphenols and betalain pigments in the Moroccan Prickly pear fruits (Opuntia ficus indica). Arab J Chem. 2016;9:S278–81.

Khorasani S, et al. Recent trends in the nanoencapsulation processes for food and nutraceutical applications. Elsevier; 2019.

King T, Osmond-McLeod MJ, Duffy LL. Nanotechnology in the food sector and potential applications for the poultry industry. Trends Food Sci Technol. 2018;72(November 2017):62–73.

Latos-Brozio M, Masek A. The application of natural food colorants as indicator substances in intelligent biodegradable packaging materials. Food Chem Toxicol. 2020;135(June 2019):110975.

Lopez de Dicastillo C, Nerin C, Alfaro P, Catala R, Gavara R, Hernandez-Munoz P. Development of new antioxidant active packaging films based on ethylene vinyl alcohol copolymer (EVOH) and green tea extract. J Agric Food Chem. 2011;59(14):7832–40.

López-de-Dicastillo C, Gómez-Estaca J, Catalá R, Gavara R, Hernández-Muñoz P. Active antioxidant packaging films: development and effect on lipid stability of brined sardines. Food Chem. 2012;131(4):1376–84.

Lucera A, Conte A, Del Nobile MA. Volatile compounds usage in active packaging systems. Elsevier Inc; 2016.

Melini V. Strategies to extend bread and gf bread shelf-life: from sourdough to antimicrobial active packaging and nanotechnology. Fermentation. 2018;4(1):9.

Moghadam M, Salami M, Mohammadian M, Khodadadi M, Emam-Djomeh Z. Development of antioxidant edible films based on mung bean protein enriched with pomegranate peel. Food Hydrocoll. 2020;104(February):105735.

Mousavi Khaneghah A, Hashemi SMB, Limbo S. Antimicrobial agents and packaging systems in antimicrobial active food packaging: an overview of approaches and interactions. Food Bioprod Process. 2018;111:1–19.

Nieva-Echevarría B, Manzanos MJ, Goicoechea E, Guillén MD. 2,6-Di-Tert-Butyl-hydroxytoluene and its metabolites in foods. Compr Rev Food Sci Food Saf. 2015;14(1):67–80.

Niu B, Shao P, Luo Y, Sun P. Recent advances of electrosprayed particles as encapsulation systems of bioactives for food application. Food Hydrocoll. 2020;99(September 2019):105376.

Oun AA, Rhim J. Carrageenan-based hydrogels and films: effect of ZnO and CuO nanoparticles on the physical, mechanical, and antimicrobial properties. Food Hydrocoll. 2017;67:45–53.

Ozer BBP, Uz M, Oymaci P, Altinkaya SA. Development of a novel strategy for controlled release of lysozyme from whey protein isolate based active food packaging films. Food Hydrocoll. 2016;61:877–86.

Paudel SK, Bhargava K, Kotturi H. Antimicrobial activity of cinnamon oil nanoemulsion against Listeria monocytogenes and Salmonella spp. on melons. LWT Food Sci Technol. 2019;111:682–7.

Pires JRA, de Souza VGL, Fernando AL. Chitosan/montmorillonite bionanocomposites incorporated with rosemary and ginger essential oil as packaging for fresh poultry meat. Food Packag Shelf Life. 2018;17(June):142–9.

Pola CC, et al. Cellulose acetate active films incorporated with oregano (Origanum vulgare) essential oil and organophilic montmorillonite clay control the growth of phytopathogenic fungi. Food Packag Shelf Life. 2016;9:69–78.

Prasad P, Kochhar A. Active packaging in food industry: a review. IOSR J Environ Sci Toxicol Food Technol. 2014;8(5):01–7.

Quesada J, Sendra E, Navarro C, Sayas-Barberá E. Antimicrobial active packaging including chitosan films with thymus vulgaris L. essential oil for ready-to-eat meat. Foods. 2016;5(3):57.

Rambabu K, Bharath G, Banat F, Show PL, Cocoletzi HH. Mango leaf extract incorporated chitosan antioxidant film for active food packaging. Int J Biol Macromol. 2019;126:1234–43.

Ramos M, Beltrán A, Peltzer M, Valente AJM, Garrigós MDC. Release and antioxidant activity of carvacrol and thymol from polypropylene active packaging films. LWT Food Sci Technol. 2014;58(2):470–7.

Ramos M, et al. Controlled release of thymol from poly(lactic acid)-based silver nanocomposite films with antibacterial and antioxidant activity. Antioxidants. 2020;9(5):395.

Regnier T, Combrinck S, Du Plooy W. Essential oils and other plant extracts as food preservatives. In: Progress in food preservation. Oxford: Wiley-Blackwell; 2012. p. 539–79.

Rezaei A, Fathi M, Mahdi S. Nanoencapsulation of hydrophobic and low-soluble food bioactive compounds within different nanocarriers. Food Hydrocoll. 2019;88:146–62.

Ribeiro-Santos R, et al. A novel insight on an ancient aromatic plant: the rosemary (Rosmarinus officinalis L.). Trends Food Sci Technol. 2015;45(2):355–68.

Ribeiro-Santos R, Andrade M, de Melo NR, Sanches-Silva A. Use of essential oils in active food packaging: recent advances and future trends. Trends Food Sci Technol. 2017;61:132–40.

Rizzolo A, et al. Volatile compound composition and antioxidant activity of cooked ham slices packed in propolis-based active packaging. Food Packag Shelf Life. 2016;8:41–9.

Rodríguez GM, Sibaja JC, Espitia PJP, Otoni CG. Antioxidant active packaging based on papaya edible films incorporated with Moringa oleifera and ascorbic acid for food preservation. Food Hydrocoll. 2020;103:105630.

Sanches Silva A, Cruz Freire JM, Sendón R, Franz R, Losada PP. Migration and diffusion of diphenylbutadiene from packages into foods. J Agric Food Chem. 2009;57(21):10225–30.

Sanches-Silva A, et al. Trends in the use of natural antioxidants in active food packaging: a review. Food Addit Contam Part A. 2014;31(3):374–95.

Santana MCCB, Machado BAS, Silva TN, Nunes IL, Druzian JI. Incorporação de urucum como aditivo antioxidante em embalagens biodegradáveis a base de quitosana. Ciência Rural. 2013;43(3):544–50.

Sganzerla WG, et al. Bioactive food packaging based on starch, citric pectin and functionalized with Acca sellowiana waste by-product: characterization and application in the postharvest conservation of apple. Int J Biol Macromol. 2020;147:295–303.

Silva AS, Freire JMC, García RS, Franz R, Losada PP. Time–temperature study of the kinetics of migration of DPBD from plastics into chocolate, chocolate spread and margarine. Food Res Int. 2007;40(6):679–86.

Silva-Weiss A, Ihl M, Sobral PJA, Gómez-Guillén MC, Bifani V. Natural additives in bioactive edible films and coatings: functionality and applications in foods. Food Eng Rev. 2013;5(4):200–16.

Sousa S, Soares NFF, Faroni LRD. Nutritional quality of common bean fortified with amino acids. Int J Dev Res. 2018;8(6):21109–14.

Souza VGL, Fernando AL. Nanoparticles in food packaging: biodegradability and potential migration to food – A review. Food Packag Shelf Life. 2016;8:63–70.

Souza VGL, Fernando AL, Pires JRA, Rodrigues PF, Lopes AASS, Fernandes FMB. Physical properties of chitosan films incorporated with natural antioxidants. Ind Crop Prod. 2017;107(February):565–72.

Souza VGL, et al. Nanomaterial migration from composites into food matrices. In: Cirillo G, Kozlowski MA, Spizzirri UG, editors. Composite materials for food packaging. 1st ed. Beverly: Scrivener Publishing LLC; 2018a. p. 401–35.

Souza VGL, Rodrigues PF, Duarte MP, Fernando AL. Antioxidant migration studies in chitosan films incorporated with plant extracts. J Renew Mater. 2018b;6(5):548–58.

Souza VGL, Pires JR, Vieira ÉT, Coelhoso IM, Duarte MP, Fernando AL. Shelf life assessment of fresh poultry meat packaged in novel bionanocomposite of chitosan/montmorillonite incorporated with ginger essential oil. Coatings. 2018c;8(5):177.

Souza VGL, Pires JRA, Vieira ÉT, Coelhoso IM, Duarte MP, Fernando AL. Activity of chitosan-montmorillonite bionanocomposites incorporated with rosemary essential oil: from in vitro assays to application in fresh poultry meat. Food Hydrocoll. 2019a;89(October 2018):241–52.

Souza VGL, et al. In vitro bioactivity of novel chitosan bionanocomposites incorporated with different essential oils. Ind Crop Prod. 2019b;140:111563.

Souza VGL, Pires JRA, Rodrigues C, Coelhoso IM, Fernando AL. Chitosan composites in packaging industry – current trends and future challenges. Polymers (Basel). 2020a;12(2):417.

Souza VGL, et al. Eco-friendly ZnO/Chitosan bionanocomposites films for packaging of fresh poultry meat. Coatings. 2020b;10(2):110.

Stoll L, Rech R, Flôres SH, Nachtigall SMB, de Oliveira Rios A. Poly(acid lactic) films with carotenoids extracts: release study and effect on sunflower oil preservation. Food Chem. 2019;281(August 2018):213–21.

Tang XZ, Kumar P, Alavi S, Sandeep KP. Recent advances in biopolymers and biopolymer-based nanocomposites for food packaging materials. Crit Rev Food Sci Nutr. 2012;52(5):426–42.

Tian F, Decker EA, Goddard JM. Controlling lipid oxidation of food by active packaging technologies. Food Funct. 2013;4(5):669.

Valdés A, Mellinas AC, Ramos M, Burgos N, Jiménez A, Garrigós MC. Use of herbs, spices and their bioactive compounds in active food packaging. RSC Adv. 2015;5(50):40324–35.

Vasile C. Polymeric nanocomposites and nanocoatings for food packaging: a review. Materials (Basel). 2018;11(10):1834.

Vilela C, et al. A concise guide to active agents for active food packaging. Trends Food Sci Technol. 2018;80(August):212–22.

Wu J, et al. The preparation, characterization, antimicrobial stability and in vitro release evaluation of fish gelatin films incorporated with cinnamon essential oil nanoliposomes. Food Hydrocoll. 2015;43:427–35.

Xie Y, et al. Active biodegradable films based on the whole potato peel incorporated with bacterial cellulose and curcumin. Int J Biol Macromol. 2020;150:480–91.

Yazgan H, Ozogul Y, Kuley E. Antimicrobial influence of nanoemulsified lemon essential oil and pure lemon essential oil on food-borne pathogens and fish spoilage bacteria. Int J Food Microbiol. 2019;306:108266.

Yong H, Wang X, Bai R, Miao Z, Zhang X, Liu J. Development of antioxidant and intelligent pH-sensing packaging films by incorporating purple-fleshed sweet potato extract into chitosan matrix. Food Hydrocoll. 2019;90(November 2018):216–24.

Yu Z, et al. Development of multifunctional nanocomposites containing cellulose nano fibrils and soy proteins as food packaging materials. Food Packag Shelf Life. 2019;21:100366.

Chapter 10
Edible Active Coating Systems for Food Purposes

Cássia H. Barbosa, Mariana A. Andrade, Fernanda Vilarinho, Ana Luísa Fernando, and Ana Sanches Silva

Abstract Currently, many foods are wasted because they are highly perishable or have not been properly packaged, and when they reach the final consumer, they do not meet minimum quality requisites. To overcome this problem, the food and packaging industries have developed several active packaging systems to increase the food's shelf life. One of the systems developed to control food spoilage and increase its shelf life are edible active coatings. Edible coatings are thin, protective, edible, and biodegradable layers deposited on a food's surface. The edible coatings are an environmentally friendly technology produced with biodegradable polymers (polysaccharides, proteins, lipids, and composites) and, can also enhance food safety, nutritional and sensory attributes by adding bioactive compounds (antioxidants, antimicrobials, or specific nutrients) to the polymeric matrix. Different studies have shown the effectiveness of edible active coatings in different foods products, particularly in fruits and vegetables. The edible active coatings can control moisture transfer, gas exchange, microbial growth, oxidation processes and other chemical reactions. This chapter summarizes the current state of knowledge on edible active

C. H. Barbosa (✉) · F. Vilarinho
Department of Food and Nutrition, National Institute of Health Dr. Ricardo Jorge, Lisbon, Portugal
e-mail: cassia.barbosa@insa.min-saude.pt

M. A. Andrade
Department of Food and Nutrition, National Institute of Health Dr. Ricardo Jorge, Lisbon, Portugal

REQUIMTE/LAQV, Faculty of Pharmacy, University of Coimbra, Coimbra, Coimbra, Portugal

A. L. Fernando
MEtRICs, Departamento de Ciências e Tecnologia da Biomassa, Faculdade de Ciências e Tecnologia, FCT, Universidade Nova de Lisboa, Caparica, Portugal

A. S. Silva
National Institute for Agrarian and Veterinary Research (INIAV), I.P., Vila do Conde, Portugal

Center for Study in Animal Science (CECA), ICETA, University of Oporto, Oporto, Portugal

S. M. Jafari, A. S. Silva (eds.), *Releasing Systems in Active Food Packaging*, Food Bioactive Ingredients, https://doi.org/10.1007/978-3-030-90299-5_10

coatings, discussing their different applications as carriers of active compounds, such as antimicrobials or antioxidants, to improve quality and extend food's shelf life.

Keywords Edible coatings · Active systems · Releasing systems · Antioxidant · Antimicrobial · Shelf life

Abbreviations

3D	Three-dimensional
a*	Redness
b*	Yellowness
BHA	Butylated hydroxyanisole (E320)
BHT	Butylated hydroxytoluene (E321)
CA	Carvacrol
CMC	Carboxymethyl cellulose
CWP	Chitosan/whey protein
EAC	Edible active coating
EC	European Commission
EFSA	European Food Safety Authority
EU	European Union
FAO	Food and Agriculture Organization of the United Nations
FDA	Food and Drug Administration
GAE	Gallic acid equivalent
Gly	Glycerol
GMP	Good manufacturing practice
GRAS	Generally regarded as safe
HPC	Hydroxypropyl cellulose
HPMC	Hydroxypropyl methylcellulose
LbL	Layer-by-layer
MAP	Modified atmosphere packaging
MC	Methylcellulose
PA	Polyamide
PE	Polyethylene
PET	Polyethylene terephthalate
PP	Polypropylene
PS	Polystyrene
PVC	Polyvinylchloride
TBARS	Thiobarbituric acid reactive substances
US	United States of America
WHO	World Health Organization
WP	Whey protein
WPC	Whey protein concentrate

WPI	Whey protein isolate
WPNFs	Whey protein isolates nanofiber

1 Introduction

Many foods deteriorate easily, especially if they are not well packed. Therefore, it is necessary to develop packaging that protects food so that it reaches the final consumer still in good condition. Food packaging has evolved over the years. Many packages are made of petrochemical-based plastics such as polyethylene terephthalate (PET), polyethylene (PE), polyvinylchloride (PVC), polystyrene (PS), polypropylene (PP), and polyamide (PA); although they provide good mechanical properties and have low cost, they may have unwanted influence on the environment as they are not completely recyclable and/or biodegradable. So, researchers and industries have been developing new packaging that besides protecting food from external factors is also biodegradable, environmentally friendly and can improve even more food shelf life (Mohamed et al. 2020; Sapper and Chiralt 2018; Siracusa et al. 2008). This has been done, among others, through the development of edible active coatings (EAC).

Edible coatings have several definitions, depending on the standpoint. There is no universal definition of edible coatings. In general, edible coatings are defined as a thin layer of edible material applied onto the food surface to extend the self-life as it provides a barrier to moisture and gases, and inhibits oxidation, weight loss, and solute movement out of the food. Various biopolymers, such as polysaccharides, proteins, lipids, or their combinations can be used to produce edible coatings (Dehghani et al. 2018; Dhall 2013; Mohamed et al. 2020; Sapper and Chiralt 2018). If the coating functions also as a carrier of an active compound, such as an antimicrobial or antioxidant compound, it is called EAC. The ingredients used to produce an EAC should be assessed and regulated before its application, once it can be further consumed with the food product (Mohamed et al. 2020; Umaraw et al. 2020; Valdés et al. 2017).

Edible coatings have been used for centuries in fruits. It started in China in the twelfth century with the application of waxes onto citrus fruits to prevent moisture loss and to create a shiny surface. Later, fats or lard were used to prolong the shelf life of meat products in England. Then, in the sixteenth century in the United States, food products were coated with lipid coatings to control moisture loss, and later paraffin wax and carnauba wax. In the early twentieth century, coatings have been used to add shine and prevent water loss in fruits and vegetables, as casings for sausages, and as some sort of sugary coating on confectionaries. In sum, edible coatings have been applied to a variety of food products. Today, EAC are successfully being applied to different food products aiming to prolong their shelf-lives, and maintaining their quality (Debeaufort et al. 1998; Dehghani et al. 2018; Dhall

2013; Galus and Kadzińska 2015; Guilbert et al. 1995; Hanssen et al. 2012; Hassan et al. 2018; Park 2003; Pop et al. 2019; Riva et al. 2020; Salgado et al. 2015).

This chapter aims to review the current state of knowledge on EAC, regarding their different application to food products as well as their ability to act as carriers of active compounds, like antioxidants or antimicrobial compounds, to improve quality and extend food's shelf life.

2 Legislation

Due to the fact that there is a wide variety of food regulations worldwide, the Food and Agriculture Organization (FAO) together with the World Health Organization (WHO) of the United Nations created the *Codex Alimentarius*, a collection of food standards and guidelines to uniform food legislation in the world (FAO and WHO 2020). EAC is normally regarded as a food additive.

The European Union (EU) regulates the materials and articles in contact with food through the Regulations No. 1935/2004 and its amendments (on materials and articles intended to come into contact with food); No. 2023/2006 (on good manufacturing practice (GMP) for materials and articles intended to come into contact with food); No. 450/2009 (active and intelligent materials); No. 10/2011 (on food plastics), and No. 282/2008 (on recycled food plastics), and with the Directives No. 84/500/EEC (ceramic articles intended to come into contact with foodstuffs), and No. 2007/42/CE (materials and articles made of regenerated cellulose film), but this Regulation does not apply to coatings materials (European Commission (EC) 1984, 2004, 2007, 2008b, 2009, 2011a). EAC are regulated by the Regulation No 1333/2008 of the European Parliament and of the Council of 16 December 2008 on food additives. And also by Regulation No 1334/2008 on flavourings and certain food ingredients with flavouring properties for use in and on foods (EC 2008a, c). According to the Regulation (EC) No 1333/2008 of the European Parliament and of the Council, food additives are "*any substance not normally consumed as a food in itself and not normally used as a characteristic ingredient of food, whether or not it has nutritive value, the intentional addition of which to food for a technological purpose in the manufacture, processing, preparation, treatment, packaging, transport or storage of such food results, or may be reasonably expected to result, in it or its by-products becoming directly or indirectly a component of such foods*" (EC 2008c). Food additives can be sorted as direct, referring to the additives directly applied to foods, and indirect additives, also recognized as substances used in food-contact articles (Andrade et al. 2020).

The EU has a community list of approved food additives that presents all substances that may be used, in this case, in the preparation of EAC. To be included in the community list, the substance must be advantageous and beneficial for the consumer, and therefore it should, besides others, enhance and/or maintain the quality or stability of a food or improve its organoleptic properties. The assessment and authorisation of food additives, food flavourings, food enzymes, and source materials of food flavourings and of food ingredients with flavouring properties used or intended for use in or on foodstuffs are regulated by Regulation (EC) No 1331/2008 of the European Parliament and of the Council of 16 December 2008. The safety of the substances used as food additives in the EU is assured by the Scientific Committee on Food and/or the European Food Safety Authority (EFSA), and only those approved by EFSA can be used in the preparation of EAC (European Commission 2008c, d, 2011b). According to this, everyone who wants to put an additive market that is not authorized or wants to expand the conditions of an authorized additive must present a request for compliance with this Regulation as well as the corresponding EFSA guide.

In the United States of America (US), the Food and Drug Administration (FDA) is responsible for the regulation of food additives as well as their approval. According to FDA, a food additive is defined as "*any substance the intended use of which results or may reasonably be expected to result, directly or indirectly, in its becoming a component or otherwise affecting the characteristic of any food including any substance intended for use in producing, manufacturing, packing, processing, preparing, treating, packaging, transporting, or holding food*" (FDA 2010). The FDA has a list of approved food additives and has a specific section of Indirect food additives with the substances that may be used in the preparation of edible coatings. According to FDA, indirect food additives are "*those that become part of the food in trace amounts due to its packaging, storage or other handling. For instance, minute amounts of packaging substances may find their way into foods during storage*" (FDA 2010). The US legislation Parts 175 to 178 of the Code of Federal Regulations (21CFR) by the FDA contain all the authorized indirect additives that can be applied in food contact materials, including adhesives and coating components (Part 175), paper and paperboard components (Part 176), polymers (Part 177), adjuvants, production aids and sanitizers (Part 178). Also, the material used for the preparation of edible coatings can be generally regarded as safe (GRAS) (Andrade et al. 2020; FDA 2010).

3 Composition of Edible Active Coatings

EAC can be composed of a single biopolymer or in combination with others to obtain the best properties. EAC carry different food additives with different purposes, such as antioxidant, antimicrobials, anti-browning, colorants, flavors, etc. Usually, plasticizers, like glycerol, sorbitol, monoglycerides, polyethylene glycol, and glucose, are added to the coating solution to improve the flexibility and elasticity of the coating, although it may affect the barrier properties (Andrade et al. 2020; Espitia et al. 2014; Galus and Kadzińska 2015; Salgado et al. 2015).

3.1 *Biopolymers*

3.1.1 Polysaccharides

Polysaccharides are complex carbohydrates that naturally exist in nature and are largely used to prepare edible films or coatings. The polysaccharides can be responsible for the plant energy storage or be the structural elements of plants and animal exoskeleton. Different polysaccharides and their derivates with different origins can be used to prepare edible coatings. The various origins of polysaccharides are plant (starch and pectin), animal (chitosan and chitin), marine (alginate and carrageenan), and microbial (pullulan and xanthan gum). Polysaccharides are wildly used to prepare edible coatings and are non-toxic materials known to originate a colourless, oily-free appearance and with a minor caloric content coating. Polysaccharides can act as a barrier to carbon dioxide and oxygen but are poor water vapour barriers. Finally, polysaccharides can be applied to prolong the shelf life of vegetables, fruits, meat products, or shellfish by significantly reducing darkening of the surface, dehydration, and oxidative rancidity (Baldwin et al. 2011; Hassan et al. 2018; Mohamed et al. 2020).

Starch

Starch is a naturally occurring polysaccharide and can be found in different sources, such as cereals (corn, wheat or rice), tubers (cassava, potato, or sweet potato), and legumes (peas). Starch is composed of several 1,4-α-*D* glucopyranosyl units, namely of amylose (Fig. 10.1), a linear chain polymer, and amylopectin (Fig. 10.2), a glucose polymer with a branched-chain structure (Hassan et al. 2018; Mohamed et al. 2020; Nesic and Seslija 2017; Sapper and Chiralt 2018). The concentration of these units varies according to the plant source and the edaphoclimatic conditions to which the plant has been exposed (Avella et al. 2005; Geigenberger et al. 1998). Starch coatings do not have good barrier properties as it has a low water vapour barrier capacity but other compounds such as emulsifiers and plasticizers or different nano-particles can be added to reduce the permeability, extensibility, flexibility, and/or the stability of the polymer matrix structure (Sahraee et al. 2019; Sapper and Chiralt 2018). Nonetheless, starch coatings are wildly used to produce biodegradable coatings as they are low cost and form transparent, colourless, odourless, and tasteless coatings and with good carbon dioxide and oxygen barriers. Starch-based coatings are mostly used to preserve fruits and vegetables (Hassan et al. 2018; Mohamed et al. 2020; Sahraee et al. 2019; Sapper and Chiralt 2018).

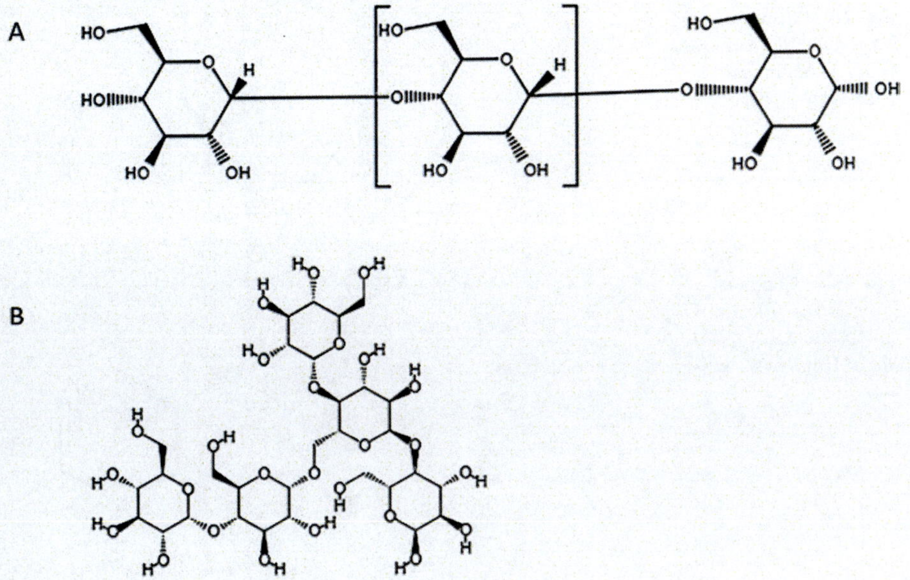

Fig. 10.1 Chemical structure of amylose (**a**) and amylopectin (**b**) (MolView 2020)

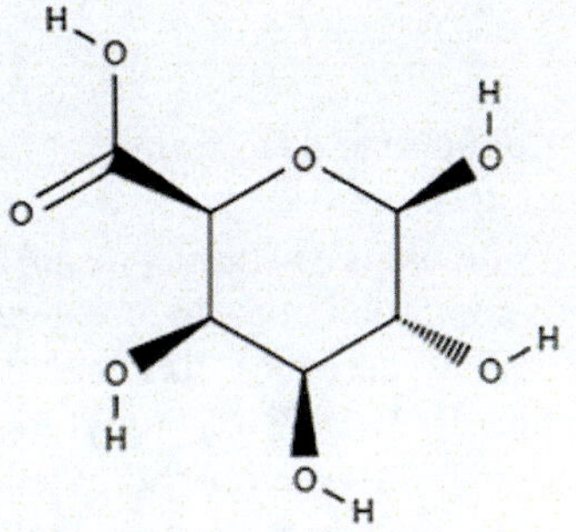

Fig. 10.2 Chemical structure of galacturonic acid (MolView 2020)

Pectin

Pectin is a complex polysaccharide composed mainly of D-galacturonic acid (Fig. 10.2). Pectin is mostly composed of linear homo-galacturonan (α-1,4-galacturonic acids) chains interspersed with branched rhamnogalacturonan (α-1,4-galacturonic acid to α-1,2- rhamnose) chains and the neutral sugar branches are connected through rhamnose residues (Dhall 2013; Hassan et al. 2018; Kurek et al. 2017; Suput et al. 2015). Pectin can be mixed with other polymers, for example, chitosan and PVA, to improve the films/coatings properties. Pectin is capable of forming gels that can be used as additive in jams, marmalades, jellies, and confectionaries, as well as edible coatings and films (Dhall 2013; Suput et al. 2015). Pectin

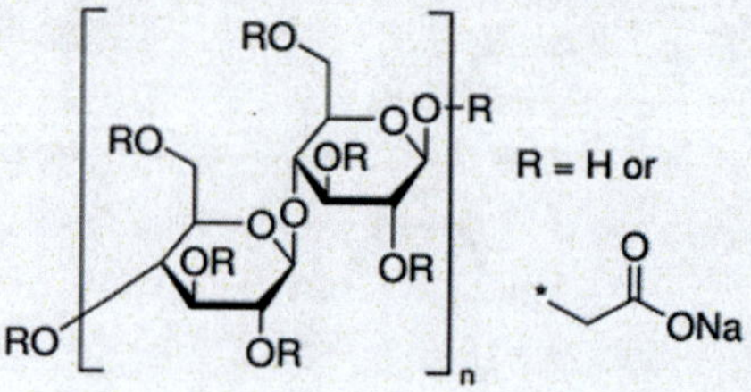

Fig. 10.3 Structure of carboxymethyl cellulose (CMC) sodium salt (MolView 2020)

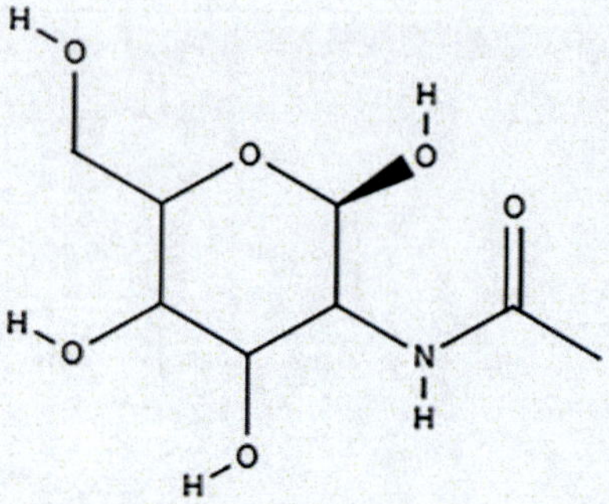

Fig. 10.4 Chemical structure of N-acetyl-D-glucosamine from shrimp shells (MolView 2020)

is suitable for food with low moisture and in the packaging of fresh fruits and vegetables (Mohamed et al. 2020).

Cellulose and Derivates

Cellulose is a naturally occurring biopolymer and it can be found in plants. Cellulose is composed of d-glucose units linked through β-1,4 glycoside bonds. Cellulose has several derivates like methylcellulose (MC), carboxymethyl cellulose (CMC; Fig. 10.3), hydroxypropyl methylcellulose (HPMC), and hydroxypropyl cellulose (HPC), that are formed by etherification of cellulose. Cellulose derivates originate films and coatings that are generally flexible, transparent, tasteless, odourless, water-soluble, resistant to oil and fat, and moderate to oxygen and water diffusion (Dhall 2013; Hassan et al. 2018; Kurek et al. 2017; Mohamed et al. 2020; Suput et al. 2015).

Chitin and Chitosan

Chitin, naturally occurring polysaccharide, is composed of N-acetyl-D-glucosamine units (Fig. 10.4) and is structurally similar to cellulose. Chitin is present in insects, invertebrates, marine diatoms, algae, and fungi. For example, it is found in the exoskeleton of crustaceans such as crab, shrimp, and crawfish. Chitin is primarily transformed to chitosan by deacetylation in concentrated alkali solutions (Campos et al.

2011; Dhall 2013; Elsabee and Abdou 2013; Hassan et al. 2018; Mohamed et al. 2020; Sabaghi et al. 2015; Sahraee et al. 2019).

Chitosan is a high molecular weight cationic polysaccharide. Chitosan is a natural, safe, non-toxic, allergen-free, biocompatible, and biodegradable polymer that has been successfully used in the food, biomedical, and chemical industries (Hassan et al. 2018; Mohamed et al. 2020; Souza et al. 2020b). Chitosan originates great coatings with good carbon dioxide and oxygen barrier properties. In addition, chitosan coatings are flexible, transparent and present good resistance to fat and oil, however are highly sensitive to moisture. Besides, chitosan coatings can delay enzymatic browning in fruits, control respiration, decline dehydration, enhance the emulsifying effect, and increase the natural flavour. Chitosan has excellent cationic properties that permit electron interactions with various compounds during processing and incorporating specific properties into the material and incorporation and/or slow release of active components, thus improving its properties (Dhall 2013; Mohamed et al. 2020; Souza et al. 2019b).

Chitosan, alongside with chitin, are known for the ability to inhibit the growth of fungi, bacteria, and yeast. Its antimicrobial activity depends on its cationic nature, degree of acetylation, concentration, the period of exposure, and the test organism. Different mechanisms have been suggested to explain the antimicrobial activity of chitosan. The intrinsic antimicrobial activity of chitosan is related to its positively charged amino groups that can react with negatively charged microbial cell membranes, resulting in permeabilization and leakage of intracellular material from the microbial cells. Also, chitosan also acts as metal chelator, forming complexes with the metals surrounding the bacteria, preventing the flow of essential nutrients (Dutta et al. 2009; Nouri et al. 2018; Verlee et al. 2017). The presence of fatty acids was also shown to enhance the antimicrobial properties of chitosan. But more research is being conducted to clarify chitosan antimicrobial action. Chitosan coatings are most commonly used for vegetables and fruits, namely cucumbers, strawberries, bell peppers, apples, peaches, pears, and plums (Elsabee and Abdou 2013; Hassan et al. 2018; Priyadarshi and Rhim 2020; Suput et al. 2015).

Carrageenan

Carrageenan is a polysaccharide that is sulphated water-soluble biopolymer extracted from various red seaweeds of the *Rhodophyceae* family, designated as λ, κ (Fig. 10.5), ι, μ, and ν-carrageenan; they consists of alternating 3-linked-β-D-galactopyranose and 4-linked-α-D-galactopyranose units. Carrageenan coatings are usually applied to fresh vegetables and fruits, but also to dry solid foods, sausage casings, oily foods, meat, poultry, and fish for preventing superficial dehydration. Carrageenan film formation includes gelation mechanism during moderate drying, leading to a three-dimensional (3D) network formed by polysaccharide double helices and to a solid film after solvent evaporation (Dhall 2013; Elsabee and Abdou 2013; Kurek et al. 2017; Mohamed et al. 2020; Suput et al. 2015).

Alginate

Alginate is a natural polysaccharide extracted from brown seaweeds of the *Phaeophyceae* family and some bacteria (Donati and Paoletti 2009; Sahraee et al. 2019). Alginate consists of units of (1 → 4)-linked-β-D-mannuronate (M) and (1 → 4)-linked-α-L-guluronate (G), in varying degrees, arrangements, and molecular weights, depending on different factors such as the age, specie, or parts of the seaweeds from which this material was obtained. Alginate coating is formed by the reaction between divalent or multivalent ions, such as calcium (Fig. 10.6), ferrum, sodium, or magnesium, which are added as gelling agents, and blocks of guluronic acid residues from two different chains resulting in a 3D network (Dhall 2013;

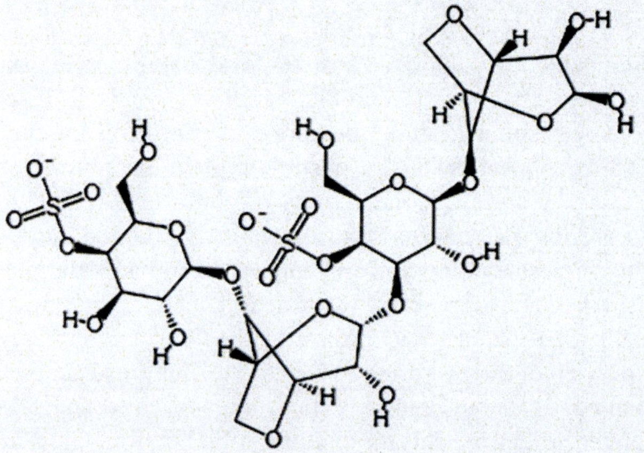

Fig. 10.5 Chemical structure of κ-Carrageenan (MolView 2020)

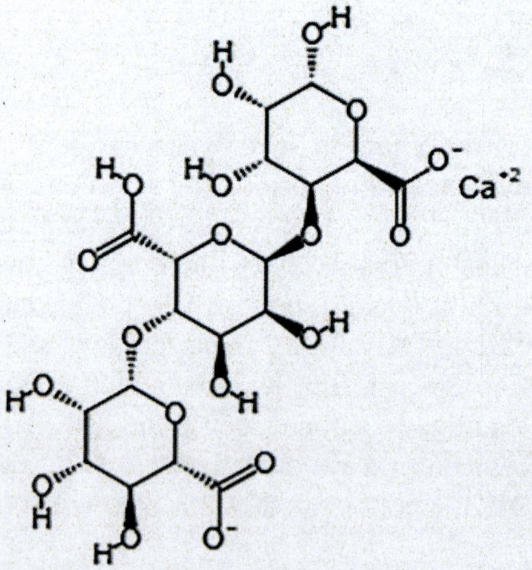

Fig. 10.6 Chemical structure of calcium alginate (MolView 2020)

Elsabee and Abdou 2013; Hassan et al. 2018; Mohamed et al. 2020; Parreidt et al. 2018). Alginate coatings are uniform, good oxygen barriers, and transparent, that can retard lipid oxidation in various fruits and vegetables, and resistant to oils and fats, but water-soluble. Nonetheless, adding calcium decreases the water vapour permeability becoming water-insoluble (Dhall 2013; Elsabee and Abdou 2013; Mohamed et al. 2020; Parreidt et al. 2018). Alginate is widely used in various industries such as pharmaceutical, food, beverage, textile, and printing, as a thickening agent, stabilizer, emulsifier, chelating agent, swelling, as suspending agent, for encapsulation, or used to form gels, films, and membranes (Parreidt et al. 2018).

Microbial Polysaccharides: Pullulan, Gellan, and Xanthan Gum

Pullulan is a microbial polysaccharide synthesized from starch by *Aureobasidium pullulans* and consists of maltotriose units interlinked by α (1, 6) glycosidic units. Pullulan is an excellent film-forming agent that is odourless, colourless, tasteless, transparent, water permeable, heat-sealable, and low oxygen and oil permeable. Pullulan-based coatings in combination with glutathione and chitooligosaccharides is an efficient way to increase the shelf-life of numerous fruits and have been effective in preserving sensory properties such as colour, odour, flavour, and texture (Hassan et al. 2018; Mohamed et al. 2020; Treviño-Garza et al. 2017).

Gellan is another microbial polysaccharide produced by the bacterium *Sphingomonas elodea* and presents unique colloidal and good capacity to form coatings. It is also capable of forming a gel in the presence of cations, like potassium, sodium, magnesium, and calcium. Gellan is usually used in the food industry as a gelling agent, texturizing, and as a carrier of food additives, such as antimicrobial and anti-browning agents colorants, nutraceuticals, and flavours. Gellan coatings have been used on fresh-cut vegetables (Mohamed et al. 2020; Sahraee et al. 2019).

Xanthan gum (Fig. 10.7) is a natural, high molecular weight, exopolysaccharide synthesized by the bacteria *Xanthomonas campestris*. Xanthan gum is formed of 1, 4-linked β-d-glucose residues and a side chain of tri-saccharide bound to an alternating d-glucose residue. Xanthan gum is usually used as a stabilizing, viscousifying, emulsifying, thickening, and temperature stability agent in the food industry. Xanthan gum edible coatings are used in fresh-cut fruits to improve their quality and shelf-life (Kurek et al. 2017; Mohamed et al. 2020; Soleimani-Rambod et al. 2018).

3.1.2 Proteins

Proteins are biopolymers containing more than 100 amino acid residues linked by peptide bonds and each amino acid contains a central carbon bonded to a carboxyl group, a hydrogen, an amino group, and R-group (Mohamed et al. 2020; Suput et al. 2015). Proteins can be separated into two groups: globular proteins and fibrous

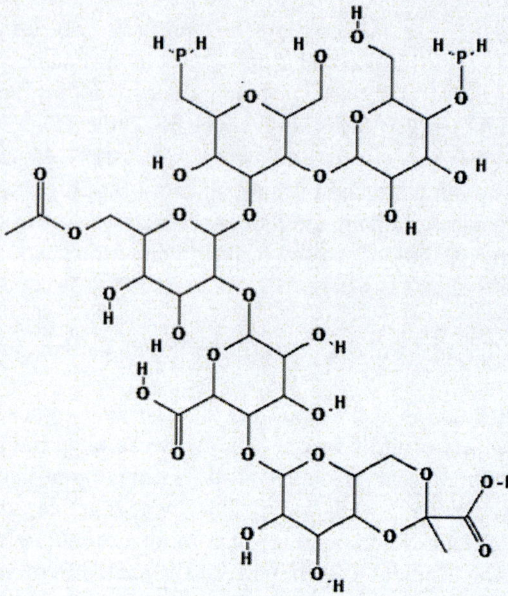

Fig. 10.7 Chemical structure of Xanthan gum (MolView 2020)

proteins. Fibrous proteins are water-insoluble and are bonded to each other on parallel through H-bonding to make fibres. Fibrous proteins are obtained from animal sources such as whey protein, casein, collagen, and gelatine. Globular proteins are soluble in water or aqueous solutions of bases, acids, or salts and have ionic, covalent, and hydrogen bonds that are folded into complex round structures. Globular proteins have a plant origin and some examples are soy protein, wheat gluten, corn zein, and peanut protein (Dhall 2013; Hassan et al. 2018; Mohamed et al. 2020).

Proteins are capable of forming coatings, depending on molecular characteristics that vary according to their biological origin and function. But first, the protein must be denatured by heat, acid, alkali, and/or solvent to form the more extended structures. Protein films and coatings are in general susceptible to moisture absorption, hydrophilic, have poor mechanical strength but good oxygen barriers at low relative humidity. Proteins are still generally better than polysaccharides in their capacity to form films with better mechanical and barrier properties, but then again the mechanical strength is poor and water vapour permeability of protein films is higher when compared to synthetic polymers (Dhall 2013; Hassan et al. 2018; Mohamed et al. 2020; Suput et al. 2015). Nonetheless, the use of protein edible coatings may be limited due to ethical or religious beliefs as well as allergenic risks (Riva et al. 2020; Valencia-Chamorro et al. 2011).

Soy Protein

Soy protein is obtained from soybeans and its protein content (38–44%) is much higher than the protein content of cereal grains (8–15%). Soy protein is available as soy flour, soy concentrate, and soy isolate. Soy protein is soluble in dilute neutral salt solutions but insoluble in water and can form coatings and films that are flexible, smooth, and clear in contrast to films formed by other proteins from plant sources. Coatings and films are normally produced from soymilk and soy isolate. Soy protein coatings and films have excellent gas barrier properties, better than lipids and polysaccharides (Dhall 2013; Mohamed et al. 2020).

Wheat Gluten

Wheat gluten is a water-insoluble protein of wheat flour that is considered to be a globular protein composed of a combination of polypeptide molecules. Wheat gluten is composed of two main components: glutelin and prolamine, also known as glutenin and gliadin, respectively. Gliadin is soluble in 70% ethanol but glutenin is not soluble. Nonetheless, wheat gluten is insoluble in water, except at low ionic strength it can be dissolved in aqueous solutions of high or low pH. Wheat gluten is capable of coatings and film formation through drying aqueous ethanol solution of wheat gluten during which the cleavage of old disulphide bonds occurs while heating the film-forming solutions and then the formation of the latest disulphide bonds during films drying occurs alongside hydrogen and hydrophobic bonds. Wheat gluten coatings and films are transparent, homogenous, strong, and with good water barriers. Plasticizers, like glycerine, are normally added to wheat gluten coatings and films to improve their flexibility, but consequently, it reduces film elasticity, strength, and water vapour barrier properties. Wheat gluten coatings and films are recommended for the conservation of fresh or minimally processed fruits and vegetables (Dhall 2013; Hassan et al. 2018; Suput et al. 2015).

Corn Zein

Corn zein is an important protein found in corn endosperm and represents up to 50% of corn's total protein content. Corn zein is obtained as a by-product of corn during the production of starch (Dhall 2013; Hassan et al. 2018; Sharaf Eddin et al. 2019). Corn zein is a prolamine protein that dissolves in 70–80% ethanol and water-insoluble. Corn zein has excellent coating and film-forming properties which are odourless, tasteless, and good barrier properties against aroma, carbon dioxide, oxygen, and lipids (Dhall 2013; Hassan et al. 2018; Kurek et al. 2017; Sharaf Eddin et al. 2019; Suput et al. 2015). Corn zein edible coatings and films are made by drying aqueous ethanol solution of zein, and plasticizers are added to enhance flexibility as they are highly brittle (Dhall 2013; Hassan et al. 2018).

Casein and Whey Protein

Casein and whey protein are milk proteins. Casein, an insoluble protein, consists of 80% of milk protein and contains α, β, and κ-casein components (Mohamed et al. 2020; Ramos et al. 2012). Casein can form coatings and films from aqueous caseinate solutions. Casein-based coatings and films are tasteless, transparent, flexible, and with good barrier properties to carbone dioxide, oxygen, and aromas. In addition, casein coatings and films are stable for a series of temperature, pH, and salt concentration and are highly stretchable. Nevertheless, casein has a relatively high price (Dhall 2013; Mohamed et al. 2020; Suput et al. 2015).

Whey is a by-product of the dairy industry and is composed mainly by water (90–94%), lactose (4.5–5%), soluble proteins (0.6–0-8%) and lipids (0.4–0.5%) (Andrade et al. 2016). For several years, the food industry considered this product as food waste and it was discarded without any treatment, becoming an environmental hazard due to its organic content and high production volume (Andrade et al. 2016; Ramos et al. 2016). The whey protein fraction is mainly constituted by β-lactoglobulin (50%), α-lactalbumin (20%), immunoglobulins (10%), bovine serum albumin (10%), lactoferrin (3%), lactoperoxidase (0.3%) (Ramos et al. 2016). Whey protein is a soluble protein obtained from cheese production or casein precipitation. Whey proteins are normally used in toddler formulae and sports activities food products.

Whey protein can be divided into whey protein isolate (WPI) and whey protein concentrate (WPC). In WPI, protein content is superior to 90% whereas in WPC, protein content varies between 20% and 80% (Campos et al. 2011; Costa et al. 2018; Hassan et al. 2018; Mohamed et al. 2020; Ramos et al. 2012). Whey protein has proven anticancer and anticarcinogenic activity (Bounous 2000; Parodi 1998), metal-chelating activity (Gad et al. 2011), antioxidant, and antimicrobial activities (Ramos et al. 2016). Whey proteins have good films and coatings forming capacity, as they are flexible, flavourless and transparent, have excellent barrier properties to aroma compounds, oil, and gas at low relative humidity, and can serve as a carrier of food additives such as colorants, antioxidants, and antimicrobial agent. Nonetheless, its barrier properties tend to decline in moisture presence due to their hydrophilic nature (Campos et al. 2011; Costa et al. 2018; Dhall 2013; Feng et al. 2019; Hassan et al. 2018; Mohamed et al. 2020; Prodpran et al. 2007; Ramos et al. 2012; Ribeiro-Santos et al. 2018).

Collagen and Gelatines

Collagen is a fibrous protein that is abundant in animal skin, tendons, and connective tissues and is highly prevalent and widely distributed fibrous protein in the animals. Collagen is pre-treated with a partial acid or alkaline solution, and further heating up to 40 °C and originates gelatine. Collagen has a complex helical and fibrous structure and therefore is insoluble and difficult to process. Collagen is mainly used in the meat industry before processing of meat products (Dhall 2013; Mohamed et al. 2020; Sharaf Eddin et al. 2019).

Gelatine is a protein derived from the degradation of bones and skin collagen, generated as waste during animal slaughtering and fish processing. (Benbettaïeb et al. 2018; Farajzadeh et al. 2016; Jridi et al. 2018; Xiong et al. 2020). Gelatine is widely used and has excellent coatings and film-forming capacity with perfect oxygen and carbon dioxide barriers at low to intermediate relative humidity but poor water vapour barrier properties. Gelatine is viscous, transparent, colourless with little flavour and dissolves easily and has good elasticity, consistency, and stability (Jridi et al. 2018; Khoshnoudi-Nia and Sedaghat 2019; Sahraee et al. 2019; Suput et al. 2015; Xiong et al. 2020). Gelatine has been indicated to have antioxidant and antimicrobial properties and has been reported as an excellent carrier of active compounds either as edible coatings or as encapsulating matrices. One of the greatest advantages of gelatine is its relatively low cost (Khoshnoudi-Nia and Sedaghat 2019; Sahraee et al. 2019; Suput et al. 2015).

3.1.3 Lipids

Lipids, commonly known as fats and oils, can be found naturally in animals, insects, and plants and are composed of phospholipids, phosphatide, mono-, di- and triglycerides, terpenes, cerebrosides, fatty alcohols, and fatty acids, among others. Lipids are naturally hydrophobic and are capable of forming edible coatings with excellent barriers against moisture migration. Yet, lipid coatings have poor mechanical properties and gas barrier properties and are thicker and brittle. Therefore, lipids are normally associated with other coatings materials like polysaccharides and proteins to improve those properties. Lipid-based coatings can be made with different compounds such as acetylated monoglycerides, natural waxes, and resins. Lipids based coatings are typically used to protect fresh fruits and vegetables (Dhall 2013; Domínguez et al. 2018; Hassan et al. 2018; Kapetanakou et al. 2014; Mohamed et al. 2020; Riva et al. 2020; Suput et al. 2015).

Waxes

Waxes are esters of long-chain aliphatic acids with long-chain aliphatic alcohols. Waxes can have different origins: animal, including insects (beeswax, shellac wax, spermaceti wax, Chinese insect wax, lanolin wax), vegetal (carnauba wax, bayberry wax, candelilla wax, rosin wood wax, palm wax, sugarcane wax, esparto wax, rice bran wax, cottonseed wax, oricury wax, Japan wax, waxol), mineral (paraffin and microcrystalline waxes) or synthetic (carbowaxes and polyethylene wax) (Dhall 2013; Galus and Kadzińska 2015; Mohamed et al. 2020). Carnauba wax, paraffin, beeswax, and candelilla wax are most commonly used and more efficient as coatings for cheese and fresh fruits and vegetables. Carnauba wax is obtained from the palm tree leaves (*Copoernica cerifera*) and has a very high melting point; beeswax is extracted from honeybees. Paraffin wax is derived from a distillate fraction of crude petroleum and consists of a mixture of solid hydrocarbons resulting from ethylene catalytic polymerization. Paraffin wax is used on raw fruits and vegetables and cheese (Dhall 2013; Hassan et al. 2018; Kapetanakou et al. 2014). Wax coatings have good water and gas barrier properties due to their low polar group content and high long-chain fatty alcohols and alkanes content. Wax coatings improve the appearance of the food as they give a glossy look but are fragile and/or brittle. Wax based coatings can be edible if applied as a thin layer but not suitable for consuming if applied a thicker layer (Dhall 2013; Galus and Kadzińska 2015; Hassan et al. 2018; Kapetanakou et al. 2014; Mohamed et al. 2020; Suput et al. 2015).

Resins

Shellac resins, an aggregate of aliphatic alicyclic hydroxyl acid polymers, are a secretion of insects like *Laccifer lacca*. Also, resins are produced by plant cells in response to injury or infection in trees and shrubs. Typically, resins are translucent with yellowish-brown tones and physically are solid or semisolid. Resins and their derivate are soluble in alcohols alkaline solutions and capable of forming coatings that are mainly used in fruits, specifically citrus. Resin-based coatings provide a lower internal oxygen and higher carbone dioxide of the coated fruit. Also, resin-coated fruits present the same glossy appearance as waxed fruits (Dhall 2013; Galus and Kadzińska 2015; Hassan et al. 2018; Kapetanakou et al. 2014; Mohamed et al. 2020).

3.1.4 Composites

Composite edible coatings are produced with at least two biopolymers aiming to obtain an improved coating with the best properties of each biopolymer and minimize their disadvantages. The possible combinations that can be made are carbohydrates and proteins, carbohydrates and lipids, proteins and lipids, or synthetic polymers and natural biopolymers. For example, the ability to form cohesive films with good gas permeability properties and no greasy texture of polysaccharides or proteins and the good water barrier properties of lipid coatings can originate a superior coating. Composite coatings can be produced through blending, extruding, or laminating and applied through the suspension, emulsion, or dispersion of the constituents, in successive layers, or as a single layer. The composite coating functional properties depend on type, preparation method, and amount of components, and their compatibility (Dhall 2013; Hassan et al. 2018; Kapetanakou et al. 2014; Riva et al. 2020; Salgado et al. 2015). Several researchers have developed composite edible coatings such as CMC and chitosan, chitosan and alginate, starch, gellan and thyme essential oil, alginate, rosemary, and oregano essential oil, chitosan and whey protein, and others (Arroyo et al. 2020; Di Pierro et al. 2011; Sapper et al. 2019; Vital et al. 2016; Yan et al. 2019).

3.2 Active Compounds

Active compounds can be obtained from natural or synthetic origins. These compounds and substances are added directly or indirectly to foods to maintain and assure the food' safety or/and for some technological purpose (European Commission 2008c).

The most used additives are from synthetic origin due to their chemical stability, economic value and, because they usually do not possess unwanted colors, flavors, or aromas and their application is quite easy (Andrade et al. 2019a). However, recent studies have reported some concerns about the prolonged exposure to these additives. The consumption of these additives, such as butylated hydroxyanisole (BHA – E320) and butylated hydroxytoluene (BHT – E321), have been linked with allergic reactions, promotion of neurodegenerative diseases, and carcinogenesis (Shahidi 2000; Pereira de Abreu et al. 2010; Sanches-Silva et al. 2014; Carocho et al. 2015). This is one of several reasons for the increased demand for natural sources of additives.

These types of additives are usually in the form of extracts or essential oils that can be extracted by several methods from plants, algae, seaweeds, and fruit sources. Industrial food by-products represent a serious problem in the modern world. Food by-products must be discarded in a responsible and environmentally friendly way, which can represent an increase in cost, with consequences in the final price of food products and food formulations. Food by-products, especially fruit by-products, usually have a high content in active compounds with powerful biological activities that can be extracted and applied in other food products and packages. This way, industries, and companies make the most of raw materials, contributing to greater sustainability and supporting the circular economy.

3.2.1 Antioxidant Compounds

Antioxidant compounds are capable of capturing free radicals, diminishing oxidative stress and, in consequence, delay, inhibit or prevent oxidation (Dai and Mumper 2010). The Regulation No. 1333/2008 defines antioxidants as "*substances which prolong the shelf-life of foods by protecting them against deterioration caused by oxidation, such as fat rancidity and color changes*" (EC 2008c). In the case of active packaging and coatings, the polymer chosen for the package or coating' matrix can have biological activities, such as antimicrobial or antioxidant activities.

Phenolic compounds, including flavonoids, tannins, and phenolic acids, are natural compounds with powerful antioxidant and antimicrobial activity, present in most edible plants, including fruits and vegetables. The plants' content in these compounds is directly related to the edaphoclimatic conditions to which the plant was exposed. These compounds are directly involved in the plants' protection against UV radiation, predators, pathogens, and parasites and are responsible for the plants' organoleptic properties (Dai and Mumper 2010; Moreno et al. 2006; Piccaglia et al.

1993; Proestos et al. 2006; Robards 2003; Sartoratto et al. 2004). The antioxidant activity of phenolic compounds can be mainly attributed to their redox properties, which translates in their capacity to absorb and neutralizing free radicals, quenching singlet and triplet oxygen, or decomposing peroxides (Zheng and Wang 2001).

Aromatic plants, such as oregano (*Origanum vulgare* L.), rosemary (*Rosmarinus officinalis* L., syn *Salvia rosmarinus* Spenn.), thyme (*Thymus vulgaris* L.), and basil (*Ocimum basilicum* L.), possess a high content in phenolic compounds (Souza et al. 2018b). Their extracts and essential oils can be incorporated into the matrix of active food packaging to inhibit or delay foods' oxidation, especially lipid oxidation. Andrade et al. (2019b) successfully incorporated an ethanolic extract of rosemary into an edible WPC based coating. The authors tested the antioxidant activity of this active coating by evaluating the lipid oxidation of salami slices for 90 days. The hexanal and malonaldehyde content of the slices packaged with the active coating was lower than the content of the salami slices packaged with the control coating (without the rosemary extract) (Andrade et al. 2019b).

In another study by Castro et al. (2019), the ability of an edible WPC based coating incorporated with a commercial green tea (*Camellia sinensis* L.) extract in inhibiting the lipid oxidation of fresh salmon (*Salmo salar* L.) was evaluated. The active coating was effective in inhibiting the salmon' lipid oxidation for, at least 14 days (Castro et al. 2019). Rastegar et al. (2019) tested the effects of several concentrations of sodium alginate coatings applied in mango fruit stored for 1 month at 15 °C. The mangos were dipped for 5 min into the coating solutions (1, 2, and 3% of sodium alginate) and the control sample into distilled water and dried at room temperature. The mangos treated with a higher percentage of sodium alginate (3%) presented the higher content in phenolic compounds and flavonoids and the highest antioxidant activity when compared with the remaining samples (Rastegar et al. 2019).

Rojas-Bravo et al. (2019) studied the effect of a mango peel powder applied in starch edible coatings and their effectiveness in preserving the antioxidant properties of apple slices, stored at 4 °C for 12 days. With the addition of the mango peel powder, the mechanical properties of the starch coating were increased and the antioxidant activity and the total content in phenolic compounds and flavonoids of the apple slices increased (Rojas-Bravo et al. 2019). The authors quantified the total content of phenolic compounds, in terms of gallic acid equivalent (GAE), in the edible films with 2% (247.70 mg GAE/100 g) and 4% (298.70 mg GAE/100 g) of mango peel powder through the Folin-Ciocalteu method. Also, the main phenolic compounds found in mango peel were gallic acid, hydroxybenzoic acid, mangiferin, quercetin, and anthocyanins (Rojas-Bravo et al. 2019).

Tannins are high molecular weight molecules, which are formed due to the polymerization of polyphenols and act in the plants' natural defense. These phenolic compounds can be classified into two groups: hydrolyzable and condensed (or proanthocyanins) tannins (Chung et al. 1998; Demarque et al. 2018; Viuda-Martos et al. 2010). Polyphenolic acids and their derivates are classified as hydrolyzable tannins. When fractionated into basic components, they can be classified as gallotannins, the simplest kind of hydrolyzable tannins, and ellagitannins, which can generate ellagic acid through hydrolyzation (de Hoyos-Martínez et al. 2019). Ellagic acid is a known and powerful antioxidant compound. Condensed tannins are chemically classified as flavonoids composed of flavan-3-ols and flavan-3,4-diols. Catechins are one of the most famous and common examples of condensed tannins (Andrade et al. 2019a; Chung et al. 1998; de Hoyos-Martínez et al. 2019).

Sáez et al. (2020) evaluated the potential of tannic acid and quebracho tannin, combined with an alginate coating, applied to rainbow trout fillets, for a 15-day cold storage period. The results showed that the fillets with the alginate coatings with the tannic acid and the quebracho tannin had a lower malonaldehyde content when compared to the control samples. The coatings with quebracho tannin presented lower malonaldehyde content than the coatings with the tannic acid (Sáez et al. 2020). In another study by Xiong et al. (2020), the antioxidant effect of a chitosan-gelatin coating with a grape seed extract was evaluated in fresh pork for 20 days. Comparing with the control samples, the fresh pork coated with the chitosan-gelatin with the grape seed extract presented the lowest malonaldehyde content at all evaluated storage times (5, 10, 15, and 20 days). The authors did not present the individual constitution of the grape seed extract but, it is general knowledge that this extract presents a great content in tannins (Andrade et al. 2019a; Xiong et al. 2020).

Phlorotannins are another group of tannins, composed by phloroglucinol units 1,3,5-trihydroxy benzene, most commonly found in brown seaweeds (Phaeophyceae) (Andrade et al. 2019a; Hermund et al. 2018; Zaragozá et al. 2008). Phlorotannins can be divided into six different classes, according to the nature of the structural linkages between the phloroglucinol units, the number and distribution of hydroxyl (OH) groups: phlorethols, fuhalols, fucols, fucophlorethols, eckols and carmalols (Lopes et al. 2012, 2016, 2018). Similar to other natural active compounds, the concentration of phlorotannins in brown seaweeds varies according to the species and the edaphoclimatic conditions to which the seaweed is exposed (Lopes et al. 2012). These compounds are known for their powerful antioxidant (Dong et al. 2019; Kirke et al. 2019; Manandhar et al. 2019), antimicrobial (Lopes et al. 2012), and antifungal (Lopes et al. 2012) activities. Sharifian et al. (2019) tested the capacity of phlorotannins in inhibiting polyphenol oxidase (also known as phenoloxidase and tyrosinase) in frozen shrimp. The shrimps were dipped in 1%, 2%, or 5% phlorotannins extracted from *Sargassum tenerimum* solutions for 10 min at 4 °C and stored in ice for 16 days and analyzed every 4 days. The shrimps treated with all the phlorotannins solutions showed lower levels of lipid oxidation and microbial contamination, being the coating with 5% phlorotannins the most efficient (Sharifian et al. 2019).

Martínez-González et al. (2020) evaluated the effectiveness of five chitosan coatings with chitosan particles and three different percentages of propolis extract in delaying the antioxidant capacity of strawberries (*Fragaria* x *ananassa*). The strawberries were immersed in the coatings' solutions for 30 s, dried at room temperature, and stored at 4 °C for 8 days. The strawberries covered with the propolis extract coatings, regardless of the propolis extract incorporated percentage, showed higher content in total phenolic compounds and flavonoids, and presented a higher antioxidant activity, at the end of the 8th storage day (Martínez-González et al. 2020).

Regarding the propolis extract, the authors did not determine the chemical composition of the propolis extract. Although, Kalogeropoulos et al. (2009) evaluated the chemical composition of 12 propolis ethanolic extracts and found that pinocembrin, abietic acid, and isopimaric acid are the most common phenolic compounds present in the ethanolic propolis extracts. Garlic extracts and essential oils have a powerful recognized antioxidant and antimicrobial activity, mostly due to their high concentration of phenolic compounds, such as 3-hydroxybenzoic acid, *p*-coumaric acid, (−) – epigallocatechin gallate (EGCG) and genistein (Emir et al. 2020). Ariviani et al. (2019) applied a chitosan coating loaded with garlic essential oil to beef meatballs. The meatballs were dipped into the coating solutions (without chitosan, chitosan, chitosan plus garlic essential oil, and garlic essential oil without chitosan) for 60 s and dried at room temperature. The meatballs were stored at 4 °C and their antioxidant activity was measured at 0, 5, 10, 14 and 18 day of storage. When compared with the control (without the addition of chitosan or garlic essential oil), all the treated meatballs with active coatings showed a significantly higher antioxidant capacity, being the meatball packaged with the chitosan plus the garlic essential oil the most powerful in terms of antioxidant capacity (Ariviani et al. 2019).

Souza et al. (2018b) also combined chitosan with essential oils and hydroethanolic extracts from several plant sources and determined the potential migration of phenolic compounds and evaluated the antioxidant capacity. The authors concluded that the chitosan film incorporated with ginger (*Zingiber officinale* Roscoe) and rosemary (*Rosmarinus officinalis* L., syn *Salvia rosmarinus* Spenn.) essential oil were the coatings that present the highest antioxidant activity and the highest phenolic compounds content (Souza et al. 2018b).

Another powerful source of phenolic compounds are fruit by-products. According to FAO, one-third of the world's food production is lost or wasted, of which 45% are fruits and vegetables (Gustavsson et al. 2011). Fruits and vegetables are often presented to the consumer in the form of food formulations such as jams, pastes, soups, and juices. These formulations produce a large quantity of by-products, which are composed of seeds, peels, leaves, and branches. These by-products still have a large number of bioactive compounds being considered products with a high biological value. On the other hand, these by-products are still seen by companies and industries as waste, which contributes to their low economic value. Besides, the companies and industries have to discard these types of products in a accountable and ecological way, which can contribute to increase the final price of the product. Apart from phenolic compounds, fruits' by-products are also rich in carotenoids and dietary fiber.

Carotenoids are natural pigments, responsible for the occurrence of the colors red, orange, and yellow in fruits, flowers, plants, and algae (Andrade et al. 2019a; Sanches-Silva et al. 2013). Due to their well-known antioxidant capacity and role in the intracellular communication and immune system, their presence in the human diet is very important. The deficiency of these compounds in the human diet can lead to visual disorders, such as night blindness and corneal ulceration (Eggersdorfer and Wyss 2018; Freitas et al. 2015; Saini et al. 2015). In a study by Sánchez-Camargo et al. (2019), mango peel composition was evaluated. Mango peel presented a good source of active compounds, specifically carotenoids, as well as crude protein and dietary fiber. Due to its richness in carotenoids, the mango peel extract was effectively applied to sunflower oil against lipid oxidation (Sánchez-Camargo et al. 2019).

3.2.2 Antibacterial Compounds

The Regulation N° 1333/2008 defines antimicrobial or preservatives as "*substances which prolong the shelf-life of foods by protecting them against deterioration caused by micro-organisms and/or which protect against the growth of pathogenic micro-organisms*" (EC 2008c).

Whey protein biological functions are directly related to its composition and the biological activities of its proteins. However, the biological properties of whey protein and its proteins are also dependent on the pH, ionic environment, pre-heat and heat treatments, and lipid presence (Walzem et al. 2002). Although its full function is still unknown, the protein present in the highest concentration, β-lactoglobulin, has proven anticancer, antiviral, and antimicrobial activities, antioxidant activity through its peptide fractions and is related with the satiety response (Ramos et al. 2016; Walzem et al. 2002). The antimicrobial activity of β-lactoglobulin can be derived from the activation of the autolytic enzyme cell system, inhibition of cell' wall, and nucleic acid synthesis, from the formation capacity of transmembrane pores and synergetic activity with other hosts innate immune molecules (Hernández-Ledesma et al. 2008). Regarding the α-lactalbumin, it is largely used in infant formulas since its structure and composition are similar to the major protein in human breast milk (Walzem et al. 2002). This whey protein has proven antimicrobial activity against *Streptococcus pneumoniae*, *Haemophilus influenza,* and *E. coli*. Similar to the β-lactoglobulin, the antimicrobial activity is most likely related to the peptides present in the α-lactalbumin (Chatterton et al. 2006).

Lactoferrin is a glycoprotein with the capacity of binding two molecules of Fe^{3+} per protein molecule. This protein has proven *in vitro* and *in vivo* antimicrobial activity against Gram-positive and Gram-negative bacteria. Logically, the antimicrobial mechanisms are attributed to its iron-sequestering capacity (Brock 1980) but other authors also observed antimicrobial activity of iron-free lactoferrin through membrane binding, subsequent release of lipopolysaccharides, and osmotic damage against *Staphylococcus mutans* and *Vibrio cholerae* (Arnold et al. 1977; Floris et al. 2003; Naidu et al. 1993; Yamauchi et al. 1993). Antimicrobial activity of this protein was also observed against *Clostridium* spp., *Micrococcus* spp., and *E. coli* (Floris et al. 2003). The inhibition of intracellular invasion of pathogenic bacteria (*E. coli* and *Yersinia pseudotuberculosis*) by the lactoferrin was also observed (Floris et al. 2003; Longhi et al. 1993). Despite its proven biological activities, most authors chose to incorporate some kind of active agent in the whey protein matrix, forming an active biodegradable coating.

Most of the active agents are essential oils or extracts obtained from plants and fruits and/or isolated phenolic compounds, such as carvacrol (Wang et al. 2019), or nanoparticles and nanofillers (Feng et al. 2019). Ribeiro-Santos et al. (2017) tested the antimicrobial activity of an optimal blend of *Cinnamomum cassia* L., Cinnamomum *zeylanicum* L., and *R. officinalis* L. essential oils incorporated in different percentages in a whey protein-based coating against *E. coli, S. aureus,* and *Penicillium* spp. The coating incorporated with 5% of the optima essential oil blend showed the highest antimicrobial activity against all the tested microorganisms (Ribeiro-Santos et al. 2017). Andrade et al. (2018) proved the antimicrobial activity of ethanolic extracts obtained from dried *R. officinalis* L. and *Thymus vulgare* L. against *Clostridium perfringens, L. monocytogenes,* and *S. aureus*. The authors also incorporated the *R. officinalis* L. extract into a whey protein-based coating which showed antimicrobial activity against *L. monocytogenes* and *S. aureus* (Andrade et al. 2018).

Chitosan, is a well-known polysaccharide with proved antimicrobial activity obtained from chitin, the second most naturally abundant polysaccharide, extracted from the crustaceans and insects exoskeleton and the cell wall of fungi (Moratti and Cabral 2017; Vunain et al. 2017). Similar to whey protein coatings, most authors chose to incorporate other biological compounds, such as phenolic compounds to increase the antimicrobial activity of chitosan. Albertos et al. (2019) tested the antioxidant and antimicrobial activity of edible chitosan films incorporated with *Himanthalia elongata* and *Palmaria palmata* seaweeds and seaweed *H. elongate* and *P. palmata* extracts, applied into aquaculture rainbow trout (*Oncorhynchus mykiss*) burgers. After the coating appliance, the burgers were kept at 4 °C for a total of 7 days, being analyzed at the end of 2, 5, and 7th day. The *H. elongata* seaweed film showed higher antimicrobial activity against psychotropic bacteria, followed by the films with the *H. elongata* extract. Regarding the mesophilic bacteria, the films incorporated with *H. elongata* seaweed, *H. elongata* extract, and *P. palmata* extract, presented the highest antimicrobial activity. In the antioxidant assays, the films incorporated with the *H. elongata* and *P. palmata* extract presented the highest antioxidant activity (Albertos et al. 2019). In another study lead by Chaparro-Hernández et al. (2019), the effectiveness of a chitosan incorporated with tomato-plant (*Lycopersicon esculentum*) from the Pitenza variety extract edible coating applied to pork loin was evaluated. The pork loins with the chitosan+0.1% of tomato plant extract and chitosan+0.3% of tomato plant extract showed higher antioxidant activity (Chaparro-Hernández et al. 2019). The authors also measured the antimicrobial activity of the chitosan edible coatings with the tomato plant extract applied in pork loins, stored for 21 days, against aerobic mesophilic bacteria, coliforms, and psychrotrophic bacteria. Comparatively to the control, all the samples with the active coatings (chitosan, chitosan +0.1% of tomato plant extract, chitosan +0.3% tomato plant extract) presented lower counts of all the tested microorganisms. The most efficient treatment against the aerobic mesophilic and the coliform bacteria was the chitosan +0.3% of tomato plant extract, and against the psychrotrophic bacteria was the treatment with chitosan +0.1% tomato plant extract (Chaparro-Hernández et al. 2019).

Also, Souza et al. (2018a, 2019a) and Pires et al. (2018) have studied the antimicrobial activity of chitosan incorporated with rosemary and ginger essential oils and concluded that chitosan films with those essential oils demonstrated good antimicrobial activity against *B. cereus* and *S. enterica*, and active films also succeeded on retarding poultry microbial deterioration.

3.2.3 Antifungal Compounds

Mycotoxins are low molecular weight compounds produced by filamentous fungi and microfungi, often referred to as molds, as secondary metabolites for their natural defense mechanism (Assunção et al. 2018; Bryden 2011; Oliveira et al. 2014). Mycotoxins present a serious health hazard since they are linked to carcinogenic and nephrotoxic occurrences, and their intake is a major risk factor for the occurrence and development of liver cancer and hepatocellular carcinoma (Bryden 2011). They cannot be eliminated through temperature and are consumed by contaminated agriculture products intake, such as cereals and fruits, or indirectly through animal product intake, such as milk and eggs (Assunção et al. 2018). Their production can occur in any part of the food chain, including pre-harvest, which can result in a major economic impact (Bryden 2011). It is estimated that approximately 25% of the world' food crops are affected by fungi. The most often encountered fungal species belong to the genera *Alternaria*, *Aspergillus*, *Cladosporium*, *Fusarium*, and *Penicillium*. Despite most fungi produce toxic compounds, not all produce toxins being the fungi belonging to the genera *Aspergillus* (aflatoxins and ochratoxins), *Fusarium* (trichothecenes and fumonisins), and *Penicillium* (ochratoxins) the most important concerning the mycotoxigenic fungi (Bryden 2011). Regarding their origins and appearance, the aflatoxins appear mostly in maize and peanuts, ochratoxin A is mostly present in cereals, wine, coffee, spices, and dried fruits, fumonisins are commonly found in maize and maize products (Bryden 2011).

Natural compounds such as plants' essential oils and extracts can help in the "battle" against the fungi responsible for the production of mycotoxins. Xing et al. (2010) proved the antifungal activity of cinnamon essential oil, whose main compound is cinnamaldehyde, against *Aspergillus flavus, Rhizopus nigricans*, and *Penicillium expansum*. The authors applied cinnamon essential oil on the surface of two fruits, Sand Sugar Orange fruit (*Shatang mandarin*) and Lingwu Long Jujube fruit (*Ziziphus jujuba* Mill cv.), proving the antifungal activity of cinnamon essential oil *in vivo* (Xing et al. 2010). Essential oils are complex mixtures of volatile secondary metabolites constituted essentially by mono and sesquiterpenes. The antimicrobial and antifungal mechanism of essential oils is still not fully understand which is generally associated with the hydrophilic or lipophilic nature of essential oils since certain essential oil compounds can interfere with proton translocation over a membrane vesicle and interrupt ADP phosphorylation. In addition, other compounds can interfere with enzyme proteins stopping their activity (Kalemba and Kunicka 2003; Knobloch et al. 1989). Other essential oils can lead to the denaturation of membrane proteins, resulting in the outer membrane disruption, respiration inhibition, and cell lysis (Adams et al. 1996; Claeson et al. 1992; de Billerbeck et al. 2001; Kalemba and Kunička 2003; Knobloch et al. 1989; Zambonelli et al. 1996).

As a bonus feature, essential oils are also able to inhibit proteins, polysaccharides, DNA, and RNA synthesis in fungal and bacterial cells (Himejima and Kubo 1993; Zani et al. 1991).

Although essential oils present important activity against fungi and mycotoxins, chitosan has also a proven antifungal activity. Wu et al. (2018a) showed production inhibition of the mycotoxins 3-acetyl- deoxynivalenol (3ADON) and 15-acetyl-deoxynivalenol (15ADON), produced by *Fusarium graminearum*, through chitosan coatings. Kazemian-Bazkiaee et al. (2020) also reported the antifungal activity of chitosan against *Aspergillus niger* in peanuts. Chitosan prevents spores' growth by inhibiting the transcription of DNA and RNA and consequently inhibiting the production of aflatoxins. The proposed mechanism of action relays on the cell wall structure whereas chitosan interacts with the membrane phospholipids, leading to the cell membrane destruction and consequently the cell death (Abd El-Hack et al. 2020; Kazemian-Bazkiaee et al. 2020).

4 Preparation of Edible Active Coatings

EAC can be prepared using various techniques, namely spraying, dipping, or spreading being these processes usually followed by drying steps. The technique has little influence on the adhesion of the coating to the food product, which is more dependent on the characteristics of the surface of the food product. However, the technique and the viscosity of the polymeric matrix affect the thickness of the coating formed around the food product (Andrade et al. 2012; Debeaufort et al. 1998; Kouhi et al. 2020; Valdés et al. 2017).

The application of coating to food products usually involves the application of a single layer of the polymeric matrix. However, lately, the layer-by-layer (LbL) technique has been applied to coatings too. The LbL technique aims to improve the coating properties by applying different layers with different polymeric matrices in a controlled manner. The LbL approach is used to deposit oppositely charged polyelectrolytes, with the objective of an efficient control of material properties and functionality. The main disadvantage of this technique is that it requires a longer period of time since after each application of the biopolymer, it is necessary to let it dry before applying another layer of biopolymer. The LbL technique is commonly used in fruits and vegetables. Chitosan, alginate, and pectin are frequently applied with this technique (Arnon-Rips and Poverenov 2018; Kouhi et al. 2020; Treviño-Garza et al. 2017; Yan et al. 2019).

Regardless of the chosen technique, all materials used, from biopolymer, plasticizer, solvent, and active compound, as well as all equipment used throughout the process, must be able to come into contact with food and fit for consumption. In addition, it is necessary to take into account the surrounding conditions, such as salt concentration, temperature, pH, and enzymatic reactions (Kouhi et al. 2020; Valdés et al. 2017).

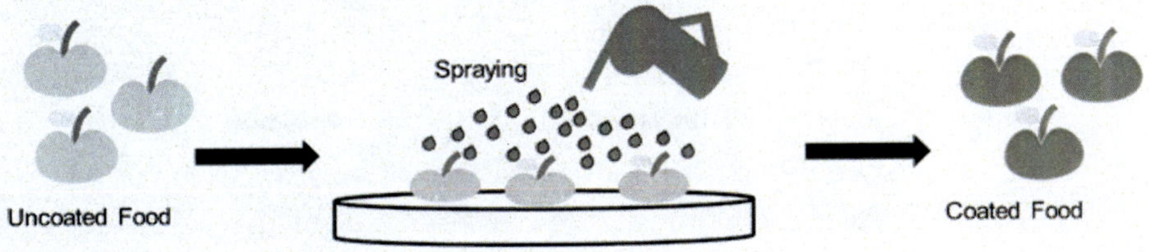

Fig. 10.8 The spraying technique

4.1 Spraying

Spraying is one of the most popular methods used to apply edible coatings, as it offers a high-quality final product (Hanssen et al. 2012; Park 2003). A spray is a collection of moving droplets consequence of atomization processes to break up bulk liquids into droplets (Valdés et al. 2017). In the spraying technique (Fig. 10.8), polymers are dispersed or solubilized into a liquid phase, cast and dried. The process of spraying consists of first the food being placed on a rotating platform at a controlled speed and sprayed with nozzles. These procedures can be repeated multiple times to ensure uniform coating on the food surface (Espitia et al. 2014; Kurek et al. 2017; Sharaf Eddin et al. 2019).

The major advantages of the spraying technique are the ability to control the process, to apply an uniform coating, use of less amount of coating material to achieve a good coverage, control the thickness of the coating, possibility of multilayer applications, temperature control of the solution and the ability to work with large surface areas. Also, this technique has a relativity low cost (Andrade et al. 2012; Parreidt et al. 2018; Valdés et al. 2017). Nonetheless, the spraying solution should not have a high viscosity once a it cannot be or is very uneasily sprayed. The spray-flow characteristics dependent on the properties of the coating solution (surface tension, density, temperature, viscosity), the operating conditions (air pressure, flow rate, etc.), and the system conditions (spray angle, nozzle design). The setback of the multilayer coating is that it requires more steps as for each layer there are two steps: application and drying processes (Parreidt et al. 2018; Skurtys et al. 2011; Valdés et al. 2017).

4.2 Dipping

Dipping is a very commonly used technique for applying coatings on foods. It is simple and relativity cheap as it only involves submerging the product into a vat containing the coating solution (Fig. 10.9) (Andrade et al. 2012; Hanssen et al. 2012; Sharaf Eddin et al. 2019; Skurtys et al. 2011). The main benefit of this method is that it allows full coating of the surface and with good uniformity, even around complex and rough surfaces (Andrade et al. 2012; Parreidt et al. 2018). The dipping

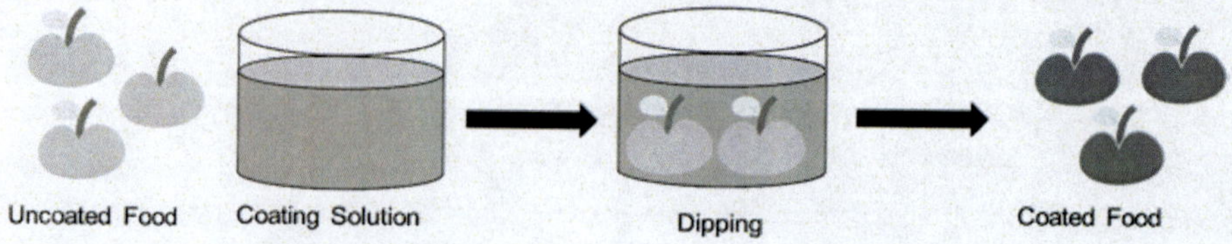

Fig. 10.9 The dipping technique

method consists of immersion of the sample into the coating solution at a constant speed followed by dwelling to guarantee that interaction of the sample with the coating solution and then the sample is withdrawn and let to drain the excessive coating-forming solution covering the product. This process is very short as it can last between 30 s and 5 min (Parreidt et al. 2018; Valdés et al. 2017). The dipping technique has several disadvantages associated. For instance, the dilution of the coating solution and the difficulty to adjust its optimal concentration. Furthermore, the dipping technique generates trash, microbial growth in the dipping tank and may cause problems with product respiration and storage characteristics, due to the thickness of the coating. Another problem that may appear is difficulty in adhesion due to the hydrophilic surface of the cut surfaces of the food products. Nonetheless, this can be avoided by applying the multi-layer technique (Andrade et al. 2012; Parreidt et al. 2018; Valdés et al. 2017).

4.3 Spreading

The spreading technique, commonly known as brushing, is the controlled spreading of a suspension onto the material surface to be than dried under controlled conditions in a sterile environment such as a laminar air-flow cabinet. This technique requires the use of instruments like a brush or spatula for direct coating on the food or indirect coating on the packaging material. The principal parameters to characterize the spreading technique are the rate of spreading and the degree of wetting, being contact angle measurements frequently used to evaluate the degree of spreading/wettability of a surface by a liquid. The spreading technique can be applied to large area food products. The spreading technique can be affected by several factors, such as system conditions such as humidity and temperature, liquid properties namely surface tension, density and viscosity, and substrate properties (its geometry like pattern and chemical make-up composition and surface roughness) (Khan and Nasef 2009; Priyadarshi and Rhim 2020; Valdés et al. 2017).

5 Application of Edible Active Coatings

5.1 *Fruits and Vegetables*

Fruits and vegetables are highly consumed, as they are an important source of nutrients. Besides, fruits and vegetables contain a high percentage of water, and consequently are highly perishable. Therefore, it is necessary to conserve the fruits and vegetables after their harvest so that they can reach the final consumer still in a good state of conservation and consumption, and this can be done through the application of EAC. EAC can extend the shelf life of fruits and vegetables by reducing gas exchange, moisture, respiration, and oxidation.

EAC have been applied to whole and fresh-cut fruits and vegetables, like orange, grapefruit, apple, passion fruit, avocado, peach, lime, lemon, fresh-cut peach, fresh-cut apple, fresh-cut pear, cucumber, bell pepper, melons, tomato, minimally processed carrot, fresh-cut tomato slices, fresh-cut cabbage, fresh-cut lettuce, minimally processed onion, fresh-cut potato (Dhall 2013). Table 10.1 contains several examples of EAC applied to fruits and vegetables.

Sessa and co-workers (2015) studied a new procedure to preserve vegetable products, in this case, rucola leaves, through the development of an edible coating based on chitosan and enriched with encapsulated essential oil nanoemulsions. Different essential oil nanoemulsions, like mandarin, oregano, lemon, and clove, were tested, and their antimicrobial activity was tested alone and combined with chitosan. The lemon essential oil nanoemulsion alone presented a higher antimicrobial activity and combined with chitosan, it presented an increase in antimicrobial activity, since chitosan alone already exhibited a significant antimicrobial activity. Therefore, the authors coated the rucola leaves with chitosan enriched with lemon essential oil nanoemulsion to prolong its shelf life for 21 days. The coating promoted a reduction of the microbial load, for 3 days of storage, while the controlled sample had an increase of the total microbial load. After the third day and until day 14 of storage, an increase of the total microbial load was observed in both samples, coated and control, but it was slower in the coated sample. The same results were observed for yeast and molds too. Regarding the colour variation, no significant variations were observed between the coated and the control samples during the storage period (Sessa et al. 2015). Whereas all samples were still in good condition on day 14, and deterioration and a yellowing of the samples were detected at day 21 of storage. Also, the authors evaluated the firmness of rucola samples through the texture analysis and observed an increase of the firmness for the control after 3 days of storage, while the coated sample was measured an increased after 7 days. Regarding these results, the authors concluded that an edible coating containing nano encapsulated lemon essential oil is an efficient way to increase the shelf life of ready-to-eat fruits and vegetables (Sessa et al. 2015).

Table 10.1 Example of different edible active coatings applied to different fruits and vegetables

Food product	Biopolymer	Active compound	Main results	Reference
Rucola leave	Chitosan	Nanoemulsified lemon essential oil	The coating was effective at prolonging the product shelf life from 3 to 7 days, compared to untreated sample.	(Sessa et al. 2015)
Tomato (*Solanum lycopersicum* L.)	Pectin and cornflour	Beetroot powder	The active coating improved shelf life and quality retention of tomato compared to control.	(Sucheta et al. 2019)
Raspberry (*Rubus idaeus* L.)	Gelatine	Ethanolic extract of Propolis	The active film presented remarkable antifungal activity allowing to extend raspberry shelf life at refrigeration temperature.	(Moreno et al. 2020)
Mango (*Mangifera indica* L.)	Guar gum	Aloe vera gel and ethanolic and aqueous extract of *Spirulina platensis*	Mango coated with the active coating presented higher firmness and reduced respiration rate. Total phenol and antioxidant activity were much higher in the coating with aqueous extract of *Spirulina platensis*.	(Ebrahimi and Rastegar 2020)
Strawberry (*Fragaria×ananassa* Duch.)	Carboxymethyl cellulose	Chitosan	Chitosan and carboxymethyl cellulose coating prevented fruit loss of firmeness and aroma volatiles but had little effect on total soluble solids and total acidity contents.	(Yan et al. 2019)
Guava (*Psidium guajava* L.)	Chitosan and alginate	ZnO nanoparticles	ZnO chitosan-alginate coating presented antibacterial activity as it delay the appearance of rot in guavas and was able to delay the ripening process up to 20 days of storage	(Arroyo et al. 2020)

(continued)

Table 10.1 (continued)

Food product	Biopolymer	Active compound	Main results	Reference
Fresh-cut pineapple (*Ananas comosus*)	Chitosan and Pullulan	Aloe mucilage, nopal mucilage, and linseed mucilage	The active coating decreased weight loss, retarded loss of total soluble solids and colour and was effective against spoilage microorganisms on fresh-cut pineapple, by 6 days compared with control.	(Treviño-Garza et al. 2017)
Apple (*Malus domestica* Borkh cv. *Golden Delicious*) and persimmon (*Diospyros kaki* Thunb. cv. *Rojo Brillante*)	Starch–gellan	Thyme (*Thymus vulgaris*) essential oil	In apples, the active coating did not reduce weight loss or firmness changes, increased the respiration rate, but reduced the severity of gray mold caused by Botrytis cinerea. In persimmon, the active coating prevented weight loss, reduced the incidence and severity of black spot caused by Alternaria alternata in persimmons, but had no significant effect on respiration rate.	(Sapper et al. 2019)

5.2 *Meat and Fish Products*

Meat, fish, and their derived products are very perishable, and a proper involvement is necessary to prevent deterioration and to prolong shelf life. Meat and fish-derived products are more susceptible to deterioration than fresh meat and fish as comminution increases the surface area and area of contact that exposes meat/fish to air (oxygen) and microorganisms (Umaraw et al. 2020). The quality of meat and fish products can be affected by intrinsic factors like pH, water activity (aw), moisture content, nutritional composition, oxidation/reduction potential, anti-microbial constituents, and biological structures, and extrinsic factors like relative humidity of the environment, temperature of storage, presence and concentration of gases, and presence or absence of other microorganisms. In order to avoid these factors, several researchers have developed EAC aiming to delay the quality decay of these foods. These coatings are carriers of antimicrobials, antioxidants, flavours, nanomaterials, and probiotics that can improve the food quality, and serve as barriers to the outer space.

For instance, Xiong et al. (2020) developed several chitosan-gelatine-based edible coatings incorporated with grape seed extract and/or nisin to evaluate its effect on fresh pork quality during 20 days stored at (4 °C). Overall, the edible coating containing 1% chitosan, 3% gelatine, and 0.5% grape seed extract had the best results on the pork preservation. Nonetheless, all coated samples had much lower pH values, at day 20, when compared to the control sample, but no significant difference was observed among all the coated samples at day 20. The same trend of results was observed regarding the colour changes of the samples, whereas the coated samples presented better results than the control sample. The pork samples coated with grape seed extract presented the highest redness (a*) and yellowness (b*) values among all groups. Regarding the lipid and protein oxidation, the edible coatings were effective in retarding the oxidation of the pork samples. Lipid oxidation was measured trough the thiobarbituric acid reactive substances (TBARS) assay and over the 20 days storage, the TBARS values significantly increased for all pork samples, but more accentuated for the control sample. Also, the TBARS value of the control sample was significantly higher than coated samples on day 5 and exponentially grow afterward but the TBARS values of all coated pork samples had a much slower increase during the storage time. The coated samples with grape seed extract presented the lowest TBARS values at day 20. In the case of protein oxidation, the control sample presented the lower values of the thiol group and therefore greater protein oxidation. On the other hand, the coated samples had greater protection against protein oxidation. Finally, regarding the microbial analysis, Xiong et al. (2020) concluded that the total viable count values of all coated and control pork samples increased significantly during the 20 days' cold storage. Yet, the total viable count of the control sample showed the fastest increase during the storage and was significantly higher than all other samples since day 5.

Concerning fish products, Choulitoudi et al. (2017) studied the potential of edible coatings made with CMC and enriched with rosemary extract and/or essential oils to extend the shelf life of smoked eel fillets during refrigerated storage. The active coatings showed moderate antimicrobial activity against *Pseudomonas* spp. and lactic acid bacteria growth and the coating with 800 ppm of extract presented the best results. Regarding the antioxidant properties, the rosemary extract was effective in retarding the oxidation of the smoked eels, and the higher the concentration of extract in the coating the higher was the protection against oxidation. Also, the combination of the extract with essential oils significantly retarded the oxidation of the smoked eels, which the authors suggested that may be indicative of possible synergistic effects.

Several researchers are studying EAC applied to meat, fish, and their derived products. So far, many studies have been carried out in which coatings are made with different biopolymers and different active substances (Chaparro-Hernández et al. 2019; Choulitoudi et al. 2017; Fang et al. 2018; Farajzadeh et al. 2016; Jasour et al. 2015; Jridi et al. 2018; Khazaei et al. 2017; Vital et al. 2016, 2018; Volpe et al. 2019; Wu et al. 2016, 2018b; Xiong et al. 2020; Yemiş and Candoğan 2017).

Table 10.2 presents other examples of EAC applied to meat, fish, and derived products.

5.3 Dairy Products

The dairy industry, specifically the cheese industry, is very prone to the application of EAC. Cheese is easily contaminated by bacteria, moulds, and yeasts, which, led to the loss of quality and consequently end of the cheese shelf life. Therefore, to ensure cheese shelf life with high quality, different packaging systems have been developed. Edible coatings have been pointed out as a good option for cheese, once it can be used as a package and as preservative, when it carries active compounds, such as antimicrobial agents. So far, many studies (Table 10.3) have been made for a better understanding of the application of EAC to cheese and other dairy products (Di Pierro et al. 2011; Saber El-Sisi 2015; Soleimani-Rambod et al. 2018; Wang et al. 2019).

For example, Wang et al. (2019) developed an edible coating with WPI nanofibers (WPNFs), combined with carvacrol (CA) as an antimicrobial agent and glycerol (Gly) as a plasticizer, to maintain the quality of fresh-cut Cheddar cheese. The WPNFs-CA/Gly film presented a smoother and continuous surfaces, lower moisture content and better transparency, when compared with the other films developed with only WPI, CA, and/or Gly. The WPNFs-CA/Gly coating was effective against two Gram-negative bacteria (*E. coli* and *S. enteritidis*) and two Gram-positive bacteria (*S. aureus* and *L. monocytogenes*). To evaluate the antioxidant activity of the WPNFs-CA/Gly, the reducing power and DPPH scavenging assays were determined. The WPNFs-CA/Gly reducing power and DPPH scavenging rates were higher than the of WPI-CA/Gly, indicating that WPNFs-CA/Gly can be applied as an edible coating to prevent food oxidation. Still, the authors evaluated the changes in the cheddar cheese and concluded that the cheddar cheese coated with WPNFs-CA/Gly had better protection capability on textural properties and lower weight loss. Therefore, Wang et al. (2019) stated that WPNFs-CA/Gly as edible coating material can be used to enhance the nutritional and overall quality properties of cheese during storage and extend the product shelf-life.

In another study, Di Pierro et al. (2011) studied fresh Ricotta cheese coated with chitosan/whey protein (CWP) film and stored under 40% CO_2/60% N_2 modified atmosphere packaging (MAP) conditions to reduce microbial growth and extend shelf-life. The effectiveness of the coating was evaluated through the measurement of the pH and the titratable acidity of the Ricotta cheese, and determination of the viable numbers of lactic acid bacteria and psychrotrophic and mesophilic microorganisms. During the 30-day storage period, the pH of control and coated Ricotta cheeses were similar. Nonetheless, the titratable acidity of the coated Ricotta cheese did not change significantly during the first 21 days, unlike the control that increased linearly during the first 2 weeks. The coated Ricotta cheese presented significantly lower viable numbers of mesophilic and psychrotrophic microorganisms different

Table 10.2 Example of different edible active coatings applied to different meat and fish products

Food product	Biopolymer	Active compound	Main results	Reference
Beef steak	Alginate	Rosemary (*Salvia rosmarinus*)and oregano (*Origanum vulgare*) essential oils	Both coatings were effective against the beef lipid oxidation and decreased colour loss, shear force and water loss compared to control. The coating with oregano essential oil presented better results with the highest antioxidant activity.	(Vital et al. 2016)
Fresh pork	Chitosan-gelatine	Grape seed extract and nisin	The incorporation of grape seed extract improved the antioxidant activity against meat oxidation but the incorporation of nisin did not improved the antioxidant and antimicrobial activities.	(Xiong et al. 2020)
Fresh pork	Chitosan	Gallic acid	The gallic acid chitosan coating inhibited microbial growth and retarded lipid and protein oxidation of fresh pork stored at 4 °C.	(Fang et al. 2018)
Beef meat	Gelatine	Henna (*L. inermis*) aqueous extract	The active coating was effective in decreasing lipid oxidation and total and phychrophilic vial bacteria counts, as well as in preserving colour properties, compared to control, at the end of storage.	(Jridi et al. 2018)
Pork	Chitosan	Tomato plant extract	The active coating was effective in prolonging the pork shelf life during 21 days, as it presented high values of antioxidant capacity and total phenolics and reduced microbial population compared to control.	(Chaparro-Hernández et al. 2019)

(continued)

Table 10.2 (continued)

Food product	Biopolymer	Active compound	Main results	Reference
Poultry meat	Chitosan	ZnO nanoparticles (NPs)	ZnO NPs, synthesized using apple peel wastes, enhanced chitosan intrinsic antimicrobial properties and improved the antioxidant activity, reducing poultry meat oxidative processes and microbiological growth.	(Souza et al. 2020a)
Fresh beef	Soy-protein	Thyme(*Thymus vulgaris* L.) or oregano (*Oreganum heracleoticum* L.) essential oils	Both active coatings presented similar antimicrobial activity against *escherichia coli* O157:H7, *listeria monocytogenes* and *staphylococcus aureus*, but thyme coating presented better results, after 14 days of storage.	(Yemiş and Candoğan 2017)
Beef	Sodium alginate	Oregano and rosemary essential oil	The authors evaluated the acceptability of the coatings loaded with oregano and rosemary essential oils. The oregano coating was preferred with high consumer acceptance and willingness too consume these products.	(Vital et al. 2018)
Shrimp (*Litopenaeus vannamei*)	Basil seed gum	Thymol	The active coating was able to significantly reduce the microbial growth without deterioration of organoleptic properties in shrimps during cold storage (4 °C).	(Khazaei et al. 2017)

(continued)

Table 10.2 (continued)

Food product	Biopolymer	Active compound	Main results	Reference
Shrimp (*Litopenaeus vannamei*)	Gelatine	Chitosan	The active coating prolonged the shrimps shelf life as it decreased the total psychotropic bacteria, lipid oxidation, and improved colour and texture properties up to 13 days compared to only 7 days of the control.	(Farajzadeh et al. 2016)
Pacific mackerel fillet (*pneumatophorus japonicus*)	Chitosan	Gallic acid	The gallic acid chitosan coating extended the Pacific mackerel shelf life up to 6 days compared to control as it was effective against microbial growth, lipid and protein oxidation and on maintaining sensory characteristics during storage.	(Wu et al. 2016)
Large yellow croaker (*Larimichthys crocea*)	Chitosan	Lysozyme	The chitosan coating enriched with lysozyme effectively inhibit staphylococcus aureus growth, retarded lipid oxidation and improved sensorial characteristics of the large yellow croaker.	(Wu et al. 2018b)
Rainbow trout	Chitosan	Lactoperoxidase system	The rainbow trout coated with chitosan and lactoperoxidase had significantly lower counts of microbial growth and lipid oxidation, and the total volatile basic nitrogen did not exceed the limits of consumption (30–35 g N 100 g $^{-1}$) compared to the control.	(Jasour et al. 2015)

(continued)

Table 10.2 (continued)

Food product	Biopolymer	Active compound	Main results	Reference
Smoked eel fillet (*Anguilla anguilla*)	Carboxyl methylcellulose	Rosemary (*Rosmarinus officinalis*, L) essential oil	The active coating enhanced the smoked eel quality by providing an antioxidant and antimicrobial protection, and retarding lipid oxidation during storage compared to control.	(Choulitoudi et al. 2017)
Trout fillet (*oncorhynchus mykiss*)	Carrageenan	Lemon essential oil	The carrageenan loaded with lemon essential oil preserved the olfactory and physicochemical characteristics and retarded lipid oxidation of fresh trout fillets stored at 4 °C during 12 days.	(Volpe et al. 2019)

to the control that reached the microbiological limit of acceptability between days 7 and 14. Still, differences in visual appearance, flavour, texture, and odour were found between the coated Ricotta cheese and the control samples, but the coated Ricotta cheese maintained its texture better than the control cheese.

5.4 Other Food Products

EAC can be applied to several other food products. They can be applied to cereals and cereal-based products to avoid moisture and improve the quality of the food product. Hydrophilic polymers are commonly used in fried products, as they are a good barrier to fats and oils, thus avoiding the absorption of fat during frying of the food product. Bakery products, such as biscuits, are often coated to reduce the hydration and therefore maintain the crispiness of the food product. EAC are also applied to nuts to avoid lipid oxidation (Galus and Kadzińska 2015; Kurek et al. 2017; Suput et al. 2015). Table 10.4 presents some examples of EAC applied to different food products.

Table 10.3 Example of different edible active coatings applied to dairy products

Food product	Biopolymer	Active compound	Main results	Reference
Fresh-cut Cheddar cheese	Whey protein isolates nanofiber	Carvacrol	The addition of carvacrol to the whey protein isolate nanofiber enhanced its antimicrobial and antioxidant activity, leading to better preservation of textural characteristics and lower weight loss of cheddar cheese during storage.	(Wang et al. 2019)
Ricotta cheese	Chitosan/whey protein	–	The active coating was able to prevent the development of undesirable acidity and of lactic acid bacteria, mesophilic and psychrotrophic microorganisms, and extended the shelf life of the ricotta cheese.	(Di Pierro et al. 2011)
Egyptian hard cheese	Chitosan	Chitosan	The chitosan coating prevented weight loss, moisture loss and effectively contribute to the flavor retention, çompared to control, during ripening.	(Saber El-Sisi 2015)
Cheddar cheese	Xanthan gum/ flaxseed (*Linum usitatissimum* L.) mucilage	–	The active coatings did not affected the growth of total mesophilic aerobic bacteria and non-starter lactic acid bacteria, but significantly affected the chemical properties with highest pH and acidity rates observed in the flaxseed mucilage and xantham gum, respectively.	(Soleimani-Rambod et al. 2018)

6 Conclusion

Lately, a vast number of food packaging research studies have been devoted to EAC. EAC are environmentally friendly and biodegradable since they are prepared with biopolymers such as polysaccharides, proteins, and lipids. These biopolymers can be used alone or in combination with others in order to obtain the best coating properties. A plasticizer is usually added to the coating formulation to improve the flexibility and elasticity of the coating. Additionally, the coatings can carry active compounds that prevent oxidation, microbial development, browning, off flavours, among others. The EAC have as main advantages to be non-toxic and digestible (since they will be consumed together with the food), have a low production cost, be easy to prepare, and have good barrier properties (control of water and gas transfer with the outside environment). Furthermore, the EAC can prevent the loss of organoleptic characteristics, flavour, colour, and appearance (important factors for the consumer), prevent and/or delay the microbiological development and oxidation, and this way maintain the quality and extend the shelf life of the food product. Coatings can be added to a variety of foods, often being applied to fresh fruits and

Table 10.4 Example of different edible active coatings applied to other food products

Food product	Biopolymer	Active compound	Main results	Reference
Walnut kernels (*Juglans regia* L., Kaghazi cultivar)	Chitosan	Green tea (*Camellia sinensis*, var. *sinensis*) extract	The chitosan incorporated with green tea extract was effective in inhibiting lipid oxidation and fungal growth during storage of walnut kernels. However, the addition of higher concentrations of green tea extract (10 g L^{-1}) was not accepted by the panelist, so the authors suggested that a formulation with 10 g L^{-1} of chitosan and 5 g L^{-1} of green tea extract is most suitable to prolong the walnuts kernels shelf life.	(Sabaghi et al. 2015)
Unsalted roasted peanut (var. "Runners")	Whey protein isolate	Ascorbic palmitate and α-tocopherol	The authors evaluated the lipid oxidation of peanuts coated with whey protein isolate, whey protein isolate loaded with ascorbic palmitate and α-tocopherol, and uncoated. The coatings were effective in delaying lipid oxidation compared to uncoated samples, however the addition of ascorbic palmitate and α-tocopherol did not have significant difference in delaying lipid oxidation.	(Han et al. 2008)
Roasted pistachio nut (*Pistacia vera* L.)	Gelatine	Ascorbic acid and/or propyl gallate	The active coating was able to delay the lipid oxidation and hydrolytic rancidity and increased hardness and sensory firmness of pistachio.	(Khoshnoudi-Nia and Sedaghat 2019)
Eggs	Sweet potato (*Ipomoea batatas Lam*) starch	Thyme (*Thymus vulgaris*) essential oil	The active coating extended the shelf life and maintained the quality of the eggs by 2 weeks compared to control. The active coating was also able to reduce *Salmonella enterica* populations to levels below the detectable level (<10 CFU/mL) after 5 weeks.	(Sharaf Eddin and Tahergorabi 2019)

vegetables, meat, fish, and cheese. In this regard, EAC are expected to be produced on a large scale by the food industry in the near future, as it uses low cost products, of natural origin and can easily be applied through effective and low-cost methods (spraying technique). So, it can be concluded that EAC have a bright future in the food industry.

References

Abd El-Hack ME, El-Saadony MT, Shafi ME, Zabermawi NM, Arif M, Batiha GE, Khafaga AF, Abd El-Hakim YM, Al-Sagheer AA. Antimicrobial and antioxidant properties of chitosan and its derivatives and their applications: a review. Int J Biol Macromol. 2020; https://doi.org/10.1016/j.ijbiomac.2020.08.153

Adams S, Kunz B, Weidenbörner M. Mycelial deformations of Cladosporium herbarum due to the application of eugenol or carvacrol. J Essent Oil Res. 1996;8:535–40. https://doi.org/10.1080/10412905.1996.9700682

Albertos I, Martin-Diana A, Burón M, Rico D. Development of functional bio-based seaweed (Himanthalia elongata and Palmaria palmata) edible films for extending the shelflife of fresh fish burgers. Food Packag Shelf Life. 2019;22:100382. https://doi.org/10.1016/j.fpsl.2019.100382

Andrade RD, Skurtys O, Osorio FA. Atomizing spray systems for application of edible coatings. Compr Rev Food Sci Food Saf. 2012;11:323–37. https://doi.org/10.1111/j.1541-4337.2012.00186.x

Andrade M, Ribeiro-Santos R, Melo NR d, Sanches-Silva A. The potential of whey protein in food packaging. EC Nutr Ed Column. 2016;ECO.01:25–6.

Andrade MA, Ribeiro-Santos R, Costa Bonito MC, Saraiva M, Sanches-Silva A. Characterization of rosemary and thyme extracts for incorporation into a whey protein based film. LWT - Food Sci Technol. 2018;92:497–508. https://doi.org/10.1016/j.lwt.2018.02.041

Andrade MA, Lima V, Sanches Silva A, Vilarinho F, Castilho MC, Khwaldia K, Ramos F. Pomegranate and grape by-products and their active compounds: are they a valuable source for food applications? Trends Food Sci Technol. 2019a;86:68–84. https://doi.org/10.1016/j.tifs.2019.02.010

Andrade MA, Ribeiro-Santos R, Guerra M, Sanches-Silva A. Evaluation of the oxidative status of salami packaged with an active whey protein film. Foods. 2019b;8:387. https://doi.org/10.3390/foods8090387

Andrade MA, Ribeiro-Santos R, Nabavi SM, Sanches-Silva A. Indirect additives. In: Food additives and human health. Sharjah: Bentham Science Publishers; 2020. p. 246–68. https://doi.org/10.2174/9789811446139120010016.

Ariviani S, Kusumawati M, Atmaka W. Chitosan-garlic essential oil incorporation on beef meatball edible coatings as antioxidant-based functional food. IOP Conf Ser Mater Sci Eng. 2019;633:012032. https://doi.org/10.1088/1757-899X/633/1/012032

Arnold RR, Cole MF, McGhee JR. A bactericidal effect for human lactoferrin. Science. 1977;197:263–5. https://doi.org/10.1126/science.327545

Arnon-Rips H, Poverenov E. Improving food products' quality and storability by using Layer by Layer edible coatings. Trends Food Sci Technol. 2018;75:81–92. https://doi.org/10.1016/j.tifs.2018.03.003

Arroyo BJ, Bezerra AC, Oliveira LL, Arroyo SJ, Melo EA d, Santos AMP. Antimicrobial active edible coating of alginate and chitosan add ZnO nanoparticles applied in guavas (Psidium guajava L.). Food Chem. 2020;309:125566. https://doi.org/10.1016/j.foodchem.2019.125566

Assunção R, Martins C, Vasco E, Jager A, Oliveira C, Cunha SC, Fernandes JO, Nunes B, Loureiro S, Alvito P. Portuguese children dietary exposure to multiple mycotoxins – an overview of risk assessment under MYCOMIX project. Food Chem Toxicol. 2018;118:399–408. https://doi.org/10.1016/j.fct.2018.05.040

Avella M, De Vlieger JJ, Errico ME, Fischer S, Vacca P, Volpe MG. Biodegradable starch/clay nanocomposite films for food packaging applications. Food Chem. 2005;93:467–74. https://doi.org/10.1016/j.foodchem.2004.10.024

Baldwin EA, Hagenmaier RD, Bai J. Edible coatings and films to improve food quality. 2nd ed. CRC Press; 2011.

Benbettaïeb N, Tanner C, Cayot P, Karbowiak T, Debeaufort F. Impact of functional properties and release kinetics on antioxidant activity of biopolymer active films and coatings. Food Chem. 2018;242:369–77. https://doi.org/10.1016/j.foodchem.2017.09.065

Bounous G. Whey protein concentrate (WPC) and glutathione modulation in cancer treatment. Anticancer Res. 2000;20:4785–92.

Brock JH. Lactoferrin in human milk: its role in iron absorption and protection against enteric infection in the newborn infant. Arch Dis Child. 1980;55:417–21. https://doi.org/10.1136/adc.55.6.417

Bryden WL. Mycotoxins: natural food chain contaminants and human health. In: Encyclopedia of environmental health. Elsevier B.V; 2011. p. 898–905. https://doi.org/10.1016/B978-0-444-52272-6.00701-7.

Campos CA, Gerschenson LN, Flores SK. Development of edible films and coatings with antimicrobial activity. Food Bioprocess Technol. 2011;4:849–75. https://doi.org/10.1007/s11947-010-0434-1

Carocho M, Morales P, Ferreira ICFR. Natural food additives: quo vadis? Trends Food Sci Technol. 2015;45:284–95. https://doi.org/10.1016/j.tifs.2015.06.007

Castro FVR, Andrade MA, Sanches Silva A, Vaz MF, Vilarinho F. The contribution of a whey protein film incorporated with green tea extract to minimize the lipid oxidation of Salmon (Salmo salar L.). Foods. 2019;8:327. https://doi.org/10.3390/foods8080327

Chaparro-Hernández S, Ruiz-Cruz S, Márquez-Ríos E, Ornelas-Paz J d J, Del-Toro-Sánchez CL, Gassos-Ortega LE, Ocaño-Higuera VM, López-Mata MA, Devora-Isiordia GE. Effect of chitosan–tomato plant extract edible coating on the quality, shelf life, and antioxidant capacity of pork during refrigerated storage. Coatings. 2019;9:827. https://doi.org/10.3390/coatings9120827

Chatterton DEW, Smithers G, Roupas P, Brodkorb A. Bioactivity of β-lactoglobulin and α-lactalbumin—technological implications for processing. Int Dairy J. 2006;16:1229–40. https://doi.org/10.1016/j.idairyj.2006.06.001

Choulitoudi E, Ganiari S, Tsironi T, Ntzimani A, Tsimogiannis D, Taoukis P, Oreopoulou V. Edible coating enriched with rosemary extracts to enhance oxidative and microbial stability of smoked eel fillets. Food Packag Shelf Life. 2017;12:107–13. https://doi.org/10.1016/j.fpsl.2017.04.009

Chung K-T, Wei C-I, Johnson MG. Are tannins a double-edged sword in biology and health? Trends Food Sci Technol. 1998;9:168–75. https://doi.org/10.1016/S0924-2244(98)00028-4

Claeson P, Rådström P, Sköld O, Nilsson Å, Höglund S. Bactericidal effect of the sesquiterpene T-cadinol on Staphylococcus aureus. Phyther Res. 1992;6:94–8. https://doi.org/10.1002/ptr.2650060209

Costa MJ, Maciel LC, Teixeira JA, Vicente AA, Cerqueira MA. Use of edible films and coatings in cheese preservation: opportunities and challenges. Food Res Int. 2018;107:84–92. https://doi.org/10.1016/j.foodres.2018.02.013

Dai J, Mumper RJ. Plant phenolics: extraction, analysis and their antioxidant and anticancer properties. Molecules. 2010;15:7313–52. https://doi.org/10.3390/molecules15107313

de Billerbeck VG, Roques CG, Bessière J-M, Fonvieille J-L, Dargent R. Effects of *Cymbopogon nardus* (L.) W. Watson essential oil on the growth and morphogenesis of *Aspergillus niger*. Can J Microbiol. 2001;47:9–17. https://doi.org/10.1139/cjm-47-1-9

de Hoyos-Martínez PL, Merle J, Labidi J, Charrier–El Bouhtoury F. Tannins extraction: a key point for their valorization and cleaner production. J Clean Prod. 2019;206:1138–55. https://doi.org/10.1016/j.jclepro.2018.09.243

Debeaufort F, Quezada-Gallo JA, Voilley A. Edible films and coatings: tomorrow's packagings: a review. Crit Rev Food Sci Nutr. 1998; https://doi.org/10.1080/10408699891274219

Dehghani S, Hosseini SV, Regenstein JM. Edible films and coatings in seafood preservation: a review. Food Chem. 2018;240:505–13. https://doi.org/10.1016/j.foodchem.2017.07.034

Demarque DP, Callejon DR, de Oliveira GG, Silva DB, Carollo CA, Lopes NP. The role of tannins as antiulcer agents: a fluorescence-imaging based study. Rev Bras Farmacogn. 2018;28:425–32. https://doi.org/10.1016/j.bjp.2018.03.011

Dhall RK. Advances in edible coatings for fresh fruits and vegetables: a review. Crit Rev Food Sci Nutr. 2013;53:435–50. https://doi.org/10.1080/10408398.2010.541568

Di Pierro P, Sorrentino A, Mariniello L, Giosafatto CVL, Porta R. Chitosan/whey protein film as active coating to extend Ricotta cheese shelf-life. LWT - Food Sci Technol. 2011;44:2324–7. https://doi.org/10.1016/j.lwt.2010.11.031

Domínguez R, Barba FJ, Gómez B, Putnik P, Bursać Kovačević D, Pateiro M, Santos EM, Lorenzo JM. Active packaging films with natural antioxidants to be used in meat industry: a review. Food Res Int. 2018;113:93–101. https://doi.org/10.1016/j.foodres.2018.06.073

Donati I, Paoletti S. Material properties of alginates. In: Rehm BH, editor. Alginates: biology and applications. Berlin/Heidelberg: Springer; 2009. p. 1–53.

Dong X, Bai Y, Xu Z, Shi Y, Sun Y, Janaswamy S, Yu C, Qi H. Phlorotannins from Undaria pinnatifida Sporophyll: extraction, antioxidant, and anti-inflammatory activities. Mar Drugs. 2019;17:434. https://doi.org/10.3390/md17080434

Dutta PK, Tripathi S, Mehrotra GK, Dutta J. Perspectives for chitosan based antimicrobial films in food applications. Food Chem. 2009; https://doi.org/10.1016/j.foodchem.2008.11.047

Ebrahimi F, Rastegar S. Preservation of mango fruit with guar-based edible coatings enriched with Spirulina platensis and Aloe vera extract during storage at ambient temperature. Sci Hortic (Amsterdam). 2020;265:109258. https://doi.org/10.1016/j.scienta.2020.109258

Eggersdorfer M, Wyss A. Carotenoids in human nutrition and health. Arch Biochem Biophys. 2018;652:18–26. https://doi.org/10.1016/j.abb.2018.06.001

Elsabee MZ, Abdou ES. Chitosan based edible films and coatings: a review. Mater Sci Eng C. 2013;33:1819–41. https://doi.org/10.1016/j.msec.2013.01.010

Emir A, Emir C, Yıldırım H. Characterization of phenolic profile by LC-ESI-MS/MS and enzyme inhibitory activities of two wild edible garlic: *Allium nigrum* L. and *Allium subhirsutum* L. J Food Biochem. 2020;44. https://doi.org/10.1111/jfbc.13165

Espitia PJP, Du WX, Avena-Bustillos R d J, Soares N d FF, McHugh TH. Edible films from pectin: physical-mechanical and antimicrobial properties - a review. Food Hydrocoll. 2014;35:287–96. https://doi.org/10.1016/j.foodhyd.2013.06.005

European Commission. COUNCIL DIRECTIVE of 15 October 1984 on the approximation of the laws of the Member States relating to ceramic articles intended to come into contact with foodstuffs (84/500/EEC). Off J Eur Communities. 1984;L277:12–6.

European Commission. REGULATION (EC) no 1935/2004 OF THE EUROPEAN PARLIAMENT AND OF THE COUNCIL of 27 October 2004 on materials and articles intended to come into contact with food and repealing Directives 80/590/EEC and 89/109/EEC. Off J Eur Union. 2004:4–17.

European Commission. COMMISSION DIRECTIVE 2007/42/EC of 29 June 2007 relating to materials and articles made of regenerated cellulose film intended to come into contact with foodstuffs. Off J Eur Union. 2007;L172:71–82.

European Commission. Regulation (EC) No 1334/2008 of the European Parliament and of the Council of 16 December 2008 on flavourings and certain food ingredients with flavouring properties for use in and on foods and amending Council Regulation (EEC) No 1601/91, Regulations (EC). Off J Eur Union. 2008a;L354:34–50.

European Commission. COMMISSION REGULATION (EC) No 282/2008 of 27 March 2008 on recycled plastic materials and articles intended to come into contact with foods and amending Regulation (EC) No 2023/2006. Off J Eur Union. 2008b;L86:9–18.

European Commission. Regulation (EC) N° 1333/2008 Of The European Parliament And Of The Council of 16 December 2008 on food additives. Off J Eur Communities. 2008c;L354:16–33.

European Commission. REGULATION (EC) No 1331/2008 OF THE EUROPEAN PARLIAMENT AND OF THE COUNCIL of 16 December 2008 establishing a common authorisation procedure for food additives, food enzymes and food flavourings. Off J Eur Union. 2008d;L354:1–6.

European Commission. COMMISSION REGULATION (EC) No 450/2009 of 29 May 2009 on active and intelligent materials and articles intended to come into contact with food. Off J Eur Union. 2009;L135:3–11.

European Commission. COMMISSION REGULATION (EU) No 10/2011 of 14 January 2011 on plastic materials and articles intended to come into contact with food. Off J Eur Union. 2011a;L12:1–89.

European Commission. Questions and answers on food additives [WWW document]; 2011b. https://ec.europa.eu/commission/presscorner/detail/en/MEMO_11_783

Fang Z, Lin D, Warner RD, Ha M. Effect of gallic acid/chitosan coating on fresh pork quality in modified atmosphere packaging. Food Chem. 2018;260:90–6. https://doi.org/10.1016/j.foodchem.2018.04.005

Farajzadeh F, Motamedzadegan A, Shahidi SA, Hamzeh S. The effect of chitosan-gelatin coating on the quality of shrimp (Litopenaeus vannamei) under refrigerated condition. Food Control. 2016;67:163–70. https://doi.org/10.1016/j.foodcont.2016.02.040

Feng Z, Li L, Wang Q, Wu G, Liu C, Jiang B, Xu J. Effect of antioxidant and antimicrobial coating based on whey protein nanofibrils with TiO2 nanotubes on the quality and shelf life of chilled meat. Int J Mol Sci. 2019;20:1184. https://doi.org/10.3390/ijms20051184

Floris R, Recio I, Berkhout B, Visser S. Antibacterial and antiviral effects of milk proteins and derivatives thereof. Curr Pharm Des. 2003;9:1257–75. https://doi.org/10.2174/1381612033454810

Food and Agriculture Organization of the United Nations, World Health Organization. About Codex Alimentarius [WWW document]; 2020. http://www.fao.org/fao-who-codexalimentarius/about-codex/en/

Food and Drug Administration (2010) Overview of food ingredients, additives & colors [WWW document]. https://www.fda.gov/food/food-ingredients-packaging/overview-food-ingredients-additives-colors

Freitas A, Moldão-Martins M, Costa HS, Albuquerque TG, Valente A, Sanches-Silva A. Effect of UV-C radiation on bioactive compounds of pineapple (Ananas comosus L. Merr.) by-products. J Sci Food Agric. 2015;95:44–52. https://doi.org/10.1002/jsfa.6751

Gad AS, Khadrawy YA, El-Nekeety AA, Mohamed SR, Hassan NS, Abdel-Wahhab MA. Antioxidant activity and hepatoprotective effects of whey protein and Spirulina in rats. Nutrition. 2011;27:582–9. https://doi.org/10.1016/j.nut.2010.04.002

Galus S, Kadzińska J. Food applications of emulsion-based edible films and coatings. Trends Food Sci Technol. 2015;45:273–83. https://doi.org/10.1016/j.tifs.2015.07.011

Geigenberger P, Geiger M, Stitt M. High-temperature perturbation of starch synthesis is attributable to inhibition of ADP-glucose pyrophosphorylase by decreased levels of glycerate-3-phosphate in growing potato tubers. Plant Physiol. 1998;117:1307–16. https://doi.org/10.1104/pp.117.4.1307

Guilbert S, Gontard N, Cuq B. Technology and applications of edible protective films. Packag Technol Sci. 1995;8:339–46. https://doi.org/10.1002/pts.2770080607

Gustavsson J, Cederberg C, Sonesson U, van Otterdijk R, Meybeck A (2011) Global food losses and food waste - extent, causes and prevention. SAVE FOOD An Initiat. Food Loss Waste Reduct.

Han JH, Hwang H-M, Min S, Krochta JM. Coating of peanuts with edible whey protein film containing α-tocopherol and ascorbyl palmitate. J Food Sci. 2008;73:E349–55. https://doi.org/10.1111/j.1750-3841.2008.00910.x

Hanssen T-ES, Mathisen TA, Jørgensen F. Generalized transport costs in intermodal freight transport. Procedia Soc Behav Sci. 2012;54:189–200. https://doi.org/10.1016/j.sbspro.2012.09.738

Hassan B, Chatha SAS, Hussain AI, Zia KM, Akhtar N. Recent advances on polysaccharides, lipids and protein based edible films and coatings: a review. Int J Biol Macromol. 2018;109:1095–107. https://doi.org/10.1016/j.ijbiomac.2017.11.097

Hermund DB, Plaza M, Turner C, Jónsdóttir R, Kristinsson HG, Jacobsen C, Nielsen KF. Structure dependent antioxidant capacity of phlorotannins from Icelandic Fucus vesiculosus by UHPLC-DAD-ECD-QTOFMS. Food Chem. 2018;240:904–9. https://doi.org/10.1016/j.foodchem.2017.08.032

Hernández-Ledesma B, Recio I, Amigo L. β-Lactoglobulin as source of bioactive peptides. Amino Acids. 2008;35:257–65. https://doi.org/10.1007/s00726-007-0585-1

Himejima M, Kubo I. Fungicidal activity of polygodial in combination with anethole and indole against Candida albicans. J Agric Food Chem. 1993;41:1776–9.

Jasour MS, Ehsani A, Mehryar L, Naghibi SS. Chitosan coating incorporated with the lactoperoxidase system: an active edible coating for fish preservation. J Sci Food Agric. 2015;95:1373–8. https://doi.org/10.1002/jsfa.6838

Jridi M, Mora L, Souissi N, Aristoy MC, Nasri M, Toldrá F. Effects of active gelatin coated with henna (L. inermis) extract on beef meat quality during chilled storage. Food Control. 2018;84:238–45. https://doi.org/10.1016/j.foodcont.2017.07.041

Kalemba D, Kunicka A. Antibacterial and antifungal properties of essential oils. Curr Med Chem. 2003;10:813–29. https://doi.org/10.2174/0929867033457719

Kalogeropoulos N, Konteles SJ, Troullidou E, Mourtzinos I, Karathanos VT. Chemical composition, antioxidant activity and antimicrobial properties of propolis extracts from Greece and Cyprus. Food Chem. 2009;116:452–61. https://doi.org/10.1016/j.foodchem.2009.02.060

Kapetanakou A, Manios S, Skandamis P. Application of edible films and coatings on food. In: Novel food preservation and microbial assessment techniques. CRC Press; 2014. p. 237–73. https://doi.org/10.1201/b16758-11.

Kazemian-Bazkiaee F, Ebrahimi A, Hosseini SM, Shojaee-Aliabadi S, Farhoodi M, Rahmatzadeh B, Sheikhi Z. Evaluating the protective effect of edible coatings on lipid oxidation, fatty acid composition, aflatoxins levels of roasted peanut kernels. J Food Meas Charact. 2020;14:1025–38. https://doi.org/10.1007/s11694-019-00352-9

Khan MI, Nasef MM. Spreading behaviour of silicone oil and glycerol drops on coated papers. Leonardo J Sci. 2009:18–29.

Khazaei N, Esmaiili M, Emam-Djomeh Z. Application of active edible coatings made from basil seed gum and thymol for quality maintenance of shrimp during cold storage. J Sci Food Agric. 2017;97:1837–45. https://doi.org/10.1002/jsfa.7984

Khoshnoudi-Nia S, Sedaghat N. Effect of active edible coating and temperature on quality properties of roasted pistachio nuts during storage. J Food Process Preserv. 2019;43:e14121. https://doi.org/10.1111/jfpp.14121

Kirke DA, Rai DK, Smyth TJ, Stengel DB. An assessment of temporal variation in the low molecular weight phlorotannin profiles in four intertidal brown macroalgae. Algal Res. 2019;41:101550. https://doi.org/10.1016/j.algal.2019.101550

Knobloch K, Pauli A, Iberl B, Weigand H, Weis N. Antibacterial and antifungal properties of essential oil components. J Essent Oil Res. 1989;1:119–28. https://doi.org/10.1080/10412905.1989.9697767

Kouhi M, Prabhakaran MP, Ramakrishna S. Edible polymers: an insight into its application in food, biomedicine and cosmetics. Trends Food Sci Technol. 2020;103:248–63. https://doi.org/10.1016/j.tifs.2020.05.025

Kurek M, Ščetar M, Galić K. Edible coatings minimize fat uptake in deep fat fried products: a review. Food Hydrocoll. 2017;71:225–35. https://doi.org/10.1016/j.foodhyd.2017.05.006

Longhi C, Conte M, Seganti L, Polidoro M, Alfsen A, Valenti P. Influence of lactoferrin on the entry process of Escherichia coli HB101(pRI203) in HeLa cells. Med Microbiol Immunol. 1993;182:25–35. https://doi.org/10.1007/BF00195948

Lopes G, Sousa C, Silva LR, Pinto E, Andrade PB, Bernardo J, Mouga T, Valentão P. Can phlorotannins purified extracts constitute a novel pharmacological alternative for microbial infections with associated inflammatory conditions? PLoS One. 2012;7:e31145. https://doi.org/10.1371/journal.pone.0031145

Lopes G, Andrade P, Valentão P. Phlorotannins: towards new pharmacological interventions for diabetes mellitus type 2. Molecules. 2016;22:56. https://doi.org/10.3390/molecules22010056

Lopes G, Barbosa M, Vallejo F, Gil-Izquierdo Á, Andrade PB, Valentão P, Pereira DM, Ferreres F. Profiling phlorotannins from Fucus spp. of the Northern Portuguese coastline: chemical approach by HPLC-DAD-ESI/MS and UPLC-ESI-QTOF/MS. Algal Res. 2018;29:113–20. https://doi.org/10.1016/j.algal.2017.11.025

Manandhar, Wagle, Seong, Paudel, Kim, Jung, Choi. Phlorotannins with potential anti-tyrosinase and antioxidant activity isolated from the marine seaweed Ecklonia stolonifera. Antioxidants. 2019;8:240. https://doi.org/10.3390/antiox8080240

Martínez-González M d C, Bautista-Baños S, Correa-Pacheco ZN, Corona-Rangel ML, Ventura-Aguilar RI, Del Río-García JC, Ramos-García M d L. Effect of nanostructured chitosan/propolis coatings on the quality and antioxidant capacity of strawberries during storage. Coatings. 2020;10:90. https://doi.org/10.3390/coatings10020090

Mohamed SAA, El-Sakhawy M, El-Sakhawy MAM. Polysaccharides, protein and lipid -based natural edible films in food packaging: a review. Carbohydr Polym. 2020;238:116178. https://doi.org/10.1016/j.carbpol.2020.116178

MolView. MolView [WWW Document]; 2020. https://molview.org/. Accessed 9 Feb 2020.

Moratti SC, Cabral JD. Antibacterial properties of chitosan. In: Jennings JA, Bumgardner JD, editors. Chitosan based biomaterials, volume 1: Fundamentals biomaterials. Matthew Deans; 2017. p. 31–44.

Moreno S, Scheyer T, Romano CS, Vojnov A a. Antioxidant and antimicrobial activities of rosemary extracts linked to their polyphenol composition. Free Radic Res. 2006;40:223–31. https://doi.org/10.1080/10715760500473834

Moreno MA, Vallejo AM, Ballester A-R, Zampini C, Isla MI, López-Rubio A, Fabra MJ. Antifungal edible coatings containing Argentinian propolis extract and their application in raspberries. Food Hydrocoll. 2020;107:105973. https://doi.org/10.1016/j.foodhyd.2020.105973

Naidu SS, Svensson U, Kishore AR, Naidu AS. Relationship between antibacterial activity and porin binding of lactoferrin in Escherichia coli and Salmonella typhimurium. Antimicrob Agents Chemother. 1993;37:240–5. https://doi.org/10.1128/AAC.37.2.240

Nesic AR, Seslija SI. The influence of nanofillers on physical–chemical properties of polysaccharide-based film intended for food packaging. In: Grumezescu AM, editor. Food packaging. Elsevier Inc.; 2017. p. 637–97. https://doi.org/10.1016/b978-0-12-804302-8.00019-4.

Nouri A, Yaraki MT, Ghorbanpour M, Agarwal S, Gupta VK. Enhanced antibacterial effect of chitosan film using Montmorillonite/CuO nanocomposite. Int J Biol Macromol. 2018;109:1219–31. https://doi.org/10.1016/j.ijbiomac.2017.11.119

Oliveira PM, Zannini E, Arendt EK. Cereal fungal infection, mycotoxins, and lactic acid bacteria mediated bioprotection: from crop farming to cereal products. Food Microbiol. 2014;37:78–95. https://doi.org/10.1016/j.fm.2013.06.003

Park HJ. Edible coatings. Food Preserv Tech. 2003:90–105. https://doi.org/10.1016/B978-1-85573-530-9.50010-3

Parodi PW. A role for milk proteins in cancer prevention. Aust J Dairy Technol. 1998;53:37–47.

Parreidt TS, Müller K, Schmid M. Alginate-based edible films and coatings for food packaging applications. Foods. 2018;7:1–38. https://doi.org/10.3390/foods7100170

Pereira de Abreu DA, Losada PP, Maroto J, Cruz JM. Evaluation of the effectiveness of a new active packaging film containing natural antioxidants (from barley husks) that retard lipid damage in frozen Atlantic salmon (Salmo salar L.). Food Res Int. 2010;43:1277–82. https://doi.org/10.1016/j.foodres.2010.03.019

Piccaglia R, Marotti M, Giovanelli E, Deans SG, Eaglesham E. Antibacterial and antioxidant properties of Mediterranean aromatic plants. Ind Crop Prod. 1993;2:47–50. https://doi.org/10.1016/0926-6690(93)90010-7

Pires JRA, de Souza VGL, Fernando AL. Chitosan/montmorillonite bionanocomposites incorporated with rosemary and ginger essential oil as packaging for fresh poultry meat. Food Packag Shelf Life. 2018;17:142–9. https://doi.org/10.1016/j.fpsl.2018.06.011

Pop OL, Pop CR, Dufrechou M, Vodnar DC, Socaci SA, Dulf FV, Minervini F, Suharoschi R. Edible films and coatings functionalization by probiotic incorporation: a review. Polymers (Basel). 2019;12:12. https://doi.org/10.3390/polym12010012

Priyadarshi R, Rhim JW. Chitosan-based biodegradable functional films for food packaging applications. Innov Food Sci Emerg Technol. 2020;62:102346. https://doi.org/10.1016/j.ifset.2020.102346

Prodpran T, Benjakul S, Artharn A. Properties and microstructure of protein-based film from round scad (Decapterus maruadsi) muscle as affected by palm oil and chitosan incorporation. Int J Biol Macromol. 2007;41:605–14. https://doi.org/10.1016/j.ijbiomac.2007.07.020

Proestos C, Boziaris IS, Nychas GJE, Komaitis M. Analysis of flavonoids and phenolic acids in Greek aromatic plants: investigation of their antioxidant capacity and antimicrobial activity. Food Chem. 2006;95:664–71. https://doi.org/10.1016/j.foodchem.2005.01.049

Ramos ÓL, Fernandes JC, Silva SI, Pintado ME, Malcata FX. Edible films and coatings from whey proteins: a review on formulation, and on mechanical and bioactive properties. Crit Rev Food Sci Nutr. 2012;52:533–52. https://doi.org/10.1080/10408398.2010.500528

Ramos OL, Pereira RN, Rodrigues RM, Teixeira JA, Vicente AA, Malcata FX. Whey and whey powders: production and uses. In: Encyclopedia of food and health. Amsterdam: Elsevier; 2016. p. 498–505. https://doi.org/10.1016/B978-0-12-384947-2.00747-9.

Rastegar S, Hassanzadeh Khankahdani H, Rahimzadeh M. Effectiveness of alginate coating on antioxidant enzymes and biochemical changes during storage of mango fruit. J Food Biochem. 2019;43:1–10. https://doi.org/10.1111/jfbc.12990

Ribeiro-Santos R, Sanches-Silva A, Motta JFG, Andrade M, Neves I d A, Teófilo RF, Carvalho MG d, Melo NR d. Combined use of essential oils applied to protein base active food packaging: study in vitro and in a food simulant. Eur Polym J. 2017;93:75–86. https://doi.org/10.1016/j.eurpolymj.2017.03.055

Ribeiro-Santos R, de Melo NR, Andrade M, Azevedo G, Machado AV, Carvalho-Costa D, Sanches-Silva A. Whey protein active films incorporated with a blend of essential oils: characterization and effectiveness. Packag Technol Sci. 2018;31:27–40. https://doi.org/10.1002/pts.2352

Riva, S.C., Opara, U.O., Fawole, O.A., 2020. Recent developments on postharvest application of edible coatings on stone fruit: a review. Sci Hortic (Amsterdam). 262, 109074. https://doi.org/https://doi.org/10.1016/j.scienta.2019.109074.

Robards K. Strategies for the determination of bioactive phenols in plants, fruit and vegetables. J Chromatogr A. 2003;1000:657–91. https://doi.org/10.1016/S0021-9673(03)00058-X

Rojas-Bravo M, Rojas-Zenteno EG, Hernández-Carranza P, Ávila-Sosa R, Aguilar-Sánchez R, Ruiz-López II, Ochoa-Velasco CE. A potential application of mango (Mangifera indica L. cv Manila) peel powder to increase the total phenolic compounds and antioxidant capacity of edible films and coatings. Food Bioprocess Technol. 2019;12:1584–92. https://doi.org/10.1007/s11947-019-02317-8

Sabaghi M, Maghsoudlou Y, Khomeiri M, Ziaiifar AM. Active edible coating from chitosan incorporating green tea extract as an antioxidant and antifungal on fresh walnut kernel. Postharvest Biol Technol. 2015;110:224–8. https://doi.org/10.1016/j.postharvbio.2015.08.025

Saber El-Sisi A. Use of chitosan as an edible coating in RAS cheese. Biolife. 2015;3:564–70. https://doi.org/10.17812/blj2015.32.32

Sáez MI, Suárez MD, Martínez TF. Effects of alginate coating enriched with tannins on shelf life of cultured rainbow trout (Oncorhynchus mykiss) fillets. LWT. 2020;118:108767. https://doi.org/10.1016/j.lwt.2019.108767

Sahraee S, Milani JM, Regenstein JM, Kafil HS. Protection of foods against oxidative deterioration using edible films and coatings: a review. Food Biosci. 2019;32. https://doi.org/10.1016/j.fbio.2019.100451

Saini RK, Nile SH, Park SW. Carotenoids from fruits and vegetables: chemistry, analysis, occurrence, bioavailability and biological activities. Food Res Int. 2015;76:735–50. https://doi.org/10.1016/j.foodres.2015.07.047

Salgado PR, Ortiz CM, Musso YS, Di Giorgio L, Mauri AN. Edible films and coatings containing bioactives. Curr Opin Food Sci. 2015;5:86–92. https://doi.org/10.1016/j.cofs.2015.09.004

Sanches-Silva A, Albuquerque TG, Finglas P, Ribeiro T, Valente A, Vasilopoulou E, Trichopoulou A, Alexieva I, Boyko N, Costea C-E, Hayran O, Jorjadze M, Kaprelyants L, Karpenko D, Filippo D'Antuono L, Costa HS. Carotenoids, vitamins (A, B 2, C and E) and total folate of traditional foods from Black Sea Area countries. J Sci Food Agric. 2013;93:3545–57. https://doi.org/10.1002/jsfa.6243

Sanches-Silva A, Costa D, Albuquerque TG, Buonocorc GG, Ramos F, Castilho MC, Machado AV, Costa HS. Trends in the use of natural antioxidants in active food packaging: a review. Food Addit Contam Part A. 2014;31:374–95. https://doi.org/10.1080/19440049.2013.879215

Sánchez-Camargo A d P, Gutiérrez L-F, Vargas SM, Martinez-Correa HA, Parada-Alfonso F, Narváez-Cuenca C-E. Valorisation of mango peel: proximate composition, supercritical fluid extraction of carotenoids, and application as an antioxidant additive for an edible oil. J Supercrit Fluids. 2019;152:104574. https://doi.org/10.1016/j.supflu.2019.104574

Sapper M, Chiralt A. Starch-based coatings for preservation of fruits and vegetables. Coatings. 2018;8:152. https://doi.org/10.3390/coatings8050152

Sapper M, Palou L, Pérez-Gago MB, Chiralt A. Antifungal Starch–Gellan edible coatings with thyme essential oil for the postharvest preservation of apple and persimmon. Coatings. 2019;9:333. https://doi.org/10.3390/coatings9050333

Sartoratto A, Machado ALM, Delarmelina C, Figueira GM, Duarte MCT, Rehder VLG. Composition and antimicrobial activity of essential oils from aromatic plants used in Brazil. Brazilian J Microbiol. 2004;35:275–80. https://doi.org/10.1590/S1517-83822004000300001

Sessa M, Ferrari G, Donsì F. Novel edible coating containing essential oil nanoemulsions to prolong the shelf life of vegetable products. Chem Eng Trans. 2015;43:55–60. https://doi.org/10.3303/CET1543010

Shahidi F. Antioxidants in food and food antioxidants. Nahrung/Food. 2000;44:158–63. https://doi.org/10.1002/1521-3803(20000501)44:3<158::AID-FOOD158>3.0.CO;2-L

Sharaf Eddin A, Tahergorabi R. Efficacy of sweet potato starch-based coating to improve quality and safety of hen eggs during storage. Coatings. 2019;9:205. https://doi.org/10.3390/coatings9030205

Sharaf Eddin A, Ibrahim SA, Tahergorabi R. Egg quality and safety with an overview of edible coating application for egg preservation. Food Chem. 2019;296:29–39. https://doi.org/10.1016/j.foodchem.2019.05.182

Sharifian S, Shabanpour B, Taheri A, Kordjazi M. Effect of phlorotannins on melanosis and quality changes of Pacific white shrimp (Litopenaeus vannamei) during iced storage. Food Chem. 2019;298:124980. https://doi.org/10.1016/j.foodchem.2019.124980

Siracusa V, Rocculi P, Romani S, Rosa MD. Biodegradable polymers for food packaging: a review. Trends Food Sci Technol. 2008; https://doi.org/10.1016/j.tifs.2008.07.003

Skurtys O, Acevedo C, Pedreschi F, Enronoe J, Osorio F, Aguilera JM. Food hydrocolloid edible films and coatings. Nova Science Publishers; 2011. p. 1–66.

Soleimani-Rambod A, Zomorodi S, Naghizadeh Raeisi S, Khosrowshahi Asl A, Shahidi S-A. The effect of xanthan gum and flaxseed mucilage as edible coatings in Cheddar cheese during ripening. Coatings. 2018;8:80. https://doi.org/10.3390/coatings8020080

Souza V, Pires JRA, Vieira ÉT, Coelhoso IM, Duarte MP, Fernando AL. Shelf life assessment of fresh poultry meat packaged in novel bionanocomposite of chitosan/montmorillonite incorporated with ginger essential oil. Coatings. 2018a;8:177. https://doi.org/10.3390/coatings8050177

Souza V, Rodrigues PF, Duarte MP, Fernando AL. Antioxidant migration studies in chitosan films incorporated with plant extracts. J Renew Mater. 2018b;6. https://doi.org/10.7569/jrm.2018.634104

Souza V, Pires JRA, Vieira ÉT, Coelhoso IM, Duarte MP, Fernando AL. Activity of chitosan-montmorillonite bionanocomposites incorporated with rosemary essential oil: from in vitro assays to application in fresh poultry meat. Food Hydrocoll. 2019a;89:241–52. https://doi.org/10.1016/j.foodhyd.2018.10.049

Souza V, Rodrigues C, Ferreira L, Pires JRA, Duarte MP, Coelhoso I, Fernando AL. In vitro bioactivity of novel chitosan bionanocomposites incorporated with different essential oils. Ind Crop Prod. 2019b;140:111563. https://doi.org/10.1016/j.indcrop.2019.111563

Souza V, Rodrigues C, Valente S, Pimenta C, Pires JRA, Alves MM, Santos CF, Coelhoso IM, Fernando AL. Eco-friendly ZnO/chitosan bionanocomposites films for packaging of fresh poultry meat. Coatings. 2020a;10:110. https://doi.org/10.3390/coatings10020110

Souza VGL, Pires JRA, Rodrigues C, Coelhoso IM, Fernando AL. Chitosan composites in packaging industry-current trends and future challenges. Polymers (Basel). 2020b; https://doi.org/10.3390/polym12020417

Sucheta, Chaturvedi K, Sharma N, Yadav SK. Composite edible coatings from commercial pectin, corn flour and beetroot powder minimize post-harvest decay, reduces ripening and improves sensory liking of tomatoes. Int J Biol Macromol. 2019;133:284–93. https://doi.org/10.1016/j.ijbiomac.2019.04.132

Suput D, Lazic V, Popovic S, Hromis N. Edible films and coatings: sources, properties and application. Food Feed Res. 2015;42:11–22. https://doi.org/10.5937/ffr1501011s

Treviño-Garza MZ, García S, Heredia N, Alanís-Guzmán MG, Arévalo-Niño K. Layer-by-layer edible coatings based on mucilages, pullulan and chitosan and its effect on quality and preservation of fresh-cut pineapple (Ananas comosus). Postharvest Biol Technol. 2017;128:63–75. https://doi.org/10.1016/j.postharvbio.2017.01.007

Umaraw P, Munekata PES, Verma AK, Barba FJ, Singh VP, Kumar P, Lorenzo JM. Edible films/coating with tailored properties for active packaging of meat, fish and derived products. Trends Food Sci Technol. 2020;98:10–24. https://doi.org/10.1016/j.tifs.2020.01.032

Valdés A, Ramos M, Beltrán A, Jiménez A, Garrigós MC. State of the art of antimicrobial edible coatings for food packaging applications. Coatings. 2017;7:56. https://doi.org/10.3390/coatings7040056

Valencia-Chamorro SA, Palou L, Delŕio MA, Pérez-Gago MB. Antimicrobial edible films and coatings for fresh and minimally processed fruits and vegetables: a review. Crit Rev Food Sci Nutr. 2011;51:872–900. https://doi.org/10.1080/10408398.2010.485705

Verlee A, Mincke S, Stevens CV. Recent developments in antibacterial and antifungal chitosan and its derivatives. Carbohydr Polym. 2017;164:268–83. https://doi.org/10.1016/j.carbpol.2017.02.001

Vital ACP, Guerrero A, Monteschio JDO, Valero MV, Carvalho CB, de Abreu Filho BA, Madrona GS, do Prado IN. Effect of edible and active coating (with rosemary and oregano essential oils) on beef characteristics and consumer acceptability. PLoS One. 2016;11:e0160535. https://doi.org/10.1371/journal.pone.0160535

Vital ACP, Guerrero A, Kempinski EMBC, Monteschio J d O, Sary C, Ramos TR, Campo M d M, Prado IN d. Consumer profile and acceptability of cooked beef steaks with edible and active coating containing oregano and rosemary essential oils. Meat Sci. 2018;143:153–8. https://doi.org/10.1016/j.meatsci.2018.04.035

Viuda-Martos M, Fernández-Lóaez J, Pérez-álvarez JA. Pomegranate and its many functional components as related to human health: a review. Compr Rev Food Sci Food Saf. 2010;9:635–54. https://doi.org/10.1111/j.1541-4337.2010.00131.x

Volpe M, Coccia E, Siano F, Di Stasio M, Paolucci M. Rapid evaluation methods for quality of trout (Oncorhynchus mykiss) fresh fillet preserved in an active edible coating. Foods. 2019;8:113. https://doi.org/10.3390/foods8040113

Vunain E, Mishra AK, Mamba BB. Fundamentals of chitosan for medical applications. In: Jennings JA, Bumgardner JD, editors. Chitosan based biomaterials, volume 1: Fundamentals. Woodhead Publishing Limited; 2017. p. 3–30.

Walzem RL, Dillard CJ, German JB. Whey components: millennia of evolution create functionalities for mammalian nutrition: what we know and what we may be overlooking. Crit Rev Food Sci Nutr. 2002;42:353–75. https://doi.org/10.1080/10408690290825574

Wang Q, Yu H, Tian B, Jiang B, Xu J, Li D, Feng Z, Liu C. Novel edible coating with antioxidant and antimicrobial activities based on whey protein isolate nanofibrils and carvacrol and its application on fresh-cut cheese. Coatings. 2019;9:583. https://doi.org/10.3390/coatings9090583

Wu C, Li Y, Wang L, Hu Y, Chen J, Liu D, Ye X. Efficacy of chitosan-gallic acid coating on shelf life extension of refrigerated Pacific mackerel fillets. Food Bioprocess Technol. 2016;9:675–85. https://doi.org/10.1007/s11947-015-1659-9

Wu D, Wan J, Lu J, Wang X, Zhong S, Schwarz P, Chen B, Rao J. Chitosan coatings on lecithin stabilized emulsions inhibit mycotoxin production by Fusarium pathogens. Food Control. 2018a;92:276–85. https://doi.org/10.1016/j.foodcont.2018.05.009

Wu T, Ge Y, Li Y, Xiang Y, Jiang Y, Hu Y. Quality enhancement of large yellow croaker treated with edible coatings based on chitosan and lysozyme. Int J Biol Macromol. 2018b;120:1072–9. https://doi.org/10.1016/j.ijbiomac.2018.08.188

Xing Y, Li X, Xu Q, Yun J, Lu Y. Antifungal activities of cinnamon oil against Rhizopus nigricans, Aspergillus flavus and Penicillium expansum in vitro and in vivo fruit test. Int J Food Sci Technol. 2010;45:1837–42. https://doi.org/10.1111/j.1365-2621.2010.02342.x

Xiong Y, Chen M, Warner RD, Fang Z. Incorporating nisin and grape seed extract in chitosan-gelatine edible coating and its effect on cold storage of fresh pork. Food Control. 2020;110:107018. https://doi.org/10.1016/j.foodcont.2019.107018

Yamauchi K, Tomita M, Giehl TJ, Ellison RT. Antibacterial activity of lactoferrin and a pepsin-derived lactoferrin peptide fragment. Infect Immun. 1993;61:719–28.

Yan J, Luo Z, Ban Z, Lu H, Li D, Yang D, Aghdam MS, Li L. The effect of the layer-by-layer (LBL) edible coating on strawberry quality and metabolites during storage. Postharvest Biol Technol. 2019;147:29–38. https://doi.org/10.1016/j.postharvbio.2018.09.002

Yemiş GP, Candoğan K. Antibacterial activity of soy edible coatings incorporated with thyme and oregano essential oils on beef against pathogenic bacteria. Food Sci Biotechnol. 2017;26:1113–21. https://doi.org/10.1007/s10068-017-0136-9

Zambonelli A, Zechini D'Aulerio A, Bianchi A, Albasini A. Effects of essential oils on phytopathogenic fungi in vitro. J Phytopathol. 1996;144:491–4.

Zani F, Massimo G, Benvenuti S, Bianchi A, Albasini A, Melegari M, Vampa G, Bellotti A, Mazza P. Studies on the genotoxic properties of essential oils with *Bacillus subtilis rec* -assay and *Salmonella* /microsome reversion assay. Planta Med. 1991;57:237–41. https://doi.org/10.1055/s-2006-960081

Zaragozá MC, López D, Sáiz MP, Poquet M, Pérez J, Puig-Parellada P, Màrmol F, Simonetti P, Gardana C, Lerat Y, Burtin P, Inisan C, Rousseau I, Besnard M, Mitjavila MT. Toxicity and antioxidant activity in vitro and in vivo of two Fucus vesiculosus extracts. J Agric Food Chem. 2008;56:7773–80. https://doi.org/10.1021/jf8007053

Zheng W, Wang SY. Antioxidant activity and phenolic compounds in selected herbs. J Agric Food Chem. 2001;49:5165–70. https://doi.org/10.1021/jf010697n

Part IV
Application of Releasing Active Packaging in Different Food Categories

Chapter 11
Application of Releasing Systems in Active Packaging of Meat Products

Mariana A. Andrade, Cássia H. Barbosa, Regiane Ribeiro Santos, Fernanda Vilarinho, Ana Sanches Silva, and Fernando Ramos

Abstract In the human diet, meat is an important source of protein, vitamin B12, omega-3 fatty acids and highly bioavailable iron. According to the Food and Agriculture Organization, around 20% of the world's meat production is lost or wasted, every year. In industrialized countries, especially in the United States and in European countries, these losses mainly occur at the end of the food supply chain, at the retail and consumer levels. The meat and meat products with lower shelf-life periods are usually fresh products with few or no additives at all. These products must be packaged with materials that could delay their natural degradation as much

M. A. Andrade (✉)
Department of Food and Nutrition, National Institute of Health Dr. Ricardo Jorge, I.P., Lisbon, Portugal

Faculty of Pharmacy, University of Coimbra, Coimbra, Portugal

REQUIMTE/LAQV, Oporto, Portugal
e-mail: mariana.andrade@insa.min-saude.pt

C. H. Barbosa · F. Vilarinho
Department of Food and Nutrition, National Institute of Health Dr. Ricardo Jorge, I.P., Lisbon, Portugal

R. R. Santos
Federal Institute of Education, Science and Technology of Pernambuco – IFPE campus Vitória de Santo Antão, Vitória de Santo Antão, PE, Brazil

A. S. Silva
National Institute for Agrarian and Veterinary Research (INIAV), I.P., Vila do Conde, Portugal

Center for Study in Animal Science (CECA), ICETA, University of Oporto, Oporto, Portugal

F. Ramos
Faculty of Pharmacy, University of Coimbra, Coimbra, Portugal

REQUIMTE/LAQV, Oporto, Portugal

S. M. Jafari, A. S. Silva (eds.), *Releasing Systems in Active Food Packaging*, Food Bioactive Ingredients, https://doi.org/10.1007/978-3-030-90299-5_11

as possible. Active packaging, both releasing/emission and absorption types, are relatively new technologies that could provide longer shelf-life periods for meat and meat products. On the other hand, most of the meat derived products are highly processed, with high levels of synthetic and natural food additives which, according to latest research studies, may represent a concern for human health. Releasing active packages appear as a solution for both cases. In a releasing active package, an active compound with biological activity (e.g. antimicrobial, antioxidant) is incorporated in the package polymer and gradually migrates to the food' surface where performs its action (e.g. inhibiting microorganisms or delaying the lipid oxidation). In this chapter, a consistent review on the application of releasing active systems to meat products will be carried out, comparing and discussing their main components, methods of production, properties and effectiveness.

Keywords Meat · Food loss · Active food packaging · Antimicrobial · Antioxidant · Releasing active systems · Shelf-life

1 Introduction

The concept of "meat" is applied to the flesh of consumed animals and often widened to the animal's musculature, organs and other edible tissues. Meat is originated from several mammalian species which can differ depending on the culture, country or continent (Lawrie and Ledward 2006). Regarding the evolution of human diet, about 5% of the diet of primates was composed by insects, eggs and small animals (Milton 1993; Eaton et al. 2002). According to paleo dental evidence, meat consumption by humankind can be traced back to 2.5 million years ago. The size of molar teeth and the robustness of the mandibular and cranium reduced, the shape of incisor teeth altered, jaws and front teeth became stronger. These alterations suggest an increase of more soft foods which requires more tearing and less grinding from the dental apparatus, which suggest the inclusion of meat in the human diet (Eaton et al. 2002; Pereira and Vicente 2013).

With the introduction of meat consumption in the humankind diet, the intake of animal origin protein and fat increased. This increase is directly related to the calorie intake rise which, in turn, is directly related to the stature increase from the australopithecines to the *Homo habilis* and *H. erectus* (Eaton et al. 2002). The current brain chemistry also supports the human consumption of meat. Brain's communication network is essentially composed by lipids (60%) which, in turn, are rich in long chain fatty acids, namely docosahexaenoic (DHA) (22:6n:3) and arachidonic (AA) (20:4n:6) fatty acids (Pereira and Vicente 2013). DHA and AA, as well as eicosapentaenoic acid (EPA, 20:5n-3) and other polyunsaturated fat acids (PUFA), perform important roles in human health, such as gene expression modulation and the regulation of the physical properties of membranes (Guesnet and Alessandri 2011). PUFA and the respective derivatives can be mainly obtained from animal sources, namely meat and fish (Anderson et al. 1990; Guesnet and Alessandri 2011; Pereira and Vicente 2013). Besides meat, there are also several derived meat products consumed by humans, such as chorizo, salami, sausages, patties, among others. In general, these products are produced using specific parts of meat carcasses, blood, fat, gut lining, and aromatic plants, salt, spices and wine.

Unlike meat, that has a short shelf-life, meat products usually undergo through several processes to increase their shelf-life and/or several additives are added to prevent spoilage and to prevent changes in color, flavor, taste and appearance of the final product. The added additives to these products are, mainly, antioxidant and antimicrobial agents from synthetic origin. The consumption of these additives has been questioned due to their still unknown long-term effects in human health and their association to carcinogenic promotion, allergies and development of neurodegenerative diseases (Sanches-Silva et al. 2007; Pereira de Abreu et al. 2010). Also, the consumer's concern for the presence of these compounds in food products are increasing, which leads for a growing demand for more natural products (Bearth et al. 2014). The application of active packaging can be a viable way to mitigate this problem. In a product packaged with an active releasing packaging, the additive can be incorporated in the package polymer being gradually released to the surface of the packaged food preventing food' spoilage and decreasing the additive' concentration in the food itself.

This chapter will focus on several releasing active packaging systems that were developed to be applied to meat and meat products. The legislation related with these systems will also be addressed.

2 Meat

The presence of meat in human diet has been increasing. The consumption of meat is related to wealth and socioeconomic status of the consumers (McAfee et al. 2010). According to Ritchie and Roser (2020), in 2017 the world's meat production was 333.59 million tones, being Asia the continent with the highest production (141.93 million tons) followed by America (Central, South and North America, 103.12 million tons), Europe, Africa and Oceania. Regarding the meat consumption, USA and Australia present the highest values per person, 124.1 kg and 121.6 kg, respectively. On the contrary, India and Bangladesh present the lowest values of meat consumption per person, 3.78 kg and 4.04 kg, respectively. Regarding the choice of the meat type, poultry is the most consumed and produced meat followed by pork, beef and buffalo (Ritchie and Roser 2020).

2.1 Composition and Lipidic Content

Meat and meat by-products are a known source of protein, fat, iron, zinc and vitamin B12. Table 11.1 compiles the composition of some common cuts of meat and meat products, according to the United States Department of Agriculture (USDA)

Table 11.1 Composition of different meats and meat products in 100 g, according to the USDA Food Database (2020)

Meat form	Protein (g)	Total fat (g)	Iron (mg)	Zinc (mg)	Vitamin B12 (µg)	Saturated fat (g)	Monounsaturated fat (g)	Polyunsaturated fat (g)	EPA (g)	DHA (g)
Chicken breast, skinless and boneless, raw	22.5	2.62	0.37	0.68	0.21	0.563	0.689	0.424	0.003	0.004
Chicken breast, meat and skin, raw	20.85	0.25	0.74	0.8	0.34	2.66	3.82	1.96	0.01	0.02
Chicken leg, skinless and boneless, raw	19.16	4.22	0.78	1.76	0.57	1.05	1.44	0.962	0.004	0.011
Chicken leg, meat and skin, raw	16.37	15.95	0.69	1.47	0.56	4.366	6.619	3.352	0.004	0.01
Chicken wing, skinless and boneless, raw	21.97	3.54	0.88	1.63	0.38	0.94	0.85	0.8	0.01	0.05
Chicken wing, meat and skin, raw	17.52	12.85	0.46	1.21	0.25	3.535	5.422	2.498	0.003	0.006
Chicken giblets, raw	17.89	9.21	5.93	3.01	10.83	2.63	2.75	2.1	0	0
Chicken heart, raw	15.55	9.33	5.96	6.59	7.29	2.66	2.37	2.71	0	0
Chicken liver, raw	16.92	4.83	8.99	2.67	16.58	1.563	1.249	1.306	0	0
Chicken gizzard, raw	17.66	2.06	2.49	2.72	1.21	0.529	0.512	0.357	0	0
Ground beef, raw	17.44	19.07	1.97	4.23	2.15	7.291	8.48	0.505	0	0
Ground beef, grass-fed, raw	19.42	12.73	1.99	4.55	1.97	5.335	4.8	0.532	0.001	0
Beef tripe, raw	12.07	3.69	0.59	1.42	1.39	1.291	1.533	0.18	0	0
Beef tongue, raw	14.9	16.09	2.95	2.87	3.79	7	7.24	0.9	0	0
Beef pastrami, cured	21.8	5.82	2.22	4.98	1.87	2.681	2.118	0.145	0	0
Dried beef, cured	31.1	1.94	2.42	4.93	1.59	0.95	0.84	0.07	0	0
Beef sausage	15.5	37.57	1.53	2.92	2.03	15.098	16.387	1.025	0	0
Pork ground, raw	16.88	21.19	0.88	2.2	0.7	7.87	9.44	1.91	–	–
Pork carcass, raw	13.91	35.07	0.69	1.59	0.61	12.44	15.93	3.8	–	–

Pork spareribs, raw	15.47	23.4	0.91	2.5	0.38	7.529	8.542	3.953	0	0
Pork loin, raw	20.65	3.53	0.97	1.87	0.52	1.181	1.355	0.562	0	0
Pork belly, raw	9.34	53.01	0.52	1.02	0.84	19.33	24.7	5.65	0	0
Pork back ribs, raw	19.07	16.33	0.76	2.54	0.56	5.783	6.861	2.676	0	0.002
Pork feet, raw	23.16	12.59	0.58	0.76	0.52	3.57	6.289	1.092	0	0
Pork kidney, raw	16.46	3.25	4.89	2.75	8.49	1.04	1.07	0.26	0	0
Pork liver, raw	21.39	3.65	23.3	5.76	26	1.17	0.52	0.87	0	0.02
Pork tongue, raw	16.3	17.2	3.35	3.01	2.84	5.96	8.13	1.78	0	0
Pork, chorizo, raw	13.63	25.1	1.41	1.68	2	8.595	10.603	4.296	0.002	0.003
Italian pork salami	21.7	37	1.52	4.2	2.8	13.1	18.2	3.6	–	–
Pork ham, cured	5.05	80.5	0.44	0.9	0.29	29.38	37.94	9.4	0	0
Pork bacon, cured	13.66	37.13	0.38	1.14	0.5	12.615	15.922	5.757	0.004	0.005
Pork sausage	18.53	27.25	1.2	2.45	0.98	8.826	11.541	5.115	0.002	0.003
Italian pork sausage, mild, raw	13.9	24.26	1.77	1.91	1	8.615	11.024	4.386	0.002	0.003
Ground turkey, raw	19.66	7.66	1.09	2.35	1	2.024	2.635	2.205	0.006	0.008
Whole turkey, meat and skin, raw	21.64	5.64	0.86	1.78	1.22	1.461	1.826	1.466	0.002	0.003
Whole turkey, meat only, raw	22.64	1.93	0.86	1.84	1.24	0.459	0.477	0.411	0	0.002
Turkey giblets, raw	18.18	5.09	5.92	3.23	13.06	1.452	1.016	1.425	0.005	0.025
Turkey heart, raw	16.7	7.44	3.7	3.21	13.3	1.923	2.045	2.153	0.002	0.006
Turkey liver, raw	18.26	5.5	8.94	3.37	19.73	1.664	0.817	1.684	0.009	0.045
Turkey gizzard, raw	18.8	3.37	2.78	3.03	3.61	0.929	0.795	0.708	0	0.007
Turkey sausage, raw	18.79	8.08	1.17	3.06	1.30	1.96	2.62	2.39	0.00	0.00

Food Data Base (USDA 2020). According to this database, besides clams, liver from beef, lamb and veal present the highest content of vitamin B12 (96.00, 90.05 and 84.60 μg/100 g, respectively). Regarding the iron content, dried thyme, basil and spearmint lead the chart with 123.60, 89.80 and 87.47 mg/100 g. According to USDA database, the meat product with highest iron content is beef spleen with 44.55 mg of Fe/100 g, followed by lamb spleen (41.89 mg/100 g), duck liver (30.53 mg/100 g) and goose liver (30.53 mg/100 g). Looking for the zinc highly rich foods, mollusks lead the chart followed by ready-to-eat cereals (64.33 to 12.50 mg/100 g), peanut butter (15.10 to 14.40 mg/100 g), babyfood (wheat vanilla biscuits) (14.15 to 12.68 mg/100 g), cottonseed meal seeds (12.32 mg/100 g) and crude wheat germ (12.29 mg/100 g). The first raw meat form is veal liver with 12.02 mg/100 g of zinc. As expected, the food with the highest EPA content is menhaden fish oil (13.17 g/100 g) followed by other fish oils (salmon, sardine, cod). The meat with the highest EPA content, with 0.13 g/100 g, is lamb liver followed by beef liver with 0.11 g/100 g and lamb kidney with 0.09 g/100 g. Regarding the content of DHA, fish oil from salmon, cod and sardine lead the chart with 18.23, 10.97 and 10.66 g/100 g, respectively. The first raw meat entry is pork brain, with 0.45 g DHA/100 g, followed by raw lamb brain with 0.36 g DHA/100 g and lamb testes with 0.18 g DHA/100 g.

Lipids are very important for human health, presenting important roles in human body. In terms of energy, lipids (9 kcal/g) provide more than twice the energy provided by proteins and carbohydrates (4 kcal/g). They also provide a feeling of satiety and are very important in the palatability of foods. Nutritionally speaking, they provide the human body with essential fatty acids, such as linoleic and arachidonic acids, and fat-soluble vitamins (A, D, E, and K). Biologically speaking, lipids act as structural elements of cell walls, form a protective barrier to vital organs and are involved in the body's heat regulation (Nawar 1996; Schmid 2010). Generally, meat and meat products are considered to have a high lipidic content and associated with health problems such as type 2 diabetes, obesity and coronary heart disease. In contrast, meat and meat products' lipidic content is highly variable from specie to specie and from product to product, being the unsaturated fats the predominant lipids (Table 11.1) (Schmid 2010). Meat' fat content is directly linked to the breeding measures, the meat cut, environmental conditions to which the animal is exposed, composition and nutritional value of the feed and the age of the animal at the moment of its death (Schmid 2010). In meat and meat products. lipids are responsible for some desirable characteristics such as flavor and aroma profile, and contribute for the tenderness and juiciness of meat (Amaral et al. 2018).

As can be observed in Table 11.1, the protein content in meat and meat products, although high, can vary. The meat product with the lowest protein value is pork ham with 5.05 g/100 g and the meat with the lowest protein value is pork belly with 9.34 g/100 g. These values contrast with the protein values from pork feet, with 23.16 g/100 g and cured dried beef with 31.1 g/100 g. When comparing food protein contents, the Protein Digestibility-Corrected Amino Acid Scores (PDCAAS) must be considered. The PDCAAS is described and recommended by the Food and Agriculture Organization (FAO) and the World Health Organization (WHO), and is based on the combination of age-related scoring patterns, calculated through the age-related amino acid requirement levels, and the digestibility measurements, to calculate the level considered safe of a diet intake or the effective intake when compared with a reference protein (WHO et al. 2002; Hughes et al. 2011).

2.2 *Meat Products*

There are several types of meat products commercialized. Looking only at Portugal, a relatively small country, several of these products can be easily found in local markets such as *farinheira, alheira, morcela, morcela de arroz, morcela de assar, salpicão de vinhais, salsichão, chouriço, chourição, chouriço de ossos, paio, salpicão, salsichão de fígado, linguiça à moda de Valpaços, tripa enfarinhada, presunto* (smoked ham), *fiambre* (ham) and bacon. Meat by-products are also part of the gastronomic world of Portugal, such as cow's hand and tongue and, pork ear and feet (Rosa-Limpo et al. 2010). These products are rich in PUFA, being highly subjected to lipid oxidation.

2.3 *Meat Degradation and Shelf-Life*

Lipid oxidation is still one of the main reasons for food spoilage. This natural occurring phenomenon leads to the development of off flavors, nutritional loss, texture and color changes, and reduced shelf-life, resulting in the rejection of these foods by the consumers (Nawar 1996; Tian et al. 2013). It can also lead to the formation of toxic compounds, which their consumption is associated with liver hypertrophy, coronary artery disease, atherosclerosis, angiotoxicity and tumor frequency (Frankel 1984; Addis 1986; Esterbauer 1993). Lipid oxidation can occur by autoxidation and photochemical reactions.

Autoxidation occurs by free radical mechanisms and it comprises three phases: initiation, propagation and termination (Fig. 11.1). Briefly, the first phase starts

$$RH + ROS \rightarrow R\cdot \quad \text{Initiation (1)}$$

$$R\cdot + O_2 \rightarrow ROO\cdot \quad \text{Propagation (2)}$$

$$ROO\cdot + RH \rightarrow ROOH + R\cdot$$

$$R\cdot + R\cdot \rightarrow Nonradical\ products \quad \text{Termination (3)}$$

$$R\cdot + ROO\cdot \rightarrow Nonradical\ products$$

$$ROO\cdot + ROO\cdot \rightarrow Nonradical\ products$$

Fig. 11.1 Lipid oxidation phases adapted from Frankel (1980) and Nawar (1996) when reactive oxygen species reacts with lipids resulting in a fatty acid radical. In propagation, the fatty acid radical, a non-stable molecule, reacts with molecular oxygen creating a peroxyl-fatty acid radical which, in turn being also a non-stable molecule, reacts with other fatty acid resulting in fatty acid radical and a lipid peroxide. The second phase continues until there is no more radical species, resulting in the third phase, termination (Frankel 1980; Nawar 1996). The first originated lipid peroxyl radicals are colorless, odorless and tasteless but, being a non-stable molecule, they will originate lipid peroxides which are a complex mixture of lower molecular weight volatile and non-volatile compounds, characterized for having off-flavors and off-odors, normally associated with food rancidity (Tian et al. 2013; Surendran Nair et al. 2020).

Regarding meat and meat products, lipid oxidation depends on the meats' (or meat product) chemical composition, oxygen and light access, storage temperature and some technological procedures which meat products and meat are subjected during processing. The meat and meat products high content in lipids, especially unsaturated fatty acids, makes these products highly vulnerable to these reactions, which can significantly affect their shelf-life. Besides, the presence of metal ions as copper and iron can triggered lipid oxidation because they can donate electrons leading to increased rate of production of free radicals. In addition, muscles with higher content of myoglobin protein and presence of lipoxygenase enzyme are more susceptible to lipid oxidation (Amaral et al. 2018).

In order to overcome this issue, food industry resorts to the addition of antioxidant compounds to foods, usually synthetic ones. The main goal of these compounds is to 'capture' the free radicals before they interact with fatty acids. These antioxidants can also be from natural origin, extracted from plants, seaweeds and fruits, in the form of extracts and essential oils.

In meat and meat products, spoilage can occur by the contamination with microorganisms, besides lipid oxidation. According to the WHO, foodborne diseases affects, every year, 600 million people and are responsible for the death of 420,000 people (WHO 2020). *Listeria* sp., *Escherichia coli*, *Salmonella* and *Staphylococcus aureus* are some of the most common bacteria present in meat and meat products. Several synthetic and natural antimicrobial agents are directly added to meat and meat products in order to assure the safety of these products. However, the safety levels of synthetic additives is constantly being updated to ensure consumer safety. A good alternative to reduce the consumption of synthetic additives is the use of natural antimicrobial agents and/or to incorporate the antimicrobial compounds in the packaging matrix. This way, the surface of the food is not contaminated with microorganisms and the concentration of the additive in the food can be reduced.

3 Legislation

3.1 Meat Legislation

To ensure food safety and quality, countries adopt several measures and laws, which may be specific or general for all food products. The WHO in cooperation with the FAO created the Codex Alimentarius, which purpose is to promote and guide the elaboration of foods definitions and requirements to assist in their harmonization and, consequently, to facilitate international trade (FAO and WHO 2020).

The European Union (EU) established several measures and legislation in this context. The EU food laws aim to achieve a high level of human and animal life protection and health and the protection of consumers' interests. These objectives should be accomplished through fair practices in food trade and the protection of the environment, animal welfare and health, and plant health,. Also, the European Commission (EC) has established an Animal Health policy that aims, besides others, to raise the health status and improve the conditions of the food-producing animals in the EU (EC 2002a, 2004a, 2020).

Meat and meat products are under several EU rules regarding food hygiene that cover all stages of the production, processing, distribution, and market placement. According to the Annex I of Regulation No. 853/2004 (EC 2004a), meat is defined by the edible parts, including blood, of poultry, domestic ungulates, lagomorphs, wild game, small wild game, farmed game, and large wild game. Still, and according to the Regulation No. 853/2004 (EC 2004a), fresh meat refers to the "*meat that has not undergone any preserving process other than chilling, freezing, or quick-freezing, including meat that is vacuum-wrapped or wrapped in a controlled atmosphere*". Offal is the fresh meat besides the carcass, including viscera and blood. While viscera are "*the organs of the thoracic, abdominal, and pelvic cavities, as well as the trachea and esophagus and, in birds, the crop*". Minced or ground meat refers to the "*boned meat that has been minced into fragments and contains less than 1% salt*". Meat preparations are known as fresh meat, including meat fragments, which has had foodstuffs, additives, or seasonings added to it or which has undergone insufficient processes to modify the internal fiber muscle structure of the meat and thus eliminating the characteristics of fresh meat. Finally, meat based products means "*processed products resulting from the processing of meat or the further processing of such processed products, so that the cut surface shows that the product no longer has the characteristics of fresh meat*" (EC 2004a). These products are under the Regulation No. 178/2002 (EC 2002a) that establishes the general principles and requirements of food law.

Also, the Regulation No. 852/2004 (EC 2004b), on the hygiene of foodstuffs, where the EU determined several rules whereas food business operators are responsible for all stages of the production, processing, and distribution of foodstuffs under their control, ensuring the compliance of the hygiene requirements, through the implementation of procedures, such as the hazard analysis and critical control point (HACCP) principles. Besides, all food businesses must comply with the

Regulation No. 853/2004 (EC 2004a), which sets hygiene rules for animal origin foods and, Regulation No. 854/2004 (EC 2004c), which specific rules for the official control organizations on products of animal origin intended for human consumption.

To ensure meat quality and safety, the EC enforce measures during animal production, ensuring animal welfare traits. The Council Directive 2002/99/EC (EC 2002b) presents the animal health rules governing the production, processing, distribution, and introduction of animal origin products for human consumption. Also, the EU establish microbiologic criteria, through Commission Regulation No. 2073/2005 (EC 2005), to provide objectives and reference points to assist competent authorities and businesses in their activities to monitor and manage the safety of foodstuffs.

In the USA, the food processing sector is not only regulated by federal agencies, but also by local and state agencies. The meat and poultry sector is regulated by the USDA specifically the Food Safety and Inspection Service (FSIS). Together they are responsible for ensuring the safety, wholesomeness, and accurate labeling of meat (FSIS 2013; USDA 2020). FSIS is responsible to ensure that the meat and poultry are within the food safety requirements and that the facilities are under HACCP principles. The US has also its codex program, based on the Codex Alimentarius which motivates stakeholders in advancing science-based international food standards to protect the consumers' health and ensure fair practices in the food trade (FSIS 2013; United States Department of Agriculture (USDA) 2020).

3.2 Food Additives Legislation

According to the WHO, food additives are substances that are added to food to maintain or improve its safety, taste, freshness,, appearance and/or texture. WHO, in cooperation with the FAO, created an international and independent scientific expert group – the Joint FAO/WHO Expert Committee on Food Additives (JECFA). JECFA is responsible for assessing the risks of food additives to human health. After the JECFA proves that the additive poses no risk to human health, national authorities can authorize the use of food additives at specified levels for specific foods, either based on the JECFA assessment or a national assessment (WHO 2018).

The EU regulates food additives through the Regulation No. 1333/2008, and respective amendments. The Regulation EC 1333/2008 defines food additive as a substance not normally consumed as a food itself and not normally used as a food' characteristic ingredient. Whether or not it has nutritive value, the intentional addition of additives to foods for a technological purpose in the stage of preparation, manufacture, treatment, processing, packaging, transport, or storage results, or can be reasonably expected to result, in it or its derivatives become, directly or indirectly, a component of such foods. The food additives can be used for several technological objectives such as colors, preservatives, and antioxidants. In the EU, all food additives are identified by an E number which is mandatory to include in the ingredient's list of the food in which they are used (EC 2008).

The safety assessment of food additives in the EU is assured by the Scientific Committee on Food and/or the European Food Safety Authority (EFSA) (EC 2008, 2011a). The annex II of the Regulation No. 1333/2008 list all food additives that are authorized in foodstuffs and their conditions of use. In the case of fresh meat, since it is an unprocessed food, no additive can be added, and no food color additive can be added in meat, poultry, and game as well as their preparations. The Regulation No. 1333/2008 divided meat and its products in eight groups, and specified the additives that are allowed, including the authorized maximum level, and the restrictions and/or exceptions that may occur (EC 2008).

In the USA, food additives are under the FDA regulation. According to Section 201(s) of the Federal Food, Drug, and Cosmetic (FD&C) Act, a food additive is "*any substance the intended use of which results or may reasonably be expected to result, directly or indirectly, in its becoming a component or otherwise affecting the characteristics of any food*". These substances can be added in the production, treatment, processing, packaging, transportation, or storage of food. This regulation does not include the substances that are generally recognized as safe (GRAS), such as sugar and salt, nor sanctioned before 1958 (FDA 2010; Office of the Law Revision Counsel of the United States House of Representatives 2020). According to the USA regulation, the FDA is the authority responsible for the approval of a food additive. The process is started through a petition that requires evidences that the new substance is safe under the conditions for which it will be applied. In the case of food additives used in meat and poultry products, the FDA consults with the USDA during the process of approval (FDA 2010).

3.3 Food Packaging Legislation

The EU regulates food contact materials through Commission Regulation No. 1935/2004. This Regulation states that materials should not release or transmit their constituents into foods at levels harmful to human health and should not change food composition, taste and odor in an unacceptable way (EC 2004d). The regulation is applied for active and intelligent materials and articles, adhesives, ceramics, cork, rubbers, glass, ion-exchange resins, metals and alloys, paper and board, plastics, printing inks, regenerated cellulose, silicones, textiles, varnishes and coatings, waxes, and wood. The Commission Regulation No. 1935/2004 defines general and specific measures for the groups of materials and articles above mentioned, general requirements for the authorization of substances, labeling and traceability (EC 2004d).

Active and intelligent materials are also regulated by the Commission Regulation No. 450/2009. Active materials and articles are "*materials and articles that are intended to extend the shelf-life or to maintain or improve the condition of packaged food; they are designed to deliberately incorporate components that would release or absorb substances into or from the packaged food or the environment surrounding the food*". On the other hand, intelligent materials and articles are "*materials and articles which monitor the condition of packaged food or the environment surrounding the food*" (EC 2004d, 2009). The EC states that only substances that are included in the Community list of authorized substances may be used in components of active and intelligent materials and articles. It also refers to the conditions for the inclusion of substances in the Community list and established additional rules on labeling (EC 2004d, 2009).

The Commission Regulation No. 10/2011 is a specific measure within the meaning of Article 5 of Regulation No. 1935/2004, that refers to the specific measures for groups of materials and articles, and establishes specific requirements for the manufacture and marketing of plastic materials and articles (EC 2004d, 2011b). Plastic materials and articles are "*materials and articles and parts thereof consisting exclusively of plastics; plastic multi-layer materials and articles held together by adhesives or by other means; or those printed and/or covered by a coating; plastic layers or plastic coatings, forming gaskets in caps and closures, that together with those caps and closures compose a set of two or more layers of different types of materials; and plastic layers in multi-material multi-layer materials and articles*" (EC 2011b). The Commission Regulation No. 10/2011 and its amendments regulate the specifications of plastic materials and articles should fulfill to be included in the union list of authorized substances and the overall and specific migration limits (EC 2011b).

The US regulation for food contact substances is under the section 409 of the FD&C Act which defines it as "*any substance that is intended for use as a component of materials used in manufacturing, packing, packaging, transporting, or holding food if such use of the substance is not intended to have any technical effect in such food*" (FDA 2017, 2018a). The FDA distinguishes food contact substances from food contact materials and food contact articles. Food contact material is made with the food contact substance and (usually) other substances and it is often (but not necessarily) a mixture, such as an antioxidant in a polymer. Food contact article is the finished film, bottle, dough hook, tray, or whatever that is formed out of the food contact material (FDA 2018b). According to the FDA, it is the responsibility of the manufacturer of a food contact substance to ensure that food contact materials comply with the specifications and limitations in all applicable authorizations (FDA 2017, 2018a).

4 Additives Applied Directly and Indirectly to Meats

Following the current legislation, as specified in section 3.2, the use of food additives must be safe and highly justified (EC 2008). They can occur in foods as the result of direct or indirect addition. Direct additives are substances added to foods to provide a desired effect, i.e. for a specific purpose. Whereas indirect additives are substances that migrate from the package into foods, becoming part of it due to its packaging, storage or other handling. Examples of indirect additives include compounds from packaging materials such as butylated hydroxytoluene (BHT). When added in the packaging, as coating or packaging additives, it is possible to reduce or eliminate the addition of large amounts of food additives that are usually integrated in foods (Ribeiro-Santos et al. 2017).

Food additives can be natural, (e.g. from plants, animals, or minerals), or synthetic. In response to recent claims that synthetic additives have the potential to cause adverse health effects and consumers' increased interest in purchasing natural products, food, pharmaceutical and cosmetic industries together with the scientific community, are seeking sources of natural additives. Thus, several studies have directed their research in the search for safe natural food additives for consumption (Polônio and Peres 2009; Shah et al. 2014).

4.1 Antioxidant Additives

Antioxidants are used to minimize the oxidative changes in food, prevent lipid rancidity, off-flavors, and to stabilize color. Regarding the particular case of meat, oxidative changes may have negative effects in the meat' quality, causing modification in their sensory properties (color, texture and flavor), and reduction of nutritional value (Shah et al. 2014). The lipid oxidation in meat is the most common chemical deterioration since lipids are one of the most chemically unstable food components that participate in oxidative reactions (Shah et al. 2014).

Antioxidant additives perform their function either as free-radical scavengers or chelators of pro-oxidant metal ions. These compounds and substances are capable of donating hydrogen radicals to the available free-radicals preventing oxidative damages, delaying or inhibiting lipid oxidation and rancidity emergence, without affecting sensory and nutritional properties (Larry Branen and Haggerty 2001).

Synthetics antioxidant additives as butylated hydroxyanisole (BHA), BHT, ethylenediamine tetraacetic (EDTA), propyl gallate (PG) and tert-butylhydroquinone (TBHQ), have been widely used in meat. They are aromatic rings that can donate hydrogen radicals to a free radical and, consequently, stop the oxidative process by forming a more stable compound (Brewer 2011).

In recent years, the processing of foods with the partial and/or total substitution of synthetic by natural additives has been increasing, with the intention of obtaining a healthier product with less risks for Human Health. Phenolic compounds are one of the major groups of natural antioxidants found in nature, among which are α-tocopherols (vitamin E), flavonoids, and phenolic acids, which can be found in plant parts (like leaves, roots, stems, fruits, seeds and bark) and extracted through different techniques (Doolaege et al. 2012; Shah et al. 2014; Fernandes et al. 2015).

For instance, Rodrigues et al. (2020) obtained an extract from male flowers (part of banana inflorescences) with a high content of phenolics, namely flavonoids, and, consequently, with a high antioxidant capacity. When applied in sausage formulation (concentration of 2%), stored under refrigeration for 28 days, they showed major potential to be used as a natural antioxidant (Rodrigues et al. 2020). In another study, rosemary extract was used in the production of a cooked meat storage for 0, 30 and 60 days, with partial replacement of animal fat by alkylglycerols rich oil. In this case, the antioxidant capacity of the extract was measured by determining the carnosic acid (major bioactive compound in the rosemary extract) content at the end of each storage time (Martin et al. 2017).

Other plant extracts have been used as natural antioxidants in meat products like extracts from thyme, green tea, oregano, lychee, marjoram and lemon balm (Fernandes et al. 2015; He and Gramza-michałowska 2016). Lychee seed water extract (LSWE) antioxidative properties in raw meat paste, during a storage period of 0–15 days, were evaluated, as well as, the sensory properties of the cooked meat paste supplemented with different amounts of LSWE (Qi et al. 2015). LSWE was effective in retarding lipid oxidation of the meat paste and it significantly improved the sensory properties of the meat paste during the late stage of the storage period (Qi et al. 2015).

Other natural antioxidants extracted from plants are essential oils. In the study carried out by Fratianni et al. (2010), fresh slices of chicken breasts were dipped, for 15 min, in sterile agar solution, with or without 0.5% of thyme and balm essential oils. The slices were then kept at 4 °C, for a 3-week period storage. The treatment with the 2 essential oils effectively limited lipid oxidation, reducing the deterioration of chicken meat and extend the shelf-life of the fresh product (Fratianni et al. 2010).

Natural antioxidants as tocopherols and polysaccharides also show potential antioxidant capacity in meat. Gadekar et al. (2014) evaluated the effect of sodium ascorbate and α-tocopherol acetate on the quality of a restructured goat meat product. Natural antioxidants significantly reduced lipid oxidation and the value of free fatty acids. Besides, sensory attributes such as appearance, color and taste were enhanced due to the use of the antioxidants. The results of the study revealed that the restructured goat meat product can be safely stored for 20 days at 4 °C with the addition of these antioxidants (Gadekar et al. 2014). In the study performed by Hamed et al. (2020) polysaccharides from pistachio (*Pistacia vera* L.) external hull were isolated. The results of thin layer chromatography showed that the carbohydrate fraction was mainly composed of rhamnose, glucose, galactose, mannose, xylose, arabinose, and galacturonic acid. Findings showed that these crude polysaccharides might have the ability to exhibit strong antioxidant capacities. Thus, the application of these polysaccharides on minced meat, reduced the lipid oxidation during chilled storage for 9 days and, also, showed significant improvement of meat color stability (Hamed et al. 2020).

Aiming of reducing the use of synthetic antioxidants, a study was carried out by Doolaege et al. (2012) who investigated the oxidative stability during the preparation of liver pate with different doses of rosemary extract and sodium nitrite (E249) before and after the cooking process, before and after exposure to light and air for 48 h at 4 °C. The addition of rosemary extract had a clear positive effect in delaying lipid oxidation and in maintaining higher levels of other antioxidants (ascorbic acid, α-tocopherol and carnosic acid). In addition, the dose of sodium nitrite added to the liver pâté was reduced when rosemary extract was added, without negative effects on lipid oxidation and color stability (Doolaege et al. 2012). In another study, several plants extracts were used to replace the synthetic antioxidant, sodium erythorbate (E316). Among the 13 plant extracts evaluated by Fernandes et al. (2015), the most promising were the extracts obtained from oregano, marjoram, lemon balm and rosemary. Therefore, the authors used these extracts in the production of lamb burgers as substitutes of the sodium erythorbate without compromising the sensory acceptance of this meat product.

Comparing synthetic and natural antioxidants, Waters et al. (2018) revealed that sorghum bran varieties delayed lipid oxidation with equal efficacy to BHA/BHT, and significantly more effective than rosemary extract in cooked pork bran and chicken patties, without negatively affecting sensory attributes. In another study performed by Yu et al. (2015), the antioxidative effect of apple phenolics extract (chlorogenic acid, phlorizin and phloretin) on lipid oxidation in Chinese-style sausage was compared with BHT and the natural antioxidant ursolic acid. At the optimum addition level, apple phenolics extract (0.5 g/kg in total fat) was more effective at inhibiting meat's lipid oxidation than BHT (0.15 g/kg in total fat) and ursolic acid (0.5 g/kg in total fat) for 120 days storage (Yu et al. 2015). Amin and Edris (2017), showed that the addition of grape seed extract significantly retarded the oxidative rancidity of minced beef, during refrigerated storage (4 °C, for 10 days) than BHT. This study introduces grape seeds extract, a natural agro-waste, as an effective alternative in extending meat's shelf-life and delaying lipid oxidation of meat without affecting sensory attributes (Amin and Edris 2017). Júnior et al. (2019), compared the effect of the direct addition of a synthetic antioxidant and curcumin microcrystals to mortadella. The addition of microcrystals of curcumin in mortadella produced a more yellowish color that negatively affected the food. However, the addition of this natural antioxidant (curcumin) can prevent lipid oxidation in meat during storage more efficiently than with synthetic antioxidant (composition: sugar; sodium isoarcorbate and citric acid) (Júnior et al. 2019).

4.2 *Antimicrobials*

Antimicrobial agents are added during meat processing to prolong their shelf-life and to control foodborne pathogens such as *Salmonella* spp., *E. coli*, *Campylobacter* spp., and *Listeria* spp. A variety of natural and synthetic antimicrobial agents are used in meat products to prevent microbial contamination against bacteria, yeast and molds, ensuring the quality and safety of meat. Synthetic antimicrobials are mostly considered safe, but several have negative and side effects that can cause serious health hazards (Molognoni et al. 2019), among theses, includes: organic acids and their derivatives, nitrites, nitrate, bacteriocins, and sulfites (Surendran Nair et al. 2020). On the other hand, studies with natural preservatives such as nisin, chitosan, plant extracts and essential oils, lysozymes and others, have increased (Surendran Nair et al. 2020).

Nitrates and nitrites are widely used preservatives, commonly used in the cured meat production, usually in the form of sodium nitrite ($NaNO_2$) and potassium nitrate (KNO_2). Between the two, nitrate is the most stable ion, although can be easily transformed in nitrite through microbial action, being usually used as a nitrite "reservation" (Dennis et al. 1990; Omar et al. 2012). These compounds are used as food additives due to their antimicrobial action, namely against *Clostridium botulinum*, and to provide cured meats, such as bacon or ham, with the pink-reddish color that the consumers are familiar with and love (Dennis et al. 1990; Govari and Pexara 2018). Nitrates and nitrites have distinct behaviors inside the human body. Nitrates are immediately absorbed by the upper gastrointestinal tract (mouth, pharynx, esophagus and stomach) and 5 to 10% of the ingested nitrates can be reduced to nitrites by the bacteria present in the tongue, stomach and intestine (Dennis et al. 1990; Hmelak Gorenjak and Cencič 2013; Carvalho 2018). On the other hand, nitrites, when they reach the stomach, can originate N-nitroso compounds which are carcinogenic (Omar et al. 2012; Hmelak Gorenjak and Cencič 2013; Bryan and Ivy 2015; Carvalho 2018). Due to health hazard, surveys from various countries indicate a decline in the nitrites content in meat products over the last years (Govari and Pexara 2018). In this context, it must be ensured that additive' residual contents do not exceed the established legislation limits. This provides microbiological safety to the final product and preserves the consumers' health (Trentini and Macedo 2019).

Accordingly, lysozyme has been used as an effective nitrite replacer in an Italian-type chicken sausage as reported by Abeyrathne (2015). This study focused on the evaluation of the antimicrobial effects of lysozyme extracted from egg white as a replacer of nitrite in cured meat products. It was concluded that the mixture of 50% nitrite +50% lysozyme was as effective as the control (100% nitrite) in suppressing the growth of *E. coli* and *Salmonella*. In addition, no negative effects were observed in color changes, and sensory characteristics were improved in the Italian type chicken sausage. Ozaki et al. (2020) tried to replace the use of nitrite, with radish powder (0.5%) and chitosan (0.25%, and 0.5%) in fermented cooked sausages. The meat was prepared and evaluated against aerobic mesophyll bacteria during the ripening process and storage time. Pure chitosan exhibited *in vitro* antimicrobial activity against *Enterobacter aerogenes*, *Listeria innocua* and *Lactobacillus rhamnosus*. Chitosan (0.5%) with radish powder (0.5%) showed promising results indicating that the products (despite the absence of nitrite) were safe for consumption. With the exception of the aroma, sensory attributes were affected by the addition of chitosan (Ozaki et al. 2020).

Bacteriocins, such as nisin, are antimicrobial peptides produced by lactic acid bacteria (LAB) such as *Lactococcus*, *Enterococcus*, *Pediococcus*, and *Lactobacillus* that inhibit the growing of pathogenic and/or deteriorating bacteria (da Costa et al. 2019). Organic acids as citric, acetic, lactic and tartaric acids, individually or in combination can result in effective shelf-life extension of meat and meat products. Dipping solutions containing organic acids (lactic and acetic acids), bacteriocins produced by *Lactobacillus curvatus* and *Lactobacillus sakei*, and nisin were tested alone and in combination against *L. monocytogenes* on vacuum-packaged frankfurters stored at 10 °C for 36 days. The organic acids mix reduced the pathogenic population during storage. Solutions containing bacteriocins prevented microbial growth, while nisin was not able to avoid its regrowth after 20 days. The combined addition of the solutions containing bacteriocins + organic acids mix was the most effective approach for pathogen reduction during refrigerated and vacuum-packaged storage (Castellano et al. 2018).

On the other hand, extracts of carvacrol and green tea showed to be weaker antimicrobial agents than sulfite, as revealed by the study of Bellés et al. (2019). The authors evaluated the use of different concentrations of carvacrol and green tea extracts and their combination in preserving lamb burger meat packaged aerobically and displayed for 8 days at 4 °C. The study tested the antimicrobial properties of extracts against total Aerobic viable counts, *Pseudomonas* spp., *Enterobacteriaceae*, lactic acid bacteria and *Brochotrix thermosphacta*, as well as, compared them with sulfite (400 ppm). It was concluded that sulfite provided a higher color stability and lower microbial counts than both natural compounds. However, carvacrol seems to be a capable alternative to replace sulfite in lamb burger meat, whereas green tea should be combined with an antimicrobial agent (Bellés et al. 2019).

Studies with cinnamon, rosemary, oregano, thyme and others have shown positive results for their capacity to act as antimicrobial in meat (Amariei et al. 2016; Ribeiro-Santos et al. 2018). Cegiełka et al. (2019) studied the effect of sage (*Salvia officinalis* L.) preparations on the storage stability of vacuum-packed low-pressure mechanically separated meat from chickens stored at −18 °C for 9 months. An aqueous extract, ethanolic extracts, and an essential oil were prepared and added to the meat. The growth of mesophilic aerobic bacteria and psychrotrophic bacteria, coliforms, *enterococci* and *enterobacteriaceae* was significantly restricted by all tested sage preparations, especially sage oil and ethanol extracts (Cegiełka et al. 2019).

Andrés et al. (2017), evaluated the effect of the addition of the aqueous extracts obtained from tomato, red grape, olive and pomegranate into lamb meat patties. In general, microbial counts were reduced by the addition of extracts, mainly grape and olive pomaces (Andrés et al. 2017). In another study, the combination of fresh herbs (*Coriandrum sativum* L. and *Rosmarinus officinalis* L.) with supercritical carbon dioxide treatment, did not increase the inactivation of either *E. coli* or natural flora (total mesophilic bacteria, yeasts, and molds) of raw chicken meat. However, the use of 0.5% v/w pure coriander essential oil, instead of the fresh herbs, showed increased inhibition. The same was not observed for the rosemary essential oil (González-Alonso et al. 2020). Thymol, cinnamaldehyde, allyl isothiocyanate, citric acid, ascorbic acid, rosemary extract, and grapefruit seed extract were tested against the spoilage bacteria: *Lactobacillus algidus*, *Leuconostoc mesenteroides*, *Leuconostoc carnosum*, *Carnobacterium maltaromaticum*, *Carnobacterium divergens*, *Brochothrix thermosphacta*, and *Serratia proteamaculans*. Single and combined antimicrobials were added to vacuum-packed pork meat to evaluate preserving effects. Although antimicrobial concentrations have showed *in vitro* activity, the same result was not observed when applied in meat at the same concentrations (Schirmer and Langsrud 2010). Badawy et al. (2020), applied chitosan nanoparticles loaded with four monoterpenes (limonene, linalool, menthol and thymol) for preservation of minced meat. *In vitro*, *E. coli* was more susceptible than *S. typhimurium* to these products. Among chitosan nanoparticles loaded monoterpenes, the chitosan nanoparticles loaded with limonene exhibited the highest *in vivo* antimicrobial activity for the minced meat samples during refrigerated storage.

Sharma et al. (2020) studied blends of essential oils as bio-preservatives to increase the shelf-life of emulsion based ready-to-eat chicken sausages, aerobically packaged and stored under refrigerated (4 °C) conditions. The microbiological analysis (total plate count, psychrophilic, yeast and mold and coliform count) was performed in the ready-to eat chicken sausages elaborated with four different treatments which were tested, separately: B-1 (clove oil: 40%, holy basil oil: 20%, cassia oil: 40%); B-2 (clove oil: 40%, holy basil oil: 20%, thyme oil: 40%); B-3 (clove oil: 30%, holy basil oil: 20%, cassia oil: 25%; thyme oil: 25%) and B-4 (clove oil: 25%, holy basil oil: 20%, cassia oil: 20%, thyme oil: 15%, ajowan oil: 10%, betel oil: 10%). The authors found that B-1, 2 and 3 (each at 0.25%) and B-4 (0.125%) enhanced the shelf-life of chicken sausages by 13–14 days, 16–17 days, 10–11 days and 6–7 days, respectively (Sharma et al. 2020).

In the study carried out by Amariei et al. (2016), the antimicrobial activity of three essential oils (thyme, rosemary and oregano), was assessed. The essential oils (0.5, 1.0 and 1.5%) were added to a mixture composed by raw minced pork and beef meat. The antimicrobial activity was monitored throughout 4 days. The results indicated that essential oils have a significant influence on the microbiological stability of meat, when compared to control samples (without oil). It was observed that the microorganism's quantity depends on the type and levels of essential oil. Besides, the pleasant taste and smell support their use as additives to prevent bacterial contamination (Amariei et al. 2016). Amentoflavone is a known bioflavonoid occurring in many natural plants. This polyphenolic compound isolated from *Nandina domestica* Thunb. significantly reduced the cell counts of *Staphylococcus aureus* and *Escherichia coli*, in minced chicken (Bajpai et al. 2019).

4.3 Coloring Agents

A color additive can be defined as a pigment or substance which, when added or applied to a food, is capable of add or restore color. They can be used in food for many purposes such as, to offset color loss that has been affected by processing (exposure to light, air, temperature extremes, moisture), storage conditions, packaging and distribution; to correct natural variations in color, making foods more visually appealing; or to provide color to colorless (EC 2008). Food coloring agents may act as emulsifying agents, stabilizers, sweeteners, antioxidants and preservatives (Shanmugasundaram et al. 2019). Coloring agents can represent a risk to human health; therefore, in consequence, there are many studies and researches for natural food coloring agents such as carotenoids, chlorophyll, and anthocyanins aiming to replace or reduce artificial coloring (Shanmugasundaram et al. 2019).

Nitrite, alone or in combination with sodium nitrate, act, among other functions, as a color fixative in cured meat and poultry products (bologna, hot dogs, bacon); however, as stated previously, nitrites have a health risk associated. In this line of though, Bolognesi and Garcia (2018) applied annatto carotenoid extracts as nitrite replacers in meat products due to their antioxidant properties and color capacity. Kim et al. (2019) conducted a study to improve the quality characteristics of pork loin cured with natural nitrite. Four treatments were tested on cured meat: nitrite free, cured meat marinated with sodium nitrite and ascorbic acid, cured meat marinated with only fermented spinach, and cured meat marinated with fermented spinach with the addition of ascorbic acid, malic acid, citric acid, and tartaric acid. Cured meat with fermented spinach presented higher redness values than sodium nitrite with ascorbic acid on cooked meat. Residual nitrite levels were lower in the presence of the added organic acids. Among various organic acid, ascorbic acid had the highest efficiency on the quality properties of cured meat (Kim et al. 2019). In the same way, Huang et al. (2020) provided a new no-added-nitrite cured meat and an innovative method for substituting nitrite with *Lactobacillus fermentum* and *Lactobacillus plantarum* as starters, beet red and Monascus color as coloring agents, and nisin as antibiotic. The results provided a good effect in the complete replace nitrite, thereby improving the quality and safety of cured meat. Thus, it is possible to ensure cured meat quality without or with low nitrite content (Huang et al. 2020).

Since organic acids have been used to improve safety and extend shelf-life by retarding food deterioration, Marcos et al. (2016) determined the effect of decanoic, malic, fumaric and octanoic acid treatments on sensory characteristic of ground beef. The authors observed that the decanoic, fumaric and octanoic acid treatments increased subjective redness, reduced discoloration. In addition, there was no differences in beef flavor between control (untreated with acids) and the rest of the treatments. The use of 3% solutions containing these organic acids on beef trimmings prior to grinding may improve or maintain sensory retail display properties without affecting beef flavor.

4.4 Other Additives

Flavor enhancers are substances which magnify, increase or intensify the existing taste and/or odor of a foodstuff (EC 2008). For example, monosodium glutamate (MSG) is a flavoring agent, basically used to develop the flavor of meat and meat based products which, at the appropriate level, could lead to consumer acceptability of the product. It comes from a common amino acid, glutamic acid. Although it is recognized as safe, several studies have questioned its due to its demonstrated negative health effects (Datta et al. 2019; Zanfirescu et al. 2019). Jůzl et al. (2019) examined differences in sensory evaluation of sliced cooked salami, manufactured according to several recipes. The monitoring factors were salt content (1.6% or 2.0%), presence (1.6% or 2.0%) or absence of MSG and age of the evaluators group. No significant negative result was found in the sensory evaluation; however, samples with MSG were rated as better, regardless of the age of the assessors. In their previous studies, Wang et al. (2019) found that consumers prefer to have MSG replaced by extracts from natural sources that are rich in umami substances, including mushroom extract, tomato extract, kelp extract, and yeast extract. Thus, the authors replaced MSG with natural umami substances in chicken soup to enhance flavor and reduce sodium chloride. It was observed that all the tested extracts (yeast, mushroom, tomato, and kelp) in different levels (0.05%, 0.1%, 0.2%, 0.4%; except yeast extract: 0.0125%, 0.025%, 0.05%, 0.1%) exhibited an enhancement effect on the overall flavor, meaty flavor, saltiness, and umami taste (Wang et al. 2019).

Although the presence of salt (sodium chloride) in high concentrations and nitrites discourage the intention of purchase of the consumer, at the same time, these additives are key factors for the taste, color and juiciness of the final product (Di Vita et al. 2019). As health strategies to reduce sodium levels (from sodium chloride) in processed foods, potassium and phosphorus from food additives may be added in meat and meat products. Sodium chloride acts to preserve and enhance the taste and texture of meat. It may be replaced by phosphorus (as potassium phosphate) and potassium (as potassium chloride, potassium lactate, and potassium phosphate) additives in sodium-reduced food (Parpia et al. 2018). The study performed by Parpia et al. (2018), showed that significantly high amounts of potassium, which confer a salty taste, are being added to meat and poultry products that can negatively influence consumers with pre-existent health conditions such as chronic kidney disease. The authors argue that the quantity of the added and total potassium should be a mandatory component of foods' labeling (Parpia et al. 2018).

Emulsifiers are substances added to foods to form or maintain a homogenous mixture of two or more immiscible phases, preventing the separation of the product components, ensuring consistency (EC 2008). Emulsified meat based products include frankfurters, chicken nuggets, bologna, spreadable sausage, surimi and other products. Actomyosin complex and, mainly, myosin protein are the most used emulsifiers, largely due to their high concentration and amphiphilic nature. The breakdown of the actomyosin complex indicates that myosin is released from the actomyosin complex and, predictably, can act as a natural emulsifier and participate in the gelation process, crucial for the fat and water stabilization in emulsified meat based products (Glorieux et al. 2017). Phosphates as food additives are used in many processed foods as stabilizers and emulsifiers. They maintain the moisture content, reduce the cooking loss and improve the textural properties and flavor protection. Furthermore, phosphates accelerate the formation of cured meat color as well as having antioxidant and antimicrobial effects (Şimşek and Kılıç 2017; Thangavelu et al. 2019). Phosphates may be added to meat products as sodium or potassium salts of phosphoric acid. However, an avoidable risk to health arises from the increased use of phosphates as food additives. Thereby, the use of an alternative natural additive, as a polysaccharide, with potential to substitute phosphates was discussed by Câmara et al. (2020). The authors evaluated the functional properties of chia (*Salvia hispanica* L.) mucilage in powder and gel format as a phosphate replacer in low-fat Bologna sausages. Chia gel provided a better emulsion stability and texture parameters (closer to the control) than chia powder. Chia mucilage gel at 2% in the total absence of phosphates and with 50% fat reduction was effective (except for the color attribute) and has proven to be a feasible strategy to substitute phosphate in low-fat Bologna sausages (Câmara et al. 2020).

Glorieux et al. (2017) also tested the possibility of phosphate reduction on cooked pork sausages, by testing the effect of seven different phosphate types (mono-, di- and trisodium phosphate; tetrasodium di- or pyrophosphate; sodium acid pyrophosphate; sodium tripolyphosphate; and sodium hexametaphosphate). The most promising phosphate type was obtained from tetrasodium pyrophosphate and sodium tripolyphosphate which caused an increase in pH, improved sausages' structural properties, presented the highest emulsion stability and lowest cooking loss and, had a very low effect on textural properties. The authors calculated, based on the structural properties results, that 0.06% of tetra sodium pyrophosphate is sufficient to obtain a quality product (Glorieux et al. 2017).

Humectants are substances added to foods to help retain moisture and improving food softness (European Commission 2008). Among several types of humectants, glycerol and sorbitol are usually used in meat products. In the study carried by Sorapukdee et al. (2016), glycerol and sorbitol, at the concentration of 0, 10, and 15%, were tested to develop a jerky product using spent hen meat. In addition, a roasting process was also applied to enhance quality of jerky product. All jerky samples showed microbial counts in undetectable levels. Moreover, the jerky containing 15% of glycerol showed better quality, indicated by low activity water, soft and springy texture than the sorbitol and control samples (no humectants added) (Sorapukdee et al. 2016). Regardless effect of humectant, a roasting process following the drying process could improve the jerky' quality by increasing color, appearance and intense flavor, leading to a positive sensory reception. Triyannanto and Lee (2015) tested the application of honey and rice syrup, natural humectants, in the replacement of sorbitol for the production of restructured duck jerky. Rice syrup (consisting of dextrin, maltose, maltotriose, and a minor amount of glucose) and honey, which is a natural humectant in a concentrated sugar form, were mixed separately, at 3%, 6%, and 10% (w/w) concentrations, with the marinating solution. The use of these humectants had a positive effect in the chemical properties of duck jerky, especially when used in higher concentrations. In addition, duck jerky samples treated with 10% honey showed the highest scores for the sensory parameters evaluated, while samples treated with rice syrup were comparable with those of samples treated with sorbitol. Therefore, honey shows a great potential to be used as a natural humectant in restricted duck jerky, replacing sorbitol (Triyannanto and Lee 2015). The addition of humectant, such as hydrocolloids, can improve the sensory properties, tenderness and decrease water activity of jerky. Effects of curing solution prepared using different combinations of humectants konjac and collagen was investigated in the study carried out by Kim et al. (2020). The authors observed that the use of combinations of collagen and konjac, 60/40 in duck jerky processing results in the best quality characteristics of the jerky.

The substances which are added to foods to thicken (increase the viscosity) or improve texture are commonly called binders. Gums, such as carrageenan, are commonly used as binders in meat due to their high hydration ability. Beyond them, the non-meat proteins as whey protein and soy protein also could be added as binders to stabilize moisture and fat and improve their moisture binding capacity in meat products (Jin et al. 2019). Thus, the combination of carrageenan and soy protein isolate was evaluated for the functional properties of chopped low-fat pork batters during heat-induced gelation by Gao et al. (2015).

Similarly, in other research lead by Zouari et al. (2012) the effect of whey powder, ι-carrageenan and fat ratio variation sand their interactions on sensory and texture properties of mechanically separated turkey meat sausages, was evaluated. Whey powder had a more notable influence than ι-carrageenan on all of the texture parameters. Sensory evaluation also indicated that whey powder increased the flavor, the firmness and the sliceability. Low-fat sausage processed with 8 g of whey powder/100 g sausage was the best evaluated (Zouari et al. 2012). The study carried out by Jin et al. (2019) tested different binders (isolated soy protein, sodium caseinate, egg white powder and pork plasma protein, at 0, 1, 1.5% concentration levels) on physicochemical and sensory properties of pork sausages. It was revealed that the sensory profiles with all treatments were indifferent, except for 1% pork plasma protein with a slightly lower overall acceptability score. The inclusion of binders lowered the pH values and cooking loss in pork sausages, besides not show any remarkable results in the textural properties. In addition, lower redness and higher yellowness values were found in pork sausages that did not contain binders (Jin et al. 2019).

5 Active Packaging

Due to the growing consumer's demand for nutritious, healthy and safe meat and meat products, along with the ever-increasing retailers demand for cost-effective technologies to extend products' shelf-life, the food packaging industry has evolved to meet and satisfy these expectations (Grunert et al. 2004). The main goals, when packaging fresh meat, are the delay of natural spoilage, avoid cross contamination, and allow some enzymatic activity to improve tenderness, decrease weight loss and preserve color and aroma. On the other hand, when it comes to package processed meat products, the main goals are to avoid or delay different phenomena, such as lipid oxidation, dehydration, loss of aroma and discoloration (Mondry 1996; Brody 1997). Meat quality and safety are highly dependent on the applied packaging materials and technologies.

Currently, several meat packaging systems are available, each of them with different characteristics and applicability. These include overwrap packaging for a short-term chilled storage; a variety of specific MAP systems for longer-term chilled storage and/or display; and a combination between vacuum packaging with MAP systems with 100% carbon dioxide for long-term chilled storage (Fang et al. 2017). Recently, different new packaging technologies and materials have been developed to ensure food safety and quality, extend shelf life, minimize environmental impact and highlight appealing aspects of the packaged product to retailers and consumers.

In this sense, and due to the diversity of characteristics of the product to be packaged and basic applications of meat packaging, a packaging technology, that can offer a better control over product quality in a more economically and diversified way, is desirable. Therefore, the interest in the use of active packaging systems for meat and meat products has increased in recent years. When a package, in addition to presenting an inert barrier to the external environment, also has any other functions, can be called active (Rooney 1995). Active packaging is a system that promotes positive interaction between the product, package and environment, extending food's shelf life and maintaining or even improving food's quality, safety or sensory properties (Miltz et al. 1995; Ahvenainen 2003). According to the Regulations No. 1935/2004/EC (EC 2004d) and No. 450/2009/EC (EC 2009), active packaging comprises packaging systems that "*deliberately incorporate components that would release or absorb substances into or from the packaged food or the environment surrounding the food*". Active packaging systems can be divided into two types: active scavenging systems (absorbers) and active-releasing systems (emitters) (Yildirim et al. 2018).

Although release packaging systems, such as antioxidant and antimicrobial active packaging, are the main goal of this chapter, oxygen scavengers, carbon dioxide and moisture scavengers will also be mentioned.

5.1 Oxygen Scavengers

High levels of oxygen in food packages can be an enabler for microbial growth, promote the off-flavors and off-odors development, color change, and nutritional losses. All these factors can induce significant reductions in the foods' shelf-life. In fresh meat, the presence of oxygen allows myoglobin oxygenation, which provides the characteristic red color (Kerry et al. 2006; Prasad and Kochhar 2014). High levels of oxygen promote the oxidation of muscle lipids, which has detrimental effects on the color of fresh meat (O'Grady et al. 1998). The decreasing of oxygen within the package helps to increase meat's shelf-life by preventing the growth of fungi and aerobic bacteria (Ozdemir and Floros 2004; Shin et al. 2009). Therefore, to limit the rate of these deteriorative and spoilage reactions in foods, it is important to control oxygen levels inside food packages.

One of the main active packaging technologies, which aims to remove any residual oxygen present in the food package, is the application of oxygen scavengers (OS) (Realini and Marcos 2014; Yildirim et al. 2018). The development of oxygen-scavenging systems includes the use of silica gel, natural clays (e.g., montmorillonite), calcium oxide, calcium chloride, molecular sieves, and modified starch, or other moisture-absorbing substances (Ozdemir and Floros 2004). The development of oxygen-scavenging systems mostly includes the use of self-adhesive labels or loose sachets in the packaging containers. Besides, the oxygen-scavenging systems are also based on the design of active substances to be included in the packaging material itself using monolayer or multilayer materials (Rooney 2005). Oxygen

Fig. 11.2 Principle of iron oxidation

$$Fe \rightarrow Fe^{2+} + 2e^-$$

$$\frac{1}{2}O_2 + H_2O + 2e^- \rightarrow 2OH^-$$

$$Fe^{2+} + 2OH^- \rightarrow Fe(OH)_2$$

$$Fe(OH)_2 + \frac{1}{4}O_2 + \frac{1}{2}H_2O \rightarrow Fe(OH)_3$$

scavenging technologies have been successfully used in the meat industry, alone or in combination with MAP (Kerry et al. 2006).

Commercially, it is more frequent to remove most of the atmospheric oxygen by MAP and then use a relatively small and inexpensive scavenger to eliminate the remaining oxygen within the food package, namely, the oxygen which permeates through the packaging film or is trapped within the meat or between meat slices (Kerry et al. 2006). Oxygen scavenging compounds react with oxygen to reduce the concentration of this inside the package (Kumar et al. 2018). Nowadays, the existing technologies use one or more of the following concepts: iron powder oxidation, ascorbic acid oxidation, photosensitive dye oxidation, enzymatic oxidation (e.g. alcohol oxidase and glucose oxidase), and immobilized yeast on a solid substrate (Floros et al. 1997; Vermeiren et al. 1999). Ferrous oxide is the most widely used scavenger. Currently, the majority of commercially available oxygen scavengers are based on the principle of iron oxidation (Fig. 11.2) (Smith et al. 1990).

The iron kept in the small sachet is oxidized to iron oxide. For the sachet to be effective, the sachet material is highly permeable to oxygen and, in some cases, to water vapor. The initial oxygen level in the package, the amount of dissolved oxygen present, the permeability of the packaging material, the water activity (size, shape, weight, etc.) and the nature of the food are several factors that determine the type and amount of absorbent that needs to be used in the sachet (Ozdemir and Floros 2004). This oxygen scavenging system was developed and introduced to the food packaging market by the Mitsubishi Gas Chemical Company, known as Ageless (Kerry et al. 2006).

5.2 *Carbon Dioxide Scavengers and Emitters*

To slow down the microbial growth on meat and poultry surfaces, high levels of CO_2 are desired (60 to 80%) to prolong the shelf-life of packaged food (Ozdemir and Floros 2004; Prasad and Kochhar 2014). Since the permeability of CO_2 is 3 to 5 times higher than that of O_2 in most plastic films, it must be continuously produced to keep the desired concentration within the package (Ozdemir and Floros 2004). Therefore, CO_2 generating system can be viewed as a technique complementary to oxygen scavenging by the impregnation of a packaging structure with a CO_2 generating system or the addition of the latter in the form of a sachet (Ha et al. 2001). These sachets can contain sodium hydroxide and calcium hydroxide or potassium hydroxide; or silica gel and calcium oxide, (Ahvenainen 2003). It should be noted that an oxygen-free environment alone is insufficient to retard the growth of *S. aureus*, *Vibrio* spp., *E. coli*, *Bacillus cereus*, and *Enterococcus faecalis* at room temperatures (Prasad and Kochhar 2014). Also, an O_2 and CO_2 absorber can inhibit the growth of *Clostridium sporogenes* (Scannell et al. 2000).

Carbon dioxide emitting sachets or labels can also be used. This innovating package consists of a standard MAP tray that has a porous sachet containing sodium bicarbonate/ascorbate under a perforated false bottom. So, CO_2 is emitted when juice exudates from the packaged meat drip onto the sachet, substituting any CO_2 absorbed by the meat and preventing package collapse (Kerry et al. 2006).

5.3 Moisture Scavenger (Absorbers)

To suppress microbial growth and prevent foggy film formation, it is important to lower the water activity of the product, which means control the excess moisture in food packages. The water accumulation inside the package is more pronounced if the package has a low permeability to water vapor (Ozdemir and Floros 2004; Kerry et al. 2006).

Several factors contribute to the excess water development inside a food package, such as temperature fluctuations in high relative humidity food packages, the respiration of a fresh product or leak of tissue fluid from cut meats and poultry (Rooney 1995). The accumulation of excess water inside the package promotes bacterial and mold growth, resulting in quality loss and shelf-life reductions (Ozdemir and Floros 2004).

The use of a moisture scavenger is a useful way of controlling excess water accumulation in a food package with a high barrier to water vapor. Moisture absorbents are commonly used as pads, sheets, or blankets for liquid water control in foods with a high water activity, like meat and fish (Kumar et al. 2018). They consist of two layers, a microporous nonwoven plastic film, like polyethylene or polypropylene, between superabsorbent polymers that is capable of absorbing moister up to 500 times its weight. Typical superabsorbent polymers include carboxymethyl cellulose (CMC), polyacrylate salts, and starch copolymers, due to their strong affinity to water. Moisture drip absorber pads are commonly placed under packaged fresh meats and fish to absorb unsighted tissue drip exudates (Kumar et al. 2018).

5.4 Antimicrobial Active Packaging

For the meat industry, antimicrobial packaging is one of the most important features to guarantee food quality once meat provides excellent nutrients for the potential growth of microorganisms. Therefore, to provide safe and healthy meat and meat products to consumers, special attention is needed to minimize microbial proliferation (Fang et al. 2017). There are spoilage microorganisms, such as bacteria, molds, yeast, and pathogenic microorganisms (such as *Salmonella* spp., *L. monocytogenes*, *S. aureus*, *Clostridium botulinum*, *Clostridium perfringens*, and *E. coli* O157: H7) leading to quality deterioration and food safety issues on meat (Jayasena and Jo 2013). In this sense, the use of antimicrobial packaging is a way to extend foods' shelf-life and to assure the food safety of meat and meat products. According to Appendini and Hotchkiss (2002) and Coma (2008), the antimicrobial packaging can be classified into four categories:

1. Inside the package, antimicrobial substances are incorporated into sachets/pads, which, during storage, are released (Otoni et al. 2016).
2. Direct incorporation of the antimicrobial compounds into the packaging film, by co-extrusion or by non-heating methods such as electrospinning, solvent compound, and casting. In both cases, the antimicrobial compounds will be gradually released from the packaging films to the packaging headspace or food surface (Sung et al. 2013).
3. Packaging coating with a matrix with a plastic film or any food derivative material, such as wax or polysaccharides. This matrix performs the transport of the antimicrobial agents, so they can be released onto the surface of food through evaporation into the headspace (volatile substances) or migration into the food (non-volatile substances) through diffusion (Fang et al. 2017).
4. The use of polymers, such as chitosan and poly-L-lysine, in food packaging and coatings, which are inherently antimicrobial. These polymers cause cellular death of microorganisms, once the charged amines of the polymers interact with negative charges on the microorganism cell membrane which causes leakage of intracellular constituents (Goldberg et al. 1990).

A large number of antimicrobial agents, including ethanol, CO_2, chlorine dioxide, silver ions, bacteriocins, antibiotics, organic acids, spices, and essential oils, are being used to inhibit the microorganisms growth in foods (Zhao et al. 2013). Table 11.2 summarizes several studies where different active packaging systems were applied. This Table compiles the changes in the physical, barrier and mechanical properties, as also the antimicrobial and/or antioxidant properties of the active film.

There is a wide variety of food packaging incorporated with antimicrobial active compounds. These indirect additives are responsible for the increasing or maintaining the quality of products. In the work of Konuk Takma and Korel (2019), active polyethylene films, assembled with antimicrobial chitosan and alginate coatings, incorporated with black cumin oil were used as packaging for chicken breast meat. These films improved the quality and shelf-life of chicken meat during refrigerated storage, exhibiting more effective against total aerobic mesophilic bacteria in comparison to psychrotrophic bacteria (Konuk Takma and Korel 2019). In another study, CMC – PVA films incorporated with citric acid as cross linked and aloe vera as the active component delayed bacterial growth of packaged minced chicken meat and thus, extended the shelf-life of product (Kanatt and Makwana 2020).

Table 11.2 Active Packaging Systems developed for meat and meat products

Polymers	Active compounds	Method	Physical, barrier and mechanical properties	Antimicrobial/ antioxidant properties	References
Starch + PVA	Nisin Z (2 to 8.70%wt.) + CNC (0 to 5%wt.) + MA (0 to 12.56%wt.)	The films were prepared by the dissolution of PVA into deionized water and stirred. Another solution prepared was plasticized corn starch (70% starch and 20% glycerol) and MA into deionized water. The solution was then mixed and stirred. CNC and NIS Z were added to the mixed solution and homogenized. The final suspension was poured into glass plates and let to dried.	The active film presented a significant improvement of the mechanical and barrier properties, with a decrease of the force to break. The compounds incorporated did not influence the swelling index and L^*, a^* and b^* color coordinates did not vary with the addition of MA, CNC, and NIS Z into starch-PVA matrices, although the b^* coordinate showed a slight yellowing in the films due to the high NIS Z addition, which is naturally yellow.	The interaction between the compounds in the active film improved the antimicrobial activity against *Listeria Monocytogenes*.	(de Oliveira et al. 2020)
LDPE	Carvacrol (10%wt.) + Nanomer I.28 (5%wt.)	The films were prepared by mixing the nanofiller and the matrix to the extruder whereas the carvacrol was added to the molten polymer using a downstream port to get antimicrobial blends at a nominal 10% content.	The addition of carvacrol to the film improved the barrier properties, through the reduction of the permeability. The mechanical properties are improved by the synergistic effects of the compounds. Carvacrol acts as a plasticizer and dispersing agent and I.28 increased strain at break value.	The antimicrobial activity of carvacrol was not affected by the presence of the Nanomer I.28, being all the indicator strains studied (*Brochotrix thermosphacta* 1R2; *Listeria innocua* ATCC 33090; *Carnobacterium* sp. 9P), sensitive to films containing carvacrol.	(Persico et al. 2009)

(continued)

Table 11.2 (continued)

Polymers	Active compounds	Method	Physical, barrier and mechanical properties	Antimicrobial/ antioxidant properties	References
CMC + PVA	Clove oil (1, 2 and 3%)	The CMC 1% solution was prepared by dissolving in distilled water at room temperature with stirring. The PVOH 5% solution was prepared in hot distilled water. The solutions of CMC and PVOH were mixed at various ratios (1:1, 1:2, and 2:1). Glycerol was added as the plasticizer in all the films. Clove oil was added to the film-forming solution at different concentrations (1%, 2%, and 3%), except the control. The film-forming mixtures were blended, and the films were prepared by the casting method. Finally, the dried films were peeled from the casting surface and preconditioned in a constant temperature humidity chamber set at 23 °C with 50% RH for 24 h.	The incorporation of clove oil into the CMC + PVA film did not affect the thickness of the films, although it has a slightly yellowish color. The addition of clove oil resulted in a decreased tensile strength and puncture force, but no significant changes in water vapor transmission rate and negligible oxygen transmission rate.	The active film showed higher antimicrobial activity against *Staphylococcus aureus* than *Bacillus cereus*, but no activity against gram-negative bacteria that was tested. The active film was also effective against reducing the natural microflora of ground chicken meat and was able to extend the shelf life of ground chicken meat during refrigerated storage.	(Muppalla et al. 2014)

LDPE	Ag (0 to 1% w/w), CuO (0 to 0.667% w/w), and ZnO (0 to 1% w/w) nanoparticles	The films were prepared through the method of extrusion, where NPs and LDPE pellets were put into a co-rotating twin-screw extruder. The LDPE and NPs were fed into the extruder from the feed hopper and the molten material was extruded as a string into a basin of cold water and cut into granules. Finally, the granules were added into another twin-screw extruder and the films obtained were chilled using a chilling roll system at room temperature.	The active films presented a homogenous distribution of NPs on the fracture surfaces and the NPs are not recrystallized into clusters after being incorporated in the LDPE matrix.	All active films presented a significant decrease of CFUs for both, *Escherichia coli* and *Staphylococcus aureus*. The containing all NPs were more effective in reducing *Escherichia coli* and *Staphylococcus aureus* compared to those containing single NP. Once again, films containing both ZnO and CuO and without Ag significantly decreased the number of colonies of both bacteria.	(Dehghani et al. 2019)
			Both tensile strength and elongation at break were improved by the incorporation of combinate NPs. LDPE-Ag presented a significant increase in tensile strength and the highest elongation at break values, which could contribute to the strength and flexibility of the films. On the other hand, films containing both ZnO and CuO and without Ag had significant effects on improving the mechanical properties of the film.		

(continued)

Table 11.2 (continued)

Polymers	Active compounds	Method	Physical, barrier and mechanical properties	Antimicrobial/ antioxidant properties	References
PP	Carvacrol (4, 6 and 8 wt.%) + thymol (4, 6 and 8 wt.%)	The different formulations were obtained by melt blending and both additives were introduced in the mixer once the polymer was already in the melt state. The films were obtained by compression-molding at 190 °C in a hot press and kept between the plates at atmospheric pressure for 5 min until melting and was successively pressed under 2 MPa for 1 min, 3.5 MPa for 1 min and finally 5 MPa for 5 min, to liberate the trapped air bubbles.	A slight modification of tensile properties and a significant decrease in elastic modulus was observed, in the result of a significant decrease in the crystallinity of the material. Also, an increase in oxygen transmission for the active film was observed.	The active films showed antioxidant activity trough differential scanning calorimetry by determining the oxidation induction parameters, i.e. oxidation onset temperature and oxidation induction time.	(Ramos et al. 2012)
			The addition of carvacrol and thymol to the polymer matrix did not significantly affect its thermal degradation profile in the inert nitrogen atmosphere.	The active film showed antimicrobial activity against *Staphylococcus aureus*, but low antimicrobial activity against *Escherichia coli*.	

LDPE + EVA	Clove leaf, Sweet basil, and Cinnamon bark oil (0.5, 2, and 4% wt.)	LDPE incorporated with 0%, 2.5%, 10%, 20%wt. EVA and neat EVA were melt-compounded in a co-rotating twin-screw extruder. After the extruded were removed from the twin-screw extruder and were ground to pellet. The same procedure was performed with the active compounds.	Grainy and porous surface was found on the active films, probably due to the evaporation of EOs.	The film incorporated with Cinnamon bark oil was found to be more effective against both microorganisms than the one incorporated with Clove leaf oil. The film with Sweet basil oil did not reveal enough antimicrobial activity.	(Wattananawinrat et al. 2014)
				The film with Clove leaf oil and Cinnamon bark oil was more active against *Escherichia coli* inhibition than *Staphylococcus aureus*.	

(continued)

Table 11.2 (continued)

Polymers	Active compounds	Method	Physical, barrier and mechanical properties	Antimicrobial/ antioxidant properties	References
LDPE + PP	Attapulgite and Allium sativum essence oil (1% w/w)	The formulations of bilayer films were precisely prepared. For the inner layer of the films, EVA (10% w/w) pellets were initially blended with active substances by using a hand whisk. Then, LDPE pellets (90% w/w) were added and mixed continually. The mixture was placed in the hopper of the inner layer of the film to manufacture the inner layer of bilayer films. Finally, the films were manufactured using a blown film extrusion process.	The active films have high transparency and no agglomeration phenomenon occurred, but the color of the films changed to yellow with the addition of AEO or AT+AEO. The incorporation of AEO and AT+AEO into the film did not affect the mechanical properties of the films and did not show significant superiority in barrier performances of water vapor and oxygen transmission rate.	AEO and AT delayed the growth of microbes on large yellow croaker, up to 9 days at 4 ± 1 °C. The yellow croakers with active films had significantly lower values of CFU of specific spoilage organisms than the control.	(Dong et al. 2019)
				The yellow croaker packaged by active films presented lower values of pH, total volatile basic nitrogen contents, K-value, and thiobarbituric acid reactive substances compared to the control group in the preservation of large yellow croaker, extending the shelf life by up to 5 days at 4 ± 1 °C.	

EVA	Anthocyanin (0 to 1% wt.); ZnO nanoparticles (0 to 1% wt.); rosemary extract (0 to 3% wt.) and modified montmorillonite (0 to 1% wt.)	EVA was dissolved in chloroform to obtain a 10 wt% solution. Then, ZnO and Fe-MMT nanoparticles were added to the above solution under ultrasonic agitation for 2 min. Required amounts of anthocyanin-MMT and rosemary extract were added to the resultant mixture subsequently. After all, mixtures were stirred for an extra 10 min to provide a homogeneous dispersion of materials inside the EVA solution. Finally, the prepared mixture was spread out on a smooth and cleaned glass plate and was dried at room temperature.	All active films presented a typical behavior of flexible plastic; films containing ZnO and modified montmorillonite exhibit higher tensile strength, and Young's modulus of these films show a similar trend.	Active films showed notable increases in antioxidant activity, compared to the control film, with a film containing anthocyanin and rosemary exhibiting significant improvement. All active films showed antibacterial activity against both *Escherichia coli* and *Staphylococcus Ureus*, and the maximum antibacterial activity belongs to the film containing rosemary extract.	(Eskandarabadi et al. 2019)

(continued)

Table 11.2 (continued)

Polymers	Active compounds	Method	Physical, barrier and mechanical properties	Antimicrobial/ antioxidant properties	References
PP	Sorbic acid (2, 4 and 6% w/w)	PP granules and SA powder were mixed well in a stainless-steel container. The mixture was introduced to a co-rotating twin-screw extruder. After the mixture of the materials melted and mixed inside the extruder as a result of the shear and pressure forces. The molten materials were left out of the extrusion in string form. They passed through cold water basin and were then cut into the granules. The granules were collected and added into another twin-screw extruder to produce the desired packaging film.	The incorporation of SA in PP matrix increases the tensile strength at the break but the elongation at break of the active films decreased by increasing the concentration of SA. Water vapor permeability was strongly influenced by the amount of SA, as it was increased significantly with higher concentrations of SA. The addition of SA to the PP matrix had no significant effect on L*, a*, and total color changes, but b*, yellowness index, and whiteness index were affected.	SA active films had a significant antimicrobial effect on *Escherichia coli* and *Staphylococcus aureus*. Increasing SA concentration in films increased the antimicrobial properties of the films. Though the inhibiting effect of SA against *Aspergillus niger* was noteworthy, the growth inhibition was more effective with the increase of the SA concentration.	(Fasihnia et al. 2018)
LDPE	Cu nanoparticles (0.5, 1, 1.5, 2, 2.5 and 3 wt. %)	The films were prepared by using solvent evaporation. For that, LDPE polymer was dissolved in xylene at a constant temperature 110 °C using an oil bath; Cu-NPs were dispersed in LDPE/xylene solution by temperature-controlled ultrasonic bath sonicator to achieve proper dispersion. Finally, the films were fabricated on glass plates and peeled off from glass plates.	Cu-NPs were uniformly dispersed on the film, an increase in melting peak temperature was observed as Cu-NPs increase in the active films, and the mechanical properties and water vapor permeability were improved.	The active film showed antimicrobial inhibition efficiency against *Escherichia coli* and *Staphylococcus aureus*.	(Lomate et al. 2018)

Whey protein isolate	Chitosan Nanofiber + Cinnamon essential oil (3–6% dry weight of WPI)	The casting method was used to prepare the films. First of all, WPI was dissolved in distilled water and mixture stirred for 2 h to obtain a homogeneous solution and heated at 80 °C for 30 min to denature the WPI, followed by the film solution cooled to the ambient temperature. The CNF was dissolved in distilled water and homogenized in the ultra-sound bath, followed by the homogenous solution of CNF added to denatured WPI solution and homogenized to obtain homogenous distribution. Then the glycerol was added to the obtained film solution and after good dispersion poured on the 15 cm plates and dried at a temperature of 25 °C.	The incorporation of CNF caused a significantly decreased water vapor permeability, while the addition of CEO caused an insignificant increase in the water vapor permeability value. The water solubility value of the WPI + CNF film decreased significantly, such as the active film were also significantly low. The WPI + CNF presented a significant increase of the tensile strength and Young's modulus and the strain to break value decreased significantly. For the active film with CEO, the tensile strength and Young's modulus decreased significantly and the strain to break value increased significantly. The active films represented the higher L* value, whiteness index and ΔE with the coincidentally lower a*, b* value and yellowness index, compared with WPI film.	The WPI and WPI-CNF film did not show any antimicrobial activity, against *Escherichia coli*, *Pseudomonas aeruginosa* and *Staphylococcus aureus*. On the other hand, the active films containing CEO presented antimicrobial activity against all tested bacteria (*Escherichia coli*, *Pseudomonas aeruginosa* and *Staphylococcus aureus*.	(Mohammadi et al. 2020)

(continued)

Table 11.2 (continued)

Polymers	Active compounds	Method	Physical, barrier and mechanical properties	Antimicrobial/ antioxidant properties	References
LDPE	Grapefruit seed extract (3%wt.) + melanin nanoparticle (0.1%wt.) + ZnO nanoparticles (3%wt.)	The extrusion blowing method was used to prepare the films. The masterbatch was prepared by mixing the active compounds with LDPE using Ca-stearate as the dispersing agent. The mixture was then extruded using a twin-screw extruder. The masterbatch compounds were blown into a film through a single screw extruder.	All the films were smooth-surfaced, homogeneous, and flexible, and the active film was dark gray due to Mel. The L* value of the active film decreased, the b* value and ΔE increased significantly and the a* value did not change significantly. Both the strength and the resiliency of the LDPE film were significantly reduced after incorporation of the GSE/Mel/ZnO NPs, so was the tensile strength and the elongation at break	The active film showed strong antibacterial activity against both bacteria, *Escherichia coli,* and *Listeria Monocytogenes.* The antimicrobial activity was mainly due to the presence of GES and ZnO NPs.	(Shankar et al. 2019)

PLA	MgO NPs (1, 2, 3, and 4% wt.)	The films were prepared using the solvent casting method. Dried PLA was dissolved in chloroform by vigorous stirring. The MgO NPs were dispersed separately in chloroform with the help of an ultrasonicator and then poured into the dissolved PLA solution and continued to stir. The mixture of MgO NPs suspended in dissolved PLA matrix was evenly spread at room temperature on a glass plate to allow the evaporation of chloroform to produce uniform thickness films, and let to dry during 2–3 days, and then were peeled out from the plate and stored in a dry place, until further use.	The mechanical properties of the film improved significantly by incorporation of MgO NPs, good compatibility of MgO NPs and PLA matrix. Also, the incorporation of MgO NPs leads to lower thermal stability, which is directly proportional to the amount of NP added. The oxygen barrier property was significantly improved by the incorporation of MgO NPs, but the water vapor permeability increased.	The active film showed significant improvement in the bactericidal properties, specifically against *Escherichia coli.*	(Swaroop and Shukla 2018)

(continued)

Table 11.2 (continued)

Polymers	Active compounds	Method	Physical, barrier and mechanical properties	Antimicrobial/ antioxidant properties	References
LDPE and PLA	Grapefruit seed extract + thermoplasticized starch	First, the TPS was prepared by mixing cornstarch with 20% wt. of glycerol and 40%wt. of GSE with a mixer. Then, the mixture was heated at 120 °C and cooled to room temperature by spreading in the room and powdered using a blender. The films were prepared by mixing the TPS and polymer. The LDPE based-films were prepared using an extrusion blowing method and the PLA based-films were prepared using an extrusion casting method.	All films were uniform and freestanding, and both active films presented a decrease in L* value and an increase in a* and b* value, and consequently an increase of ΔE. The LDPE film was very resilient with low strength, while the PLA film was tough and brittle. However, for both active films the tensile properties decreased significantly, and consequently so did the mechanical strength, flexibility, and stiffness. Water vapor permeability and water solubility value increased significantly for both active films, while water contact angle decreased significantly.	Both active films exhibited antimicrobial activity against *Escherichia coli* and *listeria Monocytogenes,* mainly due to the GSE incorporated into the TPS. However, generally, the GSE incorporated films exhibited stronger antibacterial activity against *Listeria Monocytogenes* than *Escherichia coli.*	(Wang and Rhim 2016)
PLA	CNC (1 and 3% wt.) and LNP (1 and 3% wt.)	The extrusion method was used to prepare the films. Various amounts of LNP and CNC were mixed with PLA. The films were manufactured by using a twin-screw micro-extruder.	All active films showed higher tensile strength and Young's modulus mean Values and remarkable improvements in toughness. The enhancement of the effect is more from cellulose nanocrystals than lignin nanoparticles.	The active films showed antibacterial activity with a reduction in the multiplication of the bacterial plant pathogen *Pseudomonas syringae* pv. *Tomato.*	(Yang et al. 2016)

LDPE	Curcumin (1, 2, 3, 5 and 7% wt.)	The extrusion method was used to prepare the films. The extrusion of the LDPE with and without Curcumin was carried out through a three-heating zone single-screw Rheoscam extruder. The extruded filaments were subsequently pelletized and formed pellets. Finally, the films were produced by compression molding.	The active films are transparent but have a typical orange color attributed to the curcumin filler. The active film presented higher thermal stability, an increase in the elastic modulus and a decreased in the elongation at break. The water vapor transmission rate and water vapor permeability are improved by the presence of curcumin in the matrix.	The active film presented an effective antioxidant scavenging activity against the 2,2-diphenyl-1-picrylhydrazyl free radicals (DPPH), due to the presence of curcumin.	(Zia et al. 2019)

*a** Redness value, *AEO Allium sativum* essential oil, *Ag* Silver, *AT* Attapulgite, *b** Yellowness value, *CEO* Cinnamon essential oil, *CFUs* Colony-forming unit, *CMC* Carboxymethyl cellulose, *CNC* Cellulose nanocrystals, *CNF* Chitosan nanofiber, *Cu* Copper, *CuO* Copper oxide, *EOs* Essential oils, *EVA* Ethylene vinyl acetate, *GSE* Grapefruit seed extract, *L** Lightness value, *LDPE* Low density polyethylene, *LNP* Lignin nanoparticles, *MA* Maleic anhydride, *Mel* Melanin nanoparticle, *MgO* Magnesium oxide, *NIS* Nisin Z, *NPs* Nanoparticles, *PLA* Polylactic acid, *PP* Poly propylene, *PVA* Poly vinyl alcohol, *SA* Sorbic acid, *TPS* Thermoplasticized starch, *WPI* Whey protein isolate, *ZnO* Zinc oxide, ΔE Total color changes

Clarke et al. (2017) applied two antimicrobials, sodium octanoate and auranta FV (a commercial antimicrobial composed of bioflavonoids, citric, malic, lactic, and caprylic acids) into polyethylene – polyamide packaging materials for coating elaboration. Then, these packaging materials were used to vacuum pack beef subprimal cuts and stored at 4 °C. Both films increased meat' shelf-life, however, sodium octanoate reduced microbial counts for all the tested bacteria (total viable counts, psychrotrophic bacteria, total anaerobic bacteria, Lactic acid bacteria, total coliforms and *E. coli*) (Clarke et al. 2017). Marcous et al. (2017) evaluated the antibacterial effects of low-density polyethylene (LDPE) films coated with titanium dioxide (TiO_2), zinc oxide (ZnO) and mixed TiO_2–ZnO nanoparticles. The authors concluded that the TiO_2-ZnO nanoparticle-coated LDPE films are not a suitable option to inhibit *E. coli* growth and reproduction. However, ZnO nanoparticle-coated LDPE films were identified to have improved the shelf-life and prevent *E. coli* growth in fresh calf minced meat (Marcous et al. 2017). In another study by Mulla et al. (2017), linear LDPE surface was chemically modified by chromic acid treatment and coated with clove essential oil forming, thus, an active packaging. Films exhibited high antimicrobial activity against *Salmonella typhimurium* and *L. monocytogenes* in a packed chicken sample, completely restricting their growth on the 5th day of storage and during the 21 days storage period under refrigerated (Mulla et al. 2017).

At this point, it is essential to alert for important aspects that must be considered in the use of antimicrobial packaging, namely the minimal impact that these packaging systems must have on the visual and sensory properties of the packaged product, because these greatly influence the choice of the consumer. Only in this way, the application of antimicrobial active packaging to meat and meat products will be effective.

5.5 *Antioxidant Active Packaging*

The high levels of oxygen in meat packaging increases, besides microbial growth and color changes, meats' lipid oxidation. Lipid oxidation leads to the development of rancidity and to the potential formation of toxic aldehydes due to the degradation of PUFAs (Gómez-Estaca et al. 2014). To overcome this major food industry issue, active packaging containing antioxidant compounds can be used for improving product quality and extending the shelf-life of meat and meat products, through the control of the oxygen level to which the product is exposed.

There are two different forms to apply the antioxidant active packaging system, by independent antioxidant devices, or by the incorporation of the antioxidant into the packaging material. In the first case, sachets, pads or labels, which contain oxygen scavengers, can be independent devices and added to the package (Sect. 4.1) (Gómez-Estaca et al. 2014). In the second case, the antioxidant active agent is incorporated in the polymeric matrix of the packaging with the aim of being released to the food or the headspace surrounding it (Fang et al. 2017). To select the manufacturing procedure, it is important to consider the type of polymer and the characteristics of the antioxidant agents. To produce an antioxidant packaging material, there are several ways to mix the antioxidant agent (or the reactive substances which produce the agent) with the packaging polymers, namely: (i) using coating technologies, wherein an appropriate solvent is used to dissolve the antioxidant agent and the packaging material polymer(s) together, and then this solution is applied to a substrate, (ii) using extrusion technologies, in which the polymer is melted and mixed with the antioxidant agent or (iii) immobilizing the antioxidant on the film surface (Ramos et al. 2016).

A new tendency in antioxidant active packaging for meat and meat products is reducing the use of synthetic additives, being replaced using natural antioxidants. The most common natural antioxidants are essential oils and plant extracts (e.g. rosemary, green tea, oregano), and tocopherols (Sanches-Silva et al. 2014; Barbosa-Pereira et al. 2014). Another alternative is to use the natural and synthetic antioxidants together, taking advantage of the possible synergic effect between the two components. Table 11.2 summarizes several studies in different active packaging systems, including antioxidant active packaging, applied to meat and meat products.

Many studies have been carried out in this direction with positive effects using different synthetic and natural antioxidants incorporated in films. For instance, antioxidant films exhibited a great effectiveness to protect the meat against oxidation, as showed in the study performed by Ribeiro-Santos et al. (2018) where a blend containing cinnamon (*Cinnamomum zeylanicum* L. and *Cinnamomum cassia* L.) and rosemary (*Rosmarinus officinalis* L.) essential oils was incorporated in a whey protein film aiming the reduction of salami lipid oxidation. The study showed that the active whey protein packaging can reduce the lipid oxidation phenomenon of meat products, extending their shelf-life while releasing natural antioxidants to their surface (Ribeiro-Santos et al. 2018). Alizadeh-Sani et al. (2020), applied a biopolymer-based cellulose nanofiber/whey protein matrix containing titanium dioxide particles (1% of TiO_2) and essential oil droplets (2% of rosemary oil), in lamb meat during storage which, in turn, increased its shelf-life from around 6 to 15 days (Alizadeh-Sani et al. 2020).

In another study, shelf-life extension of fresh minced meat was achieved by wrapping the fresh minced meat in a polyethylene film incorporated with an encapsulated green tea extract (Wrona et al. 2017). Gallego et al. (2020) evaluated the antioxidant potential of a freeze-dried tomato by-product in pork loin cubes. The authors found that the gelatin coating with the freeze-dried tomato by-product was able to slow the meat' lipid oxidation for 13 days. Also, the meat' quality properties, such as texture, hardness and elasticity was improved (Gallego et al. 2020). Andrade et al. (2019), evaluated the packaging of salami slices with a why protein coating incorporated with an ethanolic extract of rosemary (*Rosmarinus officinalis* L.) and stored at 5 °C for 90 days. The lipid oxidation evaluation showed that the salami slices packaged with the active film present less malonaldehyde content that the salami slices packaged with the control film (without the rosemary extract) (Andrade et al. 2019). In another study from the same authors, this rosemary extract showed antimicrobial activity against *L. monocytogenes*, *S. aureus* and *Clostridium perfringens*, which makes it an additive to be taken into account to be applied in meat and meat products (Andrade et al. 2018).

6 Conclusion

Meat and meat products are one of the major food categories consumed in the world. Their safety is assured, mainly, by synthetic food additives which are regulated by the European and FDA legislation. However, recent findings have questioned their use, as well as, their authorized concentration. Natural additives, such as fruit and plant extracts and essential oils, and their incorporation in food packaging to obtain releasing active systems, seems to be the answer for more natural and safer processed products. Essential oils present powerful antimicrobial activity against most of common microorganisms can be present in meat products, like *L. monocytogenes*, *E. coli* and *S. aureus*, presenting themselves as a safe and effective alternative to the use of synthetic additives, such as nitrites. Plant and fruit extracts, as well as their main bioactive compounds, present themselves as powerful antioxidants which can delay lipid oxidation. Their addition to the food packaging matrix and their gradual migration from the package to the packaged food can increase their antimicrobial and antioxidant activities.

New technologies and alternatives for food additives and active food packaging arise every day, which present a great challenge for the responsible authorities to test and develop legislation that enable and assure the safety of these new alternatives.

References

Abeyrathne NS. Use of lysozyme from chicken egg white as a nitrite replacer in an Italian-type chicken sausage. Funct Foods Heal Dis. 2015;5:320. https://doi.org/10.31989/ffhd.v5i9.217

Addis PB. Occurrence of lipid oxidation products in foods. Food Chem Toxicol. 1986;24:1021–30. https://doi.org/10.1016/0278-6915(86)90283-8

Ahvenainen R. Active and intelligent packaging: an introduction. In: A R, editor. Novel food packaging techniques, novel food. Cambridge: Wood head Publishing Ltd; 2003. p. 5–21.

Alizadeh-Sani M, Mohammadian E, McClements DJ. Eco-friendly active packaging consisting of nanostructured biopolymer matrix reinforced with TiO2 and essential oil: application for preservation of refrigerated meat. Food Chem. 2020;322:126782. https://doi.org/10.1016/j.foodchem.2020.126782

Amaral AB, Silva MV d, Lannes SC d S. Lipid oxidation in meat: mechanisms and protective factors – a review. Food Sci Technol. 2018;38:1–15. https://doi.org/10.1590/fst.32518.

Amariei S, et al. Rosemary, thyme and oregano essential oils influence on physicochemical properties and microbiological stability of minced meat. J Microbiol Biotechnol Food Sci. 2016;6:670–6. https://doi.org/10.15414/jmbfs.2016.6.1.670-676.

Amin R, Edris N. Grape seed extract as natural antioxidant and antibacterial in minced beef. PSM Biol Res. 2017;2:89–96.

Anderson GJ, Connoro WE, Corliss JD. Docosahexaenoic acid is the preferred dietary n-3 fatty acid for the development of the brain and retina. Pediatr Res. 1990;27:89–97. https://doi.org/10.1203/00006450-199001000-00023

Andrade MA, Ribeiro-Santos R, Costa Bonito MC, et al. Characterization of rosemary and thyme extracts for incorporation into a whey protein based film. LWT - Food Sci Technol. 2018;92:497–508. https://doi.org/10.1016/j.lwt.2018.02.041

Andrade MA, Ribeiro-Santos R, Guerra M, Sanches-Silva A. Evaluation of the oxidative status of salami packaged with an active whey protein film. Foods. 2019;8:387. https://doi.org/10.3390/foods8090387

Andrés AI, Petrón MJ, Adámez JD, et al. Food by-products as potential antioxidant and antimicrobial additives in chill stored raw lamb patties. Meat Sci. 2017;129:62–70. https://doi.org/10.1016/j.meatsci.2017.02.013

Appendini P, Hotchkiss JH. Review of antimicrobial food packaging. Innov Food Sci Emerg Technol. 2002;3:113–26. https://doi.org/10.1016/S1466-8564(02)00012-7.

Badawy MEI, Lotfy TMR, Shawir SMS. Facile synthesis and characterizations of antibacterial and antioxidant of chitosan monoterpene nanoparticles and their applications in preserving minced meat. Int J Biol Macromol. 2020;156:127–36. https://doi.org/10.1016/j.ijbiomac.2020.04.044

Bajpai VK, Park I, Lee J, et al. Antioxidant and antimicrobial efficacy of a biflavonoid, amentoflavone from Nandina domestica in vitro and in minced chicken meat and apple juice food models. Food Chem. 2019;271:239–47. https://doi.org/10.1016/j.foodchem.2018.07.159

Barbosa-Pereira L, Aurrekoetxea GP, Angulo I, et al. Development of new active packaging films coated with natural phenolic compounds to improve the oxidative stability of beef. Meat Sci. 2014;97:249–54. https://doi.org/10.1016/J.MEATSCI.2014.02.006

Bearth A, Cousin M, Siegrist M. The consumer's perception of artificial food additives: influences on acceptance, risk and benefit perceptions. Food Qual Prefer. 2014;38:14–23. https://doi.org/10.1016/j.foodqual.2014.05.008

Bellés M, Alonso V, Roncalés P, Beltrán JA. Sulfite-free lamb burger meat: antimicrobial and antioxidant properties of green tea and carvacrol. J Sci Food Agric. 2019;99:464–72. https://doi.org/10.1002/jsfa.9208

Bolognesi VJ, Garcia CER. Annatto carotenoids as additives replacers in meat products. In: Alternative and replacement foods. Elsevier; 2018. p. 355–84.

Brewer MS. Natural antioxidants: sources, compounds, mechanisms of action, and potential applications. Compr Rev Food Sci Food Saf. 2011;10:221–47. https://doi.org/10.1111/j.1541-4337.2011.00156.x

Brody AL. Packaging of food. New York: Willey; 1997.

Bryan NS, Ivy JL. Inorganic nitrite and nitrate: evidence to support consideration as dietary nutrients. Nutr Res. 2015;35:643–54.

Câmara AKFI, Vidal VAS, Santos M, et al. Reducing phosphate in emulsified meat products by adding chia (Salvia hispanica L.) mucilage in powder or gel format: A clean label technological strategy. Meat Sci. 2020;163(108085) https://doi.org/10.1016/j.meatsci.2020.108085

Carvalho CD. Determinação Analitica de Nitratos e/ou Nitritos em Amostras do Estudo de Dieta Total (TDS) por Cromatografia Líquida de Alta Resolução (HPLC): Contribuição para a Avaliação da Exposiçao da População a Nitratos/Nitritos. Instituto Superior de Engenharia de Lisboa; 2018.

Castellano P, Peña N, Ibarreche MP, et al. Antilisterial efficacy of Lactobacillus bacteriocins and organic acids on frankfurters. Impact on sensory characteristics. J Food Sci Technol. 2018;55:689–97. https://doi.org/10.1007/s13197-017-2979-8

Cegiełka A, Hać-Szymańczuk E, Piwowarek K, et al. The use of bioactive properties of sage preparations to improve the storage stability of low-pressure mechanically separated meat from chickens. Poult Sci. 2019;98:5045–53. https://doi.org/10.3382/ps/pez242

Clarke D, Tyuftin AA, Cruz-Romero MC, et al. Surface attachment of active antimicrobial coatings onto conventional plastic-based laminates and performance assessment of these materials on the storage life of vacuum packaged beef sub-primals. Food Microbiol. 2017;62:196–201. https://doi.org/10.1016/j.fm.2016.10.022

Coma V. Bioactive packaging technologies for extended shelf life of meat-based products. Meat Sci. 2008;78:90–103. https://doi.org/10.1016/j.meatsci.2007.07.035

da Costa RJ, Voloski FLS, Mondadori RG, et al. Preservation of meat products with Bacteriocins produced by lactic acid bacteria isolated from meat. J Food Qual. 2019;2019:1–12. https://doi.org/10.1155/2019/4726510

Datta A, Hossain A, Roy S. An overview on monosodium glutamate: its direct and indirect effects. Res J Pharm Technol. 2019;12:6187. https://doi.org/10.5958/0974-360X.2019.01074.6

de Oliveira TV, de Freitas PAV, Pola CC, et al. Development and optimization of antimicrobial active films produced with a reinforced and compatibilized biodegradable polymers. Food Packag Shelf Life. 2020;24:100459. https://doi.org/10.1016/J.FPSL.2019.100459

Dehghani S, Peighambardoust SH, Peighambardoust SJ, et al. Improved mechanical and antibacterial properties of active LDPE films prepared with combination of Ag, ZnO and CuO nanoparticles. Food Packag Shelf Life. 2019;22:100391. https://doi.org/10.1016/j.fpsl.2019.100391

Dennis MJ, Key PE, Papworth T, et al. The determination of nitrate and nitrite in cured meat by HPLC/UV. Food Addit Contam. 1990;7:455–61. https://doi.org/10.1080/02652039009373908

Di Vita G, Blanc S, Brun F, et al. Quality attributes and harmful components of cured meats: exploring the attitudes of Italian consumers towards healthier cooked ham. Meat Sci. 2019;155:8–15. https://doi.org/10.1016/j.meatsci.2019.04.013

Dong Z, Luo C, Guo Y, et al. Characterization of new active packaging based on PP/LDPE composite films containing attapulgite loaded with Allium sativum essence oil and its application for large yellow croaker (Pseudosciaena crocea) fillets. Food Packag Shelf Life. 2019;20:100320. https://doi.org/10.1016/j.fpsl.2019.100320

Doolaege EHA, Vossen E, Raes K, et al. Effect of rosemary extract dose on lipid oxidation, colour stability and antioxidant concentrations, in reduced nitrite liver pâtés. Meat Sci. 2012;90:925–31. https://doi.org/10.1016/j.meatsci.2011.11.034

Eaton SB, Eaton III SB, Cordain L (2002) Evolution, diet, and health. In: Ungar PS, Teaford MF (eds) Human diet: its origin and evolution. Bergin & Garvey, pp. 7–17.

Eskandarabadi SM, Mahmoudian M, Farah KR, et al. Active intelligent packaging film based on ethylene vinyl acetate nanocomposite containing extracted anthocyanin, rosemary extract and ZnO/Fe-MMT nanoparticles. Food Packag Shelf Life. 2019;22:100389. https://doi.org/10.1016/j.fpsl.2019.100389

Esterbauer H. Cytotoxicity and genotoxicity of lipid-oxidation. Am J Clin Nurs. 1993;57:779S–86S.

European Commission. Regulation (EC) No 178/2002 of the European Parliament and of the Council of 28 January 2002 laying down the general principles and requirements of food law, establishing the European Food Safety Authority and laying down procedures in matters of food safety. Off J Eur Communities. 2002a;L31:1–24.

European Commission. Council Directive 2002/99/EC of 16 December 2002 laying down the animal health rules governing the production, processing, distribution and introduction of products of animal origin for human consumption. Off J Eur Communities. 2002b;L18:11–20.

European Commission. Regulation (EC) No 853/2004 of the European Parliament and of the Council of 29 April 2004 laying down specific hygiene rules for on the hygiene of foodstuffs. Off J Eur Communities. 2004a;L139

European Commission. Regulation (EC) No 852/2004 of the European Parliament and of the Council of 29 April 2004 on the hygiene of foodstuffs. Off J Eur Communities. 2004b;L139

European Commission. Regulation (EC) No 854/2004 of the European Parliament and of the Council of 29 April 2004 laying down specific rules for the organisation of official controls on products of animal origin intended for human consumption. Off J Eur Communities. 2004c;L155

European Commission. Regulation (EU) No 1935/2004 of the European Parliament and of the Council of 27 October 2004 on materials and articles intended to come into contact with food and repealing Directives 80/590/EEC and 89/109/EEC. Off J Eur Union. 2004d:4–17.

European Commission. Commission Regulation (EC) No 2073/2005 of 15 November 2005 on microbiological criteria for foodstuffs. Off J Eur Communities. 2005;L338

European Commission. Regulation (EC) No 1333/2008 of the European Parliament and of the Council of 16 December 2008 on food additives. Off J Eur Union. 2008;L354

European Commission. Commission Regulation (EC) No 450/2009 of 29 May 2009 on active and intelligent materials and articles intended to come into contact with food. Off J Eur Union. 2009;L135

European Commission. Questions and answers on food additives; 2011a.

European Commission. Commission regulation (EU) N° 10/2011. Off J Eur Union. 2011b:1–89.

European Commission. Food safety: overview; 2020.

Fang Z, Zhao Y, Warner RD, Johnson SK. Active and intelligent packaging in meat industry. Trends Food Sci Technol. 2017;61:60–71. https://doi.org/10.1016/J.TIFS.2017.01.002

Fasihnia SH, Peighambardoust SH, Peighambardoust SJ, Oromiehie A. Development of novel active polypropylene based packaging films containing different concentrations of sorbic acid. Food Packag Shelf Life. 2018;18:87–94. https://doi.org/10.1016/j.fpsl.2018.10.001

Fernandes RPP, Trindade MA, Tonin FG, Lima CG. Evaluation of antioxidant capacity of 13 plant extracts by three different methods: cluster analyses applied for selection of the natural extracts with higher antioxidant capacity to replace synthetic antioxidant in lamb burgers. 2015. https://doi.org/10.1007/s13197-015-1994-x

Floros JD, Dock LL, Han JH. Active packaging technologies and applications. In: Food cosmetics and drug packaging; 1997. p. 10–7.

Food and Agriculture Organization of the United Nations (FAO), World Health Organization (WHO). About Codex Alimentarius; 2020.

Food and Drug Administration (FDA). Overview of food ingredients, additives & colors; 2010.

Food And Drug Administration (FDA). Packaging & Food Contact Substances (FCS); 2017.

Food And Drug Administration (FDA). Food Contact Substance Notification Program (Video); 2018a.

Food And Drug Administration (FDA). Food ingredient & packaging terms; 2018b.

Food Safety and Inspection Service (FSIS). Inspection for food safety: the basics; 2013.

Frankel EN. Lipid oxidation. Prog Lipid Res. 1980;19:1–22. https://doi.org/10.1016/0163-7827(80)90006-5

Frankel EN. Lipid oxidation: mechanisms, products and biological significance. J Am Oil Chem Soc. 1984;61:1908–17. https://doi.org/10.1007/BF02540830

Fratianni F, De Martino L, Melone A, et al. Preservation of chicken breast meat treated with thyme and balm essential oils. J Food Sci. 2010;75:M528–35. https://doi.org/10.1111/j.1750-3841.2010.01791.x

Gadekar YP, Sharma BD, Shinde AK, et al. Effect of natural antioxidants on the quality of cured, restructured goat meat product during refrigerated storage (4±1°C). Small Rumin Res. 2014;119:72–80. https://doi.org/10.1016/j.smallrumres.2014.03.005

Gallego M, Arnal M, Talens P, et al. Effect of gelatin coating enriched with antioxidant tomato by-products on the quality of pork meat. Polymers (Basel). 2020;12:1032. https://doi.org/10.3390/polym12051032

Gao X, Kang Z, Zhang W, et al. Combination of κ-carrageenan and soy protein isolate effects on functional properties of chopped low-fat pork batters during heat-induced gelation. Food Bioprocess Technol. 2015;8:1524–31. https://doi.org/10.1007/s11947-015-1516-x

Glorieux S, Goemaere O, Steen L, Fraeye I. Phosphate reduction in emulsified meat products: impact of phosphate type and dosage on quality characteristics. Food Technol Biotechnol. 2017;55:390–7. https://doi.org/10.17113/ftb.55.03.17.5089

Goldberg S, Doyle RJ, Rosenberg M. Mechanism of enhancement of microbial cell hydrophobicity by cationic polymers. J Bacteriol. 1990;172:5650–4. https://doi.org/10.1128/JB.172.10.5650-5654.1990

Gómez-Estaca J, López-de-Dicastillo C, Hernández-Muñoz P, et al. Advances in antioxidant active food packaging. Trends Food Sci Technol. 2014;35:42–51. https://doi.org/10.1016/J.TIFS.2013.10.008

González-Alonso V, Cappelletti M, Bertolini FM, et al. Microbial inactivation of raw chicken meat by supercritical carbon dioxide treatment alone and in combination with fresh culinary herbs. Poult Sci. 2020;99:536–45. https://doi.org/10.3382/ps/pez563

Govari M, Pexara A. Nitrates and nitrites in meat products. J Hell Vet Med Soc. 2018;66:127. https://doi.org/10.12681/jhvms.15856

Grunert KG, Bredahl L, Brunsø K. Consumer perception of meat quality and implications for product development in the meat sector—a review. Meat Sci. 2004;66:259–72. https://doi.org/10.1016/S0309-1740(03)00130-X

Guesnet P, Alessandri J-M. Docosahexaenoic acid (DHA) and the developing central nervous system (CNS) – implications for dietary recommendations. Biochimie. 2011;93:7–12. https://doi.org/10.1016/j.biochi.2010.05.005

Ha J-U, Kim Y-M, Lee D-S. Multilayered antimicrobial polyethylene films applied to the packaging of ground beef. Packag Technol Sci. 2001;14:55–62. https://doi.org/10.1002/pts.537

Hamed M, Bougatef H, Karoud W, et al. Polysaccharides extracted from pistachio external hull: characterization, antioxidant activity and potential application on meat as preservative. Ind Crop Prod. 2020;148:112315. https://doi.org/10.1016/j.indcrop.2020.112315

He M, Gramza-michałowska A. Effect of plant extracts on lipid oxidation and changes in nutritive value of protein in frozen-stored meat products. J Food Process Preserv. 2016:00. https://doi.org/10.1111/jfpp.12989

Hmelak Gorenjak A, Cenčič A. Nitrate in vegetables and their impact on human health. A review. Acta Aliment. 2013;42:158–72. https://doi.org/10.1556/AAlim.42.2013.2.4

Huang L, Zeng X, Sun Z, et al. Production of a safe cured meat with low residual nitrite using nitrite substitutes. Meat Sci. 2020;162:108027. https://doi.org/10.1016/j.meatsci.2019.108027

Hughes GJ, Ryan DJ, Mukherjea R, Schasteen CS. Protein Digestibility-Corrected Amino Acid Scores (PDCAAS) for soy protein isolates and concentrate: criteria for evaluation. J Agric Food Chem. 2011;59:12707–12. https://doi.org/10.1021/jf203220v

Jayasena DD, Jo C. Essential oils as potential antimicrobial agents in meat and meat products: A review. Trends Food Sci Technol. 2013;34:96–108. https://doi.org/10.1016/J.TIFS.2013.09.002

Jin S, Kim S-H, Choi J-S, Yim D-G. Effect of diverse binder materials and their addition levels on physico-chemical characteristics of sausages. J Food Meas Charact. 2019;13:1558–65. https://doi.org/10.1007/s11694-019-00071-1

Júnior MM, de Oliveira TP, Gonçalves OH, et al. Substitution of synthetic antioxidant by curcumin microcrystals in mortadella formulations. Food Chem. 2019;300:125231. https://doi.org/10.1016/j.foodchem.2019.125231

Jůzl M, Piechowiczová M, Řehůřková K. Comparison of quality parameters of the cooked salami "Gothajský" in dependence on used salt content and additives. Potravin Slovak J Food Sci. 2019;13:390–5. https://doi.org/10.5219/1117

Kanatt SR, Makwana SH. Development of active, water-resistant carboxymethyl cellulose-poly vinyl alcohol-Aloe vera packaging film. Carbohydr Polym. 2020;227:115303. https://doi.org/10.1016/j.carbpol.2019.115303

Kerry, J. P.; O'Grady, M. N.; Hogan, S. A. Past, Current and Potential Utilisation of Active and Intelligent Packaging Systems for Meat and Muscle-Based Products: A Review. Meat Sci. 2006;74(1), 113–130. https://doi.org/10.1016/j.meatsci.2006.04.024

Kim T-K, Hwang K-E, Lee M-A, et al. Quality characteristics of pork loin cured with green nitrite source and some organic acids. Meat Sci. 2019;152:141–5. https://doi.org/10.1016/j.meatsci.2019.02.015

Kim T-K, Kim H-W, Lee Y-Y, et al. Quality characteristics of duck jerky: combined effects of collagen and konjac. Poult Sci. 2020;99:629–36. https://doi.org/10.3382/ps/pez561

Konuk Takma D, Korel F. Active packaging films as a carrier of black cumin essential oil: development and effect on quality and shelf-life of chicken breast meat. Food Packag Shelf Life. 2019;19:210–7. https://doi.org/10.1016/j.fpsl.2018.11.002

Kumar V, JS W, Kumari BA, et al. Active packaging systems in food packaging for enhanced shelf life. Food Packag Shelf Life. 2018;7:2044–6. https://www.researchgate.net/publication/330010066_Active_packaging_systems_in_food_packaging_for_enhanced_shelf_life

Larry Branen A, Haggerty RJ. Introduction to food additives. In: Branen AL, Davidson PM, Salminen S, Thorngate III JH, editors. Food additives. 2nd ed. New York: Marcel Dekker, Inc.; 2001. p. 1–9.

Lawrie RA, Ledward DA, editors. Chapter 1. In: Lawrie's meat science, 7th edn. Woodhead Publishing Limited; 2006. p. 1–14.

Lomate GB, Dandi B, Mishra S. Development of antimicrobial LDPE/Cu nanocomposite food packaging film for extended shelf life of peda. Food Packag Shelf Life. 2018;16:211–9. https://doi.org/10.1016/j.fpsl.2018.04.001

Marcos J, Pohlman F, Hansen C, et al. Processing and sensory characteristics of ground beef treated with long chain organic acids. Int J Meat Sci. 2016;7:12–20. https://doi.org/10.3923/ijmeat.2017.12.20

Marcous A, Rasouli S, Ardestani F. Low-density polyethylene films loaded by titanium dioxide and zinc oxide nanoparticles as a new active packaging system against Escherichia coli O157:H7 in fresh calf minced meat. Packag Technol Sci. 2017;30:693–701. https://doi.org/10.1002/pts.2312

Martin D, Salas-Perez L, Villalva M, et al. Effect of alkylglycerol-rich oil and rosemary extract on oxidative stability and antioxidant properties of a cooked meat product. Eur J Lipid Sci Technol. 2017;119:1600412. https://doi.org/10.1002/ejlt.201600412

McAfee AJ, McSorley EM, Cuskelly GJ, et al. Red meat consumption: an overview of the risks and benefits. Meat Sci. 2010;84:1–13. https://doi.org/10.1016/j.meatsci.2009.08.029

Milton K. Diet and primate evolution. Sci Am. 1993;269:86–93.

Miltz J, Passy N, Mannhcim CH. Trends and applications of active packaging systems. In: Food and packaging materials-chemical interaction. London: The Royal Society of Chemistry; 1995.

Mohammadi M, Mirabzadeh S, Shahvalizadeh R, Hamishehkar H. Development of novel active packaging films based on whey protein isolate incorporated with chitosan nanofiber and nano-formulated cinnamon oil. Int J Biol Macromol. 2020;149:11–20. https://doi.org/10.1016/j.ijbiomac.2020.01.083

Molognoni L, Daguer H, Motta GE, et al. Interactions of preservatives in meat processing: formation of carcinogenic compounds, analytical methods, and inhibitory agents. Food Res Int. 2019;125 https://doi.org/10.1016/j.foodres.2019.108608.

Mondry H. Packaging systems for processed meat - meat quality and meat packaging. Utrecht: Ecceamst; 1996.

Mulla M, Ahmed J, Al-Attar H, et al. Antimicrobial efficacy of clove essential oil infused into chemically modified LLDPE film for chicken meat packaging. Food Control. 2017;73:663–71. https://doi.org/10.1016/j.foodcont.2016.09.018

Muppalla SR, Kanatt SR, Chawla SP, Sharma A. Carboxymethyl cellulose-polyvinyl alcohol films with clove oil for active packaging of ground chicken meat. Food Packag Shelf Life. 2014;2:51–8. https://doi.org/10.1016/j.fpsl.2014.07.002

Nawar WW. Lipids. In: Fennema OR, editor. Food chemistry. 2nd ed. Taylor & Francis; 1996. p. 139–244.

O'Grady MN, Monahan FJ, Bailey J, et al. Colour-stabilising effect of muscle vitamin E in minced beef stored in high oxygen packs. Meat Sci. 1998;50:73–80. https://doi.org/10.1016/S0309-1740(98)00017-5

Office of the Law Revision Counsel of the United States House of Representatives. 21 USC 348: Food additives; 2020.

Omar SA, Artime E, Webb AJ. A comparison of organic and inorganic nitrates/nitrites. Nitric Oxide. 2012;26:229–40. https://doi.org/10.1016/j.niox.2012.03.008

Otoni CG, Espitia PJP, Avena-Bustillos RJ, McHugh TH. Trends in antimicrobial food packaging systems: emitting sachets and absorbent pads. Food Res Int. 2016;83:60–73.

Ozaki MM, Munekata PES, Lopes A d S, et al. Using chitosan and radish powder to improve stability of fermented cooked sausages. Meat Sci. 2020;167:108165. https://doi.org/10.1016/j.meatsci.2020.108165

Ozdemir M, Floros JD. Active food packaging technologies. Crit Rev Food Sci Nutr. 2004;44:185–93. https://doi.org/10.1080/10408690490441578

Parpia AS, Goldstein MB, Arcand J, et al. Sodium-reduced meat and poultry products contain a significant amount of potassium from food additives. J Acad Nutr Diet. 2018;118:878–85. https://doi.org/10.1016/j.jand.2017.10.025

Pereira de Abreu DA, Losada PP, Maroto J, Cruz JM. Evaluation of the effectiveness of a new active packaging film containing natural antioxidants (from barley husks) that retard lipid damage in frozen Atlantic salmon (Salmo salar L.). Food Res Int. 2010;43:1277–82. https://doi.org/10.1016/j.foodres.2010.03.019

Pereira PM d CC, Vicente AF d RB. Meat nutritional composition and nutritive role in the human diet. Meat Sci. 2013;93:586–92. https://doi.org/10.1016/j.meatsci.2012.09.018

Persico P, Ambrogi V, Carfagna C, et al. Nanocomposite polymer films containing carvacrol for antimicrobial active packaging. Polym Eng Sci. 2009;49:1447–55. https://doi.org/10.1002/pen.21191

Polônio MLT, Peres F. Consumo de aditivos alimentares e efeitos à saúde: desafios para a saúde pública brasileira. Cad Saude Publica. 2009;25:1653–66. https://doi.org/10.1590/S0102-311X2009000800002

Prasad P, Kochhar A. Active packaging in food industry: A review. IOSR J Environ Sci Toxicol Food Technol. 2014;8:01–7. https://doi.org/10.9790/2402-08530107

Qi S, Huang H, Huang J, et al. Lychee (Litchi chinensis Sonn.) seed water extract as potential antioxidant and anti-obese natural additive in meat products. Food Control. 2015;50:195–201. https://doi.org/10.1016/j.foodcont.2014.08.047

Ramos M, Jiménez A, Peltzer M, Garrigós MC. Characterization and antimicrobial activity studies of polypropylene films with carvacrol and thymol for active packaging. J Food Eng. 2012;109:513–9. https://doi.org/10.1016/j.jfoodeng.2011.10.031

Ramos K, Gómez-Gómez MM, Cámara C, Ramos L. Silver speciation and characterization of nanoparticles released from plastic food containers by single particle ICPMS. Talanta. 2016;151:83–90. https://doi.org/10.1016/j.talanta.2015.12.071

Realini CE, Marcos B. Active and intelligent packaging systems for a modern society. Meat Sci. 2014;98:404–19. https://doi.org/10.1016/j.meatsci.2014.06.031

Ribeiro-Santos R, de Melo NR, Andrade M, Sanches-Silva A. Potential of migration of active compounds from protein-based films with essential oils to a food and a food simulant. Packag Technol Sci. 2017:791–8. https://doi.org/10.1002/pts.2334

Ribeiro-Santos R, de Melo NR, Andrade M, et al. Whey protein active films incorporated with a blend of essential oils: characterization and effectiveness. Packag Technol Sci. 2018;31:27–40. https://doi.org/10.1002/pts.2352

Ritchie H, Roser M. Meat and dairy production. In: OurWorldInData.org; 2020. https://ourworldindata.org/meat-production. Accessed 7 May 2020.

Rodrigues AS, Kubota EH, da Silva CG, et al. Banana inflorescences: A cheap raw material with great potential to be used as a natural antioxidant in meat products. Meat Sci. 2020;161:107991. https://doi.org/10.1016/j.meatsci.2019.107991

Rooney ML. Active packaging in polymer films. In: Rooney ML, editor. Active food packaging. Glasgow: Blackie Academic & Professional; 1995. p. 74–110.

Rooney, M. L. Oxygen-Scavenging Packaging. In Innovations in Food Packaging; Han, J. H., Ed.; Elsevier: San Diego, USA, 2005. p 123–137. https://doi.org/10.1016/B978-012311632-1/50040-1.

Rosa-Limpo B, Caetano MM, Brum do Canto J. O Livro de Pantagruel. 74th ed. Temas e Debates; 2010.

Sanches-Silva A, Cruz JM, Sendn-Garca R, Paseiro-Losada P. Determination of butylated hydroxytoluene in food samples by high-performance liquid chromatography with ultraviolet detection and gas chromatography/mass spectrometry. J AOAC Int. 2007;90:277–83.

Sanches-Silva A, Costa D, Albuquerque TG, et al. Trends in the use of natural antioxidants in active food packaging: a review. Food Addit Contam Part A. 2014;31:374–95. https://doi.org/10.1080/19440049.2013.879215

Scannell AG, Hill C, Ross R, et al. Development of bioactive food packaging materials using immobilised bacteriocins Lacticin 3147 and Nisaplin®. Int J Food Microbiol. 2000;60:241–9. https://doi.org/10.1016/S0168-1605(00)00314-7

Schirmer BC, Langsrud S. Evaluation of natural antimicrobials on typical meat spoilage bacteria in vitro and in vacuum-packed pork meat. J Food Sci. 2010;75:M98–M102. https://doi.org/10.1111/j.1750-3841.2009.01485.x

Schmid A. The role of meat fat in the human diet. Crit Rev Food Sci Nutr. 2010;51:50–66. https://doi.org/10.1080/10408390903044636

Shah MA, Bosco SJD, Mir SA. Plant extracts as natural antioxidants in meat and meat products. Meat Sci. 2014;98:21–33. https://doi.org/10.1016/j.meatsci.2014.03.020

Shankar S, Bang YJ, Rhim JW. Antibacterial LDPE/GSE/Mel/ZnONP composite film-coated wrapping paper for convenience food packaging application. Food Packag Shelf Life. 2019;22:100421. https://doi.org/10.1016/j.fpsl.2019.100421

Shanmugasundaram P, Bavenro, Rujaswini T. A review on food coloring agents-safe or unsafe? Res J Pharm Technol. 2019;12:2503. https://doi.org/10.5958/0974-360X.2019.00421.9

Sharma H, Mendiratta SK, Agarwal RK, Gurunathan K. Bio-preservative effect of blends of essential oils: natural anti-oxidant and anti-microbial agents for the shelf life enhancement of emulsion based chicken sausages. J Food Sci Technol. 2020; https://doi.org/10.1007/s13197-020-04337-1

Shin Y, Shin J, Lee YS. Effects of oxygen scavenging package on the quality changes of processed meatball product. Food Sci Biotechnol. 2009;18:73–8.

Şimşek A, Kılıç B. Et Ürünlerinde Kullanılan Fosfatların Antioksidan ve Antimikrobiyal Özellikleri. Turkish J Agric - Food Sci Technol. 2017;5:299. https://doi.org/10.24925/turjaf.v5i4.299-307.1049

Smith JP, Ramaswamy HS, Simpson BK. Developments in food packaging technology. Part II. Storage aspects. Trends Food Sci Technol. 1990;1:111–8. https://doi.org/10.1016/0924-2244(90)90086-E

Sorapukdee S, Uesakulrungrueng C, Pilasombut K. Effects of humectant and roasting on physicochemical and sensory properties of jerky made from spent hen meat. Korean J Food Sci Anim Resour. 2016;36:326–34. https://doi.org/10.5851/kosfa.2016.36.3.326

Sung S-Y, Sin LT, Tee T-T, et al. Antimicrobial agents for food packaging applications. Trends Food Sci Technol. 2013;33:110–23. https://doi.org/10.1016/J.TIFS.2013.08.001

Surendran Nair M, Nair DVT, Kollanoor Johny A, Venkitanarayanan K. Use of food preservatives and additives in meat and their detection techniques. In: Meat quality analysis. Elsevier; 2020. p. 187–213.

Swaroop C, Shukla M. Nano-magnesium oxide reinforced polylactic acid biofilms for food packaging applications. Int J Biol Macromol. 2018;113:729–36. https://doi.org/10.1016/j.ijbiomac.2018.02.156

Thangavelu KP, Kerry JP, Tiwari BK, McDonnell CK. Novel processing technologies and ingredient strategies for the reduction of phosphate additives in processed meat. Trends Food Sci Technol. 2019;94:43–53. https://doi.org/10.1016/j.tifs.2019.10.001

Tian F, Decker EA, Goddard JM. Controlling lipid oxidation via a biomimetic iron chelating active packaging material. J Agric Food Chem. 2013;61:12397–404. https://doi.org/10.1021/jf4041832

Trentini EM, Macedo REF d. The use of sodium nitrate and sodium nitrite in fresh meat product: a scientific approach. Brazilian J Technol. 2019;2:1017–41.

Triyannanto E, Lee KT. Evaluation of honey and rice syrup as replacements for sorbitol in the production of restructured duck jerky. Asian-Australasian J Anim Sci. 2015;29:271–9. https://doi.org/10.5713/ajas.15.0431

United States Department of Agriculture (USDA). The Codex Alimentarius Commission and The United States Codex Program; 2020.

USDA. USDA FoodData Central; 2020. https://fdc.nal.usda.gov/index.html. Accessed 1 May 2020.

Vermeiren L, Devlieghere F, Van Beest M, et al. Developments in the active packaging of foods. Trends Food Sci Technol. 1999;10:77–86. https://doi.org/10.1016/S0924-2244(99)00032-1

Wang LF, Rhim JW. Grapefruit seed extract incorporated antimicrobial LDPE and PLA films: effect of type of polymer matrix. LWT - Food Sci Technol. 2016;74:338–45. https://doi.org/10.1016/j.lwt.2016.07.066

Wang S, Tonnis BD, Wang ML, et al. Investigation of monosodium glutamate alternatives for content of umami substances and their enhancement effects in chicken soup compared to monosodium glutamate. J Food Sci. 2019;84:3275–83. https://doi.org/10.1111/1750-3841.14834

Waters C, Miller RK, Kerth CR, et al (2018) Sorghum bran as an antioxidant in pre-cooked ground pork and poultry products. Meat Muscle Biol 1:72. https://iastatedigitalpress.com/mmb/article/id/9189/

Wattananawinrat K, Threepopnatkul P, Kulsetthanchalee C. Morphological and thermal properties of LDPE/EVA blended films and development of antimicrobial activity in food packaging film. Energy Procedia. 2014;56:1–9. https://doi.org/10.1016/j.egypro.2014.07.125

World Health Organization (WHO). Food additives; 2018.

World Health Organization (WHO). Estimating the burden of foodborne diseases; 2020. https://www.who.int/activities/estimating-the-burden-of-foodborne-diseases. Accessed 30 May 2020.

World Health Organization (WHO), Food and Agriculture Organization of the United Nations (FAO), United Nations University (UNU). Protein and amino acid requirements in human nutrition; 2002.

Wrona M, Nerín C, Alfonso MJ, Caballero MÁ. Antioxidant packaging with encapsulated green tea for fresh minced meat. Innov Food Sci Emerg Technol. 2017;41:307–13. https://doi.org/10.1016/j.ifset.2017.04.001

Yang W, Fortunati E, Dominici F, et al. Synergic effect of cellulose and lignin nanostructures in PLA based systems for food antibacterial packaging. Eur Polym J. 2016;79:1–12. https://doi.org/10.1016/j.eurpolymj.2016.04.003

Yildirim S, Röcker B, Pettersen MK, et al. Active packaging applications for food. Compr Rev Food Sci Food Saf. 2018;17:165–99. https://doi.org/10.1111/1541-4337.12322

Yu H, Qin C, Zhang P, et al. Antioxidant effect of apple phenolic on lipid peroxidation in Chinese-style sausage. J Food Sci Technol. 2015;52:1032–9. https://doi.org/10.1007/s13197-013-1127-3

Zanfirescu A, Ungurianu A, Tsatsakis AM, et al. A review of the alleged health hazards of monosodium glutamate. Compr Rev Food Sci Food Saf. 2019;18:1111–34. https://doi.org/10.1111/1541-4337.12448

Zhao Y, Lian Z, Yue J. Recent development in food packaging, a review. J Chinese Inst Food Sci Technol. 2013;4:1–10.

Zia J, Paul UC, Heredia-Guerrero JA, et al. Low-density polyethylene/curcumin melt extruded composites with enhanced water vapor barrier and antioxidant properties for active food packaging. Polymer (Guildf). 2019;175:137–45. https://doi.org/10.1016/j.polymer.2019.05.012

Zouari N, Ayadi MA, Hadj-Taieb S, et al. Whey powder, ι-carrageenan, and fat interactions and their influence on instrumental texture and sensory properties of Turkey meat sausage using a mixture design approach. Int J Food Prop. 2012;15:1233–46. https://doi.org/10.1080/10942912.2010.517885

Chapter 12
Application of Releasing Systems in Active Packaging for Dairy Products

S. Volpe, M. Valentino, R. Muhammad, and E. Torrieri

Abstract Dairy products are a wide category of food, whose quality alteration and shelf life depend on raw materials, process, packaging, and maturation (when available). Conventional packaging provides a barrier between the packed product and its surroundings, and the durability of the product depends—to a large extent—on the effectiveness of this protection. In recent years, the use of packaging systems capable of ensuring safety, quality, and adequate prevention of the growth of post-processing contaminants in food by maintaining quality and freshness has increased considerably. This chapter presents the deteriorative reactions that limit the shelf life of the main categories of dairy foods. Different types of release active packaging used to extend the shelf life of dairy products are reviewed and assessed in terms of safety, nutritional value, and the resulting quality of dairy products. Although the application of antioxidant packaging to dairy products represents a notable strategy for future innovation, that of antimicrobial solutions have been more extensively explored and research maturity has been reached. The main innovations in this sector are related to the use of natural antimicrobial compounds, such as bacteriocin-producing lactic acid bacteria or essential oils. The application of biopolymer films or coatings represents the most studied approach for developing new active films for these kinds of packaging.

Keywords Antimicrobial · Antioxidant · Essential oil · Milk · Cheese · Bioactive coating

S. Volpe
CAISIAL-Centre of Food Innovation and Development of Food Industry, University of Naples Federico II, Portici (NA), Italy

M. Valentino · R. Muhammad · E. Torrieri (✉)
Department of Agricultural Science, University of Naples Federico II, Portici (NA), Italy
e-mail: elena.torrieri@unina.it

S. M. Jafari, A. S. Silva (eds.), *Releasing Systems in Active Food Packaging*, Food Bioactive Ingredients, https://doi.org/10.1007/978-3-030-90299-5_12

1 Introduction

The European dairy market is one of the main dairy markets worldwide, contributing to nearly 22% of the global milk production. According to the EU Milk Market Observatory,[1] European dairy production reached 162 million tons in 2020, and the EU represents the largest producer of cheese and non-fat dry milk in the world. The market is expected to grow at a compound annual growth rate of 1.2% in the forecast period 2021–2026 and reach a market value of USD 173 billion by 2026.[2] Because of the abolition of EU milk quotas in 2016, EU dairy exports have steadily increased. It is one of the leading exporters of dairy products, along with countries such as New Zealand and the United States. Cheese, skimmed milk powder, and packed milk are the primary dairy products exported by the EU (Observatory 2016). The EU is estimated to account for 49% of the global cheese exports by 2030 (European Commission 2020).

Annually, 1.3 billion tons of food globally is lost or wasted, accounting for one-third of the total food produced for human consumption. The Food and Agriculture organization of the united Nations (FAO) estimates that, in Europe alone, 29 million tons of dairy products is lost or wasted every year, representing 20% of the food losses and waste globally (Forbes et al. 2021).

Food losses during the distribution chain could be reduced by a packaging system properly designed to preserve food safety and quality, maintain natural freshness, and extend food shelf life. To this end, packaging systems must protect food from environmental factors, such as oxygen, light, humidity, and microbes, which can activate alteration processes. Ideally, packaging systems should preserve food quality by reducing the loss of nutrients, growth of microbial contaminants, and progression of oxidation reactions.

Dairy products are a wide category of food, whose quality alterations and shelf life depend on the raw materials, process, packaging, and maturation (when available). Based on the composition and manufacturing process, they can be classified into different categories, such as milk, fermented product, butter and spreads, cheese, and milk powders. The most frequently used packaging materials for dairy products are glass, plastic bottles, multi-layered materials, pouches, plastic tubs, and cans (Robertson 2009).

Although traditional packaging covers the basic needs of food protection, advances in food packaging are still required to respond to the new requirements of consumers, including high-convenience foods with long shelf life, efficient freshness preservation, high nutritional value, and enhanced environmental sustainability.

Current trends are focussed on the development of packaging systems that can actively interact with foods or the environment inside the package to preserve food

[1] https://ec.europa.eu/info/food-farming-fisheries/farming/facts-and-figures/markets/overviews/market-observatories/milk_en

[2] https://www.expertmarketresearch.com/reports/europe-dairy-market

safety and quality by reducing the rate of alteration processes. As reported by Regulation (EC) No. 1935/2004, active materials are materials intended to extend the shelf life or to maintain or improve the condition of packaged food. They are designed to incorporate components that could release or absorb substances into or from the packaged food or the environment surrounding the food (European Parliament 2004). Thus, active compounds are intentionally added to the packaging materials to improve their performance and extend the shelf life of food products.

This chapter presents the deteriorative reactions that limit the shelf life of the main categories of dairy food. Then, different types of release active packaging (AP) used to extend the shelf life of dairy products are reviewed and assessed in terms of the safety, nutritional value, and quality of dairy products.

2 Deteriorative Reactions in Dairy Products

During storage, the quality of dairy products changes because of physicochemical, chemical, or biochemical alterations. Examples of such transformations include protein gelation, lipid oxidation, microbial growth, and enzymatic degradation. Various alteration processes that tend to limit the shelf life of dairy products have been described by Muir and Banks (2000). In this section, the main deteriorative reactions that can be controlled by an AP system are discussed, focusing mainly on milk and cheese as the most studied dairy products.

2.1 Milk

Milk represents a single source of dietary elements required for the maintenance of proper health, especially in children and older citizens. Milk is a complex mix of water, proteins, lipids, carbohydrates, enzymes, vitamins, and minerals. Owing to its specific composition and pH close to neutral, it is a highly perishable product with high spoilage capacity, resulting in a rapid deterioration of its quality and safety. Microorganisms and their enzymes (such as proteases and lipases), via using the available nutrients, can deteriorate the quality and reduce the safety of milk. Moreover, milk quality can also be reduced via lipid oxidation activated by light or oxygen. Milk lipids are rapidly hydrolysed to free fatty acids because of the abundant lipoprotein lipase present in raw milk or produced by microorganisms. Usually, a product more stable than raw milk is obtained via heat treatment, which has a markedly reduced microbial load and inactivated enzymes.

As reported by Robertson (2009), based on heat treatment, fluid milk can be classified as pasteurized milk, ultra-high temperature (UHT)-treated milk, ultra-pasteurized (UP) milk, in-bottle sterilized milk, evaporated canned milk, cultured and flavoured (strawberry, cinnamon, coffee, etc.) milk, and microfiltered and bactofuged milk. The shelf life of these dairy products varies substantially because of their differing compositions, thermal processing conditions, and packaging materials. Microbiological changes in these products involve the growth of psychrotrophic bacteria (gram-negative rods such as *Pseudomonas, Enterobacteriaceae, Flavobacterium, Chromobacterium*, and *Alcaligenes*) that can produce both proteolytic and lipolytic enzymes, resulting in the formation of microbial flavours that are acidic, bitter, malty, putrid, or unclean (Muir and Banks 2000). These psychrotrophic bacteria can be almost completely destroyed with moderate heat treatment. However, under specific process conditions, some survivors of natural microflora can persist, promoting potential spoilage. Moreover, product safety may be affected through the incomplete destruction of pathogenic microorganisms during the pasteurisation process or via contamination with a particular pathogen at any stage after collection.

Among the pathogens, *Bacillus cereus* and *Clostridium* spp. may survive pasteurisation; moreover, there are some concerns over the possible survival of *Listeria monocytogenes* under similar conditions (Muriel-Galet et al. 2012), thus, refrigeration is sometimes not enough to assure the safety of the product during storage. In 1938, milk-borne outbreaks represented 25% of all infection outbreaks; recent information reveals that fluid milk products are associated with only less than 1% of these outbreaks. Despite the progress that has been made, sporadic milk-bornc outbreaks still occur, highlighting the need for continued vigilance at every stage of the production, processing, and distribution of milk products. As reported by Robertson (2009), the survival of *L. monocytogenes* due to the incorrect application of HTST pasteurisation was the cause of an outbreak in the United States. Moreover, foodborne diseases associated with pasteurised milk were associated with *Campylobacter jejuni* survival after the underprocessing of milk or batch pasteurisation. *Yersinia enterocolitica* has also been involved in outbreaks of illnesses associated with pasteurised chocolate-flavoured milk in the United States. In this case, contamination occurred during the addition of chocolate syrup to the milk after pasteurisation. Finally, *Salmonella* was also involved in two outbreaks, both of which were attributed to the contamination of pasteurised milk with raw milk.

2.2 Cheese

Cheese represents the most varied group of dairy products. Its shelf life can range from a few days to several years; thus, the generalisation of their deterioration processes is challenging. Several criteria can be used to classify cheese, such as moisture, rheological properties, texture, and various processing factors; thus, their classification is not straightforward. However, from a packaging point of view, it is convenient to classify cheese as a function of ripening time into the following categories: fresh (unripening), soft ripening (from 3 to 6 weeks of ripening), and hard ripening (from 6 to 24 months of ripening) (Robertson 2013).

The deteriorative reactions for cheese include: (1) undesirable microbial growth resulting in safety concerns or sensory defects (visible surface mould colonies, slime, putrefaction, or gas production); (2) moisture loss, (3) chemical reactions such as non-enzymatic browning and lipid oxidation, and (4) lactose crystallisation.

Fresh cheeses include ricotta, chevre, feta, cream cheese, quark and cottage cheese, fiordilatte, mozzarella, and whey cheese. Owing to their high moisture content, low salt concentration, and high pH, the shelf life of fresh cheese is mainly affected by undesirable microbial growth. They are also very sensitive to dehydration because most fresh cheeses drain slowly. In most cases, the incidence of cheese defects results from the low quality of its milk source, including problems related to the technology used or to failure during processing, transport, or storage.

For all cheeses, visual appearance failures are mostly related to mould and yeast growth, leading to colour spots inside or on the surface of the cheese. The most important spoilage mould species for cheese produced without stabilisers from several countries are *Penicillium commune* and *Pe. nalgiovense*.

In processed cheese, chemical reactions are the main causes of deterioration. In fact, pathogens are destroyed during heat treatment; therefore, specific starter cultures can be used to control the ripening processes, leading to a more stable final cheese. However, some spoilage and pathogenic microorganisms may be present in the final product, causing sporadic outbreaks (Robertson 2013).

3 Active Packaging of Dairy Products: Concepts and Application

AP is an innovative food packaging concept that combines advances in materials science, food technology, and food sciences. Generally, AP can be realised by using different methodologies that can be classified as (A) non-packaged (independent active devices) and (B) package-bound (antioxidant packaging materials and coatings). For package-bound systems, the active agent is incorporated within the container in which the product is packed or into the packaging film walls, employing its activity by absorbing undesirable agents/compounds around the product from the headspace or by releasing desirable compounds into the headspace surrounding the food products or directly into the food (Fang et al. 2017). AP-bound systems can be designed via direct incorporation of the active compound into a polymer matrix, which ensures a high concentration of the active compound and slow release to the food. For heat-sensitive active agents or those incompatible and immiscible with the polymer matrix, AP can be realised by coating an active compound onto the polymer. Different coating technologies can also be used to blend active compounds and polymers dissolved in an appropriate solvent. Moreover, immobilisation of the active compound on the polymer surface can be achieved whenever functional interacting groups are present on both the active agent and the polymer.

AP systems can also be classified based on their mechanism of action into the following types: "releasing systems" (e.g., antimicrobials, antioxidants, CO_2 or ethanol emitters, or enzymes); "absorbing systems" (e.g., oxygen, CO_2 or ethylene scavengers, moisture, or aroma absorbers); and "non-migrating systems," where full contact between the food and AP is required [antimicrobial nanoparticles (NPs), such as nanosilver, titanium dioxide (TiO_2), and zinc oxide (ZnO)]. The following sections will discuss the active releasing systems used in packaging dairy foods. These systems can be also divided into "leaching systems," where the release of the active compound is achieved via direct contact between the food and packaging material (e.g., those by bacteriocins, antibiotics, antioxidants, polyphenols, or organic acids), and "volatile systems," where the release occurs through gas-phase diffusion from the internal packaging surface to the food (Vasile and Baican 2021). Moreover, based on the active compound released, AP materials can be classified as antimicrobials or antioxidants.

4 Antimicrobial Release Systems

AP based on an antimicrobial release system could play a role in preserving food safety and extending food shelf life. It can control or even prevent the growth of undesired or spoilage microorganisms by controlling the release of antimicrobial substances into the food, extending the lag phase, and reducing the growth phase of these microorganisms (Haghighi-Manesh and Azizi 2017b). For dairy foods containing the desired microflora (e.g., ripening cheese or cottage cheese), antimicrobial packaging should ensure the development of these microorganisms within the foreseeable shelf life of the product.

To be effective, the antimicrobial substances need to be present at the food surface at concentrations above the minimum inhibitory concentration. Moreover, their efficacy depends on the mechanism of action of the active compound and the way it is introduced into the packaging system. The most commonly investigated antimicrobial substances for use in dairy products are bacteriocins—mainly nisin (NI), organic acids, and their derivatives. Recently, the use of natural compounds has mostly been investigated, and the most studied solutions are derived from lactic acid bacteria (LAB) strains that produce bacteriocin, calcinated corals, and essential oils (EOs). Moreover, antimicrobial packaging has mostly been applied to milk and cheese products (Table 12.1).

Haghighi-Manesh and Azizi (2017a) evaluated the possibility of incorporating cinnamaldehyde, a natural antimicrobial substance, directly into a multi-layer active film for packaging pasteurised milk. They showed that the AP they created can prolong the shelf life of milk from 4 to 6 days under storage at 4 °C. However, the inhibition of bacterial growth was highly dependent on the thickness of the middle layer containing the active compound and the composition of the multi-layer film. In fact, after the antimicrobial substance is integrated into the film, its release is governed by its diffusion coefficient in the film, which is characterised by a rapid initial release, followed by a progressively slow release over time. If not properly designed, the release of the antimicrobial substance could not present sufficient

Table 12.1 Application of antimicrobial releasing packaging to dairy products

Bioactive compound	Film or coating	Dairy product	Impact of antimicrobial packaging	Reference
Cinnamaldehyde	Low-density polyethylene (LDPE) at different concentrations of cinnamaldehyde (0.5, 1, 1.5, and 2%).	Pasteurised milk	Reduction of total CFU growth rates with a consequent shelf-life extension of 2 days at 4 °C	Haghighi-Manesh and Azizi (2017a)
Nisin	Polymer coating onto paper	Pasteurised milk	3% nisin-coated paper could delay the onset and the rate of growth of total aerobic bacteria and yeast	Kim et al. (2002)
Nisin	Coating on LDPE	Milk (raw, pasteurized, ultra-high temperature-treated milk)	Nisin-coated films were effective in inhibiting the bacterial flora with a reduction of 0.9 log CFU^{-1} and 1.3 log CFU^{-1} for raw milk and pasteurized milk	Mauriello et al. (2005)
Nisin (NI), natamycin (NA) and NI + NA	Cellulose polymer	Mozzarella	Films containing NI and NA showed an in vitro antimicrobial effect against *S. aureus* and *Listeria monocytogenes* and against *Penicillium* sp. and *Geotrichum* sp., respectively. Film containing NA could reduce the count of yeasts and moulds on cheese of 2 log units. NI film was active against psychrotrophic bacteria after 6 days of storage.	Pires et al. (2008)
NI-producing *Lactococcus lactis* subsp. *Lactis*	Polyvinyl alcohol (PVOH)	Pasteurised milk	Reduction of 2 Log of *L. monocytogenes* growth	Settier-Ramírez et al. (2020)

(continued)

Table 12.1 (continued)

Bioactive compound	Film or coating	Dairy product	Impact of antimicrobial packaging	Reference
Lactic acid bacteria+ grapefruit seed extract (GSE)	LDPE, biodegradable polybutylene adipate-co-terephthalate (PBAT)	Fresh soft cheese	The most effective antimicrobial effect was observed in soft cheese made with LAB and packaged with PBAT + GSE. The lag time of *L. monocytogenes* in soft cheese packed with antimicrobial film at 15 °C was 2.7 times longer than that in the control.	Lim et al. (2020)
Bacteriocin-producing *lactobacillus curvatus* P99	Starch	"Prato" cheese	The film containing the minimal bactericidal concentration of *Lactobacillus curvatus* P99 was effective in controlling of *L. monocytogenes* Scott A, demonstrating maintenance of activity for 30 days.	Lima et al. (2017)
Calcined corals (calcium oxide)	Nylon/linear low-density polyethylene	Pasteurised milk	Antimicrobial performance against *Escherichia coli* and *Staphylococcus aureus*. Extension of 50% of milk shelf-life (from 12 to 24 days)	Thanakkasaranee et al. (2020)

efficacy to protect the food. At a slow release rate, an insufficient amount of the antimicrobial substance is released to retard food deterioration; thus, the product may lose its quality before the AP can interact with the foods. Conversely, at a fast release rate, an excessive amount of the antimicrobial substance is released, and the active compound may not be present anymore when it is needed. The best packaging design requires a controlled release such that the kinetics of alteration is consistent with the kinetics of release. Consequently, the antimicrobial substance compound will be available when it is needed.

An alternative to the incorporation of antimicrobial compounds during extrusion is to apply antimicrobial additives as coatings. As explained in the previous section, this has the advantage of placing a specific antimicrobial additive in a controlled manner without subjecting it to high temperature or shearing forces. In addition, the coating can be applied at a later step, minimising the exposure of the product to contamination.

Recently, Haghighi-Manesh and Azizi (2017b) reviewed the applications of antibacterial packaging in dairy products. Their review discussed how the coefficient of desorption depends on the solubility of the migrant in the food and concluded that antimicrobial films are highly efficient when the packaging material and the food are in direct contact, but it is not always possible.

Among the available antibacterial packaging, NI derived from *Lactococcus lactis* was one of the earliest bacteriocins to be described and is the only one that has been recognised as a safe biological food preservative (Kumariya et al. 2019). The first application of NI-incorporated polymer coatings for dairy products was reported by Kim et al. (2002). They showed that antimicrobial packaging based on NI-incorporated polymer coatings could control the development of bacterial flora in pasteurised milk. Similar results were also reported by Mauriello et al. (2005), showing that NI-activated films retarded microbial growth and lowered the maximum growth levels of bacteria in raw milk, pasteurized milk, and UHT milk. However, the release of NI from the plastic film was unpredictable and temperature- and pH-dependent.

Pires et al. (2008) developed antimicrobial films through the incorporation of NI, natamycin (NA), and their combination (NI + NA) into a cellulose polymer. NI films showed an antimicrobial effect *in vitro* against *S. aureus* and *L. monocytogenes*, whereas NA films showed *in vitro* activity against *Penicillium* sp. and *Geotrichum* sp. isolated from sliced mozzarella. The films containing NA and NI + NA extended the shelf life of the cheese for 6 days compared with the control films. However, they did not show any synergistic effect on antimicrobial activity.

A novel and natural alternative preservation method is the incorporation of viable LAB strains into a film or coating matrix for bacteriocin production during food storage (La Storia et al. 2020). Few applications of this method to dairy products have been reported.

De Lima et al. (2017) evaluated the efficacy of a biodegradable film based on the antimicrobial metabolites produced by *Lactobacillus curvatus* P99, aiming to control *L. monocytogenes* in sliced Prato cheese. The minimum inhibitory concentration and minimum bactericidal concentration of the cell-free supernatant against *L. monocytogenes* Scott A were 15.6 μL/mL and 62.5 μL/mL, respectively. Prato cheese slices packed with films containing cell-free supernatants (at bactericidal concentrations) comprising bacteriocins produced by *Lactobacillus* displayed lower bacterial counts (<3 CFU/g) at the end of storage than those packed in other materials.

Similarly, Moraes et al. (2020) studied the effect of antimicrobial starch films coupled with pulsed light on *Listeria* content in sliced cheddar cheese. Cheese slices packed in active starch films without light treatment (and treated with sodium benzoate and citric acid) displayed a decrease in the concentration of *Listeria* (~2 log CFU/slice).

Settier-Ramírez et al. (2020) showed that polyvinyl alcohol (PVOH) based antimicrobial films, supplemented with proteins, protein hydrolysates, or yeast extract incorporating NI-producing *L. lactis* subsp. *Lactis* active against *L. monocytogenes*, could significantly reduce the number of *L. monocytogenes* in refrigerated pasteurised milk. They showed that PVOH-based films can act as effective carriers of living bacterial cells that could be used as a competitive culture against *L. monocytogenes*. Lim et al. (2020) assessed the combined effect of LAB isolated from kimchi and packaging films containing grapefruit seed extract (GSE) on *L. monocytogenes* growth in fresh soft cheese. Their results showed significantly reduced maximum population density and prolonged lag time values of *L. monocytogenes* in soft cheese packed with both LAB and GSE, showing that cheese manufacturing plants may be able to use this approach as a hurdle technology to control the growth of *L. monocytogenes* in the cheese products they sell to the retail market.

Thanakkasaranee et al. (2020) proved that the shelf life of refrigerated pasteurised milk (4 °C) could be extended using composite films based on nylon (NY), low-density polyethylene, and 1% calcined corals (NY/LL-CORALS), owing to their antimicrobial activity. Coral is a marine organism and a rich natural source of $CaCO_3$. In the presence of $Ca(OH)_2$, $CaCO_3$ is converted to CaO, which can act as a natural antimicrobial agent. NY/LL-CORALS films showed very strong antibacterial activity against *S. aureus* and *E. coli*, highlighting their strong antimicrobial activity against both gram-positive and gram-negative bacteria. Furthermore, these packaging films significantly extended the lag time of bacteria and suppressed the bacterial growth cycle in the pasteurised milk stored in refrigerated (4 °C) and ambient temperature (20 °C). In particular, the packaging that contained 1% NY/LL-CORALS could sufficiently maintain the bacterial number below the maximum population limit in refrigerated pasteurised milk, extending its shelf life by 100%—i.e., from 12 to 24 days (maximum storage time studied)—when stored at 4 °C.

5 Antioxidant Release Systems

The influence of antioxidant releasing packaging systems on milk-based products has been widely reported in the literature since approximately 15 years; however, researchers mainly focussed on the development of antioxidant packaging for different cheese types and whole milk powders (Table 12.2). Generally, for milk powders, volatile compounds such as pentanal, hexanal, and heptanal are used as indicators of oxidation. For instance, Granda-Restrepo et al. (2009b) studied the influence of α-tocopherol on the lipid oxidation rate of whole milk powder. The authors observed a slow increase in the concentration of volatile compounds during lipid oxidation (i.e., pentanal, hexanal, and heptanal) for whole milk powder packed in multi-layer AP, especially for samples stored at a temperature range of 30–40 °C. The heptanal concentration remained below 0.25 μg mL^{-1} at the end of the storage period for samples packed in AP. However, this type of system is temperature-dependent, as the rate of oxidation is much higher at higher temperatures. Peroxide values were also used to determine the oxidative stability of different types of cheese (Yang and Bin 2016; Pérez et al. 2021). Pérez et al. (2021)

Table 12.2 Application of antioxidant releasing packaging to dairy products

Antioxidant compound	Film or coating	Dairy product	Impact of antioxidant packaging	Reference
α-tocopherol	Multilayer film (high-density polyethylene, ethylene vinyl alcohol, and low-density polyethylene)	Whole milk powder	Better release values of antioxidant (3.06–3.11 × 10^{-11} cm^2 s^{-1}) observed at higher temperatures (30–40 °C); thus, antioxidant releasing systems delayed oxidation in the milk powder at elevated temperatures.	Granda-Restrepo et al. (2009a)
α-tocopherol, butylated hydroxyanisole (BHA), and butylated hydroxytoluene (BHT)	Multilayer film (high-density polyethylene, ethylene vinyl alcohol, and low-density polyethylene)	Whole milk powder	45% decrease in vitamin A content of whole milk packed in active antioxidant films as compared to 50% with control packaging.	Granda-Restrepo et al. (2009b)
α-tocopherol, BHA, and BHT	Poly (lactide-co-glycolide) film	Dry whole and butter milk powder	Reduced pentanal concentration for whole milk powder and butter milk powders packed in α-tocopherol- and BHA/BHT-releasing films, respectively.	Aardt et al. (2007)
BHT	Low-density polyethylene (LDPE) film	Asadero cheese	The active packaging maintained the same levels of oxidation odour for cheese samples from day 20 until the end of storage.	Soto-Cantú et al. (2008)
Catechin and gallic acid	Zein film	Kashar cheese	Lower thiobarbituric acid values (0.47–0.66 mg MDA/kg cheese) were observed for treated cheese than for control.	Ünalan et al. (2013)
Black carob extract	Carrageenan film	Semi-hard cheese	Coated samples displayed low peroxide values.	Perez et al. (2021)
Moringa Oleifera leaf extract	Gelatine film	Gouda cheese	Lowest peroxide value (3.04 meq peroxide/kg sample) observed	Yang and Bin (2016)
Pomegranate peel extract	Zein film	Himalayan cheese	Reduced lipid oxidation.	Mushtaq et al. (2018)
Green tea and Pu-erh extracts	Furcellaran-whey protein isolate film	Acid-curd cheese	Negative impact on sensory properties.	Pluta-Kubica et al. (2021)
Myrtle and rosemary extract	Polypropylene film	Mayonnaise	Significant reduction in peroxide level in treated cheese compared with that in control.	Mousapour Balegh and Yassini Ardakani (2020)

observed lower peroxide values for cheese slices packed in all-antioxidant packaging materials (<4 meq O_2 kg^{-1}) than for the control slices (>4 meq O_2 kg^{-1}) at the end of the storage period, thus indicating the antioxidant-releasing activity of the packaging. Similarly, Yang and Bin (2016) observed low peroxide values for gouda cheese samples stored in *Moringa oleifera* leaf extract-based antioxidant films, indicating the inhibition of the formation of primary oxidation products in the cheese samples. In contrast, a negative influence of antioxidant packaging on cheese has also been observed in the literature. For example, Pluta-Kubica et al. (2021) developed active whey protein (WPI) isolate films with green tea and Pu-erh extracts and observed their negative influence on the sensory parameters of the cheese samples. All samples preserved in edible films were less desirable than the control because of the staining of films on the sample surface and introduction of a sour taste to the samples.

6 Active Films and Coatings Based on Essential Oils

EOs have gained increasing attention because of their safety, high efficiency, and activity against many pathogenic bacteria (Mishra et al. 2020). The antimicrobial activity of EOs is probably associated with their hydrophobicity and is responsible for the increase in permeability of cells and loss of cellular components, leading to lysis and death (Bevilacqua et al. 2017). EOs are widely used for their antioxidant activity, which may be determined through their complex phenolic structures represented mainly by flavonoids, anthocyanins, and coumarins (Vintila 2017) (Table 12.3).

EOs have been shown to have a high efficacy in restricting the growth and survival of microorganism in cheese. Seydim et al. (2020) studied the effect of oregano EO (OEO) and garlic EO (GEO) incorporated in edible WPI films on the reduction of microbial contamination in sliced Kasar cheese. NA and NI were also included in the WPI films to compare the efficacies of these films. The results showed that WPI films incorporated with OEO showed superior antimicrobial activity against *E. coli* O157:H7 in Kasar cheese slices than the other films and that WPI films containing EOs or NI revealed a similar microbial inactivation pattern for *Salmonella enteritidis*.

Moringa oil (MO) was loaded into chitosan nanoparticles (MO@CNPs) to improve their stability and prolong their activity (Lin et al. 2019). MO@CNPs were embedded in a gelatine solution at a concentration of 9 mg mL^{-1} to produce electrospun nanofibers with antimicrobial activity. The results showed the strong antibacterial effect of the nanofibers against *L. monocytogenes* and *S. aureus* when applied to cheese stored at 4 °C and 25 °C.

An AP material composed of different concentrations of clove leaf oil incorporated in foxtail millet starch (FMS)-based films was applied to queso blanco cheese (Yang et al. 2018). These films containing clove oil showed antioxidant and antimicrobial activities against *L. monocytogenes*. In addition, FMS film packaging

Table 12.3 Application of active film and coating based on essential oil to dairy products

Active compound	film or coating	Dairy product	Impact of the active packaging	Reference
Oregano oil, garlic oil, nysin (NI), niatamicin (NA)	Whey protein (WPI), candelilla wax	Kasar cheese	WPI films containing essential oils or NI showed similar microbial inactivation pattern for *Salmonella enteritidis*. Film with NI had the highest microbial inactivation against *L. monocytogenes*, whereas WPI films with NA showed the highest reduction of *Penicillium* spp. Application of WPI films containing plant essential oils to sliced Kasar cheese caused significant microbial reductions during storage.	Seydim et al. (2020)
Moringa oil	Chitosan, gelatine nanofibers	Cheese	The nanofibers possessed strong antibacterial effect against *L. monocytogenes and S. aureus* on cheese at 4 °C and 25 °C and negligible effect on the surface colour and sensory quality of cheese during 4 days of storage.	Lin et al. (2019)
Grape seed oil	Poly (vinyl alcohol), nanofibers (gsN)	Kashar cheese	Encapsulation was effectively provided within nanofibers, which were used as coating materials in Kashar cheese. The growth of total mesophilic bacteria, yeast, and mould in food coated with electrospun gsN was effectively delayed (~1.50 log CFU/g) at the same storage conditions. Furthermore, rapid oxidation was limited (up to 0.85 mg MDA/kg) by using grape seed oil-loaded nanofibers.	Ceylan et al. (2021)

(continued)

Table 12.3 (continued)

Active compound	film or coating	Dairy product	Impact of the active packaging	Reference
Pink pepper essential oil	Cellulose acetate	Mozzarella cheese	The concentrations of 2, 4 and 6% of essential oil added to the films made them active against *L. monocytogenes and S. aureus*. The *in situ* tests demonstrated that the affinity between nonpolar molecules of essential oil and the lipid components of cheese allows the essential oil to migrate to food by direct contact, indicating favourable characteristics for its use as active packaging.	Dannenberg et al. (2017)
Oregano essential oil (OEO)	Sodium alginate, mandarin fibres	Low fat cheese	Edible coatings with at least 2.0% w/w OEO improved the microbial stability of the cheese pieces, resulting in effective decontamination of external pathogens such as *S. aureus* and preservation of cheese outward appearance during the time. Consequently, the incorporation of nanoemulsion-based edible coatings containing OEO onto low-fat cut cheese extended the shelf-life of this product	Artiga-Artigas et al. (2017)
Origanum vulgare essential oil (OR); Ethyl lauroyl arginate HCl (LAE)	PP and PET	Spanish ripened sheep cheese (Zamorano)	For inoculated cheese slices, OR and LAE PP films did not effectively decrease *E. coli* O157:H7 counts after 7 days of cold storage. PET films incorporating 6 and 8% of OR and LAE significantly ($p < 0.05$) decreased the numbers of both strains and also did 4% for the reference and wild strain depending on the antimicrobial. LAE PP, OR PET and LAE PET did not significantly ($p > 0.05$) affect sensorial characteristics of Zamorano cheese.	Otero et al. (2014)
Cymbopogon citratus essential oil	Galactomannan	Coalho cheese	A reduction in microbial growth, especially that of total coliforms was observed.	Lima et al. (2021)

(continued)

Table 12.3 (continued)

Active compound	film or coating	Dairy product	Impact of the active packaging	Reference
Clove leaf oil	Foxtail millet starch	Queso blanco cheese	The films containing essential oil showed antimicrobial activity against *L. monocytogenes*. During the storage of queso blanco cheese that was packed with the FMS film containing 1% of clove leaf oil, a decrease in microbial growth and lipid oxidation in the cheese was observed.	Yang et al. (2018)
Carum copticum essential oil	Pectin, nanoclay, beta carotene	Butter	Active film by controlling butter oxidation as well as by controlling the growth of bacteria in packaged butter can delay the oxidative and microbial corruption of local butter.	Asdagh and Pirsa (2020)
Zataria multiflora essential oil	Zein	Feta cheese	Addition of *Z. multiflora* Boiss essential oil resulted in lower count of viable *Salmonella enteritidis*, *L. monocytogenes*, *E. coli*, and *S. aureus* in the final cheese.	Sara et al. (2015)
Rosemary and thyme essential oil	Microcellular foam starch	Mozzarella cheese	The sachet containing either rosemary or thyme oil alone or both oils (1% w/w oil total) did not eliminate *L. monocytogenes* in mozzarella cheese after inoculation at approximately 3 log CFU/g. Nonetheless, the sachets retarded the growth of *L. monocytogenes* and produced a significant difference in *Listeria* counts by day 15.	Han et al. (2014)

containing 1% clove leaf oil reduced the *L. monocytogenes* population in queso blanco cheese by 1.19 log CFU g^{-1} after 24 days of storage.

The addition of *Zataria multiflora* Boiss EO to zein-based edible films for packing cheese could prevent the growth of *S. enteritidis*, *L. monocytogenes*, *E. coli*, and *Staphylococcus aureus* and preserve the overall quality and sensory attributes of treated cheese during 70 days of storage (Sara et al. 2015). Edible coatings composed of *Caesalpinia pulcherrima* seeds, galactomannan, *Cymbopogon citratus* leaf EO, and polysorbate 80 positively affected the shelf life of coalho cheese (Lima et al. 2021). Samples coated with 1% galactomannan, 0.2% EO, and 0.1% polysorbate 80 showed high inhibition of mesophilic aerobes with values below 10 CFU/g.

In addition, the coating reduced the water loss in cheese stored for 20 days and did not alter the melting capacity and instrumental texture of cheese stored for 30 days. Dannenberg et al. (2017) studied the effect of pink pepper EO as an antimicrobial component in a cellulose acetate film on sliced mozzarella cheese. The authors evaluated the action of these films via diffusion in a solid medium (agar), dispersion in liquid medium (broth), volatilisation (micro-atmosphere), and *in situ* (sliced mozzarella cheese) against *S. aureus*, *L. monocytogenes*, *E. coli*, and *Salmonella typhimurium*. Films with EO concentrations of 2%, 4%, and 6% were active against *L. monocytogenes* and *S. aureus* in all the evaluated media. *E. coli* was sensitive in the liquid medium, micro-atmosphere, and *in situ*, whereas *S. typhimurium* showed sensitivity to the films in the liquid medium and *in situ*. The *in situ* test demonstrated that the EO could migrate to food because of the affinity between the lipid components of mozzarella cheese and nonpolar molecules of the EO. Nanoemulsion edible coatings based on sodium alginate, mandarin fibre, Tween 80, and OEO at different concentrations have been successfully applied to low-fat cut cheese to extend the shelf life (Artiga-Artigas et al. 2017). In particular, edible coatings with at least 2.0% w/w OEO improved the microbial stability of cheese, resulting in the effective decontamination of external pathogens such as *S. aureus*.

Rosemary oil and thyme oil, alone or in combination (1% w/w oil total), were added into Nylon/EVOH/PE bags to evaluate whether they could improve the microbiological safety of shredded mozzarella cheese (Han et al. 2014). The combination of rosemary oil and thyme oil was more effective in retarding the growth of *L. monocytogenes* up to 2.5 log CFU/g on day 9 at 10 °C than the package containing one oil alone. Although they did not prevent the growth of LAB and total aerobic bacteria (TAB), the oils could restrict their growth. After 15 days at 10 °C, the numbers of LAB and TAB in the samples containing a sachet with both oils showed a 1.2 and 1.4 log CFU/g reduction, respectively.

Origanum vulgare essential oil (OR) was used to coat two packaging films (PP and PET), and its antimicrobial activity against two *E. coli* O157:H7 strains was evaluated using *in vitro* systems and a raw milk sheep cheese (Zamorano) (Otero et al. 2014). The results showed that the inhibitory activity of the PP and PET films coated with OR depended on the active compound concentration (above 6%), the target strain, and the packaging material, with PET being more effective than PP. For the inoculated cheese slices, films with OR moderately affected *E. coli* O157:H7 counts after 7 days of cold storage. In a study by Asdagh and Pirsa (2020), a pectin/nanoclay (montmorillonite)/*Carum copticum* essential oil/β-carotene (Pec/Clay/CCE/βC) composite film was used to package a local butter. The film was more effective against *B. cereus* than against *E. coli* and showed the highest oxidative stability, lowest microbial load, and least colour change during storage.

In a study conducted by Melo et al. (2020), lemongrass essential oil (MEO) was macro-encapsulated into microcapsules made of Arabic gum and maltodextrin and then applied to cohalo cheese. Both MEO-macro-encapsulated and non-macro-encapsulated samples could inhibit the growth of coliforms at 45 °C after 21 days of storage.

7 Conclusions

Packaging technologies in the dairy industry have been increasingly changing to match the needs of consumers asking for healthy products containing as many natural components as possible and with long shelf-lives. Novel concepts of dairy packaging include numerous possibilities that enable not only the protection of the products but also the extension of their shelf life.

The application of AP is an innovative solution for improving the safety and extending the shelf life of dairy products, such as pasteurised milk, fresh cheese, ripened cheese, and butter. Active antimicrobial packaging can be a solution for reducing dairy product-borne microbial outbreaks. However, the use of AP in the food industry, mainly in cheeses, presents some challenges. These challenges include: (i) the specific mechanism of each active compound against specific microorganisms, (ii) the impact on sensory properties, and (iii) the safety of new natural compounds that must be confirmed. Furthermore, the results, albeit promising, are not sufficient to achieve the desired objectives of safety or shelf-life extension. The possible success and future application of AP in the dairy product sector depend on the broadening of the spectrum of action, owing to the combination of multiple active compounds or with other technologies (as a form of hurdle technology). Active films could be applied as one of several hurdles in the design of active packages for dairy products, with the purpose of inhibiting the growth of psychrophilic pathogenic microorganisms such as bacteria of the genus *Listeria*. However, by broadening the spectrum of antimicrobial activity, product safety and shelf life extension can be guaranteed. Edible films and coatings represent an opportunity for cheese preservation because of their ability to act as carriers of active compounds in addition to their capacity to act as barriers to gases and moisture. Active films are promising materials for preserving the safety of cheese and extending the shelf life and are likely to have additional applications in the future when combined with new and emerging technologies.

References

Aardt MV, Duncan SE, Marcy JE, Long TE, Keefe SFO, Sims SR. Original article release of Antioxidants from poly (lactide- co -glycolide) films into dry milk products and food simulating liquids (1988). Int J Food Sci Technol. 2007:1327–37. https://doi.org/10.1111/j.1365-2621.2006.01329.x

Artiga-Artigas M, Acevedo-Fani A, Martín-Belloso O. Improving the shelf life of low-fat cut cheese using nanoemulsion-based edible coatings containing oregano essential oil and mandarin fiber. Food Control. 2017;76:1–12. https://doi.org/10.1016/j.foodcont.2017.01.001

Asdagh A, Pirsa S. Bacterial and oxidative control of local butter with smart/active film based on pectin/nanoclay/Carum copticum essential oils/β-carotene. Int J Biol Macromol. 2020;165:156–68. https://doi.org/10.1016/j.ijbiomac.2020.09.192

Bevilacqua A, Speranza B, Perricone M, Sinigaglia M, Corbo MR. Bioactivity of essential oils towards fungi and bacteria. Essent Oils Food Process Chem Safety Appl. 2017:231–46. https://doi.org/10.1002/9781119149392.ch7

Ceylan Z, Kutlu N, Meral R, Ekin MM, Kose YE. Protective effect of grape seed oil-loaded nanofibers: limitation of microbial growth and lipid oxidation in kashar cheese and fish meat samples. Food Biosci. 2021;42(March):101076. https://doi.org/10.1016/j.fbio.2021.101076

Dannenberg G d S, Funck GD, Cruxen CE d S, Marques J d L, Silva WP d, Fiorentini ÂM. Essential oil from pink pepper as an antimicrobial component in cellulose acetate film: potential for application as active packaging for sliced cheese. LWT Food Sci Technol. 2017;81:314–8. https://doi.org/10.1016/j.lwt.2017.04.002

De Lima J, Daiana G, Dannenberg S, Eduardo C, Lisie S, El M, et al. Bacteriocin-like substances of Lactobacillus curvatus P99: characterization and application in biodegradable films for control of Listeria monocytogenes in cheese. 2017;63. https://doi.org/10.1016/j.fm.2016.11.008

European Commission. (2020). EU agricultural outlook for markets, income and environment 2020 - 2030. https://doi.org/10.2762/252413.

European Parliament. Regulation EC 1935/2004 on materials and articles intended to come into contact with food. Off J Eur Communities. 2004;1935(4):1–20.

Fang Z, Zhao Y, Warner RD, Johnson SK. Active and intelligent packaging in meat industry. Trends Food Sci Technol. 2017;61(2):60–71. https://doi.org/10.1016/j.tifs.2017.01.002

Granda-Restrepo DM, Soto-valdez H, Peralta E, Troncoso-rojas R, Vallejo-córdoba B, Gámez-meza N, Graciano-verdugo AZ. Migration of a -tocopherol from an active multilayer film into whole milk powder. Food Res Int. 2009a;42(10):1396–402. https://doi.org/10.1016/j.foodres.2009.07.007

Granda-Restrepo D, Peralta E, Troncoso-rojas R, Soto-valdez H. Release of antioxidants from co-extruded active packaging developed for whole milk powder. Int Dairy J. 2009b;19(8):481–8. https://doi.org/10.1016/j.idairyj.2009.01.002

Haghighi-Manesh S, Azizi MH. Active packaging systems with emphasis on its applications in dairy products. J Food Process Eng. 2017a;40(5) https://doi.org/10.1111/jfpe.12542

Haghighi-Manesh S, Azizi MH. Production and evaluation of type of multi-layer active film for packaging of pasteurized milk. J Food Process Eng. 2017b;40(3) https://doi.org/10.1111/jfpe.12442

Han JH, Patel D, Kim JE, Min SC. Retardation of listeria monocytogenes growth in mozzarella cheese using antimicrobial sachets containing rosemary oil and thyme oil. J Food Sci. 2014;79(11):E2272–8. https://doi.org/10.1111/1750-3841.12659

Hamish Forbes, Tom Quested, Clementine O´Connor. (2021). Food Waste Index United Nations Environment Programme. In Wrap

Kim YM, An DS, Park HJ, Park JM, Lee DS. Properties of nisin-incorporated polymer coatings as antimicrobial packaging materials. Packag Technol Sci. 2002;15(5):247–54. https://doi.org/10.1002/pts.594

Kumariya R, Garsa AK, Rajput YS, Sood SK, Akhtar N, Patel S. Bacteriocins: classification, synthesis, mechanism of action and resistance development in food spoilage causing bacteria. Microb Pathog. 2019;128(October 2018):171–7. https://doi.org/10.1016/j.micpath.2019.01.002

La Storia A, Di Giuseppe FA, Volpe S, Oliviero V, Villani F, Torrieri E. Physical properties and antimicrobial activity of bioactive film based on whey protein and Lactobacillus curvatus 54M16 producer of bacteriocins. Food Hydrocoll. 2020;108(April):105959. https://doi.org/10.1016/j.foodhyd.2020.105959

Lim JY, Lee CL, Kim GH, Bang YJ, Rhim JW, Yoon KS. Using lactic acid bacteria and packaging with grapefruit seed extract for controlling Listeria monocytogenes growth in fresh soft cheese. J Dairy Sci. 2020;103(10):8761–70. https://doi.org/10.3168/jds.2020-18349

Lima, JD, Daiana, G., Dannenberg S, Eduardo C, Lisie S, El M, Dias G. (2017). Bacteriocin-like substances of Lactobacillus curvatus P99 : characterization and application in biodegradable fi lms for control of Listeria monocytogenes in cheese. 63. https://doi.org/10.1016/j.fm.2016.11.008.

Lima AEF, Andrade PL, de Lemos TLG, Uchoa DE d A, Siqueira MCA, do Egito AS, et al. Development and application of galactomannan and essential oil-based edible coatings applied to "coalho" cheese. J Food Proces Preserv. 2021;45(1):1–12. https://doi.org/10.1111/jfpp.15091

Lin L, Gu Y, Cui H. Moringa oil/chitosan nanoparticles embedded gelatin nanofibers for food packaging against Listeria monocytogenes and Staphylococcus aureus on cheese. Food Packag Shelf Life. 2019;19(December 2018):86–93. https://doi.org/10.1016/j.fpsl.2018.12.005

Mauriello G, De Luca E, La Storia A, Villani F, Ercolini D. Antimicrobial activity of a nisin-activated plastic film for food packaging. Lett Appl Microbiol. 2005;41(6):464–9. https://doi.org/10.1111/j.1472-765X.2005.01796.x

Melo AM d, Turola Barbi RC, Souza WFC d, Luna LC, de Souza HJB, Lucena GL, et al. Microencapsulated lemongrass (Cymbopogon flexuosus) essential oil: a new source of natural additive applied to Coalho cheese. J Food Proces Preserv. 2020;44(10) https://doi.org/10.1111/jfpp.14783

Mishra AP, Devkota HP, Nigam M, Adetunji CO, Srivastava N, Saklani S, et al. Combination of essential oils in dairy products: a review of their functions and potential benefits. LWT. 2020;133(March):110116. https://doi.org/10.1016/j.lwt.2020.110116

Moraes JOD, Hilton ST, Moraru CI. The effect of pulsed light and starch films with antimicrobials on Listeria innocua and the quality of sliced cheddar cheese during refrigerated storage. Food Control. 2020;112(August 2019):107134. https://doi.org/10.1016/j.foodcont.2020.107134

Mousapour Balegh S, Yassini Ardakani SA. Antimicrobial and antioxidant effects of polypropylene films containing Myrtle and Rosemary extract on mayonnaise packaging. Food Sci Technol. 2020;16(97):113–26. https://doi.org/10.29252/fsct.16.97.113

Muir DD, Banks JM. Milk and milk products. In: Kilcast D, Subramanian P, editors. The stability and shelf-life of food. Boca Raton: CRC Press; 2000. p. 197–219.

Muriel-Galet V, López-Carballo G, Gavara R, Hernández-Muñoz P. Antimicrobial food packaging film based on the release of LAE from EVOH. Int J Food Microbiol. 2012;157(2):239–44. https://doi.org/10.1016/j.ijfoodmicro.2012.05.009

Mushtaq M, Gani A, Gani A, Punoo HA, Masoodi FA. Use of pomegranate peel extract incorporated zein fi lm with improved properties for prolonged shelf life of fresh Himalayan cheese (Kalari / kradi). Innov Food Sci Emerg Technol. 2018;48(April):25–32. https://doi.org/10.1016/j.ifset.2018.04.020

Observatory, M. M. (2016). EU dairy exports to third countries.

Otero V, Becerril R, Santos JA, Rodríguez-Calleja JM, Nerín C, García-López ML. Evaluation of two antimicrobial packaging films against Escherichia coli O157: H7 strains invitro and during storage of a Spanish ripened sheep cheese (Zamorano). Food Control. 2014;42:296–302. https://doi.org/10.1016/j.foodcont.2014.02.022

Pérez JM, Moreno A, Martínez-Abad A, Cattaneo F, Zampini C, Inés Isla M, et al. Interest of black carob extract for the development of active biopolymer films for cheese preservation. Food Hydrocoll. 2021;113. https://doi.org/10.1016/j.foodhyd.2020.106436

Perez MJ, Moreno MA, Martínez-abad A, Cattaneo F, Zampini C, Isla MI, Lopez-Rubio A, Fabra MJ. Interest of black carob extract for the development of active biopolymer films for cheese preservation. Food Hydrocoll. 2021;113:106436. https://doi.org/10.1016/j.foodhyd.2020.106436

Pires ACDS, De Fátima Ferreira Soares N, De Andrade NJ, Da Silva LHM, Camilloto GP, Bernardes PC. Development and evaluation of active packaging for sliced mozzarella preservation. Packag Technol Sci. 2008;21(7):375–83. https://doi.org/10.1002/pts.815

Pluta-Kubica A, Jamróz E, Juszczak L, Krzyściak P, Zimowska M. Characterization of furcellaran-whey protein isolate films with green tea or pu-erh extracts and their application as packaging of an acid-curd cheese. Food Bioprocess Technol. 2021;14:78–92.

Robertson G. Food quality and indices of failure. Food Packag Shelf Life. 2009; https://doi.org/10.1201/9781420078459-c2

Robertson GL. Food packaging: principles and practice. 3rd ed. Taylor & Francis Group; 2013.

Sara G, Nasrin HSJ, Mehran M, Kianoush KD. Application of zein antimicrobial edible film incorporating Zataria multiflora boiss essential oil for preservation of Iranian ultrafiltered Feta cheese. Afr J Biotechnol. 2015;14(24):2014–21. https://doi.org/10.5897/ajb2014.13992

Settier-Ramírez L, López-Carballo G, Gavara R, Hernández-Muñoz P. PVOH/protein blend films embedded with lactic acid bacteria and their antilisterial activity in pasteurized milk. Int J Food Microbiol. 2020;322(July 2019):108545. https://doi.org/10.1016/j.ijfoodmicro.2020.108545

Seydim AC, Sarikus-Tutal G, Sogut E. Effect of whey protein edible films containing plant essential oils on microbial inactivation of sliced Kasar cheese. Food Packag Shelf Life. 2020;26:100567. https://doi.org/10.1016/j.fpsl.2020.100567

Soto-Cantú CD, Graciano-Verdugo AZ, Peralta E, Islas-Rubio AR, González-Córdova A, González-León A, Soto-Valdez H. Release of butylated hydroxytoluene from an active film packaging to asadero cheese and its effect on oxidation and odor stability. J Dairy Sci. 2008;91(1):11–9. https://doi.org/10.3168/jds.2007-0464

Thanakkasaranee S, Sadeghi K, Lim IJ, Seo J. Effects of incorporating calcined corals as natural antimicrobial agent into active packaging system for milk storage. Mater Sci Eng C. 2020;111:110781. https://doi.org/10.1016/j.msec.2020.110781

Ünalan IU, Arcan I, Korel F, Yemenicioğlu A. Application of active zein-based films with controlled release properties to control Listeria monocytogenes growth and lipid oxidation in fresh Kashar cheese. Innov Food Sci Emerg Technol. 2013;20:208–14. https://doi.org/10.1016/j.ifset.2013.08.004

Vasile C, Baican M. Progresses in food packaging, food quality, and safety—controlled-release antioxidant and/or antimicrobial packaging. Molecules. 2021;26:1263.

Vintila I. Basic structure, nomenclature, classification and properties of organic compounds of essential oil. Essent Oils Food Proces Chem Safety Appl. 2017:173–90. https://doi.org/10.1002/9781119149392.ch5

Yang KLH, Bin K. Application of a puffer fish skin gelatin film containing Moringa oleifera Lam. leaf extract to the packaging of Gouda cheese. J Food Sci Technol. 2016;53(11):3876–83. https://doi.org/10.1007/s13197-016-2367-9

Yang SY, Cao L, Kim H, Beak SE, Song KB. Utilization of foxtail millet starch film incorporated with clove leaf oil for the packaging of Queso Blanco cheese as a model food. Starch/Staerke. 2018;70(3–4):1–7. https://doi.org/10.1002/star.201700171

Chapter 13
Application of Releasing Packaging in Beverages

P. Garcia-Oliveira, A. G. Pereira, M. Carpena, A. Carreira-Casais, M. Fraga-Corral, M. A. Prieto, and J. Simal-Gandara

Abstract Nowadays, packaging is a crucial step to ensure the preservation of beverages during storage, transport, distribution, retailing and consumer handling. The drawbacks of traditional materials and the increasing demand for higher quality and safe products by consumers has led the packaging industry to develop new packaging systems. Many of the advances have been focused on active packaging, which aims to prolong shelf life, ensure food quality and safety and reduce the environmental impact of the packaging industry. This chapter will be focused on the main applications of the releasing active packaging systems in different beverages. Antioxidant and antimicrobial applications stand out, since oxidative reaction and microorganisms' growth are the major causes of beverage deterioration and spoilage. Regarding antioxidant packaging, several extracts and natural antioxidants have been effectively included and intentionally released from packaging. Similarly, a great variety of compounds have been employed in antimicrobial packages, like metal nanoparticles, bacteriocins or natural antimicrobials. In addition, other type of applications could be described. Functional applications, which aim to improve the organoleptic characteristics of the product and/or bring health benefits to the consumers. Some examples are the gas releasing in beer or the nutrient release in different drinks. Finally, self-cooling packages for beer and soft drinks and self-heating packages for chocolate, soup and coffee have been developed, but some of these advances have not obtained a great commercial reception. This great variety of applications show that active packaging is a promising industrial sector, which is expected to continue growing an incorporating new scientific advance.

P. Garcia-Oliveira · A. G. Pereira · M. Fraga-Corral
Nutrition and Bromatology Group, Analytical and Food Chemistry Department, Faculty of Food Science and Technology, University of Vigo, Ourense, Spain

Centro de Investigação de Montanha (CIMO), Instituto Politécnico de Bragança, Bragança, Portugal

M. Carpena · A. Carreira-Casais · M. A. Prieto (✉) · J. Simal-Gandara (✉)
Nutrition and Bromatology Group, Analytical and Food Chemistry Department, Faculty of Food Science and Technology, University of Vigo, Ourense, Spain
e-mail: mprieto@uvigo.es; jsimal@uvigo.es

S. M. Jafari, A. S. Silva (eds.), *Releasing Systems in Active Food Packaging*, Food Bioactive Ingredients, https://doi.org/10.1007/978-3-030-90299-5_13

Keywords Packaging · Beverages · Active systems · Deterioration · Preservation

Abbreviations

General

EC	European Commission
EFSA	European Food Safety Authority
EU	European Union
FCMs	Food contact materials
FDA	Food and Drugs Administration
UHT	Ultra-high temperature
US	United States

Materials & Compounds

ABTS	2,2′-azino-bis(3-ethylbenzothiazoline-6-sulfonic acid
DPPH	2,2-diphenyl-1-picrylhydrazyl radical
EDTA	Ethylenediaminetetraacetic acid
EVOH	Ethylene–vinyl alcohol
LDPE	Low-density polyethylene
PET	Polyethylene terephthalate
PHB	Polyhydroxybutyrate

1 Introduction

1.1 History, Role and Other Implications of the Beverage Packaging

Food and beverage industry are continuously growing and developing new products thanks to the new advances in food technologies and also to the increasing demands of consumers, whose interests change depending on their preferences, publicity and progressively more on product's health properties and their environmental implications (Hao et al. 2019). Although packaging is not the industry most striking at first sight, it is a powerful and competitive sector as every item produced by food and beverage industry needs to be packaged in order to store and protect the product (Chen et al. 2017). From a simple point of view, packaging needs to fulfil three objectives: food protection, identification and labelling, and marketing purposes. However, consumer demands are evolving to other interests. Nowadays, they seek healthier foods, more sustainable, improved food safety and quality, and waste reduction (Borah and Dutta 2019). This change on their preferences has led companies to face the continuous challenge of satisfying consumers' requirements. Currently, companies are forced to explore new shapes, sizes, materials and functions of packaging to boost their productivity, improve their processes and the end of the process, find the most convenient products for consumers. In general terms, packaging can be defined as a way for presenting, protecting, identifying, containing and providing compliance for a product through the different processes until they reach consumer hands (Borah and Dutta 2019). Food supply chain is usually formed by independent processes that could be group in five steps: raw material collection, food or beverage processing, distribution, retailing and consumer handling. Packaging procedure occurs during the second step and so, the product must pass through the remaining sections once it is packaged. Therefore, packaging is a crucial step as it must ensure the preservation of the product during storage, transport, distribution, retailing and consumer handling (Hao et al. 2019). By packaging, products are closer to reach higher food quality and prolong their shelf life. Food quality is modified by different parameters, fundamentally, biological factors (i.e. spoilage microorganisms), time and environmental conditions (i.e. temperature, humidity, light, contaminants presence, etc.) that can lead to product degradation, and this can be avoided by the type of packaging (Van Der Vorst et al. 2010).

The history of beverage packaging first started linked to prehistoric times when humans left their nomadic lifestyle to become a sedentary society, thus packaging arose as a technique for preserving food. Firstly, beverages were storage into wooden barrels, due to its low cost manufacturing. Later, glass containers were developed (Butler and Joseph 2008). Particularly, glass has a long history as packaging material as it guarantees healthy taste, it offers great protection due to its inflexibility and it is recyclable, among other properties. This material was one of the first used for beverage packaging and is still used today. Then, metal came and since its appearance, it has been widely employed and sometimes even categorized as the most adaptable material for its blending properties (Boarca et al. 2019). At last, plastic arrived hand in hand with the carbonated refreshment industry, particularly, in 1970 by Pepsi® who presented the 2 L bottle made of polyethylene terephthalate (PET). From this moment, plastics became the most employed materials for packaging due to their properties, as they are very malleable and synthetically safe and therefore, they have low production cost. However, plastic has shown a series of drawbacks as its variability in terms of vapors, light, gases and low molecular size particles penetrability (Borah and Dutta 2019). But most important, its widespread use has been questioned during the last years because of the ecological implications of its excessive use. For this reason, different organizations such as World Economic Forum or Ellen MacArthur Foundation have claimed that innovation systems could solve this problem (Cordier and Uehara 2019). At this context, biopolymers have been presented as an alternative for plastics as they are biodegradable, ecofriendly and an excellent vehicle for carrying substances with active properties as antimicrobial, antioxidant or food supplements. Unfortunately, these polymers do not show the key properties that plastic offers and thus, their use has been limited (Borah and Dutta 2019).

Beverages market is constantly growing and therefore, companies compete for distinguish their products, among others. New packaging approaches and trends are a promising option for increasing the products attractiveness (Borah and Dutta 2019). Traditional packaging acts as a passive barrier, but now manufacturers are intended to find modern and safe packaging systems (Boarca et al. 2019). Active packaging is aimed to prolong shelf life, ensure food quality and safety, reduce environmental impact and increase product attractiveness (Fang et al. 2017). The term refers to an active system that can influence and improve the product's quality, solving the interactions among the packaging, the product and the environment. In general, active packaging allows to control and modify the product atmosphere by means of two strategies: absorbing (scavengers) or releasing (emitters) molecules. For instance, it can act as oxygen (O_2) scavenger, carbon dioxide (CO_2) scavenger or emitter, moisture regulator, flavor or odor liberator or absorber and emitter of other compounds with bioactive properties such as antioxidant or antimicrobial (Borah and Dutta 2019; Boarca et al. 2019). A schematic picture of how molecules are absorbed or released is presented in Fig. 13.1. In summary, it is responsible for releasing or absorbing substances in a controlled way to add extra functions and improve product functionality at the same time that product deterioration is decreased and spoilage microorganisms growth is inhibited (Boarca et al. 2019). In addition, a third category has been noted: immobilized, where active agents linked

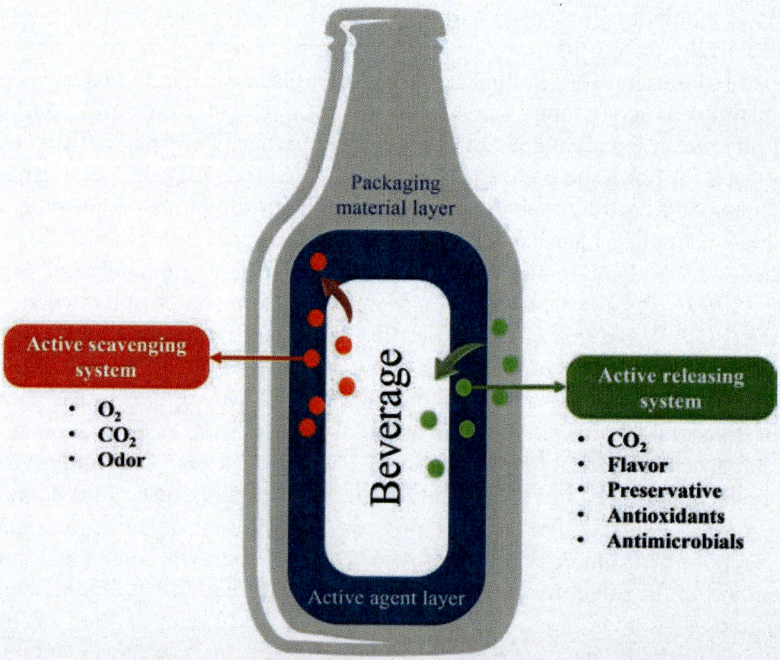

Fig. 13.1 Schematic picture of how molecules are absorbed or released in active packaging systems. (Modified from Ahmed et al. 2017)

by covalent bounds interact with the packaged beverage (Romani et al. 2020). Among active packaging releasing strategies, the most studied have been the controlled release of antioxidant and antimicrobial substances, since oxidative reactions and spoilage microorganisms are the main causing agents of degradation in beverages (Ho et al. 2004; Gómez-Estaca et al. 2014).

1.2 Legislation

As active and intelligent packaging are relatively new in food and beverage industry, legislation is needed to accomplish the requirements of safety and quality. In this regard, a brief revision of the regulations in Europe and United States (US) is presented in the following lines. In general terms of the European Union (EU), containers must fulfil the Directive 2001/95/EC known as the General Product Safety Directive that provides the guidelines for safety requirements and the related responsibilities of ensuring products safety by the producers (Cushen et al. 2012). There are also general regulations for classification, labelling and packaging of substances and mixtures, regulated by Regulation (EC) No 1272/2008 (Cushen et al. 2012). In EU, the first regulation that explicitly mentions active and intelligent packaging is the European Regulation on Materials and Articles Intended for Food Contact (EC) No 1935/2004. This regulation gives the legal frame for food packaging, also including active food contact materials (FCMs). This is a new type of material that maintains or improves the condition of the product. It regulates the use of active and intelligent packaging due to its beneficial properties on food safety, quality and shelf life. It also refers to the risk of migration of constituents that could damage human health or deteriorate food quality. In addition, products should accomplish with the Food Additive Directive (89/109/EEC) in relation to labelling of the authorized compounds allowed to be released from the active FCMs (Cushen et al. 2012; Dainelli 2015). More recently, the specific regulation (EC) No 450/2009 on Active and Intelligent Materials and Articles Intended to Come in Contact with Food was adopted in active packaging legislation. In this case, it regulates that, if the product is submitted to a legislation that limits the total quantity of a substance in a product, the released substance by the active FCMs must appear in the ingredients list. It also points out the migration issue that must undergo the same safety assessment that plastic materials, according to the European Food Safety Authority (EFSA) evaluation and an European Community (EC) authorization (Cushen et al. 2012). At this regulation is also stated the definition of active packaging system as "deliberately incorporate components that would release or absorb substances into or from the packaged food or the environment surrounding the food" (Yildirim et al. 2018). It is worth to mention that depending on the material where the active or intelligent system is incorporated, packaging must accomplish with the corresponding associated legislation. For instance, if this material is plastic, it should fulfil Regulation (EC) No. 10/2011 and its amendments (Dainelli 2015). Moreover, legislation can vary among different countries. For instance, in the case of US there is no specific regulation about active packaging, although their use is regulated by the normative provided by the FD&C Act, Section 201(s) (21 U.S.C. 321) (1958) as they act as food indirect additives and their approval is submitted to the Food and Drugs Administration (FDA) (Dainelli 2015). However, even though the regulations on the EU and in the US have been developed independently, general requirements are quite similar. Moreover, there is still an open via for further regulations and in general, future challenges in terms of legislation, innovation and risk assessment of the use of these new active packaging systems (Restuccia et al. 2010).

2 Packaging Technology

As stated before, packaging industry moves huge amounts of money annually and is subject to continuous technologic progress with the aim of reducing costs, improving preservation or being a distinctive brand, making a product recognizable just by looking at the package. This progress allows us to distinguish between traditionally used packages and those developed in recent decades.

2.1 Typical Materials Used in Beverages Packaging

Several packaging materials can be used in beverages such as plastic, paper, paperboard, cardboard based packaging, wood, metal or glass, having all been used over many years. However, despite this great variety of materials not all can be used with all beverages. A suitable choice of packaging material is essential to better preserve the food during its shelf life. The different advantages and disadvantages of each type of packaging material, summarized in Table 13.1, are of great interest for the choice of material. In many cases, to avoid the inconveniences, several materials are combined to obtain better packaging characteristics. In this way, protection is increased due to the application of a multilayer system.

One of the containers with the worst opinion by the consumer is plastic. Despite the risks present in plastic, it is still the most widely used material in beverages not only for being economical, but also for its mechanical properties that make it easy to manipulate (heat seal, optical properties, moldable) (Farmer 2013). In fact, if the most common way to see plastic packaging is in a bottle, there are also packaging in the form of bags consisting of several heat-sealed layers, used to package beverages such as milk, wine or teas (Ghoshal 2019). This material has the advantage of being able to be applied in multilayers, which reduces part of the drawbacks it presents. Its main drawbacks are a variable permeability to light, gases, vapors, and low molecular weight molecules. Different types of plastic have been developed and each one has its own characteristics and properties that will determinate its application. The most used are PET, high-density polyethylene, polycarbonate, low-density polyethylene (LDPE), polyvinyl chloride and polystyrene, being all polymeric

Table 13.1 Advantages and disadvantages of different conventional packaging materials

Material	Advantages	Disadvantages
Plastic	Recyclable	Variable permeability
	Break resistance	Limited reuse and recycling properties
	Moldable, flexible	Use of nonrenewable resources
	Lightweight	Risk of carcinogenic diseases
	Economical	May possess electrostatic charges
	Integrated into production processes	High CO_2 emissions during production
		High environmental impact during if not recycled or if incinerated
Paper o paperboard	Recyclable	Brittleness
	Lightweight	Susceptibility to breakages
	Economical	Poor barrier properties
		Not suitable for long periods
		It is usually necessary to combine it with other materials which makes it difficult to recycle
		Tears easily
Cardboard based packaging	Recyclable	Use of many materials (paper, aluminum, plastic) CO_2 emission during the production
	Lightweight	Use of some nonrenewable resources (oil and bauxite)
	Economical	
Wood	Reusable and recyclable	Expensive
	Break resistance	Poor barrier properties
		Not suitable for long periods
Metal	Recyclable	Depending on the material it can be expensive. E.g. aluminum
	Break resistance	Corrosion
	Impermeable to gases and vapors	Use of nonrenewable resources
	Good food preservation	
	Useful for heat sterilization	
	Consumer acceptance	
Glass	Reusable and recyclable	Heavy
	Odorless and chemically inert	Expensive
	Impermeable to gases and vapors	Brittleness
	Good food preservation if color is added to the glass	Susceptibility to breakages
	Useful for heat sterilization	CO_2 emission during the production
	Can see the content	Use of nonrenewable resources

materials (Núñez et al. 2015) (Table 13.2). Among all these types, the most suitable is usually PET, since it allows hot filling, a technique that permits sterilization and increases the shelf life of the packaged product. In addition, it is cheaper than other

Table 13.2 Applications of the different plastics used to make beverage containers

Polymer	Code	Uses	Symbol
PET	1	Microwavable packages.	♳
		Take-out packages.	
		Single-serving food trays.	
		Bottles: soft drinks, sport drinks, water, sauces, vegetable oil.	
HDPE	2	Juice, milk, vinegar, syrups.	♴
PVC	3	Bubble foils and foils.	♵
LDPE	4	Cans: coffee, soda, etc.	♶
PP	5	Containers: yogurt, syrup, creams.	♷
PS	6	Cups.	♸
Others	7	Depends on the polymer.	♹

Adapted from Cooper (2013)
PET polyethylene terephthalate, *HDPE* high-density polyethylene, *PVC* polyvinyl chloride polycarbonate, *LDPE* low-density polyethylene, *PP* polypropylene, *PS* polystyrene

types of packaging. PET have less barrier toward O_2, which may be corrected by using flexi composite combination such as metalized polyester/polyester/polyethylene. Thermoplastic along with paperboard and aluminum foils are used in aseptic packaging, which also act as a strong barrier to O_2 and light. These packages are characterized by being stable at room temperature (Welle 2011).

Other common materials in beverage packaging are glass and metal. The first one is used in the production of bottles for the packaging of fruit drinks, especially for those of higher quality. Furthermore, the drinks packaged with this material are usually gourmet or alcoholic. However, more and more juice is produced in bricks (Ghoshal 2019). This material has the advantage that it can be melted and re-molded as many times as desired, and glass can also be reused. Fusion process is carried out at relatively low temperatures, so it does not entail a high energy expenditure either. The latest advances in this field have also reduced the thickness of the material, reducing not only the weight of the containers, but also CO_2 emissions (Vinci et al. 2019). Regarding the second one, several metal materials have been used in beverage packaging. For example, steel is frequently used to make beverage bottles, while aluminum is typical used in carbonated refreshments to make coatings or layers. Aluminum presents the advantage of being moldable and light (Borah and Dutta 2019). Nevertheless, this material cannot be used in all beverages. For example, in wine, it develops sulfur smells. Corrosion problems can be avoided by making containers of several pieces that also prevent the release of aromas (Deshwal and Panjagari 2020). The case of alcoholic beverages is especially interesting because consumers believe that the type of packaging has a great influence on its flavor. However, different studies have not shown significant differences (Barnett et al. 2016).

2.2 New Technologies and Materials for Beverage Packaging

The increasing demand for higher quality and safe products by consumers has led the packaging industry to make some significant changes in the use of ingredients and processing systems. This last section has generated great interest since consumer's demand minimally processed, natural and fresh products, a challenge that has increased with the current globalization. Different advances have allowed to satisfy many of these needs as well as lengthening the useful life of the product (Dávila-Aviña et al. 2015). The most recent advances in packaging are not only focused on the application of multilayer systems, but also an in-depth investigation on active or intelligent packaging and materials with lower environmental impacts is being carried out (Marsh and Bugusu 2007). Therefore, the new advances are focused on changing the structure of the packaging material and the development of new systems (active or intelligent). In Table 13.3, the advantages and disadvantages of the emergent packaging systems have been summarized. The hot filling technique is based on filling the container with the hot drink (70–93 °C), and then sealed and kept it at that temperature for about 10 min, which sterilizes the product. This method is used especially for acid or acidified beverages and it has been described to improve their shelf life up to 6 months (Grumezescu and Holban 2019). Another packaging technology consists in the use of modified atmospheres. This method entails the exchanging of the gas surrounding by introducing a gas mixture in the container, optimizing the preservation of the beverage. There are different forms, being the most applied in drinks the gas-flushing and the compensated vacuum (Emblem 2013). But the systems that are grabbing the most attention are active and intelligent packaging. As mentioned before, active packaging is intended to extend the shelf life of a food, mainly achieved by controlling the permeability to O_2 and moisture as these factors facilitate microbial growth, increase oxidative reactions and induce the development of off-flavor and color changes. Thus, it allows to improve the preservation of beverages due to interactions with the product or its environment (Dombre et al. 2015). This packaging system can be applied to different products, like concentrated milk, fermented milk, probiotic-rich fermented beverage, fruit juice, wine, beer, tea, coffee and others (Ghoshal 2019). Several examples of active packaging technologies are O_2, CO_2 and ethylene scavengers, antimicrobial component releasers, moisture, flavor and odor absorbers, and ethanol emitters (Sahu 2016). The addition of active substances through the packaging instead of directly to the food may be more efficient, thus reducing the amount of compound necessary, since most degradative processes take place on the surface of food. In addition, when added to the packaging, interactions with food compounds that may inhibit or reduce their activity or be lost during food processing are avoided (Yildirim et al. 2018). Other emergent technology is the use of intelligent systems,

Table 13.3 Advantages and disadvantages of the emergent packaging systems

Packaging	Advantages	Disadvantages
Active	Increase storage stability	Can hide product defects
	Extend shelf life	Extra cost
	Slow down metabolism of food	There may be unwanted migration processes that can induce toxicity
	Control of organoleptic characteristics	Lack of recyclability
	Necessity of use lesser amounts of preservatives	Active agent should not act until packaging
	Reduce food loss and waste	Active agent must be compatible with the packaging manufacturing processes
	Packaging can be carried out with traditional equipment	The active compound may be included in another container that separates it from the food (bags), but it presents the risk of being manipulated
		Inadequate concentrations can have adverse effects not only on health, but also on food preservation
Intelligent	Provides reliable and correct information about the conditions of the food and packaging integrity	Older products can be rejected despite being in good condition
		Possible mistrust/confusion of information
		Extra cost
Hot fill	Sterilization of the interior of the packaging	Cannot be used for carbonated beverages
	Reduction of cost (PET bottles)	
	More natural	
Modified atmosphere	Increase shelf life	Does not inhibit the growth of some harmful bacteria
	Better control of food quality	Need of other preservation techniques such as refrigeration
	Longer freshness cycles	Once open the package, the protective effect is lost
	No need of chemicals	

based on the use of components that monitor the condition of packaged food or the environment surrounding the product during transport and storage (Commission E 2011). Therefore, this system allows the users to obtain reliable and accurate information about food conditions, the environment and/or the packaging integrity (Vanderroost et al. 2014).

Regarding new advancements in packaging materials, numerous biopolymers have been developed, as well as nanoparticles, used to improve the mechanical and barrier properties of packaging materials (Borah and Dutta 2019). Other challenge for packaging industry is the reduction of the waste generated. As a possible solution, edible films and coatings are being developed. However, this type of coating still requires some of the other packaging (Malhotra et al. 2015). All these new advances mean that the standards and the requirements to be met by packaging are constantly changing and updating. It is important to point out that, before its application, toxicological studies should be carried out to demonstrate the safety of the packaging used. Different surveys carried out among consumers showed that they are willing to change the packaging so that it is more sustainable as long as neither the taste nor the price of the product is changed. This last factor may be due to the belief that individual actions are not enough to contribute to a greener world (van Birgelen et al. 2009).

3 Releasing Active Packaging Applications in Beverages

Nowadays, active packaging is one of the most promising fields in beverage packaging. Examples of active packaging technologies include absorbing systems (*i.e.* oxygen scavengers, humidity absorbers, etc.) and releasing systems. The last one is a new generation of materials that possess the ability to release active compounds, such as antioxidants or antimicrobials, at controlled rates. The packaging material acts as reservoir of the active compound, released into the food in a controlled manner during the expected storage time of the product. These active molecules need to be encased appropriately, to assure the release is not too slow or too fast (Farris and Piergiovanni 2012). Numerous applications and studies based on the release of compound have been developed. Some examples are described in Table 13.4. Generally, these applications are focused in the preservation of beverage against oxidation and the action of foodborne microorganisms, but several applications search enhancing the organoleptic qualities (flavor, taste or color of the product) or provide beneficial properties for the consumers.

3.1 Antioxidant

Antioxidant packaging aims to prevent or slow down the oxidation of beverage compounds, which causes the deterioration of physical characteristics (such as flavor and color). Generally, two basic methodologies to obtain antioxidant packaging systems have been described: (1) Independent devices like sachets, pads or labels, which contain the antioxidant agent separately from the food product, added to a conventional package, and (2) Antioxidant compounds included (directly or in combination with a releasing system, such as emulsions or encapsulated) in the manufacture of the package, so these compounds further migrate into the beverage (Gómez-Estaca et al. 2014). However, sachet scavengers loss their activity rapidly when they get wet (Butler and Joseph 2008). Thus, the second option is the most used among the bibliography. The chosen antioxidants must provide no color or odor and should be thermally resistant, so it do not degrade during the processing (Lagarón and Busolo 2012).

Table 13.4 Current approach of active packaging in beverages

Active packaging	Application	Principle	Material/System	Ref
Antioxidant	Fruit juices	Release of encapsulated antioxidants	Plastic	(Gómez-Estaca et al. 2014)
	Fruit juices	Antioxidant compounds from fruit by-products	Plastic	(Ramos et al. 2015)
	Aqueous simulants	Release of flavonoid-rich cocoa extract	Plastic	(Calatayud et al. 2013)
	Aqueous simulants	Release of carvacrol and thymol	Plastic	(Ramos et al. 2014)
	Mango juice	Release of Vitamin E	Nanoemulsion	(Dasgupta et al. 2015)
	Fruit juices	Release of a-tocopherol:	Encapsulation using zein and cyclodextrins	(Saldanha do Carmo et al. 2017)
	Apple juice	Trans-cinnamaldehyde release	Chitosan-alginate nanoparticles	(Loquercio et al. 2015)
Antimicrobial	Orange juice	Chitosan and/or nisin as coating	Paperboard	(Ho et al. 2004)
	Apple and orange juice	Silver and Zinc oxide nanoparticles	Plastic	(Cushen et al. 2012)
	Aqueous simulants	Release of flavonoid-rich cocoa extract	Plastic	(Calatayud et al. 2013)
	Mango juice	Release of Vitamin E	Nanoemulssion	(Dasgupta et al. 2015)
	Kiwi juice and melon juice	Cellulose and silver nanocomposites	Plastic	(Lloret et al. 2012)
	Melon and pineapple juice	Cellulose and copper composites	Plastic	(Llorens et al. 2012)
	Orange juice, liquid egg white	Nisin bacteriocin as a polymer coating	Plastic	(Jin and Zhang 2008)
	Tomato juice	Nisin release	Nanocapsules	(Chopra et al. 2014)
	Apple juice	Nisin release	Nanofibers	(Soto et al. 2019)
	Grape juice	Natamycin release	Nanohydrogels	(Fuciños et al. 2015)
	Carrot, apple and orange juices	Release of carvacrol	Emulsion	(Char et al. 2016)
	Cantaloupe juice	Thymol and nisin release	Emulsion	(Sarkar et al. 2017)
	Carrot juice	Isoeugenol release	Emulsion encapsulation	(Krogsgård et al. 2016)
	Meat broth	Thymol, carvacrol in combination with acetic acid, lactic acid	–	(Rocha et al. 2017)
	Apple juice	Carvacrol, trans-cinnamaldehyde and thymol	–	
	Apple juice	Orange essential oil	Nanoemulsion	(Sugumar and Singh 2016)

(continued)

Table 13.4 (continued)

Active packaging	Application	Principle	Material/System	Ref
Functional	Beer	Gas emission	Plastic, metal	(Butler and Joseph 2008; Boarca et al. 2019)
	Water, other drinks	Flavor release		
	Health, wellness, and sports drinks	Nutrient release		
	Sugarcane juice	β-glucosidase release	Calcium-alginate beads	(Ephrem et al. 2018)
	Grape juice	Naringinase release	κ-carrageenan beads	
	Chocolate shake	Enhance the flavor of shakes without having to add sugars	Nanoparticles	(Ríos-Corripio et al. 2019)
	Fruit juice, tea	Fortification with nutrients by nanoencapsulation	Nanoencapsulation	
	Isotonic beverage	Color enhancement	Encapsulation with cyclodextrins	(Lobo et al. 2017)
	Pineapple juice	Enzymatic clarification	Encapsulation	(Speranza et al. 2017)
	Berry-pomegranate juice	Enzymatic clarification	Encapsulation	(Gassara-chatti et al. 2013)
	Pineapple juice	Enzymatic clarification	Inmobilization on clay	(Mohammadi et al. 2020)
	Pomegranate juice	Microencapsulation of probiotic bacteria	Alginate beads coated with chitosan	(Nualkaekul et al. 2012)
	Mandarin juice	Fortification with vitamin C	Nanoliposomes	(Liu et al. 2017)
	Apple juice	Improvement of vitamin C stability	Liposōsomes	(Wechtersbach et al. 2012)
	Functional beverages	Fortification with refined kenaf seed oil	Encapsulation with cyclodextrins	(Chin et al. 2018)
	Fruit juices	Fortification with fatty acids	Nanoencapsulation with sodium caseinate and gum Arabic	(Assadpour and Jafari 2018)
	Model beverage	Fortification by nanoencapsulation of anthocyanins	Chitosan nanoparticles	(He et al. 2017)
Self-heating	Chocolate drink, soup, coffee	Glycerol and potassium salt reaction	Plastic	(Boarca et al. 2019)
	Coffee, tea, clear soup, soup, chocolate	Calcium monoxide reaction	Metal	(Palomero et al. 2016)

(continued)

Table 13.4 (continued)

Active packaging	Application	Principle	Material/System	Ref
Self-cooling	Beer, soft drinks	Water and desiccant reaction	Plastic, metal	(Boarca et al. 2019)
	Soft drinks	Vacuum heat pump technology	Metal	(Palomero et al. 2016)

Adapted from Boarca et al. (2019)

Nowadays, there is a tendency to substitute chemical synthetized antioxidant compounds used in food and beverages, such as butylated hydroxyl anisole and butylated hydroxyl toluene, by natural ones, because the first are suspected of damaging the health of consumers. Vitamins C and E and also phenolic compounds have been used as antioxidant additives. These antioxidants may prevent the oxidation of the product by different ways, such as scavenging reactive oxygen species or chelation of transition materials (Ephrem et al. 2018). This switch could have advantages to the packaging producer's awareness and favor strategies more sustainable. For example, the project PHBOTTLE aims to develop polyhydroxybutyrate (PHB)-based materials with antioxidant properties made from sugars and carbon, nitrogen and O_2-rich residues found in the waste water from juice bottling industries (Ramos et al. 2015). Numerous scientific studies have been focused on the development of new systems that effectively release antioxidants to protect different beverage product. Although some studies have not tested the systems in beverages, some of them employed aqueous models. The positive results suggest that they could be really useful for the beverage industry. For instance, green tea extracts were added to ethylene–vinyl alcohol (EVOH) copolymer films (López De Dicastillo et al. 2011). Green tea extract is rich in flavonoids and is currently considered a safe food additive. Flavonoids, like other phenolic compounds, have gained a great attention in the last years, due to their wide range of activities, including antioxidant and antimicrobial, among others. Furthermore, they present beneficial effects on the consumers' health. The antioxidant properties of the material were evaluated by ABTS (2,2′-azino-bis(3-ethylbenzothiazoline-6-sulfonic acid) and DPPH (2,2-diphenyl-1-picrylhydrazyl radical) scavenging assays. The results showed that the extract was effectively released into the aqueous model and the material was able to reduce DPPH and ABTS radicals (López De Dicastillo et al. 2011). A similar study incorporated flavonoid rich cocoa extracts into ethyl EVOH copolymer (Calatayud et al. 2013). The antioxidant properties of the released compounds were evaluated by ABTS scavenging activity assay, using an aqueous model. The results showed that the films exerted antioxidant properties directly proportional to flavonoid content. In addition, the cocoa extract also protected Caco-2 human epithelial colorectal adenocarcinoma cell against oxidative stress induced by H_2O_2, without any cytotoxic effect. These results revealed that the developed films may be applied in the package of beverage as effectively antioxidant packaging (Calatayud et al. 2013). Novel active packaging materials have been developed by the covalent bonding of polyacrylates to tyrosol and hydroxytyrosol, two natural antioxidants compounds.

The antioxidant properties of the materials were evaluated by DPPH assay and also in fresh orange juice. The last analysis indicated the ability of both polymers to inhibit the oxidation of ascorbic acid. However, more studies are needed, to assess the safety of the migrating compounds prior to its application in the beverage industry (Fazio et al. 2017). Other example is the incorporation of two natural antioxidant compounds, carvacrol and thymol, into polypropylene films. The migration pattern was evaluated in aqueous models and the antioxidant activity was assessed by DPPH assay. The results showed that both compounds incorporated into the films (at 80 g/kg) were effectively released in the models. Furthermore, they still remaining in the material after 15 days, showing a great potential for antioxidant packaging (Ramos et al. 2014).

The diverse application of nanotechnology in the food and beverage sector have been widely recognized. Some examples are the enhanced encapsulation of selected compounds or development of innovative active packaging systems. Recently, the interest of the food and beverage industry in nanoemulsions and nanoencapsulation has increased, since they can carry and release functional lipophilic compounds. This is the case of vitamin E acetate nanoemulsions, produced using edible mustard oil and surfactant Tween-80. The antioxidant activity of these material was evaluated by DPPH, to check its potential use in beverages. The positive results showed that the nanoemulsions could be used as beverage preservative (Dasgupta et al. 2015). Other studied have employed zein and cyclodextrins to encapsulate α-tocopherol (a type of vitamin E) to improve the shelf life of fruit juices (Saldanha do Carmo et al. 2017). The antioxidant activity was estimated by the oxygen radical absorbance capacity assay and shelf-life stability tests were conducted in strawberry juice. Despite the reduced antioxidant capacity observed, the α-tocopherol-zein-cyclodextrin nanoparticles enhanced significantly the shelf life of the juice. In addition, the particles do not induced cytotoxicity, thus, they are safe to be used in the beverage industry (Saldanha do Carmo et al. 2017). Trans-cinnamaldehyde has been recognized as a safe substance and has been applied in different food applications. This compound has been encapsulated, using chitosan-alginate nanoparticles and the antioxidant activity of the nanoparticles was evaluated by DPPH assay. The nanoparticles showed a concentration dependent antioxidant activity in apple juice, compared with free trans-cinnamaldehyde (Loquercio et al. 2015).

3.2 *Antimicrobial*

The antimicrobial active packaging is an emerging market sector in beverage packaging, principally because it faces the primary responsible factor of food spoilage: microbial spoilage. Studies in active packaging have focused in the incorporation of antimicrobial compounds to eliminate or reduce the growth of pathogens, extending the shelf life of the beverage. The release rate of the antimicrobial compounds should be controlled to maintain them at a specific range, providing an equilibrium between effective inhibition of microorganisms and preservation of the quality and safety of beverages (Boarca et al. 2019; Haghighi-Manesh and Azizi 2017). Most releasing systems with antimicrobial properties include metal nanoparticles (silver, gold, copper and zinc), natural antimicrobial compounds (nisin, chitosan, essential oils, etc.) and their combinations (Palomero et al. 2016).

Nanomaterials may improve the packaging properties and provide interesting properties, including antimicrobial (Ríos-Corripio et al. 2019). In the past years, several products whose package contain nanoparticles have arrived at the market. For instance, silver and zinc oxide nanoparticles have been demonstrated to extend the self life of apple and orange juice (Cushen et al. 2012; Yildirim et al. 2018). In numerous studies, nanoparticles have been employed in combination with cellulose. For instance, silver/cellulose composite were used in the packaging of melon and kiwi juice. The results demonstrated that this system reduces the microbial growth of selected bacteria, molds and yeast during manipulation and storage (Lloret et al. 2012). Copper/cellulose composites have shown antifungal activity in pineapple and melon juice, obtaining four log cycle reductions of spoilage-related yeasts and molds, so it could be an interesting matrix for beverage packaging (Llorens et al. 2012). In addition, sub-lethal concentrations of copper inhibited the growth of *Salmonella* spp. and *Escherichia coli* if combined with lactic acid in carrot juice. Nevertheless, worldwide regulatory agencies should study the potential migration ions into the beverages and the possible health risks for the consumers before the industrial application of nanomaterials (Ríos-Corripio et al. 2019).

Several natural compounds have been used in active packaging to kill and inhibit the growth of microorganisms, being two of the most used nisin and chitosan. Bacteriocins are antimicrobial peptides synthesized from bacteria that have gained attention in the last years and several studies have demonstrated their potential in active packaging. Nisin is the most used bacteriocin, employed in packaging applications in numerous countries and has been approved by the FDA. The antimicrobial activity of nisin has been tested against different food pathogens. For example, this bacteriocin has been used in a soy protein film to inhibit the growth of *Lactobacillus plantarum* (Boarca et al. 2019). Nisin has been employed in the package of egg white and orange juice, showing inhibitory effects against bacteria and yeast and the consequent improvement of the self-life of the product (Boarca et al. 2019; Yildirim et al. 2018). Nisin/polylactic acid films have been designed for their use as packaging materials or coating on the surface for bottles to reduce the growth of microorganisms in fruit juice packaging (Jin and Zhang 2008). The effectivity of chitosan and carrageenan nanocapsules containing nisin has been evaluated in tomato juice. The releasing studies conducted *in vitro* showed that the particles released the nisin slowly and continuously during 2 weeks. Also, the antibacterial properties of the nisin were remained intact, since it inhibited the growth of *M. luteus, P. aruginosa, S. enterica and Enterobacter aerogenes, even in the sixth month of storage* (Chopra et al. 2014). A recent study has developed nanofibers using amaranth (*Amaranthus hypochondriacus)* protein isolate and pullulan with the ability to encapsulate nisin. The releasing kinetics of the nisin and the antimicrobial properties were assessed in apple juice. The fibbers effectively released the bacteriocin into the juice. In addition, the nisin maintained its antimicrobial properties, showing activity against *L. monocytogenes, Leuconostoc mesenteroides* and *S.* Typhimurium (Soto et al. 2019). The promising results suggest that this innovative material can be used in future active packaging applications. Nevertheless, other bacteriocins could be suitable for beverage protection. For example, pimaricin (natamycin) nanohydrogels have been applied in grape juice to control the microbial spoilage. The gels effectively released the piramicin and protected the juice against fungal degradation. Authors have stated the possibility of using this system, incorporating it into packaging materials to extend the shelf life of beverages (Fuciños et al. 2015).

Natural compounds and extracts with proven antimicrobial properties have been employed in active packaging applications. In the previously mentioned study of a new material based on the incorporation of cocoa extracts into EVOH polymer, the antimicrobial activity was assessed against the foodborne pathogens *S. aureus*, *L. monocytogenes*, *E. coli* and *S. enterica*. The results showed that films containing a 10%, 15%, and 20% of cocoa extract totally inhibited the growth of all microorganisms tested (Calatayud et al. 2013). The antimicrobial activity of the vitamin E acetate nanoemulsions, earlier mentioned, has been also assessed in mango juice. The nanoemulsion enhanced significantly the shelf life of the product, which was attributed to the antibacterial properties (Dasgupta et al. 2015). For years, essential oils have been used in the nutraceutical, cosmetic and pharmaceutic applications, due to its antimicrobial, antioxidant and flavoring properties. However, their inclusion in aqueous food is difficult, since they poorly soluble in water (Char et al. 2016). To solve this problem, different systems have been developed. Some examples which have proven their efficacy in different fruit juices will be described below. A nanoemulsion of orange essential oil has been formulated to prevent the spoilage of beverage produced by the yeast *Saccharomyces cerevisiae*. Its antimicrobial activity was tested in apple juice. In the highest concentrations of the nanoemulsion (8–2 μg/mL), no growth was observed during incubation. Carvacrol, a component from the essential oil of different plants, has been incorporated into carrot, apple and orange juices by emulsification. The antimicrobial activity of this compound were not altered by the process, so authors considered that the emulsions could be interesting for juice packaging (Char et al. 2016). In another study, thymol, a phenolic compound found in the essential oil of thyme, was stabilized in combination with nisin in oil-in-water emulsions, using starch octenyl succinate. The antimicrobial activity of the system was assessed in cantaloupe juice. The emulsion retained and protected the compounds effectively during the storage time. In addition, the antimicrobial properties of the compounds were not altered, since the growth of *L. monocytogenes* and *S. typhimurium* was inhibited (Sarkar et al. 2017). The antibacterial activity of chitosan-encapsulated and unencapsulated emulsions of the essential oil isoeugenol has been evaluated in carrot juice. Both systems presented antibacterial efficacy, quantified as the minimal bactericidal concentration against *E. coli*, being the encapsulated emulsion the most effective (Krogsgård et al. 2016). Other interesting compounds, chitosan and its derivatives, have been proposed for application in the beverage industry due to its antioxidant and antimicrobial, film-forming capacity, biodegradability, and recyclability. Paper-boards covered with a mixture of chitosan and/or nisin and a thermoplastic material have demonstrated to suppress the growth yeast on orange juice at low temperatures (<10 °C) (Ho et al. 2004). Recently, chitosan nanofibers were incorporated into poly(butylene adipate-co-terephthalate) films, which showed antibacterial activity against foodborne pathogens. Thus, this material is a promising option for active packaging for beverage industry (Rocha et al. 2017).

3.3 Functional

Functional packaging has gain attention in the recent years in the beverage industry. It consists mainly in the maintenance or improvement of characteristics of the packed beverage through different mechanisms. Sometimes, if a certain active packaging system implies that the beverage has beneficial properties for the health of consumers, some authors speak of the term "bioactive packaging" (Lopez-Rubio et al. 2006; Majid et al. 2018). Research and progress in this field include the development of functional beverages, nutrient delivery systems and methods for optimizing beverage characteristics, such as color, flavor and consistency (Butler and Joseph 2008). Some examples of releasing functional packaging are the gas release in beer; flavor releasing in chocolate or bottled water; nutrient release in health, wellness and sports drinks; or probiotics encapsulation into fruit juices (Ramos et al. 2015; Nualkaekul et al. 2012).

Numerous active packaging applications are based in the release of compounds to enhance the organoleptic beverage properties. Gas releasing "widget" is one of the most known applications, developed initially for canned and bottled beer products, such as Guinness®. The widget releases CO_2 to create a creamy and thick head, similar to the draught beer experience. Other breweries have adopted this system, and also other beverages, like coffee or shakes. Furthermore, the widget could prolong bubble release and generate other 'theatre effects' at the time of consumption. Beverages like fruit juices and coffee are expected to have a fresh aroma that replicates or evokes memory of the likely flavor of the product. Incorporating aromas into the packaging material may be employed to attract consumers when the container is open and balance any detrimental aroma loss. For instance, a chocolate aroma has been used in PET packaging of chocolate-flavored products. This strategy has been also applied in bottled water, to give different fruit smell to the product (Butler and Joseph 2008). Another example is the WO patent 2013032631 A1, which involves the encapsulation of aromatic compounds from essential oils in gelatin capsules. The gelatin capsules are broken when the package is opened, so the aroma compounds are released, causing a positive response from consumers (Zanetti et al. 2018). Regarding flavor improvement, several systems have been developed. In the beverage industry, enzymes are used to improve the yield and the organoleptic characteristics of the product, to appeal the consumer (Speranza et al. 2017). However, the inclusion of these compounds before packaging can be detrimental to the product. Thus, the idea of producing systems that allow the controlled release of these enzymes into the product has raised (Ramos et al. 2015). In the juice processing, the enzyme β-glucosidase is commonly use to hydrolyze the glycoside precursors of the aroma compounds, what enriches the aroma and flavor of the juice (Speranza et al. 2017). Calcium-alginate beads have been employed to encapsulate β-glucosidase and its stability was investigated in sugarcane juice. The results showed that the enzyme was more stable to temperature and pH variations and its ability to hydrolyze the glyosidic linkage between sugars and phenolic compounds remained intact. Other enzyme widely used in the beverage industry is the naringinase, which eliminates the bitterness of juices, specially of the citrus one. This enzyme has been encapsulated into κ-carrageenan beads and added to grape juice, decreasing the bitterness of the juice (Ephrem et al. 2018). Finally, some beverages contain nanoclusters that allow to enhance the taste of the product without having to add sugars to the drink (Pradhan et al. 2015).

Cyclodextrins have been considered useful compounds for the beverage industry and numerous studies have demonstrated that they protect interesting bioactive compounds, specially phenolic compounds, the antioxidant properties and, in some cases, the color and the organoleptic characteristic of beverages (Astray et al. 2020). In this aspect, a recent study evaluated the use of β-cyclodextrin to encapsulate the pigments of yellow bell pepper to provide color to isotonic beverages. Analysis demonstrated that the color provided by the encapsulated pigments were more stable than the pigment extract without cyclodextrins (Lobo et al. 2017). Other important characteristic that influences consumer's appealing is the turbidity. The turbidity and cloudy appearance of some products, specially fruit juices, is caused by the presence of polysaccharides, that usually settle during storage, leading to a decrease of product quality. Enzymes are also used to clarify juices and appeal the consumer (Speranza et al. 2017). For example, the use of hydrogels to encapsulate ligninolytic enzymes has been considered. The results showed that the encapsulation improved the thermal stability of the enzyme in a wide range of temperatures (4–75 °C). In addition, the clarification of berry-pomegranate juice was higher with encapsulated enzymes than with the free ones (Gassara-chatti et al. 2013). Similarly, encapsulated xylanase was more efficient to clarify pineapple juice, obtaining a product less viscous and with few suspended solids (Speranza et al. 2017). Recently, a study immobilized a pectinase on modified montmorillonite clay by covalent binding. The immobilization did not alter the stability of the enzyme. Furthermore, the immobilized pectinase achieved a higher clarification rate, compared with the free enzyme. Thus, this system could be helpful for the fruit juice industry (Mohammadi et al. 2020).

As it is generally known, the interest in products more natural and with beneficial properties has growth in the last years. Numerous "bioactive packaging" applications have been developed and implemented at industrial level in recent years. This packaging can be performed using diverse mechanisms: (1) The regulation of the controlled release of compounds with bioactive properties; (2) Encapsulation of bioactive agents; and (3) Use of enzymes to transform particular beverage components. To our knowledge, most of these types of applications in beverage focus on the first two mechanisms. The first one allows to release nutrients which cannot be preserved in liquids, discharging them in the product just before consumption (Majid et al. 2018). The controlled release of the bioactive compound presents several advantages such as less degradation by light and O_2 of the compound or less variation in the organoleptic properties of the products. For instance, the enterprise Atlantic Multipower Germany, Europe's leading supplier in the sports food sector, has developed a ready-to-drink creatine drink. The product has 4.6 g creatine citrate, a natural dietetic supplement that enhances the performance in sports involving intensive muscle workout. With wedge support, the dried creatine is freshly mixed with the drink when the can is opened. This system is a more comfortable way to consume creatinine, since, traditionally, this compound is sold in the form of powder or tablets that must be dissolved to be ingested. Portola company has produced the "Fusion cap". By twisting the cap, the consumers add flavors or vitamins to the bottled beverage. This two-piece, resealable cap is designed to keep the compounds powdered, tablet, or liquid separate from the beverage until the consumption time. Many different products with similar systems have been developed, controlling the release of vitamins, minerals and other compounds (Butler and Joseph 2008). Probiotics could be also released in a controlled way into the product. This is the case of a fruit drinkable yogurt packaged in a regular single-serve carton with straws. The interesting active system of this product is that the probiotic bacteria are attached inside the straw, so they are only releases into the product when liquid passes through the straw at the moment of consumption (Butler and Joseph 2008).

Encapsulation is an attractive tool when adding compounds to a beverage. This technique allows to introduce low soluble or non-stable compounds in the product, control their release and prevent their deterioration. In addition, they avoid the possible changes in the organoleptic properties associated with the incorporation of the desired compounds (Ur Rahman et al. 2019). In the literature, diverse delivery systems have been developed, including liposomes, emulsions, cyclodextrins, or nanocapsules, among others. Microencapsulation is useful for the fortification of beverage with probiotics, since it helps to protect the microorganisms against bacteriophages and contaminant yeasts and also during the processing, the storage and further passage through the gastrointestinal tract. Different polymers could be used, such as alginate, pectin, casein or starch derivatives. This technique has been employed in both in dairy and non-dairy products, mainly in fruit juices. For example, alginate beads coated with chitosan have been demonstrated to increase the survival of the probiotic bacteria *L. plantarum* during the simulated digestion of pomegranate juice. In addition, microencapsulation allowed cells to be detected until the sixth week of storage, while in the absence of the beads, the probiotic bacteria died by the fourth week (Nualkaekul et al. 2012). Other probiotic bacteria, such as *Bifidobacterium adolescentis, L. casei or L. acidophilus*, have been effectively encapsulated in fruit and vegetable juices (Ephrem et al. 2018). Nanoencapsulation has been employed to fortify different beverages, mainly fruit juices, by transporting fat-soluble compounds, such as vitamins, carotenoids, fatty acids, steroids, minerals and other compounds of interest (Ríos-Corripio et al. 2019; Musthaba et al. 2009). Diverse vitamins and lipophilic compounds have been encapsulated and the suitability of the encapsulates has been evaluated. For example, chitosan-sodium alginate based nanoliposomes have been employed to encapsulate vitamin C and fortify mandarin juice. Compared with uncoated nanoliposomes, the first ones produced less changes in organoleptic characteristics, like color, pH or acidity. Furthermore, the rate of lipid peroxidation was reduced and the vitamin C was protected after 90 days of storing (Liu et al. 2017). A similar study has stabilized vitamin C using liposomes, reducing its oxidation rate significantly in apple juice (Wechtersbach et al. 2012). Vitamin D has been encapsulated using potato proteins as protective nanovehicle. The nanoencapsulation significantly protected vitamin D and reduced its loss during pasteurization and also during simulated shelf life tests, imitating storage conditions. Thus, this system could be useful to fortify beverages and promote human health (David and Livney 2016). Different studies have developed systems to encapsulate vitamin E that could be employed in the beverage industry, using nanocapsules, nanoemulsions and cyclodextrins (Dasgupta et al. 2015; Saldanha do Carmo et al. 2017; Katouzian and Mahdi 2016; Hategekimana et al. 2015). Cyclodextrins, in combination with Arabic gum and sodium caseinate, have been also employed to encapsulate kenaf seed oil, rich in beneficial monounsaturated and polyunsaturated fatty acids, to develop functional beverages (Chin et al. 2018). Finally, the fatty acids eicosapentaenoic acid and docosahexaenoic acid, obtained from fish oils, were successfully encapsulated in sodium caseinate and gum Arabic nanoparticles to fortify fruit juices. According to the authors, this enrichment did not cause changes in the organoleptic characteristic (Assadpour and Jafari 2018). Several studies have been focused in the encapsulation of phenolic compounds. As mentioned before, these compounds present a wide range of interesting biological properties, thus, their use in functional packaging is of great interest. For example, anthocyanins have been demonstrated to possess interesting bioactive properties to combat neurodegenerative disorders, diabetes or cancer, among other diseases. However, these compounds present low stability and bioavailability. To protect these compounds, they were encapsulated into chitosan nanoparticles and its stability was studied in simulated gastrointestinal fluid and during the storage of a model beverage. The results showed that the nanoparticles significantly protected the anthocyanins (He et al. 2017). A similar strategy has been employed with the flavonoid quercetin. This compound has been demonstrated to possess beneficial properties, such as antioxidant, anti-inflammatory, cancer prevention and cardio-protective effects. Nevertheless, this compound presents a low solubility, stability and little absorption in the gastrointestinal tract, which limits its activity when consuming. Thus, nanoencapsulation of quercetin is an interesting option to improve its stability before consumption and also its gastrointestinal absorption. Lecithin and chitosan-based nanoparticles have been employed successfully to encapsulate quercetin. The stability evaluation demonstrated that the encapsulated compound was stable between 5 to 70 °C and pH 3.3 to 5.0. Furthermore, the antioxidant activity of the encapsulated compound was assessed by DPPH assay, showing better results that free quercetin. The results suggests that the nanoencapsulation of quercetin could be used to develop new nutraceutical beverages (Souza et al. 2014). Finally, cyclodextrins have been used to encapsulate caffeic acid, enhancing its solubility and antioxidant capacity. This system could be used in the development of functional food enriched in this phenolic compound (Astray et al. 2020).

3.4 *Self-Cooling and Self-Heating*

The use of packages that causes physical alterations is a curious case in the active packaging of beverage. Self-cooling packages for beer and soft drinks and self-heating packages for chocolate, soup and coffee have been one of the first developments of this field. However, some of these advances have not achieved a great commercial reception. The first self-cooling packaging was the Instant Cooling Can, created for soft drinks by Tempra Technologies and Crown Cork and Seal in 2006. It is based on vacuum heat pump technology. It uses the latent heat of evaporating water to produce the cooling effect. The water is bound in a gel layer that lines a separate container inside the beverage can and is in close thermal contact with the beverage. The system is activated when the consumer twists the base of the can to open a valve, exposing the water to the desiccant held in a separate evacuated external chamber. The water evaporates at room temperature, achieving a cooling effect as the heat is removed from the system (Ramos et al. 2015). Nowadays, self-cooling packages in larger volume formats have been designed, mostly based in endothermic reactions of ammonium nitrate (NH_4NO_3) or ammonium nitrite (NH_4NO_2). When dissolved in water, these compounds absorb heat from the system (Palomero et al. 2016). On the other hand, the self-heating packages use the exothermic hydration reactions of calcium monoxide (CaO) or magnesium monoxide (MgO). When dissolved in water, these monoxides are transformed into calcium hydroxide ($Ca(OH)_2$) and magnesium hydroxide ($Mg(OH)_2$), respectively. When using CaO, the container must show high barrier properties against water vapor passage, to avoid the hydration of lime and the rendering of the package useless. In addition, there should be enough space to contain the entire device. Another self-heating systems are based in the reaction between glycerol and potassium salt. In these case, it is necessary to adapt the heat generation, in order to control the rate of the reaction, to introduce a lag before it initiates, and to regulate the final temperature reached by the product (Ramos et al. 2015). Regarding self-heating packages, an example is a Dutch star-up, which allows to contain a dehydrated or dried drink (like coffee or tea) to infuse in a separate compartment. This system improves the shelf life of the product and the sensory characteristics of the final product, while reduces the use of preservatives (Palomero et al. 2016). In Fig. 13.2, a schematic representation of the self-cooling and self-heating packages is presented.

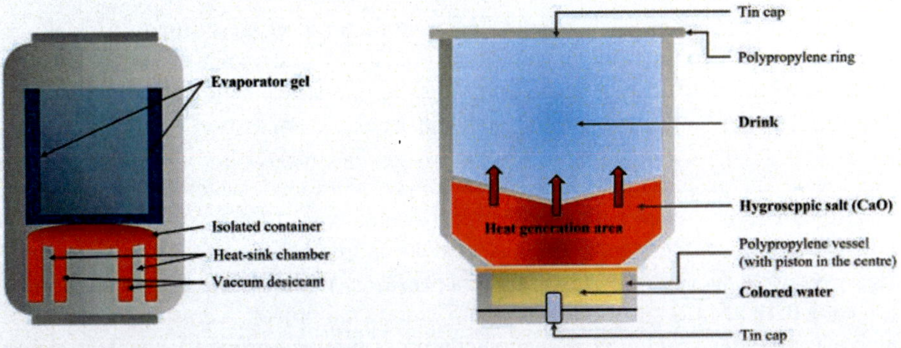

Fig. 13.2 Schematic picture of the self-cooling and the self-heating systems. (Adapted from Palomero et al. 2016)

4 Future Trends in Active Packaging

The field of beverage packaging technologies is continuously increasing, with the main goal of developing more effective systems to preserve beverage quality. Recent developments have focused on two main types: (1) packaging systems, mainly plastics and metallic materials for bottles and cans, with scavenging compounds located in their crowns and (2) new active materials, usually natural compounds or synthetic plastic films (Palomero et al. 2016). Several active materials have gaining attention in the last years, such as stimuli-responsive polymeric materials, edible films or biodegradable packages. The former ones present several advantages, such as unique properties and compliance with the current packaging regulation. Polymeric materials can release specific targets through external stimulus, which could be temperature, pH, or adjustments carried out in the chemical composition. These type of molecular structures have been only recently manufactured, due to their complex structure (Boarca et al. 2019). The second ones, edible films, are also an interesting advance in active packaging. This technology may offer future possibilities to satisfy the consumers' demands for eco-friendly, green foods. Edible films should present certain characteristics: (1) act as a barrier against CO_2, O_2, water and oil leakages, (2) microbial, biochemical, and physicochemical stability, (3) safe for the consumers and (4), reduced-cost production. In addition, edible films should be compatible vehicles for antioxidants, antimicrobials, flavor, color and nutritional additives. It has been proposed that edible films may increase the time of storage and improve the shelf life of beverage, acting as barrier against gases and moisture. In addition, they can enhance microbial resistance, by incorporation of antimicrobial compounds. However, it should be highlighted that edible films do not eliminate the necessity of traditional packaging, but they could work as a complement. Edible coatings or films are manufactured from agricultural wastes produced by industrial food production, thus, they increase the value of biomass and favor the circular economy. Several polysaccharides such as cellulose, chitosan or alginates can be used to make edible coatings. Moreover, the use of these films and coatings may help to reduce the price of traditional packaging (Boarca et al. 2019). Finally, several projects have been focused on the development of efficient biodegradable packages, that retain the freshness and nutritional composition of the beverage, while also reduce the amount of waste produced by the packaging industry. This is the case of the project PHBOTTLE, previously mentioned, which develops biodegradable materials using fruit by-products (Ramos et al. 2015).

Application of nanotechnology is also a leading theme in active packaging. However, for a successful application it is necessary to deliberate about regulations, focusing in their safety/toxicology and environmental impact. Currently, there is not any international regulation of nanotechnology or nano-products and just a few organizations from particular countries have established standards and regulations to define and control the use of nanotechnology. In the US, nanofoods and most of the food nanopackages are regulated by the FDA. Regarding EU; since 2010, the EFSA created a network between member states for risk assessment of nanotechnologies in food and feed. Some nanoforms have been approved to be used in the manufacture of plastic materials (amendments of Regulation N° 10/2011 on plastic materials and articles intended to come into contact with food) but their use must respect restrictions. In 2011, the EC suggested a definition of "nanomaterial". Later, in 2014, the FDA elaborate guidance documents for industry, recommending premarket safety assessments of the FDA-approved products that either apply nanotechnology or if the engineered product shows properties attributable to the nanoscale. The US also implemented mandatory changes to their food labeling law in 2014, which requires that all nanomaterials have to be included on the list of ingredients, followed by the term "nano" in parenthesis (Ríos-Corripio et al. 2019).

5 Conclusions

The global packaging industry is continuously introducing new technical advances, while also meeting consumer demands and the regulations established. From traditional materials, such as wood, glass or metal, this industry has evolved towards more efficient strategies to package beverage, such as the hot filling technology or the use of modified atmospheres. Nowadays, active packaging is considered as innovative field in beverage packaging, which allows preserving beverages in a more efficient way, maintaining their properties and increasing their shelf life. Currently, numerous applications of releasing active packaging have been developed and employed in different beverage (beer, shakes, fruit juices, etc.) being the most prominent the packages with antioxidant and antimicrobial activity. Antioxidant strategies are focused in the use of natural extracts and compounds with antioxidant properties, which are included in different materials and have demonstrated to prevent oxidative reactions that cause the deterioration of the product. Regarding antimicrobial applications, numerous compounds have been used to prevent microorganisms' spoilage, such as metal nanoparticles, enzymes, nisin and other natural compounds with antimicrobial properties. Several functional applications have been developed. Some of them aim to improve the organoleptic characteristics of the product, like the gas release in beer or the flavor releasing in bottled water. Besides, the objective of other applications, also known as "bioactive packaging", is bringing health benefits to consumers. This is case of the fortification of beverage with different Innutrients, such as vitamins or phenolic compounds. Probiotics could be also encapsulated and released fruit juices. Finally, several self-cooling and self-heating packages have been developed, but do not always have a good commercial reception. Considering the great variety of applications developed and their excellent results protecting beverages, the active packaging industry is a sector on the rise. It is expected to continue growing in the following years, exploring new natural compounds and towards materials such as edible films or a greater inclusion of nanotechnology. Nevertheless, since active packaging is a relatively new field, a global legislation should be established to regulate the requirements of safety and quality.

Acknowledgements The research leading to these results was funded by FEDER under the program Interreg V Spain-Portugal by POPTEC, IBERPHENOL Project (ref. 0377-Iberphenol-6-E) and by NANOEATERS Project (ref. 0181-NANOEATERS-01-E) that supports the pre-doctoral work of C. Jimenez-Lopez; by MICINN supporting the Ramón&Cajal grant for M.A. Prieto (RYC-2017-22891); by Xunta de Galicia and University of Vigo supporting the post-doctoral grant for M. Fraga-Corral (ED481B-2019/096), and the pre-doctoral grants for A.G. Pereira (ED481A-2019/0228) and P. García-Oliveira (ED481A-2019/295); by the company AlgaMar (www.algamar.com) and Axudas Conecta Peme (Xunta de Galicia) supporting the IN852A 2018/58 NeuroFood Project that financed the pre-doctoral grant for C. Lourenço-Lopes; EcoChestnut Project (Erasmus+ KA202) for supporting the work of M. Carpena; Ibero-American Program on Science and Technology (CYTED – AQUA-CIBUS, P317RT0003) and UP4HEALTH Project (H2020-BBI-JTI-2019) for financial support.

References

Ahmed I, Lin H, Zou L, Brody AL, Li Z, Qazi IM, Pavase TR, Lv L. A comprehensive review on the application of active packaging technologies to muscle foods. Food Control. 2017;82:163–78. https://doi.org/10.1016/j.foodcont.2017.06.009

Assadpour E, Jafari SM. A systematic review on nanoencapsulation of food bioactive ingredients and nutraceuticals by various nanocarriers. Crit Rev Food Sci Nutr. 2018;59:3129–51. https://doi.org/10.1080/10408398.2018.1484687

Astray G, Mejuto JC, Simal-gandara J. Latest developments in the application of cyclodextrin host-guest complexes in beverage technology processes. Food Hydrocoll. 2020;106:105882. https://doi.org/10.1016/j.foodhyd.2020.105882

Barnett A, Velasco C, Spence C. Bottled vs. canned beer: do they really taste different? Beverages. 2016;2(25) https://doi.org/10.3390/beverages2040025

Boarca B, Lungu I, Holban AM. Bioactive packaging for modern beverage industry. In: Trends in beverage packaging. Elsevier Inc.; 2019. p. 51–71.

Borah H, Dutta U. Trends in beverage packaging. In: Trends in beverage packaging. Elsevier Inc.; 2019. p. 1–19.

Butler P, Joseph K. Smart packaging technologies for fast moving consumer goods. Chichester: Wiley; 2008.

Calatayud M, López-de-dicastillo C, López-carballo G, Vélez D, Hernández P, Gavara R. Active films based on cocoa extract with antioxidant, antimicrobial and biological applications. Food Chem. 2013;139:51–8. https://doi.org/10.1016/j.foodchem.2013.01.097

Char C, Cisternas L, Pérez F, Guerrero S. Effect of emulsification on the antimicrobial activity of carvacrol. CYTA - J Food. 2016;14:186–92. https://doi.org/10.1080/19476337.2015.1079558

Chen JC, Chen TL, Harianto H. Capacity planning for packaging industry. J Manuf Syst. 2017;42:153–69. https://doi.org/10.1016/j.jmsy.2016.12.007

Chin S, Ping C, Lin K. Microencapsulation of refined kenaf (*Hibiscus cannabinus* L.) seed oil by spray drying using β -cyclodextrin/gum arabic/sodium caseinate. J Food Eng. 2018;237:78–85. https://doi.org/10.1016/j.jfoodeng.2018.05.016

Chopra M, Kaur P, Bernela M, Thakur R. Surfactant assisted nisin loaded chitosan-carageenan nanocapsule synthesis for controlling food pathogens. Food Control. 2014;37:158–64. https://doi.org/10.1016/j.foodcont.2013.09.024

Commission E. EU guidance to the commission regulation (EC) No 450 / 2009 of 29 May 2009 on active and intelligent materials and articles intended to come into contact with food. 2011. p. 1–26.

Cooper TA. Developments in bioplastic materials for packaging food, beverages and other fast-moving consumer goods. In: Trends in packaging of food, beverages and other fast-moving consumer goods (FMCG). Woodhead Publishing; 2013. p. 108–52.

Cordier M, Uehara T. How much innovation is needed to protect the ocean from plastic contamination? Sci Total Environ. 2019;670:789–99. https://doi.org/10.1016/j.scitotenv.2019.03.258

Cushen M, Kerry J, Morris M, Cruz-Romero M, Cummins E. Nanotechnologies in the food industry - recent developments, risks and regulation. Trends Food Sci Technol. 2012;24:30–46. https://doi.org/10.1016/j.tifs.2011.10.006

Dainelli D. Global legislation for active and intelligent packaging materials. In: Global legislation for food contact materials. Elsevier Ltd.; 2015. p. 184–99.

Dasgupta N, Ranjan S, Mundra S, Kumar A. Fabrication of food grade vitamin E nanoemulsion by low energy approach, characterization and its application. Int J Food Prop. 2015;19:700–8. https://doi.org/10.1080/10942912.2015.1042587

David S, Livney YD. Potato protein based nanovehicles for health promoting hydrophobic bioactives in clear beverages. Food Hydrocoll. 2016;57:229–35. https://doi.org/10.1016/j.foodhyd.2016.01.027

Dávila-Aviña JE, Solís-Soto LY, Rojas-Verde G, Salas NA. Sustainability and challenges of minimally processed foods. In: Siddiqui MW, Rahman MS, editors. Food engineering series. Springer; 2015. p. 279–95.

Deshwal GK, Panjagari NR. Review on metal packaging: materials, forms, food applications, safety and recyclability. J Food Sci Technol. 2020;57:2377–92. https://doi.org/10.1007/s13197-019-04172-z

Dombre C, Rigou P, Chalier P. The use of active PET to package rosé wine: changes of aromatic profile by chemical evolution and by transfers. Food Res Int. 2015;74:63–71. https://doi.org/10.1016/j.foodres.2015.04.026

Emblem A. Modified atmosphere packaging and other active packaging systems for food, beverages and other fast-moving consumer goods. In: Trends in packaging of food, beverages and other fast-moving consumer goods (FMCG); 2013. p. 22–34.

Ephrem E, Najjar A, Charcosset C, Greige-gerges H. Encapsulation of natural active compounds , enzymes, and probiotics for fruit juice fortification , preservation, and processing: an overview. J Funct Foods. 2018;48:65–84. https://doi.org/10.1016/j.jff.2018.06.021

Fang Z, Zhao Y, Warner RD, Johnson SK. Active and intelligent packaging in meat industry. Trends Food Sci Technol. 2017;61:60–71. https://doi.org/10.1016/j.tifs.2017.01.002

Farmer N. Trends in packaging of food, beverages and other fast-moving consumer goods (FMCG): markets, materials and technologies. Elsevier; 2013.

Farris S, Piergiovanni L. Emerging coating technologies for food and beverage packaging materials. In: Emerging food packaging technologies: principles and practice. Woodhead Publishing Limited; 2012. p. 274–302.

Fazio A, Caroleo MC, Cione E, Plastina P. Novel acrylic polymers for food packaging: synthesis and antioxidant properties. Food Packag Shelf Life. 2017;11:84–90. https://doi.org/10.1016/j.fpsl.2017.01.002

Fuciños C, Fuciños P, Fajardo P, Amado IR, Pastrana LM, Rúa ML. Evaluation of antimicrobial effectiveness of pimaricin-loaded thermosensitive nanohydrogels in grape juice. Food Bioprocess Technol. 2015;8:1583–92. https://doi.org/10.1007/s11947-015-1499-7

Gassara-chatti F, Brar SK, Ajila CM, Verma M, Tyagi RD, Valero JR. Encapsulation of ligninolytic enzymes and its application in clarification of juice. Food Chem. 2013;137:18–24. https://doi.org/10.1016/j.foodchem.2012.09.083

Ghoshal G. Recent development in beverage packaging material and its adaptation strategy. In: Trends in beverage packaging. Elsevier; 2019. p. 21–50.

Gómez-Estaca J, López-de-Dicastillo C, Hernández-Muñoz P, Catalá R, Gavara R. Advances in antioxidant active food packaging. Trends Food Sci Technol. 2014;35:42–51. https://doi.org/10.1016/j.tifs.2013.10.008

Grumezescu AM, Holban AM. Preservatives and preservation approaches in beverages. In: Volume 15: The science of beverages. Academic; 2019.

Haghighi-Manesh S, Azizi MH. Active packaging systems with emphasis on its applications in dairy products. J Food Process Eng. 2017;40 https://doi.org/10.1111/jfpe.12542.

Hao L, Li Y, Gong P, Xiong W. Material, process and business development for 3D chocolate printing. In: Fundamentals of 3D food printing and applications. Elsevier Inc.; 2019. p. 207–55.

Hategekimana J, George K, Ma J, Zhong F. Encapsulation of vitamin E: effect of physicochemical properties of wall material on retention and stability. Carbohydr Polym. 2015;124:172–9. https://doi.org/10.1016/j.carbpol.2015.01.060

He B, Ge J, Yue P, Yue X, Fu R, Liang J, Gao X. Loading of anthocyanins on chitosan nanoparticles influences anthocyanin degradation in gastrointestinal fluids and stability in a beverage. Food Chem. 2017;221:1671–7. https://doi.org/10.1016/j.foodchem.2016.10.120

Ho C, Jin H, Lee DS. Influence of antimicrobial packaging on kinetics of spoilage microbial growth in milk and orange juice. J Food Eng. 2004;65:527–31. https://doi.org/10.1016/j.jfoodeng.2004.02.016

Jin T, Zhang H. Biodegradable polylactic acid polymer with nisin for use in antimicrobial food packaging. J Food Sci. 2008;73 https://doi.org/10.1111/j.1750-3841.2008.00681.x.

Katouzian I, Mahdi S. Nano-encapsulation as a promising approach for targeted delivery and controlled release of vitamins. Trends Food Sci Technol. 2016;53:34–48. https://doi.org/10.1016/j.tifs.2016.05.002

Krogsgård C, Kjems J, Mygind T, Snabe T, Schwarz K, Serfert Y, Louise R. Enhancing the antibacterial efficacy of isoeugenol by emulsion encapsulation. Int J Food Microbiol. 2016;229:7–14. https://doi.org/10.1016/j.ijfoodmicro.2016.04.002

Lagarón J-M, Busolo MA. Active nanocomposites for food and beverage packaging. In: Emerging food packaging technologies. Woodhead Publishing Limited; 2012. p. 55–65.

Liu W, Tian M, Kong Y, Lu J, Li N, Han J. Multilayered vitamin C nanoliposomes by self-assembly of alginate and chitosan: long-term stability and feasibility application in mandarin juice. LWT - Food Sci Technol. 2017;75:608–15. https://doi.org/10.1016/j.lwt.2016.10.010

Llorens A, Lloret E, Picouet P, Fernandez A. Study of the antifungal potential of novel cellulose/copper composites as absorbent materials for fruit juices. Int J Food Microbiol. 2012;158:113–9. https://doi.org/10.1016/j.ijfoodmicro.2012.07.004

Lloret E, Picouet P, Fernández A. Matrix effects on the antimicrobial capacity of silver based nanocomposite absorbing materials. LWT - Food Sci Technol. 2012;49:333–8. https://doi.org/10.1016/j.lwt.2012.01.042

Lobo FATF, Silva V, Domingues J, Costa V, De LKG. Inclusion complexes of yellow bell pepper pigments with β -cyclodextrin: preparation, characterisation and application as food natural colorant. J Sci Food Agric. 2017;98:2665–71. https://doi.org/10.1002/jsfa.8760

López De Dicastillo C, Nerín C, Alfaro P, Catalá R, Gavara R, Hernández-Muñoz P. Development of new antioxidant active packaging films based on ethylene vinyl alcohol copolymer (EVOH) and green tea extract. J Agric Food Chem. 2011;59:7832–40. https://doi.org/10.1021/jf201246g

Lopez-Rubio A, Gavara R, Lagaron JM. Bioactive packaging: turning foods into healthier foods through biomaterials. Trends Food Sci Technol. 2006;17:567–75. https://doi.org/10.1016/j.tifs.2006.04.012

Loquercio A, Castell-perez E, Gomes C, Moreira RG. Preparation of chitosan-alginate nanoparticles for trans -cinnamaldehyde entrapment. J Food Sci. 2015;80:N2305–15. https://doi.org/10.1111/1750-3841.12997

Majid I, Thakur M, Nanda V. Innovative and safe packaging technologies for food and beverages: updated review. In: Innovations in technologies for fermented food and beverage industries. Cham: Springer; 2018. p. 257–87.

Malhotra B, Keshwani A, Kharkwal H. Antimicrobial food packaging: potential and pitfalls. Front Microbiol. 2015;6:611. https://doi.org/10.3389/fmicb.2015.00611

Marsh K, Bugusu B. Food packaging - roles, materials, and environmental issues: scientific status summary. J Food Sci. 2007;72:R39–55. https://doi.org/10.1111/j.1750-3841.2007.00301.x

Mohammadi M, Rezaei R, Shahvalizadeh R, Sarabandi K. Immobilization and stabilization of pectinase on an activated montmorillonite support and its application in pineapple juice clarification. Food Biosci. 2020;36:100625. https://doi.org/10.1016/j.fbio.2020.100625

Musthaba SM, Baboota S, Ahmed S, Ahuja A, Ali J. Status of novel drug delivery technology for phytotherapeutics. Expert Opin Drug Deliv. 2009;6:625–37. https://doi.org/10.1517/17425240902980154

Nualkaekul S, Lenton D, Cook MT, Khutoryanskiy VV, Charalampopoulos D. Chitosan coated alginate beads for the survival of microencapsulated *Lactobacillus plantarum* in pomegranate juice. Carbohydr Polym. 2012;90:1281–7. https://doi.org/10.1016/j.carbpol.2012.06.073

Núñez O, Gallart-Ayala H, Martins CPB, Lucci P. Fast liquid chromatography-mass spectrometry methods in food and environmental analysis. World Scientific; 2015.

Palomero F, Morata A, Suárez-Lepe J, Calderón F, Benito S. Antimicrobial packaging of beverages. Elsevier Inc.; 2016.

Pradhan N, Singh S, Ojha N, Shrivastava A, Barla A, Rai V, Bose S. Facets of nanotechnology as seen in food processing, packaging, and preservation industry. Biomed Res Int. 2015;2015. https://doi.org/10.1155/2015/365672

Ramos M, Beltrán A, Peltzer M, Valente AJM, Garrigós C. Release and antioxidant activity of carvacrol and thymol from polypropylene active packaging films. LWT - Food Sci Technol. 2014;58:470–7. https://doi.org/10.1016/j.lwt.2014.04.019

Ramos M, Valdés A, Mellinas A, Garrigós M. New trends in beverage packaging systems: a review. Beverages. 2015;1:248–72. https://doi.org/10.3390/beverages1040248

Restuccia D, Spizzirri UG, Parisi OI, Cirillo G, Curcio M, Iemma F, Puoci F, Vinci G, Picci N. New EU regulation aspects and global market of active and intelligent packaging for food industry applications. Food Control. 2010;21:1425–35. https://doi.org/10.1016/j.foodcont.2010.04.028

Ríos-Corripio RCMA, López-Díaz AS, Ramírez-Corona N, López-Malo A, Palou E. Metallic nanoparticles: development, applications, and future trends for alcoholic and nonalcoholic beverages. Elsevier Inc.; 2019.

Rocha MAM, Coimbra MA, Nunes C. Applications of chitosan and their derivatives in beverages: a critical review. Curr Opin Food Sci. 2017;15:61–9. https://doi.org/10.1016/j.cofs.2017.06.008

Romani VP, Martins VG, Goddard JM. Radical scavenging polyethylene films as antioxidant active packaging materials. Food Control. 2020;109:106946. https://doi.org/10.1016/j.foodcont.2019.106946

Sahu JK. Introduction to advanced food packaging processing technologies. CRC Press; 2016. p. 302–9.

Saldanha do Carmo C, Maia C, Poejo J, Lychko I, Gamito P, Nogueira I, Bronze MR, Serra AT, CMM D. Microencapsulation of a-tocopherol with zein and b-cyclodextrin using spray drying for colour stability and shelf-life improvement of fruit beverages. RSC Adv. 2017;7:32065–75. https://doi.org/10.1039/c6ra25946d

Sarkar P, Bhunia AK, Yao Y. Impact of starch-based emulsions on the antibacterial efficacies of nisin and thymol in cantaloupe juice. Food Chem. 2017;217:155–62. https://doi.org/10.1016/j.foodchem.2016.08.071

Soto KM, Hernández-Iturriaga M, Loarca-Piña G, Luna-Bárcenas G, Mendoza S. Antimicrobial effect of nisin electrospun amaranth: pullulan nanofibers in apple juice and fresh cheese. Int J Food Microbiol. 2019;295:25–32. https://doi.org/10.1016/j.ijfoodmicro.2019.02.001

Souza MP, Vaz AFM, Vicente AA, Carneiro-da-cunha MG. Quercetin-loaded lecithin / chitosan nanoparticles for functional food applications. Food Bioprocess Technol. 2014;7:1149–59. https://doi.org/10.1007/s11947-013-1160-2

Speranza B, Petruzzi L, Bevilacqua A, Gallo M, Campaniello D, Sinigaglia M, Corbo MR. Encapsulation of active compounds in fruit and vegetable juice processing: current state and perspectives. J Food Sci. 2017;82:1291–301. https://doi.org/10.1111/1750-3841.13727

Sugumar S, Singh S. Nanoemulsion of orange oil with non ionic surfactant produced emulsion using ultrasonication technique: evaluating against food spoilage yeast. Appl Nanosci. 2016;6:113–20. https://doi.org/10.1007/s13204-015-0412-z

Ur Rahman U, Sahar A, Ishaq A, Khalil AA. Design of nanoparticles for future beverage industry. In: Nanoengineering in the beverage industry. Academic; 2019. p. 105–36.

van Birgelen M, Semeijn J, Keicher M. Packaging and proenvironmental consumption behavior: investigating purchase and disposal decisions for beverages. Environ Behav. 2009;41:125–46. https://doi.org/10.1177/0013916507311140

Van Der Vorst JGAJ, Van Der Zee DJ, Tromp SO. Simulation modelling for food supply chain redesign; integrated decision making on product quality, sustainability and logistics. In: Delivering performance in food supply chains. Woodhead Publishing Limited; 2010. p. 387–415.

Vanderroost M, Ragaert P, Devlieghere F, De Meulenaer B. Intelligent food packaging: the next generation. Trends Food Sci Technol. 2014;39:47–62. https://doi.org/10.1016/j.tifs.2014.06.009

Vinci G, D'Ascenzo F, Esposito A, Musarra M. Glass beverages packaging: innovation by sustainable production. Trends Beverage Packag. 2019;16:105–33. https://doi.org/10.1016/b978-0-12-816683-3.00005-0

Wechtersbach L, Ulrih NP, Cigic B. Liposomal stabilization of ascorbic acid in model systems and in food matrices. LWT - Food Sci Technol. 2012;45:43–9. https://doi.org/10.1016/j.lwt.2011.07.025

Welle F. Twenty years of PET bottle to bottle recycling - an overview. Resour Conserv Recycl. 2011;55:865–75. https://doi.org/10.1016/j.resconrec.2011.04.009

Yildirim S, Röcker B, Pettersen MK, Nilsen-Nygaard J, Ayhan Z, Rutkaite R, Radusin T, Suminska P, Marcos B, Coma V. Active packaging applications for food. Compr Rev Food Sci Food Saf. 2018;17:165–99. https://doi.org/10.1111/1541-4337.12322

Zanetti M, Carniel TK, Dalcanton F, dos Anjos RS, Gracher Riella H, de Araújo PHH, de Oliveira D, Antônio Fiori M. Use of encapsulated natural compounds as antimicrobial additives in food packaging: a brief review. Trends Food Sci Technol. 2018;81:51–60. https://doi.org/10.1016/j.tifs.2018.09.003

Chapter 14
Application of Releasing Active Packaging in Cereals and Cereal Based Products

Sara Hedayati, Vahid Baeghbali, and Seid Mahdi Jafari

Abstract Cereal products are used as staple food items around the world. Many of these products such as breads, cakes, muffins, cookies and fresh pasta are perishable and have a short shelf life. Nevertheless, the internationalization of markets has increased distribution distances of these products. Therefore, longer storage periods are needed to meet the requirements of market. Active substances such as antimicrobials can be incorporated to the bulk of cereal products, but, microbial growth and degradation generally occur at their surface. Packaging is an important part of food production systems which can extend their shelf life. Active packaging is the system in which a certain material is incorporated into the packaging to preserve or improve the quality of packaged foods or prolong their shelf life. They may have releasing, scavenging or removing properties. Releasing active packaging systems such as essential oil releasing, ε-Polylysine releasing, ethanol releasing, metal ions releasing and chemical preservatives releasing active packaging have been used for cereal products packaging and has shown promising effects in prevention or retarding microbial spoilage, staling and oxidation.

Keywords Releasing active packaging · Cereal products · Shelf-life · Preservation

S. Hedayati
Nutrition Research Center, Shiraz University of Medical Sciences, Shiraz, Iran

V. Baeghbali
Nutrition Research Center, Shiraz University of Medical Sciences, Shiraz, Iran

Department of Food Hygiene and Quality Control, Shiraz University of Medical Sciences, Shiraz, Iran

S. M. Jafari (✉)
Faculty of Food Science and Technology, Gorgan University of Agricultural Sciences and Natural Resources, Gorgan, Iran
e-mail: smjafari@gau.ac.ir

S. M. Jafari, A. S. Silva (eds.), *Releasing Systems in Active Food Packaging*, Food Bioactive Ingredients, https://doi.org/10.1007/978-3-030-90299-5_14

1 Introduction

Cereals are the most important food crops cultivated around the world. A large amount of cereal grains is milled to produce products such as breads, buns, cakes, cookies and many other perishable commodities. These products remain fresh for a short time after leaving the oven. Microbial growth and staling are two key factors in deteriorating the quality of cereal products. Intervening on formulation and applying different packaging systems such as active packaging and modified atmosphere packaging (MAP) strategies have been suggested to improve the shelf life of these products by retarding staling and microbial growth (Pasqualone 2019). Staling occurs due to physical and chemical changes of crust and crumb during storage. Crumb hardening is the main consequence of staling which can be reduced by the addition of anti-staling agents to the formulation of baked goods. Throughout baking operation most of the microorganisms are destroyed but, during cooling, slicing and packaging, recontamination may occur and cause microbial growth (Cauvain 2017). The microbial spoilage in cereal products is mainly due to mold growth which led to the formation of mycotoxins, off-flavor and an unpleasant appearance (Nielsen and Rios 2000). Molds are aerobic thus, MAP systems with low O_2 and high CO_2 levels seem to be effective in mold growth prevention. However, several studies have revealed that mold growth can occur in MAP with high CO_2 concentrations when O_2 is present (Suhr and Nielsen 2005; Tabak and Cooke 1968). Therefore, complete O_2 elimination is required to extend the shelf life of baked goods. Many of the bakery products have highly porous texture which does not allow the complete O_2 elimination (Soares et al. 2002). Thus, application of MAP without incorporation of antimicrobial agents is not very effective in controlling mold growth. *Rhizopus stolonifer* (black bread mould), *Aspergillus* and *Penicillium* genera are the most predominant microorganisms in bakery products spoilage however certain types of yeast and bacteria may also cause spoilage of cereal products (Sachdeva et al. 2017; Pateras 2007). Incorporation of preservatives such as propionates, organic acids or essential oils to the formulation of cereal products can be effective in preventing the microbial spoilage. Nevertheless, when the active substances are directly added to food matrix, the interactions between food components and the active substances, decrease or inhibit their desired activity. On the other hand, microbial growth or degradation occurs at the surface of products. Therefore, the addition of active materials to packaging system instead of direct incorporation to food, decreases the required amount of these substances while increasing their effectiveness (Yildirim et al. 2018). Active Packaging is an innovative type of packaging system in which the interaction between package, environment and the product leads to shelf life extension, sensory and safety improvement of the product (Suppakul et al. 2003). In active packaging a certain material is incorporated into the packaging to impart an active function. Active packaging can improve or preserve the quality of foods or prolong their shelf life. They may have releasing or scavenging properties to control food quality and microbial growth (Day and Potter 2011). In scavenging systems, the undesired compounds such as moisture, oxygen, carbon dioxide, ethylene or odor are removed from food to its environment while in releasing systems, antimicrobial compounds, antioxidants, carbon dioxide, flavors, ethanol or ethylene are added to the packed food or the headspace (Yildirim et al. 2018). Releasing active packaging systems have been extensively used for shelf-life extension of cereal based products which will be discussed in this chapter.

2 Releasing Active Packaging Systems for Cereal Products

2.1 Essential Oil and Plant Extract Releasing Active Packaging

In recent years, consumers have become more interested in decreasing the consumption of synthetic food preservation due to the health benefits of individuals and the environmental issues. This trend has resulted in increasing their demand for mildly preserved food products and application of natural additives such as essential oils (EOs) (Alves-Silva et al. 2013). EOs are highly concentrated aromatic compounds produced by plants as secondary metabolite. EOs are volatile and easily evaporated which are extracted from seeds, flowers, stems, bark, resin, leaves, roots or fruit rinds and cause a wonderful scent (Hanif et al. 2019). In addition to odoriferous, therapeutic, insecticidal and antioxidant properties, EOs have strong antimicrobial activity against yeast, molds and bacteria. Nevertheless, Gram positive bacteria are more sensitive than Gram negative bacteria because their membrane don't have hydrophilic lipopolysaccharides (Nikaido 2003). EOs contain active antimicrobial compounds such as terpenes, terpenoids, phenylpropenes, isothiocyanates, allicin etc. (Hyldgaard et al. 2012). Due to their lipophilic nature, EOs pass the cell wall and cytoplasmic membrane and damage the microbial cells (Seow et al. 2014). The antimicrobial properties of EOs are affected by their concentration, composition and structure (Burt 2004). Several mechanisms such as cytoplasm coagulation, cell wall damage, membrane lesion, increasing cell permeability, hydrolysis of ATP, destruction of membrane proteins and reduction of ergosterol have been suggested for antimicrobial activity of EOs and their constituents (Khorshidian et al. 2018; Pinto et al. 2009). Most of EOs are categorized as GRAS (Moleyar and Narasimham 1992) thus, EOs or their constituents have the capacity to be used in releasing active packaging. Cinnamon, clove, thymol, oregano, star anise EOs or their active constituents have been widely used in edible coatings and films for cereal products packaging. Cinnamaldehyde is an active component of cinnamon EOs. It can inhibit the microbial cell division and prolong the shelf-life of cereal products (Sachdeva et al. 2017). Balaguer et al. (2013) investigated the antifungal activity of gliadin films incorporated with 1.5, 3 and 5% cinnamaldehyde on sliced bread. They tested the active food packaging system against *Aspergillus niger* and *Penicillium expansum*. Incorporation of cinnamaldehyde into gliadin films was very effective against fungal growth. The results revealed that films containing 3% cinnamaldehyde completely prevent the growth of both fungi during 10 days of storage while mold growth was observed after 4 days in control bread slices. Increasing the concentration of cinnamaldehyde to 5% in gliadin films increased the shelf-life of sliced bread and molds appeared on bread slices after 27 days. Figure 14.1 shows the influence of cinnamaldehyde on fungal growth in bread samples. In vitro studies confirmed that the cinnamaldehyde content of films after storage at 20 °C for 45 days was adequate to prevent the growth of *P. expansum* and *A. niger*.

Cashew gum (CG) and gelatin (G) films incorporated with *Cymbopogon citratus* essential oil (EO) as an antimicrobial agent and ferulic acid (FA), as a cross linker were used for bread packaging by Oliveira et al. (2020). They reported that incorporation of EO and FA did not have substantial effects on opacity and water vapor permeability (WVP) of films. Whereas the solubility of the films decreased significantly owing to cross-linking. Scanning electron microscopy (SEM) images revealed films have smooth surfaces without irregularities. The presence of EO in cross-linked films decreased the rupture stress but increased elongation-at-break. The permeability of films was higher than polyethylene as a commercial packaging material and caused more harshness in bread samples. However, the preservation period of bread was extended to 6 days while it was 3 days for the polyethylene

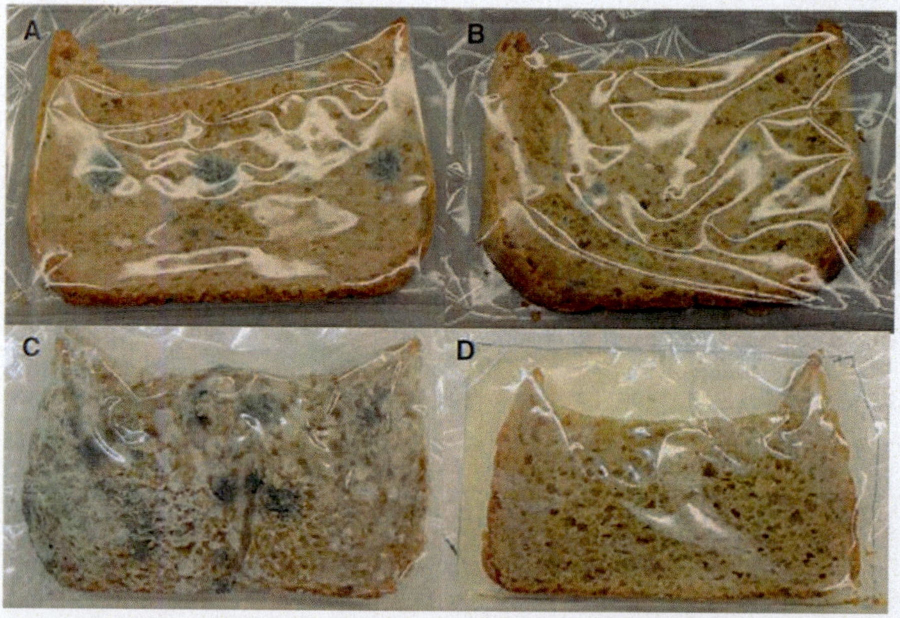

Fig. 14.1 Influence of packaging systems on growth of inoculated *Penicillium expansum* and naturally available fungi in bread. (**a**) inoculated bread sample in polypropylene bag after 14 days of storage, (**b**) uninoculated bread in polypropylene bag after 14 days of storage, (**c**) uninoculated bread in polypropylene bag after 1 month of storage, (**d**) inoculated bread in active packaging (polypropylene bag + cinnamaldehyde incorporated gliadin film) after 1 month of storage (Balaguer et al. 2013)

bags. Heras-Mozos et al. (2019) produced an antimicrobial film containing garlic extract to extend the shelf-life of preservative-free sliced pan loaf. They added 0.25 and 0.5% of garlic extract to ethylene-vinyl alcohol copolymer (EVOH), polyethylene (PE) aqueous emulsion and zein hydroalcoholic solutions to produce PE/EVOH, PE/PE and PE/zein active composite films. The antimicrobial activity of films was investigated in vitro against *Penicillium expansum*, and also in sliced bread. They found that all the composite films have antimicrobial properties. However, PE/zein films with 0.5% garlic extract showed the best results and breads remained mold free for 30 days. In another study, they added garlic extract to low density polyethylene (LDPE) coated with ethylene vinyl alcohol (EVOH) copolymer. They reported that garlic extract release offers antimicrobial properties and delayed mold development on bread slices from 6 to 12 days. However, breads have an inappropriate aroma which was masked by the addition of bread aroma and sensory results showed that breads were acceptable for consumers (Heras-Mozos et al. 2018).

Encapsulation of EOs for their controlled release is effective to increase the antimicrobial activity of EOs during storage. Ju et al. (2020) produced sachets containing microencapsulated essential oils (ASEOs-CM) to prolong bread shelf life. They used high density polyethylene (HDPE), low density polyethylene (LDPE) and polypropylene (PP) bags for bread packaging reported that ASEOs-CM was highly effective against mold and yeast growth *in vitro* and in sliced bread. Breads were placed into different packages (HDPE, LDPE and PP) had a shelf life of 15, 13, and 13 days, respectively, whereas control bread samples had a shelf life of 5 days. The release of EOs from ASEOs-CM fitted to first-order kinetics and the amount of eugenol and citral EOs increased in bread samples during storage. Application of electrospinning for fabrication of EOs loaded fibrous films increases the efficiency of active substances in antimicrobial films. In a study by Altan et al. (2018) carvacrol loaded fibrous films were prepared from poly(lactic acid) and zein by electrospinning and the influence of this releasing active food packaging on shelf life extension of whole wheat bread was investigated. They used different concentrations (0, 5, 10 and 20%) of carvacrol and found that morphology and size of fibers were affected by carvacrol content. The Fourier transform infrared (FTIR) spectroscopy and thermogravimetric analysis (TGA) results revealed that carvacrol was encapsulated in fibrous films. The antioxidant activity of zein films was significantly higher than PLA films. A sustained controlled release of carvacrol was observed during 7 days of storage and bread shelf life was significantly enhanced. As Figs. 14.2 and 14.3 illustrate that the microbial growth inhibition rate was increased with carvacrol concentration and mold growth was not observed in samples with high levels of carvacrol.

In another study, Fonseca et al. (2020) used electrospinning to produce starch/carvacrol nanofibers and evaluated their antifungal activity *in vitro* or in bread samples. The nanofibers containing carvacrol were significantly effective against *Aspergillus flavus* and *Penicillium* sp. Thus, nanofibers are promising options for bread packaging in food industry.

Fungi such as *Aspergillus flavus, Penicillium roqueforti, P. commune* and *Endomyces fibuliger* can grow at very low oxygen levels and *E. fibuliger* can grow

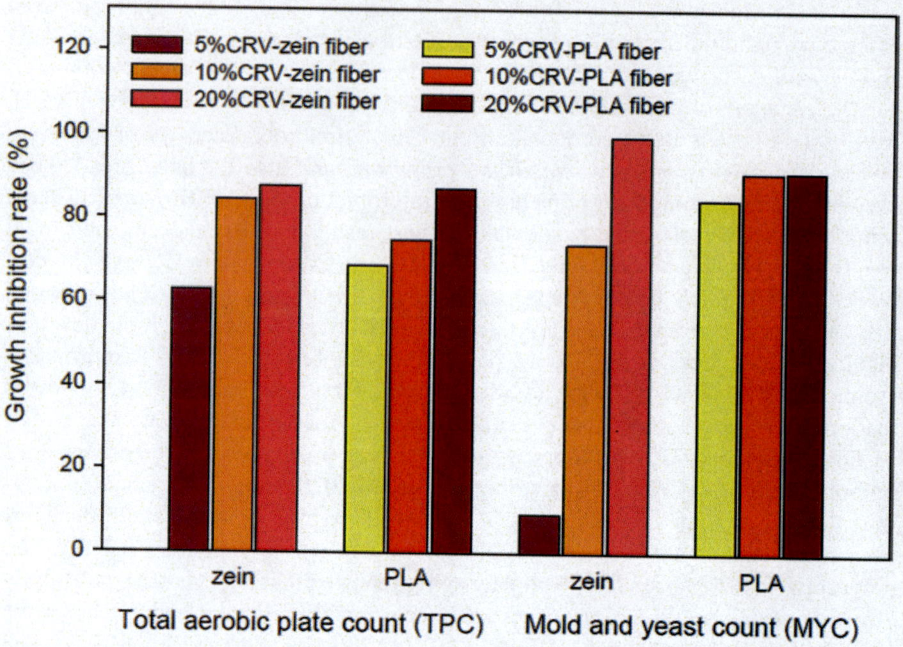

Fig. 14.2 Influence of carvacrol concentration and packaging material ((poly(lactic acid) or zein)) on growth inhibition rate (%) of aerobic bacteria and mold and yeast (Altan et al. 2018)

even in the presence of oxygen absorbers. High levels of CO_2 in modified atmosphere packaging (MAP) can retard the growth of these microorganisms but not completely. Application of EOs and oleoresins (OL) from herbs and spices is an alternative to MAP. Nielsen and Rios (2000) tested different concentrations (1, 10 or 100 μl) of EOs (mustard and garlic) and OL (cinnamon, clove, oregano and vanilla) against bread spoiling fungi and compared them with MAP. They inoculated the Petri dishes with the common fungi found in bread, added OL or EO to a filter paper placed in the lid and hermetically sealed the Petri dishes. Mustard had the most effective EO. Garlic, clove and cinnamon, also had strong effects, whereas oregano OL weakly inhibited fungi growth and vanilla didn't have inhibitory effect at the applied levels. *A. flavus* was the most resistant and *P. roqueforti* was the most sensitive microorganism. Allyl isothiocyanate (AITC) is the active component of mustard EO. The minimal inhibitory concentration (MIC) values for AITC against different molds and yeasts was between 1.8 and 3.5 mg/ml. They found that AITC can have fungistatic or fungicidal properties depending on AITC and spore concentration. The MAP could not prevent fungi growth completely and yeasts were able to grow even in the presence of oxygen absorber in MAP systems. These findings are in agreement with the results described by Gutiérrez et al. (2011). They compared the effect of MAP and cinnamon EOs releasing active packaging on shelf life extension of gluten-free bread. The results revealed that active packaging was more

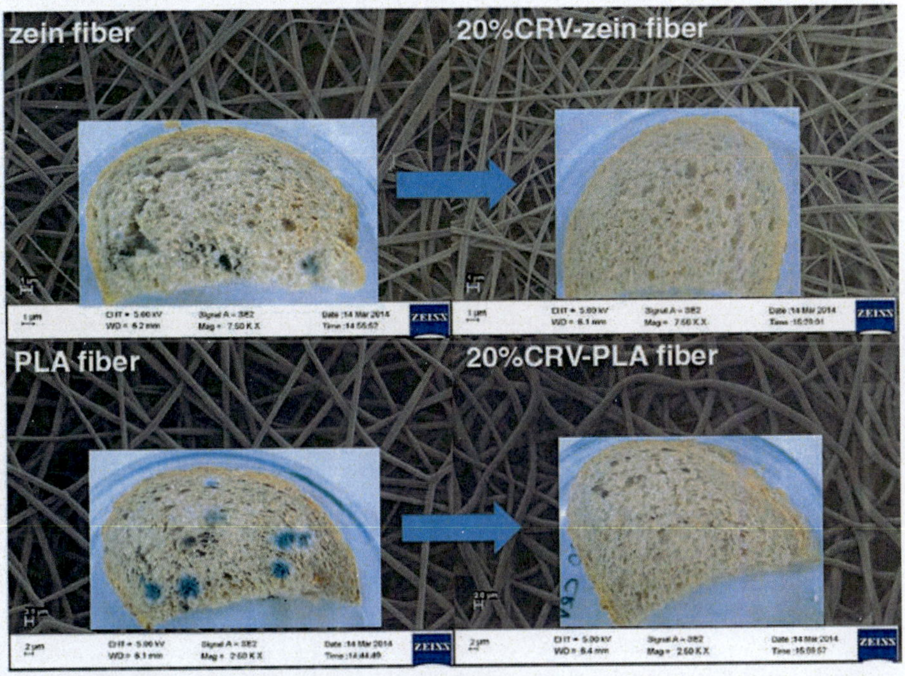

Fig. 14.3 Images of bread samples after 7 days of storage at 25C in electrospun PLA or zein films with and without carvacrol (Altan et al. 2018)

effective in increasing bread shelf life and provided better sensory attributes than MAP.

Some other studies incorporating EOs or their constituents to cereal products are summarized in Table 14.1.

2.2 *Ethanol Releasing Active Packaging*

Ethanol is naturally produced in dough during fermentation which helps its preservation before baking. But throughout baking operation alcohol is evaporated and removed from bakery products. Direct or indirect post-baking ethanol treatment can be used to prolong shelf life of baked goods. Spraying ethanol on the surface of products such as bread, cake and pizza has been shown to increase their shelf life by preventing or hindering microbial spoilage, oxidative changes and staling in these products. (Suppakul et al. 2003). Ethanol can be used for packaging of different types of baked goods. In a study by Latou et al. (2010) the impact of ethanol emitter (EE) or combination of ethanol emitter and oxygen absorber (EE OA) packaging systems on shelf life of sliced wheat bread was compared with commercial preservatives. Physicochemical, microbiological and sensory attributes of bread samples

Table 14.1 Influence of different EOs and their active constituents on shelf life of cereal products

Product	Packaging material	Active compound	Target microorganism	Result	Reference
Moon cake	Temperature-sensitive polyurethane (TSPU) films	Cinnamyl aldehyde and carvacrol	Total plate count and coliform	Coliforms were in standard range for all samples but total plate count was much lower in TSPU active film compared with commonly used PE film during 35 days of storage.	Dong et al. (2020)
Fresh pasta	Chitosan	Chestnut extract	Total bacterial count, yeast and molds	Active components prevented microbial growth of pasta during 60 days	Kõrge et al. (2020)
Bread	Poly(butylene adipate-co-terephthalate) (PBAT) and poly(lactic acid) (PLA) film	Trans-cinnamaldehyde	Aspergillus niger, Penicillium sp. and Rhizopus sp	Films exhibited high antifungal activity against Penicillium sp. and Aspergillus niger but low efficacy against Rhizopus sp. and the bread shelf-life extended to 21 days	Srisa and Harnkarnsujarit (2020)

(continued)

Table 14.1 (continued)

Product	Packaging material	Active compound	Target microorganism	Result	Reference
Bread	Poly(lactic acid) and poly(butylene-succinate-co-adipate) film	Thymol EO	*Penicillium spp* and *Aspergillus spp*	Mold growth was observed after 9 days of storage in EOs containing film while neat film showed visible mold growth after 6 days.	Suwanamornlert et al. (2020)
Sliced bread	Carboxymethyl cellulose (CMC)-polyvinyl alcohol (PVA) based films	Cinnamon EOs	*P. digitatum*	In CMC-PVA film mold growth was observed after 6 days while in samples with 1.5 and 3% EOs no mold was detected after 60 days.	Fasihi et al. (2019)
Sliced bread	PP/PET/LDPE multilayer film	Thymol and star anise EOs	*S. aureus* and *P. roqueforti*	Active film had limited inhibitory effect against *S. aureus* but mold growth was not observed after 14 days.	Lee et al. (2019)
Bread	Chitosan film	Apricot kernel EOs	……	Fungal growth was not observed in bread slices packed in active films after 10 days.	Priyadarshi et al. (2018)

(continued)

Table 14.1 (continued)

Product	Packaging material	Active compound	Target microorganism	Result	Reference
Sliced bread	Pectin film	Cinnamon and clove EOs	Aspergillus and Penicillium	Mold growth was not detected after 8 days of storage at 30 °C in EOs containing films while mold growth in control sample was severe.	Sachdeva et al. (2017)
Sliced bread	Chitosan	Cinnamaldehyde	Rhizopus Stolonifer	Mold growth was not detected after 21 days in EOs containing films whereas, mold was observed after 7 days in control sample	Demitri et al. (2016)
Sliced bread	Methylcellulose	Clove and oregano EOs	Yeasts and molds	Both EOs decreased the yeasts and molds count in bread slices during 15 days, and size reduction of EOs droplets improved their antimicrobial properties.	Otoni et al. (2014)
Sliced bread	Porous, polypropylene-based resin sachets	Oregano EOs	Thermotolerant coliforms, *Salmonella* sp., molds and yeasts,	The micobial count results meet the standard for sliced bread after 15 days of storage.	Passarinho et al. (2014)
Sliced bread	Paper package	Cinnamon EOs	*Rhizopus stolonifer*	Complete fungal growth inhibition was observed after 3 days of storage.	Rodriguez et al. (2008)

were monitored during 30 days of storage. They reported that EE and EE OA were more effective than commercial preservatives in reducing the counts for yeasts, molds and *Bacillus cereus* after 30 days of storage at 20 °C. The aroma deterioration and off-flavor development was observed in bread slices during storage. Aroma is formed due to enzymatic activity during kneading of dough, fermentation of yeasts and also caramelization and Maillard reactions during baking (Pozo-Bayon et al. 2006). Throughout storage of bread at ambient temperature lipid oxidation, loss of volatile compounds and accumulation of hexanal cause negative changes on aroma and flavor of bread. Nevertheless, these negative changes significantly decreased in EE and EE OA samples. The results of sensory evaluation revealed that EE and EE OA samples got higher scores compared to chemical preservatives in sensory tests. Microbiological and sensory results revealed that the shelf life was 4 days for control bread; 6 days for commercial preservatives containing samples; 24 days for EE containing samples and more than 30 days for EE OA containing breads and EE OA was the most effective treatment for bread packaging. The influence of an ethanol emitter active packaging (Ethicap1) on microbiological quality and shelf life of pre-baked buns, revealed that the total mesophilic count of buns significantly increased to an unacceptable level in samples without ethanol within 1 week. Whereas, the total mesophilic count remained at a consumable level (10^5–10^6 cfu g) in the presence of ethanol. Mold growth was observed on the surface of pre-baked buns within 4–6 days and included *Penicillium solitum, P. corylophilum, P. commune, Cladosporium herbarum* and *C. sphaerospermum*. Addition of Ethicap1 delayed the mold growth for 13 days. The shelf life extension is due to the ethanol absorption of buns which suppress growth of *Bacillus* spp. and molds (Franke et al. 2002). Hempel et al. (2013) used a MAP system with 10% CO_2 and 90% N_2 in combination with ethanol emitters (EE) or ethanol spray on the surface of breads (ES) and investigated the effects of these packaging systems on shelf life of bread in comparison with samples held in air. The O_2 level in MAP samples was constant during storage but in samples packed in air O_2 level reduced by day 6, while samples held in air with EE and ES showed O_2 reduction over time. The complete O_2 depletion was occurred by day 14 in ES, whereas the O_2 reduction was slower in EE and the complete O_2 depletion occurred after 35 days of storage. The mycological counts decreased in bread samples incorporated with ES, but predominantly EE, in control and MAP systems. The results of sensory evaluation confirmed that incorporation of ethanol in bread packs didn't cause negative effects on organoleptic properties.

The influence of ethanol vapor on growth and toxin production of *Cl. Botulinum* type A and B in English style crumpets was investigated by Daifas et al. (2000). Crumpet samples were inoculated with 500 spores/g, packed in plastic bags with Ethicap® 2, 4 or 6G (commercial ethanol vapor generators) or cotton wool pads containing 2, 4 or 6 g of 95% ethanol and kept at 25 °C. In the inoculated control samples (0% ethanol) toxin was identified at day 5 day of storage. The toxicity was delayed for 10 days in Ethicap® 2 whereas complete toxin inhibition (more than 2 1 days) was found in samples packed with 4 or 6G Ethicap® or with 2, 4 or 6 g of ethanol in cotton wool pads. Nevertheless, all of the crumpet samples were obviously spoiled by this time. Therefore, ethanol vapor is an effective barrier against growth and toxin production by *C. botulinum* in crumpets at room temperature. Salminen et al. (1996) used ethanol emitters and oxygen absorbers for shelf-life extension of whole grain rye bread samples within **6** weeks of storage. They incorporated small and large ethanol emitters reported that small ethanol emitters were not effective in extending the microbial shelf life of breads while in the presence of large ethanol emitters the shelf life of packed bread slices extended from 8–12 to 26–27 days. They also found that ethanol emitters and oxygen absorbers did not affect the texture and moisture changes of bread slices within storage period.

Janjarasskul et al. (2016) used an ethanol emitter (EE) and an O2 scavenger (OS) to extend the shelf life of sponge cakes. The combination of EE and OS had synergistic effects in delaying staling, lipid oxidation and microbial growth. They reported that the mold-free shelf life of control cake was 1 day and increased to more than 42 days in cakes with active packaging.

2.3 *ε-Polylysine Releasing Active Packaging*

ε-Polylysine, is an emerging antimicrobial cationic peptide composed of 5 to 100 amino acids. It is obtained from microbial metabolism and has a wide antibacterial spectrum against Gram-positive and Gram-negative bacteria, molds, yeasts, viruses, etc. The interactions between ε-Polylysine and plasma membrane damage the integrity of fungal cells. Plasma membrane is vital for normal activity of living cells. Thus, the damage caused by ε-Polylysine inhibit the spore germination, mycelial growth, germ tube elongation, interrupt cellular homeostasis, disturb the ionic balance of membrane, increase reactive oxygen species (ROS) and contribute to cell death (Luz et al. 2016; Liu et al. 2017). ε-Polylysine is edible, biodegradable, water-soluble, nontoxic and heat resistant. It is a FDA approved GRAS material thus can be used in food products such as baked goods as a natural preservative or as an antimicrobial in active packaging. Luz et al. (2018) investigated the influence of ε Poly-L-lysine (ε-PL) against fungal growth and reducing the production of aflatoxins (AFs). They produced starch films containing different levels of ε-PL and determined its antifungal activity against *Penicillium expansum* and *Aspergillus parasiticus* (AFs producer) in solid medium. Afterwards, films were used for packaging of breads inoculated with *A. parasiticus* CECT 2681 and *P. expansum* CECT 2278 and the influence of films on shelf life extension and AFs content of breads was determined. The ε-PL concentration less than 1.6 mg/cm^2 didn't have antifungal activity in solid medium, while in films with ε-PL higher than 1.6 mg/cm^2 the antifungal activity was dose dependent. Application of films containing 1.6–6.5 mg ε-PL/cm^2, increased the shelf life of *A. parasiticus* inoculated bread by 1 day with the, whereas packaging of bread inoculated by *P. expansum* with films incorporating 6.5 mg E-PL/cm2 increased their shelf life by 3 days. As Fig. 14.4 shows by increasing the concentration of ε-PL in packaging, the mold growth was decreased in bread samples. The ε-PL films greatly (93–100%) inhibited the production of AFs. Thus, they have the potential to be used for bread packaging as a natural preservative.

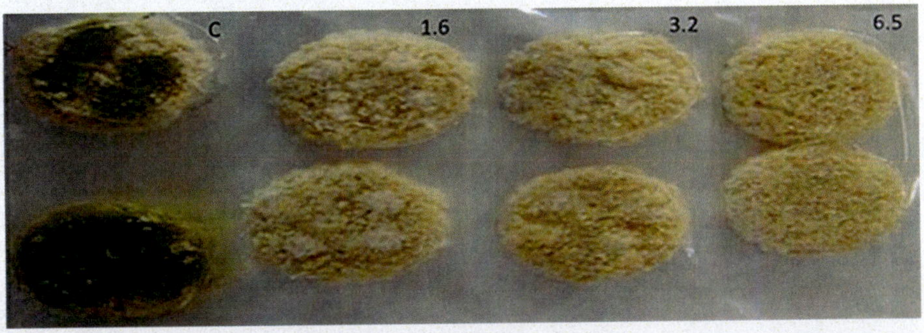

Fig. 14.4 Influence of ε-PL concentration on Growth inhibition of *P. expansum* and *A. parasiticus* in breads (Luz et al. 2018)

Luz et al. (2016) selected three mycotoxigenic fungi including and *Penicillium expansum*, *Aspergillus parasiticus* and *Fusarium verticilloides* and determined the fungal growth as well as aflatoxin, patulin and fumonisin content respectively by Kirby–Bauer test. Subsequently, they inoculated bread loaves with *P. expansum* and *A. parasiticus* and investigated the influence of starch based films containing different levels of ε-Polylysine on shelf life extension, fungal growth and mycotoxin content in bread samples. Films showed fungicide properties against *A. parasiticus* and F. *verticilloides*, and fungistatic properties against *P. expansum*. Antimicrobial films reduced the fungal growth 54 to 99% and aflatoxins content 93–99% and significantly increased the shelf-life of breads.

2.4 Metal Ions Releasing Active Packaging

Metal ions have antimicrobial activity by penetrating into microorganisms, inactivating their enzymes or generating hydrogen peroxide. Generally, when the size of metal ions decreases their antibacterial activity increases and a small dose of them can provide high antibacterial activity due to the large surface area (Kim et al. 1998; Beyth et al. 2015). Metal or metal oxide nanoparticles can be applied in releasing active packaging to increase the shelf life of cereal products. Incorporation of metal oxide nanoparticles into packaging materials not only increases their mechanical strength, light blocking properties and oxygen and moisture barrier properties, but also imparts antimicrobial activity. Metal oxide nanoparticles such as aluminum oxide, titanium dioxide, zinc oxide and silicon dioxide are used for fresh food packaging. Migration of nanomaterial into food matrices is a consequence of dissolution, diffusion and abrasion of packaging surface but it is negligible. Thus, the health risks caused by consumer exposure would be low (Garcia et al. 2018). Direct contact of microorganisms' cell walls with ZNO nanoparticles and release Zn^{2+} ions, generate reactive oxygen species (ROS) and destroy bacterial cell integrity (Jones et al. 2008). Zinc oxide nanoparticles (ZnO NPs) (0.5, 1 and 2%) were added to carboxymethyl cellulose-chitosan-oleic acid (CMC-CH-OL) based film by Noshirvani et al. (2017). They used this film for *Aspergilus niger* inoculated sliced bread packaging and investigated its influences on microbial quality and staling of bread samples. They stated that the shelf life of control bread was 3 days while it increased in other samples. The active coatings decreased the fungal growth in sliced bread. However, after 15 days of storage, all samples except those with 2% ZnO NPs presented fungal growth. CMC-CH-OL-ZnO NPs 2% packed bread samples offered the best results and fungal development was detected after 22 days Fig. 14.5 shows the images of bread slices in different packages during storage.

Incorporation of oleic acid and ZnO NPs (2%) significantly decreased the water vapor permeability (WVP) of the CMC-CH film and water activity (aw) and moisture content of breads in active coatings were higher than the control bread. Differential scanning calorimetry (DSC) results of control bread revealed an endothermic peak which is attributed to retrogradation of amylopectin and firmness of control sliced bread sample was higher after 15 days of storage.

SiO2 nanoparticles are usually added to packaging materials in order to improve their mechanical strength, gas and water barrier properties and thermal stability. Nevertheless, it has been proven that SiO2 has antimicrobial and enzyme-inhibiting activity and can be added to active packaging systems (Garcia et al. 2018). An antimicrobial film based on poly (vinyl alcohol) (PVA), chitosan (CS) incorporating zinc oxide nanoparticles and silicon dioxide nanoparticles (ZnO-SiO2) (0.50, 1.0, 3.0 and 5.0%) were fabricated by Al-Tayyar et al. (2020). The PVA/CS/ZnO-SiO2 films were tested against Gram-negative (*Escherichia coli*) and Gram-positive

Fig. 14.5 Images of *Aspergilus niger* inoculated bread slices in different packages during storage (Noshirvani et al. 2017)

(*Staphylococcus aureus*) bacteria and showed great antibacterial activity Also, the bionanocomposite films were used for bread packaging. The visual appearance of breads in PVA/CS/ZnO-SiO2 films was greatly improved, shelf life was extended, and food-borne pathogens count reduced. Silver nanoparticles can anchor and penetrate to the bacterial cell wall and cause changes in cell membrane. Accumulation of silver nanoparticles on cell surface can form pits on cell membrane and lead to cell death (Prabhu and Poulose 2012; Sondi and Salopek-Sondi 2004). Also, silver nanoparticles can form free radicals which can make pores, damage the cell membrane and cause cell death (Danilczuk et al. 2006; Kim et al. 2007). The interaction of silver nanoparticles with phosphorus and sulfur of DNA is another mechanism suggested for antimicrobial activity of silver nanoparticles. These interactions cause destruction of DNA or problems in DNA replication which finally result in cell death (Hatchett and White 1996). The release of silver ions from silver nanoparticles can inhibit several cell functions such as respiratory enzyme and generate reactive oxygen species lead to cell damage (Prabhu and Poulose 2012). TiO_2 is generally used in food products as a white pigment however, it has photocatalytic antibacterial activity (Garcia et al. 2018). TiO_2 is usually doped with other metals or metal oxide nanoparticles to increase its antimicrobial properties in a synergistic way. Mihaly Cozmuta et al. (2015) tested the effects of Ag and TiO_2 nanocomposite (Ag/TiO_2-P) based on high density polyethylene on microbiological properties and shelf life of bread in comparison with high density polyethylene packaging (HDP-P) and no packaging (CS). Microbial and chemical stability of bread was examined for 6 days and the results indicated that Ag/TiO_2-P reduced lipid hydroperoxides, bacteria, yeasts and moulds counts and significantly prolonged the bread shelf life compared to HDP-P and CS. Mihaly-Cozmuta et al. (2017) studied the effect of different active cellulose-based papers comprising TiO2 (P–TiO2), Ag–TiO2 (P–Ag–TiO2) and Ag–TiO2–zeolite (P–Ag–TiO2–Z) nanocomposites on bread shelf life and compared to plain paper (PP) as control. The microbiological tests revealed that P–Ag–TiO2–Z was the most effective treatment in improving the microbial quality of bread followed by P–Ag–TiO2 and P–TiO2. The molds and yeasts counts were within the admitted limit for 10 and 12 days at 20 and 4 °C for P–Ag–TiO2–Z packed breads while it was 7 and 8 days for PP packed breads.

In a study by Metak (2015) 0.1% titanium dioxide (TiO_2) nanoparticles and 1% silver nanoparticles (Ag) were incorporated to polyethylene (PE) and used for packaging of different food products including bread to determine its antimicrobial activity and found that the packaging materials exhibited great antimicrobial properties. TiO_2 and Ag nanoparticles are effective against a wide range of bacterial strains and combination of them is a suitable method for improving the characteristics of nanocomposites. Peter et al. (2016) used paper-packages modified with Ag/N–TiO_2, Ag/TiO_2–SiO_2, or Au/TiO_2 for white bread and investigated microbiological and chemical properties during storage period. They found that Ag/TiO_2–SiO_2-paper was more effective followed by Ag/N–TiO_2 in preventing bread spoilage and water retention while Au/TiO_2-paper coated bread spoiled after 15 days of refrigerated storage. The higher effectiveness in Ag/TiO_2–SiO_2 is due to its large specific surface area and photoactivity which generate active charge carriers and head to cytoplasm release and cell death. Migration assessments didn't detect traces of gold, silver or titanium because the amount of nanomaterials in contact with bread was relatively low, thus the bread slices were safe for consumers. The water retention and color stability.

2.5 Chemical Preservatives Releasing Active Packaging

Weak lipophilic acids and their salts such as sorbate, acetate, benzoate propionate are used as preservative agents in a wide range of food products due to their stability, solubility in water and ease of handling (Brul and Coote 1999; Jideani and Vogt 2016). These preservative agents exhibit their antimicrobial activity through a variety of mechanisms including membrane disruption, preventing metabolic reactions, accumulation of toxic anions and increasing the proton concentration (Brul and Coote 1999). Sodium propionate (SP) is one of the most common preservatives which can be used in active packaging systems. SP generates sodium and propionate ions when dissociates in water. SP reduces pH and forms high proton concentration that enhances the acid diffusion through cell membrane and intracellular acidification of microorganisms. The preservation of intracellular pH homeostasis reduces the available energy that inhibit growth and essential metabolic functions of microorganisms, bringing about cell death. Thanakkasaranee et al. (2018) incorporated different levels (0.5, 1, 2, 3, and 5%) of sodium propionate (SP) into polypropylene (PP) and fabricated an antimicrobial composite film (PP/SP) to extend bread shelf life. The feasibility of films for bread packaging was investigated by physical, chemical and antimicrobial properties. Increasing SP concentration enhanced the hydrophilicity, thermal and mechanical stability as well as antimicrobial activity of the PP/SP films. Which preserve bread freshness and extend its shelf life. The antimicrobial activity of SP is due to the interaction of bread moisture and SP changes the bread pH and destroy the fungal cell structure and also decreases the growth rate of bacteria. In another study cellulose acetate films incorporating sodium propionate (0, 2 and 4%) were produced by Soares et al. (2002) and used for bread packaging. They sandwiched bread slices between this film, packed them in LDPE bags and kept them at ambient temperature (25 ± 2 °C) for 15 days. Bread slices in LDPE bags served as control. The microbiological tests during storage period revealed that mold growth decreased with increased propionate concentration. Sorbic acid and its salts particularly potassium sorbate are used as antimicrobial agents in cereal products or packaging materials. Sorbic acid or sorbate inhibit the vegetative cell division and spore germination by mechanisms such as inhibition of key enzymes and transport systems, alteration of cell membranes and formation of a proton flux into microbial cells (Sofos et al. 1986). Antimicrobial active films of 70 and 25 μm thickness containing 3 and 7% sorbic acid respectively were implemented for pastry dough packaging by Silveira et al. (2007a). Pastry doughs were refrigerated at 8 °C for 40 days and their sorbic acid migration was evaluated. Furthermore, the antimicrobial properties of films were compared with a pastry dough containing potassium sorbate packed in LDPE film as control. The migration of sorbic acid from films into pastry dough was considerably lower than the maximum allowed limit for both samples. A decrease of 2 log cycles was detected in aerobic mesophilic count; while, psychotropic and Staphylococcus spp. counts were not significantly reduced for dough wrapped in 25 μm thickness film with 7% sorbic acid. For doughs wrapped with 70 μm thickness films with 3% sorbic acid mesophilic and psychotropic counts, reduced 2 and 1.5 log cycles respectively. In control sample, the Staphylococcus spp., aerobic mesophilic and psychotropic counts increased 2, 1.5 and 1 log cycles, respectively. In another study Silveira et al. (2007b) used antimicrobial films with 0–6% of sorbic acid for pastry dough packaging and reported that microbial growth in samples wrapped in 70 μm thickness films containing 3–6% sorbic acid was considerably lower than control. The psychrotrophic count in films of 25 μm-thick with 6% sorbic acid was1 log cycle lower than that of the control film. While the *Staphylococcus* spp. and aerobic mesophilic counts were 1 log cycle higher than the count of control dough. Figure 14.6 shows the images of pastry doughs in active films with sorbic acid.

Fialho et al. (2011) implemented an active film for pastry dough packaging. They added sorbic acid to films and evaluated the antimicrobial properties against *Penicillium* sp. They also added sorbic acid to pastry dough directly and found that active films were more effective for mold growth inhibition. Sangsuwan et al. (2015) produced antimicrobial films by blending chitosan and potassium sorbate (PS) or vanillin and compared their influence on shelf life extension of butter cake with commercial stretch film as control. Incorporation of PS into film increased oxygen permeability and flexibility but decreased their water permeability, transparency and strength. Addition of vanillin decreased film's strength and permeability and increased their yellowness. Molds were detected on cake surface on day 7 and 4 for butter cakes packed in chitosan film and stretch film respectively, whereas mold was not detected on cake packed in PS or vanillin incorporated chitosan films during 8 days of storage. The results of sensory analysis revealed that these samples were

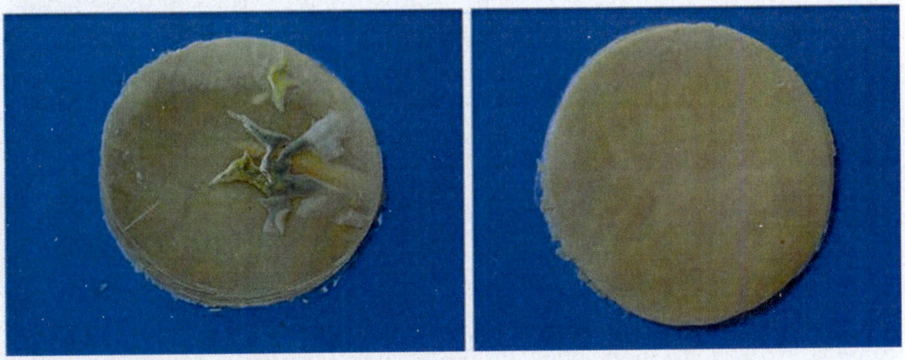

Fig. 14.6 Images of pastry doughs wrapped in sorbic acid incorporated active films (**a**) 1% sorbic acid after 13 days of storage at 8 °C, (**b**) 2% sorbic acid after 40 days of storage at 8 °C (Silveira et al. 2007b)

acceptable for consumers at the end of storage period. de Camargo Andrade-Molina et al. (2013) prepared a biodegradable film based on thermoplastic starch, poly(butylene adipate-co-terephthalate) (PBAT) containing potassium sorbate (1.5, 3 and 4.5%) as a preservative for fresh pasta packaging. Different types of fresh pasta are perishable products and packaging in active films can prevent or retard their spoilage. Pasta sheets were covered with composite films, packed in LDPE bags and kept at 10 °C. The microbiological quality and migration of potassium sorbate (PS) to pasta as well as film properties, were evaluated throughout storage period. The mechanical properties of films were suitable for packaging of fresh pasta. The microbial growth was controlled by active films and shelf-life of pasta samples increased and the amount of PS in pasta was much lower than the allowed concentrations. The composite film incorporated with 4.5% potassium sorbate was the most effective in microbial growth inhibition. Fresh lasagna is another type of perishable pasta product. Sousa et al. (2013) used poly(butylene adipate co-terephthalate) (PBAT), rice flour, glycerol and PS to produce a biodegradable film for active packaging of fresh lasagna. They characterized the water vapor barrier, mechanical, optical and microstructural properties of films during 45 days of storage for fresh pasta packaging. The results revealed that when the PS concentration was greater than or equal to 3% the opacity was reduced but water vapor permeability increased. They found that PS has plasticizing effects and incorporation of 1 to 5% PS modify the mechanical properties of biodegradable films and these films are easy to handle after application for lasagna. In a study by Ahmadi et al. (2011) different levels (2, 6, 10, 14 and 18 w/w dry matter) of potassium sorbate (PS) was added to edible films based on starch and hydroxypropylmethylcellulose (HPMC). They investigated the staling of bread and found that staling of baguette bread was retarded. Incorporation of benzoic acid or its derivatives to packaging materials has been investigated by several researchers. These antimicrobial agents cause a general energy decline, ATP depletion and restricting growth by inhibiting the phosphofructokinase production in glycolysis (Krebs et al. 1983; Warth 1991). Dobiáš et al. (2000) added 5 and 10 g/kg of benzoic anhydride (BA), ethyl paraben and propyl paraben as preservative agents to LDPE and investigated their influences on film properties as well as microbiological quality of food matrices including toasted bread. They found that incorporation of antimicrobials into films significantly changed their gas and water permeability, tensile strength, friction coefficient and transparency. Mold growth in toasted bread packed in films containing 10 g/kg of BA was significantly reduced during storage period at 6 °C.

3 Conclusion

Considering the high consumption of perishable cereal-based products around the world, increasing their shelf life is required. Incorporation of antimicrobial agents can be considered as an effective solution to extend the shelf-life of cereal products. However, application of active packaging in which antimicrobials are added into packaging material is more effective than direct incorporation to food matrix and decreases the required amount of active compounds. Therefore, releasing active packaging systems have gained attention and are known as promising options for cereal products packaging. Different types of active materials such as essential oils, oleoresins, plant extracts, ethanol, ε-polylysine, propionic acid, sorbic acid, benzoic acid and their salts, titanium dioxide, zinc oxide, silicon dioxide and aluminum oxide nanoparticles have been added to packaging films. The results indicate that by selection of appropriate type and concentration of active substances shelf life of cereal products increases significantly. However, more research is needed to develop more effective active films for cereal products packaging.

References

Ahmadi E, Azizi MH, Abbasi S, Hadian Z, Sareminezhad S. Extending bread shelf-life using polysaccharide coatings containing potassium sorbate. J Food Res (University of Tabriz). 2011;21:209–17.

Altan A, Aytac Z, Uyar T. Carvacrol loaded electrospun fibrous films from zein and poly (lactic acid) for active food packaging. Food Hydrocoll. 2018;81:48–59.

Al-Tayyar NA, Youssef AM, Al-Hindi RR. Antimicrobial packaging efficiency of ZnO-SiO2 nanocomposites infused into PVA/CS film for enhancing the shelf life of food products. Food Packag Shelf Life. 2020;25:100523.

Alves-Silva JM, dos Santos SMD, Pintado ME, Pérez-Álvarez JA, Fernández-López J, Viuda-Martos M. Chemical composition and in vitro antimicrobial, antifungal and antioxidant properties of essential oils obtained from some herbs widely used in Portugal. Food Control. 2013;32(2):371–8.

Balaguer MP, Lopez-Carballo G, Catala R, Gavara R, Hernandez-Munoz P. Antifungal properties of gliadin films incorporating cinnamaldehyde and application in active food packaging of bread and cheese spread foodstuffs. Int J Food Microbiol. 2013;166(3):369–77.

Beyth N, Houri-Haddad Y, Domb A, Khan W, Hazan R. Alternative antimicrobial approach: nano-antimicrobial materials. Evid Based Complement Alternat Med. 2015:2015.

Brul S, Coote P. Preservative agents in foods: mode of action and microbial resistance mechanisms. Int J Food Microbiol. 1999;50(1–2):1–17.

Burt S. Essential oils: their antibacterial properties and potential applications in foods—a review. Int J Food Microbiol. 2004;94(3):223–53.

Cauvain SP. Baking problems solved. Woodhead Publishing; 2017.

Daifas DP, Smith JP, Tarte I, Blanchfield B, Austin JW. Effect of ethanol vapor on growth and toxin production by Clostridium botulinum in a high moisture bakery product. J Food Saf. 2000;20(2):111–25.

Danilczuk M, Lund A, Sadlo J, Yamada H, Michalik J. Conduction electron spin resonance of small silver particles. Spectrochim Acta A Mol Biomol Spectrosc. 2006;63(1):189–91.

Day BPF, Potter L. Active packaging. In: Food beverage packaging technology; 2011. p. 251–62.

de Camargo Andrade-Molina TP, Shirai MA, Grossmann MVE, Yamashita F. Active biodegradable packaging for fresh pasta. LWT - Food Sci Technol. 2013;54(1):25–9.

Demitri C, De Benedictis VM, Madaghiele M, Corcione CE, Maffezzoli A. Nanostructured active chitosan-based films for food packaging applications: effect of graphene stacks on mechanical properties. Measurement. 2016;90:418–23.

Dobiáš J, Chudackova K, Voldrich M, Marek M. Properties of polyethylene films with incorporated benzoic anhydride and/or ethyl and propyl esters of 4-hydroxybenzoic acid and their suitability for food packaging. Food Addit Contam. 2000;17(12):1047–53.

Dong H, He J, Xiao K, Li C. Temperature-sensitive polyurethane (TSPU) film incorporated with carvacrol and cinnamyl aldehyde: antimicrobial activity, sustained release kinetics and potential use as food packaging for Cantonese-style moon cake. Int J Food Sci Technol. 2020;55(1):293–302.

Fasihi H, Noshirvani N, Hashemi M, Fazilati M, Salavati H, Coma V. Antioxidant and antimicrobial properties of carbohydrate-based films enriched with cinnamon essential oil by Pickering emulsion method. Food Packag Shelf Life. 2019;19:147–54.

Fialho AR, Vidigal MCTR, Soares NDFF, de Moraes LP, de Melo NR, Gonçalves MPJ. Development and evaluation of antimicrobial and flavored film for using on pastry dough. Ciência Rural. 2011;41:537–42.

Fonseca LM, Souza EJD, Radünz M, Gandra EA, da Rosa Zavareze E, Dias ARG. Suitability of starch/carvacrol nanofibers as biopreservatives for minimizing the fungal spoilage of bread. Carbohydr Polym. 2020;252:117166.

Franke I, Wijma E, Bouma K. Shelf life extension of pre-baked buns by an active packaging ethanol emitter. Food Addit Contam. 2002;19(3):314–22.

Garcia CV, Shin GH, Kim JT. Metal oxide-based nanocomposites in food packaging: applications, migration, and regulations. Trends Food Sci Technol. 2018;82:21–31.

Gutiérrez L, Batlle R, Andújar S, Sánchez C, Nerín C. Evaluation of antimicrobial active packaging to increase shelf life of gluten-free sliced bread. Packag Technol Sci. 2011;24(8):485–94.

Hanif MA, Nisar S, Khan GS, Mushtaq Z, Zubair M. Essential oils. In: Essential oil research. Cham: Springer; 2019. p. 3–17.

Hatchett DW, White HS. Electrochemistry of sulfur adlayers on the low-index faces of silver. J Phys Chem. 1996;100(23):9854–9.

Hempel AW, O'Sullivan MG, Papkovsky DB, Kerry JP. Use of smart packaging technologies for monitoring and extending the shelf-life quality of modified atmosphere packaged (MAP) bread: application of intelligent oxygen sensors and active ethanol emitters. Eur Food Res Technol. 2013;237(2):117–24.

Heras-Mozos R, Muriel-Galet V, Lopez-Carballo G, Catalá R, Hernandez-Munoz P, Gavara R. Active EVOH/PE bag for sliced pan loaf based on garlic as antifungal agent and bread aroma as aroma corrector. Food Packag Shelf Life. 2018;18:125–30.

Heras-Mozos R, Muriel-Galet V, López-Carballo G, Catalá R, Hernández-Muñoz P, Gavara R. Development and optimization of antifungal packaging for sliced pan loaf based on garlic as active agent and bread aroma as aroma corrector. Int J Food Microbiol. 2019;290:42–8.

Hyldgaard M, Mygind T, Meyer RL. Essential oils in food preservation: mode of action, synergies, and interactions with food matrix components. Front Microbiol. 2012;3:12.

Janjarasskul T, Tananuwong K, Kongpensook V, Tantratian S, Kokpol S. Shelf life extension of sponge cake by active packaging as an alternative to direct addition of chemical preservatives. LWT - Food Sci Technol. 2016;72:166–74.

Jideani VA, Vogt K. Antimicrobial packaging for extending the shelf life of bread—a review. Crit Rev Food Sci Nutr. 2016;56(8):1313–24.

Jones N, Ray B, Ranjit KT, Manna AC. Antibacterial activity of ZnO nanoparticle suspensions on a broad spectrum of microorganisms. FEMS Microbiol Lett. 2008;279(1):71–6.

Ju J, Xie Y, Yu H, Guo Y, Cheng Y, Qian H, Yao W. A novel method to prolong bread shelf life: sachets containing essential oils components. LWT. 2020;131:109744.

Kim TN, Feng QL, Kim JO, Wu J, Wang H, Chen GC, Cui FZ. Antimicrobial effects of metal ions (Ag+, Cu2+, Zn2+) in hydroxyapatite. J Mater Sci Mater Med. 1998;9(3):129–34.

Kim JS, Kuk E, Yu KN, Kim JH, Park SJ, Lee HJ, et al. Antimicrobial effects of silver nanoparticles. Nanomedicine. 2007;3(1):95–101.

Khorshidian N, Yousefi M, Khanniri E, Mortazavian AM. Potential application of essential oils as antimicrobial preservatives in cheese. Int J Food Sci Emer Technol. 2018;45:62–72.

Kõrge K, Bajić M, Likozar B, Novak U. Active chitosan-chestnut extract films used for packaging and storage of fresh pasta. Int J Food Sci Technol. 2020;

Krebs HA, Wiggins D, Stubbs M, Sols A, Bedoya F. Studies on the mechanism of the antifungal action of benzoate. Biochem J. 1983;214(3):657–63.

Latou E, Mexis SF, Badeka AV, Kontominas MG. Shelf life extension of sliced wheat bread using either an ethanol emitter or an ethanol emitter combined with an oxygen absorber as alternatives to chemical preservatives. J Cereal Sci. 2010;52(3):457–65.

Lee JS, Park MA, Yoon CS, Na JH, Han J. Characterization and preservation performance of multilayer film with insect repellent and antimicrobial activities for sliced wheat bread packaging. J Food Sci. 2019;84(11):3194–203.

Liu K, Zhou X, Fu M. Inhibiting effects of epsilon-poly-lysine (ε-PL) on Pencillium digitatum and its involved mechanism. Postharvest Biol Technol. 2017;123:94–101.

Luz C, Calpe J, Saladino F, Quiles J, Ruiz J, Franzón MF, Meca G. Antifungal and antimycotoxigenic activity of polylysine biofilms. Toxicol Lett. 2016;258:S170.

Luz C, Calpe J, Saladino F, Luciano FB, Fernandez-Franzón M, Mañes J, Meca G. Antimicrobial packaging based on ε-polylysine bioactive film for the control of mycotoxigenic fungi in vitro and in bread. J Food Process Preserv. 2018;42(1):e13370.

Metak AM. Effects of nanocomposite based nano-silver and nano-titanium dioxide on food packaging materials. Int J Appl Sci Technol. 2015;5(2):26–40.

Mihaly Cozmuta A, Peter A, Mihaly Cozmuta L, Nicula C, Crisan L, Baia L, Turila A. Active packaging system based on Ag/TiO2 nanocomposite used for extending the shelf life of bread. Chemical and microbiological investigations. Packag Technol Sci. 2015;28(4):271–84.

Mihaly-Cozmuta A, Peter A, Craciun G, Falup A, Mihaly-Cozmuta L, Nicula C, Vulpoi A, Baia M. Preparation and characterization of active cellulose-based papers modified with TiO 2, Ag and zeolite nanocomposites for bread packaging application. Cellulose. 2017;24(9):3911–28.

Moleyar V, Narasimham P. Antibacterial activity of essential oil components. Int J Food Microbiol. 1992;16(4):337–42.

Nielsen PV, Rios R. Inhibition of fungal growth on bread by volatile components from spices and herbs, and the possible application in active packaging, with special emphasis on mustard essential oil. Int J Food Microbiol. 2000;60(2–3):219–29.

Nikaido H. Molecular basis of bacterial outer membrane permeability revisited. Microbiol Mol Biol Rev. 2003;67(4):593–656.

Noshirvani N, Ghanbarzadeh B, Mokarram RR, Hashemi M. Novel active packaging based on carboxymethyl cellulose-chitosan-ZnO NPs nanocomposite for increasing the shelf life of bread. Food Packag Shelf Life. 2017;11:106–14.

Oliveira MA, Gonzaga ML, Bastos MS, Magalhães HC, Benevides SD, Furtado RF, et al. Packaging with cashew gum/gelatin/essential oil for bread: release potential of the citral. Food Packag Shelf Life. 2020;23:100431.

Otoni CG, Pontes SF, Medeiros EA, Soares NDF. Edible films from methylcellulose and nanoemulsions of clove bud (Syzygium aromaticum) and oregano (Origanum vulgare) essential oils as shelf life extenders for sliced bread. J Agric Food Chem. 2014;62(22):5214–9.

Pasqualone A. Bread packaging: features and functions. In: Flour and breads and their fortification in health and disease prevention. Academic; 2019. p. 211–22.

Passarinho ATP, Dias NF, Camilloto GP, Cruz RS, Otoni CG, Moraes ARF, Soares NDFF. Sliced bread preservation through oregano essential oil-containing sachet. J Food Process Eng. 2014;37(1):53–62.

Pateras IM. Bread spoilage and staling. In: Technology of breadmaking. Boston: Springer; 2007. p. 275–98.

Peter A, Mihaly-Cozmuta L, Mihaly-Cozmuta A, Nicula C, Ziemkowska W, Basiak D, Danciu V, Vulpoi A, Baia L, Falup A, Craciun G, Ciric A, Begea M, Kiss C, Vatuiu D. Changes in the microbiological and chemical characteristics of white bread during storage in paper packages modified with Ag/TiO2–SiO2, Ag/N–TiO2 or Au/TiO2. Food Chem. 2016;197:790–8.

Pinto E, Vale-Silva L, Cavaleiro C, Salgueiro L. Antifungal activity of the clove essential oil from Syzygium aromaticum on Candida, Aspergillus and dermatophyte species. J Med Microbiol. 2009;58(11):1454–62.

Pozo-Bayon MA, Guichard E, Cayot N. Flavor control in baked cereal products. Food Rev Intl. 2006;22(4):335–79.

Prabhu S, Poulose EK. Silver nanoparticles: mechanism of antimicrobial action, synthesis, medical applications, and toxicity effects. Int Nano Lett. 2012;2(1):32.

Priyadarshi R, Kumar B, Deeba F, Kulshreshtha A, Negi YS. Chitosan films incorporated with Apricot (Prunus armeniaca) kernel essential oil as active food packaging material. Food Hydrocoll. 2018;85:158–66.

Rodriguez A, Nerín C, Batlle R. New cinnamon-based active paper packaging against Rhizopusstolonifer food spoilage. J Agric Food Chem. 2008;56(15):6364–9.

Sachdeva A, Vashist S, Chopra R, Puri D. Antimicrobial activity of active packaging film to prevent bread spoilage. Int J Food Sci Nutr. 2017;2:29–37.

Salminen A, Latva-Kala K, Randell K, Hurme E, Linko P, Ahvenainen R. The effect of ethanol and oxygen absorption on the shelf-life of packed sliced rye bread. Packag Technol Sci. 1996;9(1):29–42.

Sangsuwan J, Rattanapanone N, Pongsirikul I. Development of active chitosan films incorporating potassium sorbate or vanillin to extend the shelf life of butter cake. Int J Food Sci Technol. 2015;50(2):323–30.

Seow YX, Yeo CR, Chung HL, Yuk HG. Plant essential oils as active antimicrobial agents. Crit Rev Food Sci Nutr. 2014;54(5):625–44.

Silveira MFA, Soares NFF, Geraldine RM, Andrade NJ, Goncalves MPJ. Antimicrobial efficiency and sorbic acid migration from active films into pastry dough. Packag Technol Sci Int J. 2007a;20(4):287–92.

Silveira MFA, Soares NFF, Geraldine RM, Andrade NJ, Botrel DA, Gonçalves MPJ. Active film incorporated with sorbic acid on pastry dough conservation. Food Control. 2007b;18(9):1063–7.

Soares NFF, Rutishauser DM, Melo N, Cruz RS, Andrade NJ. Inhibition of microbial growth in bread through active packaging. Packag Technol Sci Int J. 2002;15(3):129–32.

Sofos JN, Pierson MD, Blocher JC, Busta FF. Mode of action of sorbic acid on bacterial cells and spores. Int J Food Microbiol. 1986;3(1):1–17.

Sondi I, Salopek-Sondi B. Silver nanoparticles as antimicrobial agent: a case study on E. coli as a model for Gram-negative bacteria. J Colloid Interface Sci. 2004;275(1):177–82.

Sousa GM, Júnior MS, Yamashita F. Active biodegradable films produced with blends of rice flour and poly (butylene adipate co-terephthalate): effect of potassium sorbate on film characteristics. Mater Sci Eng C. 2013;33(6):3153–9.

Srisa A, Harnkarnsujarit N. Antifungal films from trans-cinnamaldehyde incorporated poly (lactic acid) and poly (butylene adipate-co-terephthalate) for bread packaging. Food Chem. 2020;333:127537.

Suhr KI, Nielsen PV. Inhibition of fungal growth on wheat and rye bread by modified atmosphere packaging and active packaging using volatile mustard essential oil. J Food Sci. 2005;70(1):M37–44.

Suppakul P, Miltz J, Sonneveld K, Bigger SW. Active packaging technologies with an emphasis on antimicrobial packaging and its applications. J Food Sci. 2003;68(2):408–20.

Suwanamornlert P, Kerddonfag N, Sane A, Chinsirikul W, Zhou W, Chonhenchob V. Poly (lactic acid)/poly (butylene-succinate-co-adipate)(PLA/PBSA) blend films containing thymol as alternative to synthetic preservatives for active packaging of bread. Food Packag Shelf Life. 2020;25:100515.

Tabak HH, Cooke WB. The effects of gaseous environments on the growth and metabolism of fungi. Bot Rev. 1968;34(2):126–252.

Thanakkasaranee S, Kim D, Seo J. Preparation and characterization of polypropylene/sodium propionate (PP/SP) composite films for bread packaging application. Packag Technol Sci. 2018;31(4):221–31.

Warth AD. Mechanism of action of benzoic acid on Zygosaccharomyces bailii: effects on glycolytic metabolite levels, energy production, and intracellular pH. Appl Environ Microbiol. 1991;57(12):3410–4.

Yildirim S, Röcker B, Pettersen MK, Nilsen-Nygaard J, Ayhan Z, Rutkaite R, Radusin T, Suminska P, Marcos B, Coma V. Active packaging applications for food. Compr Rev Food Sci Food Saf. 2018;17(1):165–99.

Khorshidian, N., Yousefi, M., Khanniri, E., & Mortazavian, A. M. (2018). Potential application of essential oils as antimicrobial preservatives in cheese. *Innovative Food Science & Emerging Technologies*, *45*, 62-72.

Chapter 15
Releasing Active Systems Applied to Fruits and Vegetables

Letricia Barbosa-Pereira, Antía Lestido-Cardama, Patricia Vázquez-Loureiro, Ana Rodríguez Bernaldo de Quirós, and Raquel Sendón

Abstract Fruits and vegetables are high preference products of human consumption due to their several health benefits. However, their short shelf life requires innovative strategies of preservation such as active packaging systems to maintain the quality and enhance the organoleptic and nutritional properties of the food. This chapter provides insights on release active packaging systems, including ethanol and carbon dioxide emitters, antimicrobial releasers agents of essential oils, or other natural agents such as natural extracts and other active compounds into the package headspace. The active release systems can be in several formats, such as sachets, bags, filters, films, wrapping papers, or paper bags. A brief section dedicated to the ethylene scavengers is also included due to the extended application and commercialization of these systems in this food sector. Recent studies have been dedicated to active nanocomposites with different mechanisms of action to enhance packaging barriers against gas/moisture/vapor and improve their mechanical and thermal properties, as well as to improve the controlled release of active substances to increase the shelf-life of fruits and vegetables. However, some aspects still require special attention from researchers and regulators, such as the interactions between antimicrobial-loaded nanocarriers and food components, the migration of nanostructures into foodstuffs, and the levels of toxicity for humans and the environment.

Keywords Active release systems · Active packaging · Essential oils · Ethanol emitters · Ethylene scavengers · Nanocarrier systems

L. Barbosa-Pereira · A. Lestido-Cardama · P. Vázquez-Loureiro
A. Rodríguez Bernaldo de Quirós · R. Sendón (✉)
Department of Analytical Chemistry, Nutrition and Food Science, Faculty of Pharmacy, University of Santiago de Compostela, Santiago de Compostela, Spain
e-mail: raquel.sendon@usc.es

S. M. Jafari, A. S. Silva (eds.), *Releasing Systems in Active Food Packaging*, Food Bioactive Ingredients, https://doi.org/10.1007/978-3-030-90299-5_15

1 Introduction

The fruits and vegetables production has increased during the last decade (FAOSTAT 2021) and it is of great importance to ensure the high quality of these foods products from the farm to the fork. Furthermore, it is assumed that approximately one third of the fruits and vegetables production will not be consumed as food. Usually, post-harvest losses are related with storage conditions (e.g. temperature, relative humidity, decontamination treatments, atmosphere composition) and packaging technologies (Shinde et al. 2018). Additionally, the long-term storage of a product inside a packaging can have a negative impact on the taste and quality, favouring the natural colonisation and multiplication of microorganisms (Ju et al. 2019; Patrignani et al. 2015). Therefore, there is a clear need for new methods of protecting food, such as antimicrobial active packaging (Bajpai and Baek 2016). Thus, active packaging should preserve fruits and vegetables for longer periods of time without altering both its quality and its organoleptic characteristics.

With the new lifestyles, the consumption of minimally processed foods, as fresh-cut fruit and vegetables, is also increased as they are healthy and also easy to eat or use. But its main drawback is that even they are minimally processed, usually its shelf life is shorter than the whole fruit or vegetable. Moreover, the deterioration of these products is accelerated after its processing. In these cases, active packaging can also play an important role in extending the self-life (Mehyar and Han 2011).

When focus on fruits and vegetables, the most important active systems that can be applied are oxygen scavengers, moisture absorbers, ethylene absorbers, carbon dioxide emitters, and antimicrobial releasers agents (Mehyar and Han 2011). All these systems involve the absorption and/or release of compounds. Between the releasers systems, the most described in the bibliography are those intended to release an antimicrobial compound to control the spoilage and foodborne pathogenic microorganisms (Sung et al. 2013). These antimicrobials, such as bacteriocins, enzymes, organic acids and nanosized metal oxides have different chemical properties (e.g., volatile or non-volatile); lately bacteriophages have been also included. They are integrated into the polymer matrix to provide to the packaging material the desired antimicrobial properties. Moreover, volatile antimicrobials such as essential oils, plant extracts or ethanol, are easily released to the headspace of the packaging and the releaser system can be the polymer matrix or other device as sachets resulting in antimicrobial packaging systems. The essential oils more used for fruits and vegetables packaging applications are reviewed in Sect. 2.1.

The mode of action will depend on the nature of the antimicrobial compound (Pereira de Abreu et al. 2012), and it will be based on evaporation in the headspace of the packaging or diffusion (Kapetanakou and Skandamis 2016).

An innovative technology is the controlled-release active packaging (CRP) that regulates the release of active antimicrobial compounds during the storage to maintain the quality of food and extend its shelf life (Hadi et al. 2020). In CRP the active compound is released over time slowly and in a controlled way, which makes it appropriate for enhancing both quality and safety of foods during long-time storage (Mastromatteo et al. 2010). Several examples are described along this chapter.

There is special interest in the development of new food packaging materials based on nanotechnology. The recent advances in nanomaterials have led to new possibilities in the active packaging of fruits and vegetables. The main applications based in the use of release systems are reviewed in Sect. 4.

A particular case of active systems that are almost exclusive for using with fruits and vegetables are the ethylene absorbers that are the active systems most mentioned in the literature and for which more alternatives are available commercially. For this reason, a brief description of these systems is also included in Sect. 3.

2 Releasing Systems Applied to Fruits and Vegetables

2.1 Essential Oils

Fruits and vegetables require protection during their storage from microbial spoilage because they are perishable. Therefore, there is a clear need for new methods of protecting them, such as antimicrobial active packaging (Bajpai and Baek 2016). An antimicrobial active packaging is designed to incorporate an antimicrobial intended to be released into the food to retard or prevent microbial growth and contamination during its useful life (Jung and Zhao 2016).

Antimicrobial active packaging is intended for make the fruits and vegetables shelf-life longer maintaining its quality and to enhance consumers' safety since it allows to reduce, inhibit, and/or retard the growth of pathogenic organisms in packaged foods (Patrignani et al. 2015).

Although the role of active packaging is to reduce these negative effects in some measure, there are some drawbacks associated with synthetic food additives; in general, natural additives get a better acceptance from the consumers (Ju et al. 2019). Antimicrobial agents extracted from natural sources are capable of inactivating pathogenic and spoilage bacteria even at low concentrations having no or little effect on the sensory properties of foods and being safe for human consumption (Jung and Zhao 2016).

Essential oils (EOs) are a complex combination of highly volatile compounds with strong sensory effects. They are natural secondary metabolites obtained from all the organs of many plants and accumulated in secretory cells, canals, cavities, glandular trichomes, or epidermis cells (Patrignani et al. 2015). To date, thousands of essential oils are available in the market for their multipurpose uses due to their unique flavours and fragrances (Bajpai and Baek 2016). Initially, EOs have been used to improve the aroma of foods, but in these past two decades, EOs and their components have attracted widespread attention from several researchers due to the extension of the shelf-life of different food systems such as to the replacement of less ideal synthetic additives because they have a variety of properties such antioxidant, antimicrobial, and anticancer activities (Patrignani et al. 2015).

EOs have a strong broad antibacterial spectrum. The antibacterial mechanism of EOs is not yet fully understood, and it seems to be related to the presence of some major biologically active compounds. It has been proven that the main antimicrobial components in EOs are phenols, terpenes, acids, aliphatic alcohols, aldehydes, ketones, and isoflavonoids (Deng et al. 2020). These biologically active compounds are usually found in bulbs (garlic and onions), leaves (rosemary, sage, basil, oregano, thyme and marjoram), fruits (cardamom and pepper), seeds (caraway, fennel, nutmeg), flowers or buds (clove) (Aziz and Karboune 2018). These compounds have the ability to interact with cell membrane lipids, increasing their fluidity and permeability, and causing the leaching out of important cell components. They can also bind to proteins, leading to the inactivation of enzymes. These modifications can disrupt stages of the microbial metabolism, affecting to the transport of electrons for energy production, the proton motive force, the translocation of proteins, and the synthesis of cellular components, ultimately leading to cell lysis and death (Deng et al. 2020).

The antimicrobial effectiveness of EOs is intimately connected to chemical species and concentration, storage time, microorganism and types of fruits and vegetables, among others. Generally, the microbial reduction increases with the EOs concentration and storage time. It has been shown that Gram-positive bacteria are more sensitive to EOs than Gram-negative bacteria because the latter have lipopolysaccharides in their outer membrane which could act as a barrier for phenolic compounds (Deng et al. 2020). Nevertheless, numerous EOs have shown broad spectrum antimicrobial activity in fruits and vegetables, as can be seen in Tables 15.1 and 15.2, respectively.

Most EOs are classified as "generally recognized as safe" (GRAS) by the Food and Drug Administration (FDA, 2021). However, the regulation of some EO components varies depending on the country, such as *Artemisia annua*, which has been approved by the FDA but is explicitly banned in the European Union. Newly discovered natural aromatic substances can be approved for use after toxicological and *in vivo* metabolic assessment (Ju et al. 2019).

The most used active release systems for fruits and vegetables, besides antimicrobial edible coatings, are the antimicrobial sachets inside the package that release the antimicrobial compound into the headspace of the packaging in order to interact with the microorganisms, or the antimicrobial films made by inserting the antimicrobial agent into a polymer matrix to release it into the surface of food. Antimicrobial sachets have received great success in commercial applications because they can release the antimicrobial compound without direct contact with the food. However, consumers´ acceptance of sachets is low because of the possible accidental ingestion of their contents (Jung and Zhao 2016). In the study of Marques et al. (2019), the sachets were the least accepted by consumers with scores between 5 and 6 for overall impression attribute. In the case of antimicrobial films, the key point is trying to extend the release of the antimicrobial agent from the film along with the continuous migration into the product throughout its shelf life (Jung and Zhao 2016).

As can be seen in the Tables 15.1 and 15.2, the use of biodegradable packaging systems such as alginate or starch that are eco-friendly alternatives to conventional

Table 15.1 Application of essential oils into the packaging for fruits

Essential oil	Fruit	Material	Target microorganism	Effect	Reference
Grapefruit Grape seed	Grape	Biodegradable alginate coating	*Penicillium digitatum*	Reduce decay incidence in inoculated fruits	Aloui et al. (2014)
Oregano oil	Cherry tomatoes	Polyvinyl alcohol film	*Salmonella enterica*, yeast, molds and mesophilic aerobic bacteria	Strong antimicrobial effects within changes in the physical characteristics of cherry tomatoes	Kwon et al. (2017)
Eugenol and thymol	Table grapes	On sterile gauze inside the non-perforated oriented polypropylene bag	Mesophilic aerobics and mold and yeast	Reduction in the microbial, higher for yeast and mold than for mesophilic aerobic	Valero et al. (2006)
Oregano, cinnamon and lemongrass	Papaya	Sachets made of a high-absorption polymeric resin and nonwoven fabric.	Mesophilic aerobic bacteria, yeasts and mold	Significant reduction in the growth, cinnamon the greatest reduction	Espitia et al. (2012)
Eugenol, thymol, menthol and eucalyptol	Sweet cherry	Polypropylene bags	Molds, yeasts and total aerobic mesophilic	Reduced grown of molds and yeasts and total aerobic mesophilic grown compared with control	Serrano et al. (2005)
Cinnamon	Chinese bayberry	Ethyl cellulose pads modified	*Aspergillus niger* and *Penicillium* sp.	A significantly lower decay incidence compared to the control group	Niu et al. (2018)
Cinnamon	Banana	Sodium alginate/ carboxymethyl cellulose films	*Staphylococcus aureus and Escherichia coli*	Extend the shelf life of bananas	Han et al. (2018)

(continued)

Table 15.1 (continued)

Essential oil	Fruit	Material	Target microorganism	Effect	Reference
Cinnamon, toronjil, rosemary, laurel, albaca, thyme and propolis ethanol extract	Strawberries	Bio-active packaging, made up of native starch, essential oil and natural extract	*Botritys cinerea, Escherichia coli* and *Staphylococcus aureus*	Rosemary, cinnamon and extract ethanol from propolis, showed greater inhibitory capacity	Beltran and Florez (2018)
Carvacrol and thymol	Strawberries	Nanocomposite films based on thermoplastic starch/ montmorillonite	*Botrytis cinerea*	Antimicrobial effectiveness without altering neither quality parameters nor organoleptic properties	Campos-Requena et al. (2017)
Cinnamon	Apple	Multi-layer film of chitosan and sodium alginate	*Penicillium*	Significant and lasting inhibition of *penicillium* expansion	Zhang et al. (2019)

plastic packaging, have called the food packaging industry attention (Issa et al. 2017). Among the variety of essential oils, oregano is one of the most tested in the group of vegetables and cinnamon for fruits.

Unfortunately, due to several factors which could influence the overall antimicrobial activity, such as the variability of the composition of EOs depending on the geographic origin, extraction methods, season, etc., the possible application of EOs in some foods continues to be limited. Moreover, the food matrix can interact limiting the contact between these molecules and the microbial cells, thus reducing the impact on cell viability. In addition, the mechanisms by which these molecules have antimicrobial activity is poorly understood (Patrignani et al. 2015); as well as there is little knowledge about safety, toxicity, and effect of EOs on the human intestinal tract; or the antimicrobial activity of trace components (Deng et al. 2020). Furthermore, they present unsuitable mechanical and permeability properties and/or strong, and sometimes unpleasant, odours and flavours. EOs are difficult to stabilize in the antimicrobial system due to their insolubility in water and strong lipophilicity and volatility (Ju et al. 2019; Jung and Zhao 2016). As demonstrated in several studies, *in vivo* tests require higher concentration of EOs to obtain the same microbial reduction as *in vitro* tests, which can cause the acceptable sensory level of these antimicrobials to be exceed (Deng et al. 2020). Therefore, to overcome these limitations, more research is needed to design antimicrobial packaging that allow inactivation of a broad spectrum of pathogenic microorganisms, provide a good barrier to moisture and gas, without effect on the organoleptic quality of foods. Encapsulation

Table 15.2 Application of essential oils into the packaging for vegetables

Essential oil	Vegetable	Material	Target microorganism	Effect	Reference
Oregano	Package mixed vegetables	Films made of poly(lactic acid)–cellulose nanocrystals	*Listeria monocytogenes*	Strong antimicrobial effect for 14 days at 4 °C	Salmieri et al. (2014)
Oregano	Packed salad	Ethylene-vinyl alcohol copolymer coated polypropylene films	*Salmonella Enterica, Escherichia coli, Listeria monocytogenes* and natural microflora	Antibacterial effect greater against gram-negative bacteria	Muriel-Galet et al. (2012)
Oregano and citral	Salad	Ethylene-vinyl alcohol copolymer coated polypropylene films	Enterobacteria, total aerobic counts, yeasts and molds	Reduction of the bacterial growth, citral appeared to be the most effective	Muriel-Galet et al. (2013)
Nutmeg, lemongrass, citral, oregano, and pimento berry	Fresh broccoli	Two different films: (i) Methylcellulose and (ii) a blend of polycaprolactone/alginate	*Escherichia coli, Salmonella typhimurium* and *Listeria monocytogenes*	Significant reduction of *S. typhimurium* concentration and good ability to inhibit the growth of *L. monocytogenes* and *E. coli*	Takala et al. (2013)
Oregano	Fresh-cut iceberg lettuce	Sachet containing oregano microcapsules	*Dickeya chrysanthemi*, molds and yeasts (MY), and total mesophilic aerobic bacteria (MAB)	Inhibition of the growth of *D. chrysanthemi*, and significant growth inhibitory effect against MY and total MAB	Chang et al. (2017)
Eugenol, carvacrol and *trans*-anethole	Ready-to-eat iceberg lettuce	Sachets in cellulose and polypropylene pillow packages	Coliforms, yeast and mold, and total bacterial count	All sachets demonstrated antimicrobial activity	Wieczyńska and Cavoski (2018)
Thyme essential oil	Baby spinach leaves	Film of sweet potato starch and montmorillonite nanoclays	*Salmonella typhimurium* and *Escherichia coli*	The incorporation of thyme in the film significantly reduces the population of *S. Typhi and E. coli*	Zhang et al. (2019)

technology can be employed to capture volatile compounds to minimize their organoleptic impact, control the release of the antimicrobial agents, and reduce the possible interactions with other food compounds. Moreover, multiple antimicrobial systems may be required to gain synergistic effects (Jung and Zhao 2016). For example, developing effective delivery systems to micro or nano encapsulate EOs together with other conservation techniques such as high pressure, ozone, low temperature, and modified atmosphere techniques to make a synergistic alternative can enhance the suppression of microbial growth maintaining the quality of the product (Ju et al. 2019; Jung and Zhao 2016). Other alternative can be to study the synergistic effect among EOs and/or their components to optimize their antibacterial activity and thus be able to decrease their needed dose to achieve a particular antibacterial effect (Patrignani et al. 2015).

2.2 *Ethanol*

Fruits and vegetables are susceptible to microbial contamination and decay. Therefore, reducing the spoilage is essential to maintain the quality and extend the shelf-life. Ethanol is a well-recognized broad-spectrum antimicrobial and widely applied in the preservation of fruits and vegetables (Mu et al. 2017). Different methods of ethanol treatment have been used including ethanol dip, vapors and ethanol emitters allowing to achieve a controlled release (Dao and Dantigny 2011; Mu et al. 2017). The response of fruits to ethanol treatment depends on several factors such as species, maturity, the concentration of ethanol and time of exposure, the method applied, among others. (Bai et al. 2011). Herein, some examples selected from the literature about the ethanol treatments in the preservation of fruits and vegetables are commented on.

Lichter et al. (2002) investigated the ethanol dips as a treatment to control the postharvest decay in table grapes. In the experiment, before packaging, grape bunches were dipped in ethanol, dried in the shade for 30–60 min and packed in ventilated polyethylene bags. An inhibition of the berry decay was observed. In another study the application of sulphur dioxide treatment in combination with ethanol vapor generated from a pre-soaked paper pad proved to be effective in improving the shelf life of table grapes (Chervin et al. 2005).

The combination of ethanol dip with modified atmosphere using a special film with high water conductance (Xtend) for packaging table grapes was effective in controlling decay (Lichter et al. 2005).

In other study reported in the literature, different treatments with ethanol were applied to climacteric fruits to evaluate the capacity of ethanol of inhibiting the ripening. Fruits including bananas, honeydews, muskmelons, nectarines, pears, peaches, plums and avocados were exposed to ethanol vapors. In the other treatment, ethanol was directly injected into the seed cavity of muskmelons and honeydew melons. The exposure to ethanol vapors did not show a significant effect on the inhibition of the ripening in bananas, honeydews, muskmelons, nectarines, pears, peaches and plums, however the ripening was inhibited when ethanol was directly injected (Ritenour et al. 1997). Jin et al. (2013) found that the effectiveness of treatment with ethanol vapors to maintain the quality of oriental sweet melons depends on the dose of ethanol.

One of the main disadvantages of liquid ethanol is that it volatilizes quickly. Therefore, in order to achieve an effective concentration of ethanol during a certain period of time, controlled release technology is needed. In a more recent study Mu et al. (2017) propose a system consisted of a gel prepared by a reaction with ethanol and sodium stearate. The gel was loaded on to diatomite which was selected as the adsorbent. The concentration of sodium stearate plays a key role in the control of the release, thus depending on the final application (fruit selected) and the storage period to be achieved the sodium stearate concentration should be adjusted. The release was strongly influenced for the temperature, lower temperatures lead to lower rates of release of ethanol. The active system was applied to preserve Chinese bayberry stored at 4 °C and good results were achieved, in fruits stored with the active system the decay rate was reduced and the fruit firmness was maintained.

Within the fruits and vegetable category the preservation of fresh cut fruit products is a great challenge for the food industry since these products are more susceptible to spoilage than intact fruits.

Several strategies have been developed to address this issue including, surface treatments with solutions and coatings, low temperatures, ethanol vapors and so on (Bai et al. 2004). An interesting approach involves the application of the treatment before cutting. Bai et al. (2004) evaluated the effect among other treatments of ethanol vapor pre-treatment in intact apples ("Gala") at 25 °C for 24 h. They founded that in treated apples the shelf-life was increased by reducing the ethylene production of the cut slices and inhibiting the decay on the cut surface.

In a similar approach intact mango were exposed to ethanol vapor after subjecting intact fruits to heat treatment to increase the shelf life of fresh-cut mango. The results showed that the ethanol treatment did not contribute to delay ripening but inhibited the microbial growth. Besides, the exposure to ethanol vapor during extended periods (20 h) lead to formation of off-flavors (Plotto et al. 2006).

More recently the use of an ethanol vapor release pad to preserve the quality of intact and fresh-cut sweet cherries was explored. Basically, the pad consisted of silica gel powder containing ethanol protected with a film laminated with ethylene-vinylacetate and Japanese paper with a high gas permeability allowing ethanol vapor diffusion gradually. Intact fruits and fresh-cut cherries were packed in clamshells with and without the ethanol pad and stored at different temperatures. Different quality parameters such as color, firmness and sensory evaluation among others were monitored. The ethanol-treated intact fruits showed a better firmness and a better visual quality compared with the control, similarly the ethanol-treated fresh cut cherries resulted in a better visual quality. From the sensorial point of view, no difference in the flavor was found with the ethanol treatment (Bai et al. 2011). A similar approach, that is ethanol vapor from alcohol powder was used for the preservation of broccoli during the storage in order to extend the shelf-life. Treated and untreated broccoli were packed in polyethylene bags and stored at 20 °C. The system showed to be effective in preventing yellowing of broccoli and reducing the ethylene formation (Suzuki et al. 2004).

A different active system that allows controlled release of ethanol vapor was used to delay microbial spoilage in fresh peeled shallots. The system was a sachet made of low-density polyethylene or nylon/polyethylene on one side and aluminium foil laminated on the other. And 1 g of silica gel adsorbent with ethanol inside the sachet. The system was placed in a tray which was covered with a LDPE film. The trays contained the shallots were stored at 10 °C for 10 days. The results on the efficacy of ethanol were not conclusive. It is suggested that more research is needed to understand the interactions of ethanol with fresh peeled shallots (Utto et al. 2018).

2.3 CO_2

Other postharvest technologies to delay the ripening of fruits and vegetables involves the exposure to CO_2. Elevated levels of CO_2 and reduced levels of O_2 have demonstrated to be effective for extending the shelf-life of fruits (Beaudry 1999).

In a study conducted by Ueda and Bai (1993) strawberries exposed to 20% CO_2 for 12, 24 or 48 h and stored at 1 °C, presented higher firmness than those exposed to air.

Sometimes the packaging material used has different permeability to O_2 and CO_2 (being higher in this last case) so it is interesting the use of dual systems that combine an oxygen scavenger with a CO_2 emitter thus helping to maintain a high concentration of CO_2 to delay the respiration rate of fruits and vegetables (Ozdemir and Floros 2004).

2.4 Other Volatile and Non-volatile Compounds Used as Active Release Systems

In addition of essential oils as an example of volatile compounds (Sect. 2.1) released by antimicrobial sachets or edible coatings, and nanoparticles as an example of non-volatile compounds (Sect. 3), there are some other compounds that can be used to preserve fruits and vegetables. These are summarized in Table 15.3. Sachets are the most used form to contain these active compounds that can be encapsulated inside this type of systems as i.e., antimicrobial sachet, for example allyl isothiocyanate to inactivate *Escherichia coli O157:H7* on spinach leaves (Hyun-Sun et al. 2012) or chlorine dioxide gas sachets for enhancing the microbiological quality and safety of blueberries (Popa et al. 2007). Equilibrium-modified atmosphere packaging (EMAP) in combination with a sachet release system were developed by Eva Almemar et al. using 2-nonanone as antimicrobial active compound, since it is an antifungal volatile compound naturally present in strawberries. It inhibits fungal

Table 15.3 Fruits and vegetables preservation by releasing active volatile compounds (excluding EO and edible coatings) and non-volatile compound (without nanoparticles)

Active compound	Format	Food matrix	Target microorganism	Reference
Allyl isothiocyabate (AIT)	Sachets	Spinach leaves	*Escherichia coli O157:H7*	Hyun-Sun et al. (2012)
			Yeasts and molds	
Chlorine dioxide (ClO_2)	Sachets	Blueberry	*Escherichia coli O157:H7*	Popa et al. (2007)
			L. *monocytogenes*	
			Salmonella	
			Yeasts and molds	
			Total aerophic mesophilics	
			Coliforms and E. *coli*	
Chlorine dioxide (ClO_2)	Ethylene moisture sachets	Strawberries	Shelf life	Mehmet and Cengiz (2010)
Eugenol, thymol and menthol	Polypropylene bag	Table grapes	Yeasts and molds	Valverde et al. (2005)
2-nonanone	Metallocene polypropylene sachets	Wild strawberries	*Botrytis cinerea*	Almenar et al. (2009)
Hexanal	Flushing hexanal vapor in packaging headspace	Tomato	*Botrytis cinerea*	Weerawate et al. (2008)
β-cyclodextrin-hexanal	Hexanal vapor	Berries	*Colletotrichum acutatum*	Almenar et al. (2007)
			Alternaria alternata	
			Botrytis cinerea	
Sweet basil extract	Coatings (Pullulan films)	Apples	*Rhizopus arrhizus*	Synowiec et al. (2014)
			Total mesophilic bacteria	
Layered Double Hydroxide (LDH) hosting 2-acetoxybenzoic anion (salicylate)	PET	Table grapes	*Pseudomonas*	Gorrasi et al. (2020)
			Listeria	
			Lactobacillus	
Carrier of nisin, natamycin, pomegranate and grape seed extract	Chitosan coatings	Fresh strawberries	Aerobic mesophilic bacteria	Merve et al. (2016)
			Yeast and mold	

(continued)

Table 15.3 (continued)

Active compound	Format	Food matrix	Target microorganism	Reference
1-methylcyclopropene (1-MCP)	Cellulose paper packaging	Fruits and vegetables		Zhijun et al. (2017)
Allyl isothiocyabate (AIT) entrapped in alpha and beta cyclodextrin inclusion complexes (ICs)		Fresh cut onions	*Penicillium expansum*	Piercey et al. (2012)
			Escherichia coli	
			Listeria monocytogenes	
Potassium permanganate	Polyethylene film lining	Banana		Abugoukh and Elamin (2009)

decay and delays the senescence of highly perishable wild strawberry (Almenar et al. 2009). The application of a coating over other material is also used, as Giuliana et al. reports the antimicrobial effect of PET coated with a Layered Double Hydroxide (LDH) hosting 2-acetoxybenzoic anion (salicylate) as antimicrobial molecule inhibiting the grown of Pseudomonas, Listeria and Lactobacillus in table grapes (Gorrasi et al. 2020).

Another way to incorporate the active compound into the packaging is by encapsulation. An example of that is the release of allyl isothiocyanate (AITC), encapsulating in poly(lactic) (PLA) fibers of submicron sizes and electrospun onto the surfaces of PLA films for food application in which the authors studied the growth of *Listeria innocua* and *Escherichia coli k12* on packaged foods (Hasan et al. 2015) or S-nitroso-N-acetyl-D-penicillamine (SNAP) encapsulated into nanocellulose fibrils studied by Jaya S. (Jaya et al. 2016). These biodegradable composite membranes have antimicrobial properties and inhibit bacterial strains of *Enterococcus faecalis*, *Staphylococcus aureus*, and *Listeria monocytogenes*.

Films produced with modified starch-based polymers can also be used to package fruits and vegetables. This type of materials is often expensive and more difficult to process in comparison with synthetic polymers (Kuorwel et al. 2011).

1-methylcyclopropene (1-MCP) is an excellent eco-friendly inhibitor of ethylene which retard the ripening of fruits and vegetables, thus, blocking the ethylene-induced respiration process. So, it is used as active substance in paper products to ship, wrap or decorate and in this way, also to preserve fruits and vegetables Zhijun et al. (2017). Active paper packaging using 1-MCP encapsulated will be maybe an interesting coating to prolong the shelf life and improve the quality of fruits and vegetables by the gradual release of 1-MCP. This substance was also used in fresh cut apple as has been reported by Linchun Mao et al. (2007).

Active packaging will be increasingly developed in the future to improve the quality and safety of the food, preferably minimally processed and naturally preserved foods. Active packaging systems of a very different nature such as those mentioned above are in development due to the consumer demands and market trends. Edible films are the most used as active packaging applied to fruits and vegetables, but these are out of the scope of these chapter.

3 An Example of Success: Ethylene Scavengers

Ethylene is a phytohormone present in fruits and vegetables, this volatile compound besides other specific roles in plants, is responsible to accelerate fruit ripening. It is considered one of the main causes of the fruits and vegetables spoilage. Thus, removing ethylene is essential to retard the ripening and so maintain the quality and extend the shelf-life (Wei et al. 2021; Awalgaonkar et al. 2020; Martínez-Romero et al. 2007). Ethylene scavengers have been successfully used to approach this issue. Ethylene scavengers can be classified according to the mechanism of action: (1) oxidation and (2) adsorption/absorption. In the first group are included those systems based on potassium permanganate, sodium permanganate and titanium dioxide whereas the second group comprises activated carbon, zeolites, clays, metal-organic frameworks (Awalgaonkar et al. 2020).

In the systems that use potassium permanganate the chemical is immobilized onto an inert substrate such as, alumina or silica gel (Gaikwad et al. 2020). On the other hand, sodium permanganate has been effectively impregnated on diatomaceous earth, zeolites, silica gel, perlite, etc. (Awalgaonkar et al. 2020).

With respect the formats, the most popular presentation of ethylene scavengers are the sachets made of porous materials that contain the ethylene scavenger in form of granules, beads or powder (Gaikwad et al. 2020). Recently, the incorporation of ethylene scavengers into the packaging films is being investigated. Some examples including the incorporation of nano-Ag, nano-TiO_2 into polyethylene (Wei et al. 2021).

Nowadays there are available many commercial ethylene scavengers. In the Table 15.4 are summarized some examples.

4 Nanotechnology Applied to Packaging Release Systems

Nanotechnology applied to food packaging is a broad field of study that has been provided new improved materials to extend the shelf life of foods. The development of polymer materials filled with nano-sized (1–100 nm) particles can lead to significant modifications on the physicochemical properties of the packaging materials: improved packaging, active packaging, intelligent packaging and edible packaging (Enescu et al. 2019). The use of nanomaterials in food packaging for fruits and vegetables has been extensively studied in recent years. The most common applications of nanomaterials in fruits and vegetables packaging are (i) improved food packaging to enhance packaging barriers against gas/moisture/vapor and improve their mechanical and thermal properties, and (ii) active food packaging with controlled release of active substances such as antibacterial to increase shelf-life of food products (Ma et al. 2017).

This section will be focused on the latest innovations in active packaging systems based on polymers nanomaterials designed to deliberately incorporate

Table 15.4 -Commercial ethylene scavengers for food packaging applications

Commercial brand	Format	Mechanism of action	Active agent
Neupalon™	Sachet	Adsorption/ Absorption	Activated carbon
Hatofresh System	Paper bag or corrugated box	Adsorption/ Absorption	Activated carbon impregnated with bromine-type inorganic chemicals
Extend-A-life™, Produce Saver™	Sachet	Adsorption/ Oxidation	Zeolite coated with potassium permanganate
EvertFresh	Bags	Adsorption/ Absorption	Oya clay
Nichem's Freshness Plus Bags	Bags	Removal	Nano-silver additive
Debbie Meyer® Bags	Bags	Adsorption/ Absorption	Zeolite
Purafil	Sachet	Oxidation	Sodium permanganate
Dry Pak's linePeak Fresh®	Sachet, filter and films	Adsorption/ Absorption	Zeolite
PEAKfresh®	Bags	Removal	Natural mineral
Biopac	Sachet	Adsorption/ Oxidation	Porous beads mixed with potassium permanganate
Keep-it-Fresh	Bags	Adsorption/ Oxidation	Special minerals and oxidizing agent
Erisfilter	Filter	Oxidation	Potassium permanganate

components that would release substances into or from the packaged food or the environment surrounding the food. Nowadays, active packaging has been mainly developed for antimicrobial packaging applications, but the use of other functional additives such as antioxidants and browning inhibitors into the active packaging of fresh fruits and vegetables has successfully enhanced the shelf-life by absorbing the ethylene gas and inhibiting microbial growth (Ma et al. 2019; Zanetti et al. 2018). However, these functional additives may interact with food or food packaging components, reducing their biological activity. This effect may be avoided by using carrier systems, such as nanocomposites, nanofillers, nanofibers, nanotubes, etc., to protect and release the additives to target sites in a controlled approach (Zanetti et al. 2018). A nanocomposite is a multiphase material that comprises two or more components, including a continuous phase (matrix) and a discontinuous nanodimensional phase with at least one nanoscale filler (Bahrami et al. 2020).These nanofillers can be classified into organic including proteins and polysaccharides (e.g. chitosan, cellulose), and inorganic such as metals (e.g. silver (Ag)), metal oxides (e.g. zinc oxide (ZnO), titanium dioxide (TiO_2)) and clays (e.g. montmorillonite (MMT)) (Azeredo et al. 2019). The most current nanocarrier systems developed with antimicrobial agents for food packaging purposes are nanoparticles (NPs), nanofibers, nanoemulsions, and nanoliposomes. The incorporation of antimicrobials and other bioactive compounds to NPs have the following technical advantages: (i) increase of the surface area to interact with substracts; (ii) enhance the solubility and miscibility; (iii) protection of antimicrobials against environmental, especially volatiles such as essential oils (OE); and (iv) contribute to allowing a target and controlled release of bioactive compounds from packaging materials (Zanetti et al. 2018; Bahrami et al. 2020; Azeredo et al. 2019).

Tables 15.5 and 15.6 presents a summary of the most recent studies on the application of packaging nanomaterials to extend the shelf life of several fruits (Table 15.5) and vegetables (Table 15.6) with the description of the nanocomposites, polymeric materials, their nanofillers composition, and particle sizes, methods of production, and the main positive effects for the extension of the food shelf life.

The incorporation of the nanoparticles is commonly performed by extrusion, solvent evaporation (casting) as part of the polymeric material, or by a coating applied on the packaging material of fruits and vegetables to enhance their shelf-life. The polymeric materials used are varied from petroleum-based materials (e.g. low-density polyethylene (LPDE), polyethylene (PE), polypropylene (PP), polyvinyl chloride (PVC) among others), and biopolymers for packaging such as polylactic acid (PLA), poly(vinyl alcohol) (PVOH), ethylcellulose pads, and Nano-fibrillated cellulose (NFC), among other biobased materials such as gelatin (GL) and chitosan (CS), as described in above mentioned Tables.

Among the numerous nanoparticles used for functionalizing polymeric materials, metal nanoparticles and metal oxide nanomaterials are the most used nanoparticles to develop antimicrobial active packaging. These particles function on direct contact, but they can also migrate slowly and react preferentially with organics present in the food.

Silver nanoparticles (AgNPs) have been widely studied, mainly due to their antimicrobial properties. Several authors have studied the AgNPs application in different polymeric matrices to extend the shelf life of fruits such as strawberries (Kowsalya et al. 2019; He et al. 2020; Li et al. 2017b), grapes (Kumar et al. 2018; Deng et al. 2019), lemons (Kowsalya et al. 2019), apples (Bhople et al. 2016; Li et al. 2017b). The paper coated with AgNPs developed by Bhople et al. (2016) was used to wrapped apples. The active system showed higher antibacterial capacity against *S. aureus*, *S. typhi*, *E. coli*, *K. pneumoniae*, and *P. aeruginosa* and antifungal activity against *A. niger* and *F. oxysporum*. The active paper used increased the shelf life of apples up to 15 days in a shelf-life study performed at room temperature. This remarkable antimicrobial activity of silver nanoparticles has been related by several authors to the capacity of the silver ions to promote electrostatic interactions with negatively charged bacterial membranes (biomacromolecular components and nucleic acids). The AgNPs might interfere with membrane functions and generate reactive oxygen species (ROS) that lead to structural changes and bacterial cell wall deformation, increasing permeability and cell death (Azeredo et al. 2019).

Metal oxide NPs have been used as photocatalysts in antimicrobial food packaging. Nanoparticles of zinc oxide (ZnONPs) and titanium dioxide (TiO_2NPs) activities are recognized to numerous mechanisms, including the release of antimicrobial ions, the interaction of nanoparticles with the microorganism that damage the integrity of the bacterial cell, and the generation of reactive oxygen species (ROS) by the effect of light irradiation resulting in the oxidation of cytoplasm of bacterial cells

Table 15.5 Recent studies on the application of nanotechnology-based packaging in fruits and vegetables

Nanofiller/ concentration	Particle size	Material base/Film	Methodology	Food application	Effects	Ref
AgNPs (0, 1, 5, 10% w/w)	–	PLA	Casting	Strawberries	Effective reduction of vitamin C loss, delaying the decline of total phenols in strawberries. The strawberries had a higher antioxidant capacity in the later period of storage, delaying the aging of the fruit and providing a better preservation effect. It also showed better physical properties. Reduce the weight loss rate of strawberries during storage at 4 °C for 10 days, and delay the drop in hardness, soluble solids, and titratable acid content.	Zhang et al. (2018)
AgNPs (0.05% and 1% w/w) (*Mimusops elengi* fruit extract and $AgNO_3$ solution)	25–45 nm	CS/GL	Casting	Red grapes	Enhanced mechanical properties (mechanical strength) and decrease in light transmittance. The shelf life of the fruit extended for additional 2 weeks (14 days) at 37 °C, in case of the hybrid film CS-GL-AgNPs at 0.1%.	Kumar et al. (2018)
AgNPs (37.5 and 75 mg/L)	–	PVOH	Casting	Grapes	Controlled the sensory score, mass loss, decay rate, ascorbic acid and titratable acid in grapes. Remarkable effect in prolonging grapes' shelf life due to its breathability and antifungal of nano-silver capability to *Aspergillus niger*, during storage at room temperature for 10 days.	Deng et al. (2019)
AgNPs (0.5% w/v) with fruit peels extracts of *Vitis vinifera*, *Carica papaya*, and *Citrullus lanatus*	30 nm	PVOH electrospun nanofibers	Coating	Lemon (Citrus limon) and strawberry (*Fragaria ananassa*)	The active AgPVOH nanofibers enabled the inhibition of the microorganisms (*Bacillus cereus*, *Staphylococcus aureus*, *Escherichia coli*, and *Pseudomonas aeruginosa*) in fruits for a period of 10 days of storage at room temperature.	Kowsalya et al. (2019)

AgNPs	85 nm	Wrapping Paper	Coating	Apples	Antibacterial potential against *Staphylococcus aureus*, *Salmonella typhi*, *E. coli*, *Klebsiella pneumoniae* and *Pseudomonas aeruginosa* and antifungal activity against *A. niger* and *F. oxysporum*. Paper impregnated with AgNPs used for wrapping of apples increases their shelf life up to 15 days (shelf-life study for 45 days at room temperature).	Bhople et al. (2016)
(1 mM)						
CMC/CNC-AgNPs	AgNPs 10–20 nm	Wrapping Paper	Coating	Strawberries	CMC/CNC-AgNPs 7% coated paper exhibited 1.26 times increase in tensile strength, 45.4% decrease in WVP, 93.3% reduction in air permeability as well as the best antibacterial activities against *E. coli* and *S. aureus*. Coated paper could maintain better strawberries quality and extend the shelf-life of strawberries to 7 days at room temperature.	He et al. 2020
(1, 3, 5, and 7 wt.% CMC)	CNC needle- like crystals: 81–286 nm (L), 8–21 nm (D), 16 nm (L/D ratio)					
Ag^{+}/Zn^{2+}-permutite/ CEO	≤10 μm	Ethyl cellulose pads	2 mL dispersion to absorbent paper	Chinese bayberry	Ag^{+}/Zn^{2+}-permutite/CEO has a sustained release effect and inhibit mold growth up to 5 days of storage *Aspergillus niger* and *Penicillium*.	Niu et al. (2018)
(313.07 μL/g)					Ethyl cellulose pads modified by composite antimicrobial particles were applied in the preservation of Chinese bayberry significantly with lower decay incidence compared to the control group during 3–4 days at room temperature.	

(continued)

Table 15.5 (continued)

Nanofiller/ concentration	Particle size	Material base/Film	Methodology	Food application	Effects	Ref
ZnONPs	200–400 nm	PVC	Coating	Fresh-cut 'Fuji' apple	Coating of nano-ZnO film reduced fruit decay rate, reduced the accumulation of MDA and ethylene content, maintained TSS and TA levels and inhibited the activity of PPO and POD during storage at 4 °C for 12 days. Easy processing and feasibility to be industrialized with lower time-consuming, costs and off-flavor changing.	Li et al. (2011)
(5.0 g/L)						
ZnONPs	30 nm	CS/CAP	Casting	Black grape	Thermal stability and barrier properties of the fabricated films increased with increasing amount of nano ZnO in the range of 2–7.5% (w/w). CS-CAP film loaded with 5% (w/w) nano ZnO showed the most optimal tensile strength and stiffness. Low surface wettability and high contact angle value up to 90°. Extend the shelf life of black grapefruits up to 9 days. Antimicrobial activity against *S. aureus* and *E. coli*.	Indumathi et al. (2019)
(2.5, 5, 7.5 wt.%)						
ZnONPs	30 nm	LPDE	Extrusion	"Hujingmilu" peaches	Rapid formation of the gas atmosphere with low O_2 and high CO_2, successfully applied in the cold storage of peaches at 2 °C.	Li et al. (2017a)
(30 wt.%)					Increase chilling tolerance of peaches during cold storage, showed higher fruit firmness with lower browning index, electrolyte leakage, relative viscosity, and decay rate during storage at 2 °C for 40 days.	
					Inhibition of pectin esterase and enhance polygalacturonase and β-galactosidase.	

ZnONPs (1, 3 and 5 wt.%)	10–30 nm (ZnONPs)	LPDE	Hot press machine	Fresh Strawberries	Effective antimicrobial activity against yeast and molds up to 16 days of storage at 4 °C. The addition of PE-g-MA into the polymer matrix resulted in a good dispersion of nanoparticles in the polymer, increased antimicrobial activity, improved mechanical properties of nanocomposite polymers, reduce the degradation of ascorbic acid content, and loss of acidity.	Emamifar and Mohammadizadeh (2015)
ZnONPs-PE-g-MA (3 wt.% ZnO + 10 wt.% PE-g-MA)	10–20 nm (PE-g-MA)					
ZnONPs	14–48 nm	Agar	Casting	Green grape	Loading of ZnONPs in the agar matrix improved thermal stability, elongation and film thickness, whereas tensile strength and transparency decreased. Grapes packaged in composite films showed fresh appearance up to 14 and 21 days in ambient conditions for 2% (w/w) and 4% (w/w) ZnONPs in films, respectively. The green grapes wrapped in the films remained acceptable in appearance for extended periods in ambient storage for 25 days.	Kumar et al. (2019)
(2 and 4 wt.%) with *Mimusops elengi* fruit extract						
ZnONPs with cinnamaldehyde (C6)	–	PLA	Casting	Fresh-Cut Apple	Nano-blend packaging films had the highest weight loss higher water vapor permeability and opacity, and a lower oxygen permeability.	Li et al. (2017b)
(C6/ZnO 1 wt.%, C6/ZnO 3 wt.%)					Maintenance of tissue firmness, total phenolic content, and the sensory quality, and in the reduction of the activity of PPO, as well as in the inhibition of the browning index (BI) and the microbial growth (total bacteria, mold and yeasts) of the fresh-cut apple at 4 °C for 14 days.	

(continued)

Table 15.5 (continued)

Nanofiller/ concentration	Particle size	Material base/Film	Methodology	Food application	Effects	Ref
TiO_2NPs (0.5, 1.0, 1.5 and 2.0 wt.%)	40 nm	LDPE	Extrusion	Strawberry (Fragaria ananassa Duch.)	TiO_2NPs-LDPE packaging contributed to forming a high CO_2 and low O_2 atmosphere in the package. Beneficial effects of TiO_2NPs-LDPE packaging on postharvest overall quality of strawberry fruit (weight loss, firmness and titratable acid), including maintaining higher organoleptic and nutritional quality, and antioxidant capacity of strawberry fruits. Correct air composition and suppressed ROS accumulation and ROS catalysis-related antioxidant enzyme activities. ROS balance between O_{-2} and H_2O_2 was maintained prolonging the shelf-life of strawberry fruit at 4 °C for 17 days.	Li et al. (2017c)
M-TiO_2 nanocomposite (2 wt.%)	20–80 nm	LDPE	Melt blending method – extrusion	Fresh pears	The number of mesophilic bacteria, *Pseudomonas spp.* and *Rhodotorula mucilaginosa*, and yeast cells in fresh pears packaged in TiO_2 nanocomposite film decreased significantly during storage under fluorescent light lamp at 5 °C for 17 days.	Bodaghi et al. (2013)

TiO_2NPs and clay (Cloisite 20A) NC (5 wt.% clay), NT (3 wt.% TiO2), NTC (5 wt.% clay and 3 wt.% TiO2)	<100 nm	LDPE	Extrusion	Pear fruit (Prunus communis L. cv. Williams)	Weight loss, firmness and the total solids content of pear in clay/TiO_2-nanocomposite film decreased at the rates of 50%, 39% and 13%, respectively due to the modified atmosphere. LDPE films, clay/TiO_2-nanocomposite increaser ascorbic acid content, decrease ethylene production, reduce sugar content and PPO activity and increase POD of the packed fruit. Clay/TiO_2-nanocomposite film reduce the risk of pathogens, contamination and improve the maintenance quality of pear fruit, reducing degradation of nutrients during the storage during 48 days of storage at 4 °C. Simultaneous application of three nanoparticles in clay-TiO_2 nanocomposite film could be considered as a new packaging strategy to retard ripening and to maintain the quality of pear fruit.	Bodaghi and Hagh (2019)

(continued)

Table 15.5 (continued)

Nanofiller/ concentration	Particle size	Material base/Film	Methodology	Food application	Effects	Ref
BEO/TiO_2NPs/ AgNPs	TiO_2NPs <100 nm	PLA	Casting	Mangoes	Nanocomposite films decrease the weight loss from day 9 to day 12 and delay the loss of mango firmness during the entire storage period.	Chi et al. (2019)
BEO (9 wt.%);	AgNPs <150 nm				The PLA nanocomposite films reduce the changes in color, total acidity, vitamin C, and microbial property when compared with PLA and PLA/BEO films. PLA nanocomposite films could retard the growth of microorganisms on mango pericarp. The sensory evaluation indicated the superior quality of mangoes in PLA/BEO/Ti and PLA/BEO/Ti + Ag groups. The overall acceptability of mangoes packed by the PLA nanocomposite films was still higher than 5 scores and within limit of marketability at the end of storage period. The result indicated that the PLA nanocomposite films could be used to maintain the quality of fresh mango and extend its postharvest life to 15 days of storage at 20 °C.	
BEO (9 wt.%) / TiO_2NPs (3 wt.%);						
BEO (9 wt.%) / TiO_2NPs (2 wt.%) / AgNPs (1 wt.%)						

AgNPs (5 wt.%)/ TiO_2NPs (40 wt.%)/ MMT (25 wt.%)	40–80 nm	LDPE	Extrusion	Kiwifruit (Actinidia deliciosa)	The weight loss, softening, color variation and soluble solid content of kiwifruit were significantly inhibited by 22.67%, 124.84%, 23.46% and 14.42%, respectively. Increase of ascorbic acid and total phenols contents in NCP-treated fruit. 7.44% lower headspace ethylene concentration, 29.44% for malondialdehyde (MDA), lower polyphenol oxidase (PPO) activity and higher peroxidase (POD) activity during storage at 4 °C for 42 days.	Hu et al. (2011)
SiO_2NPs (0.1 wt.%)	40–60 nm	LDPE	Extrusion	Loquat fruit	LDPE- SiO_2NPs material inhibited the internal browning, retarded the decline of total soluble solids, titratable acidity, ascorbic acid content of fruits during storage at room temperature for 12 days. LDPE- SiO_2NPs enhanced the contents of individual phenolic compounds and soluble sugar compounds, induced higher antioxidant enzyme activities, and increased the antioxidant capacity.	Wang et al. (2020)
CuNPs-PEO (1 wt.%)	12 nm	PVA	Coating	Fruit salad made of Apples (Granny Smith) and pears (Abate Fétel)	CuNPs-PEO composites prevented quality decay of fruit salad. The active films preserved color and texture of fruit that remained acceptable for 5–6 days more than the control sample. Microbiological quality, pH and sensory properties retained on samples in contact with the antimicrobial active composite, for more than 2 weeks (18 days) at 4 °C. Browning was delayed in the active samples up to 14 days of storage.	Sportelli et al. (2019)

(continued)

Table 15.5 (continued)

Nanofiller/ concentration	Particle size	Material base/Film	Methodology	Food application	Effects	Ref
Se-AgNPs 5% (2.5% SeNPs +2.5% AgNPs); 10% (5% SeNPs +5% AgNPs); 15% (7.5% SeNPs +7.5% AgNPs) (v/v)	SeNPs 50 nm AgNPs 20 nm	FUR/GL films	Casting	Mini kiwi (Actinidia arguta)	The incorporation of Se-AgNPs in the FUR/GL films improved the UV-light barrier, mechanical and thermal stability attributes. The increase in the concentration of Se-AgNPs results in change of elasticity and elongation at break. Antibacterial activity against *S. aureus*, Multi Resistant *S. aureus* (MRSA) and *E. coli*. Biopolymer film had longer shelf life of kiwis up to 8 days at room temperature.	Jamróz et al. (2019)
Nano-Carrier of Salicylate (25 wt.%)	1–2 micron	PE	Extrusion multilayer	Blueberries	Active film delayed the rate of respiration of fruits and limited weight loss, preserving the fruit firmness. The release of salicylate on blueberries preserved the nutritional quality without alterations the fruits' sensory traits. Antimicrobial effect (mold and yeasts) starting from the 7th day of storage, a sampling time in which the salicylate concentration was released by more than 80%. The combination of active packaging and modified atmosphere resulted a valid solution to extend blueberries' shelf-life (50% shelf-life extension respect to the Control) up to 13 days at 8 °C.	Bugatti et al. (2020)
Nanotubes HNTs: HNTs/ 2 wt.% carvacrol HNTs/ wt.% carvacrol	HNTs tubular: <100 nm (e.d.), 20 nm (i.d.), 0.2–2 mm (L)	Polyamide (Nylon 6)	Cast extrusion	Cherry tomatoes, lychee and grapes	PA/(HNT/Carvacrol) films exhibit fungicidal and/or fungistatic effects against a wide range of postharvest fungal molds: *A. alternata*, *B. cinerea*, *P. digitatum*, *P. expansum* and *A. niger*. Inhibit the development of decay and extending shelf life under storage conditions (cherry tomatoes: 51 days at 10 °C; Lychee: 29 days at 4 °C; table grapes: 38 days at 1 °C).	Shemesh et al. (2016)

CEO/β-CD	Fibers: 240 nm (D)	Electrospun nanofibrous PVOH film (6% (w/w))	Electrospinning	Strawberry	Antimicrobial activity against *S. aureus* and *E. coli.*	Wen et al. (2016)
(2:1; 2:2; 2:3 wt.%)					Mild electrospinning process is favorable for maintaining greater CEO (volatile antimicrobial agent) in the film resulting in an improved antimicrobial activity compared with that of casting film.	
CEO (80% trans-cinnamaldehyde)					PVOH/ CEO/β-CD nanofibrous film can effectively prolong the shelf-life of strawberry at 4 °C for 18 days.	
Nanofibers Pul-CMC-TP	Fibers 127 nm (D)	Pul-CMC	Electrospinning	Strawberry	The pullulan-CMC-TP nanofibers significantly decreased weight loss and maintained the firmness of the strawberries and improved the quality of the fruit during storage at 4 °C for 10 days.	Shao et al. (2018)
Tea polyphenols (TP) (0.5%, 1%, 1.5%, (w/v))						

Silver nanoparticles (AgNPs); Zinc oxide nanoparticles (ZnONPs); Titanium dioxide nanoparticles (TiO_2NPs); Cinnamon essential oil (CEO); Cinnamaldehyde (C6); clay-nanocomposite (NC), TiO2-nanocomposite (NT), clay-TiO2 nanocomposite (NTC), Bergamot essential oil (BEO); montmorillonite (MMT); malondialdehyde (MDA); polyphenol oxidase (PPO); peroxidase (POD); polyethylene oxide (PEO); copper nanoparticles (CuNPs); Selenium nanoparticles (SeNPs); Halloysite nanotubes (HNTs); β-cyclodextrin (β-CD); external diameter (e.d.); internal diameter (i.d.); length (L), diameter (D); oregano essential oil (EO); tea polyphenols (TP); modified TiO_2 (M-TiO_2). Polymers: (PLA); Low density polyethylene (LDPE); Chitosan (CS); Gelatin (GL); Poly(vinyl alcohol) (PVOH); Carboxymethyl cellulose (CMC); Cellulose nanocrystals (CNC); Cellulose acetate phthalate (CAP); Polyvinyl acetate (PVA); Polyethylene (PE); Pullulan-Carboxymethylcellulose sodium (Pul-CMC); Furcellaran (FUR); Polyvinylpyrrolidone (PVP); polyvinyl chloride (PVC); Nano-fibrillated cellulose (NFC); Polyethylene-grafted maleic anhydride (PE-g-MA)

Table 15.6 Application of nanotechnology in vegetables

Nanocomposite/ concentration	Particle size	Material base	Methodology	Food application	Effects	Ref
C/G-AgNPs based glycerosomes	G-AgNPs 37 nm, G-PVP-AgNPs 123 nm	PVP	Coating	Fresh-cut bell pepper (Capsicum annuum L. var. grossum (L.) Sendt)	Antibacterial activity: G-PVP-AgNPs > C-PVP-AgNPs against Salmonella enterica, P. aeruginosa, E. coli, B. cereus, S. aureus.	Saravanakumar et al. (2020)
(0.01 M)	C-PVP-AgNPs 45 nm				Nano coating of G-PVP-AgNPs extended the shelf life of the red or yellow fresh cut bell pepper for 12 d at 4 °C.	
G-AgNPs (Diospyros kaki L (persimmon) pedicel extracts)						
AgNPs-SiO_2 AgNPs-TiO_2	–	LDPE	Extrusion	Fresh-Cut Carrots	AgNPs-TiO_2-LDPE showed improved performance (lower permeability coefficient) compared to AgNPs-SiO_2-LDPE. Films with lower concentration of AgNP (SiO_2 2.5 and TiO_2 2.5) showed higher antimicrobial activity. Carrots packaged with TiO_2 films presented lower mesophilic aerobic and total coliforms count, weight loss, indicating better physicochemical and microbiological properties. TiO_2-LDPE films leads to lower soluble solids and maintain the higher levels of ascorbic acid in packed carrots and, therefore, to a better postharvest quality conservation during storage at 10 °C for 10 days. Migration of AgNPs from the packages to fresh-cut carrot was not observed.	Becaro et al. (2016)
(2.5, 50 and 100 wt.% AgNPs)						

NFC (4 wt.%)	80–105 nm	PVOH/PAA (5% (v/v) / 5% (v/v))	Casting	Tomato	Nanofilm was UV protectant, strong cross-linking, high oxygen barrier capacity, thermally stable up to 365 °C, high tensile strength compared to control. NFC/PVOH/PAA films with high capacity to preserve the tomato during storage at room temperature for 15 days, extending the shelf life 9 days more than unwrapped tomato and 7 days compared to control film.	Ponni et al. (2020)
ZnONPs (1, 3 and 5 wt.%)	30 nm	PU/CS (75/25% (w/w))	Casting	Carrot	PU/CS with 5% ZnONPs composite film improved thermal stability, tensile strength and stiffness. The incorporation of ZnONPs enhanced the antibacterial properties, against S. aureus and E. coli, barrier properties and hydrophobicity of the film. Carrot pieces wrapped in composite films PU/CS-5%ZnONPs extended shelf-life up to 9-days at room temperature.	Indumathi and Rajarajeswari (2019)
CS-ZnONPs (0.147 mg/dm^2)	35–45 nm	LPDE	Coating	Okra (Abelmoschus esculentus)	CS-ZnONPs nanocomposite coatings did not affect the quality attributes of the okra, such as pH, total soluble solids, moisture content, and weight loss. CS-ZnONPs nanocomposite reduced fungal and bacterial growth in the okra samples after 12 storage days. CS-ZnONPs nanocomposite coating increased shelf-life of Okra from 5-days to 12 days under room temperature storage.	Al-Naamani et al. (2018)

(continued)

Table 15.6 (continued)

Nanocomposite/ concentration	Particle size	Material base	Methodology	Food application	Effects	Ref
TiO_2/Ag-NPs (0, 1, 3, 5, 8 and 10 wt.%)	30–40 nm	PBAT	Casting	Cherry tomato	TiO_2/AgNPs enhanced the physical properties PBAT matrix, such as mechanical, UV blocking, and barrier properties, thermal stability, and food preservation properties. PBAT-TiO_2/AgNPs biofilms exhibited antibacterial activity against E. coli and S. aureus. The optimal performance was achieved with the formulation developed at 5% (w/w) TiO_2/Ag NPs, which reduced the weight loss and improved conservation of cherry tomatoes during 25 days of storage.	Cao et al. (2020)
TiO_2NPs (1 wt.%)	21 nm	CS	Casting	Tomato fruit	The CS-TiO_2 nanocomposite film exhibited better tensile strength and barrier properties as well as ethylene photodegradation, delay the ripening process and extend the storage life of the tomato fruit up to 14 days at 20 °C.	Kaewklin et al. (2018)
$CaCO_3$NPs (30 wt.%)	60–100 nm	LDPE	Extrusion	Fresh-cut sugarcane	$CaCO_3$NPs-LDPE decrease transmission rate of O_2 and CO_2 that lead to the more rapid formation of gas environment with low O_2 and high CO_2 concentration in the package. $CaCO_3$NPs reduced the total bacterial, yeast and mold counts of fresh-cut sugarcane during storage at 10 °C for 5 days. Fresh-cut sugarcane packed with $CaCO_3$NPs-LDPE exhibited significantly lower activities of PAL, PPO, POD. $CaCO_3$NPs-LDPE improved overall quality by the reduction of browning index and the total phenolic content.	Luo et al. (2014)

$CaCO_3NPs$ (30 wt.%)	60–100 nm	LDPE	Extrusion	Fresh-cut Chinese yam	$CaCO_3NPs$ reduced the growth of bacterial, yeast and mold of fresh-cut Chinese yam and inhibit respiration and ethylene production. $CaCO_3NPs$-LDPE packaged fresh-cut Chinese yam exhibited significantly lower activities of PAL, PPO, and POD compared to the control yam samples. $CaCO_3NPs$-LDPE delay the browning process, decrease the total phenolic and malondialdehyde contents, and maintained overall quality, titratable acid, and ascorbic acid during storage at 10 °C for 5 days.	Luo et al. (2015)
CNC/EO (3 wt.%)	–	PLA	Casting	Mixed vegetables	EO addition increased elongation at break of films and reduced their tensile strength and tensile modulus (increase of elasticity). Antimicrobial activity of PLA–CNC–oregano films against L. monocytogenes after 14 days of storage at 4 °C, as a promising bioactive packaging to preserve fresh food products against food borne pathogens. PLA–CNC films allowed a slow controlled release of the total phenolic compounds during storage, increased continuously from day 0 to day 4, up to 16.6% at day 14.	Salmieri et al. (2014)

Silver nanoparticles (AgNPs); Zinc oxide nanoparticles (ZnONPs); Titanium dioxide nanoparticles (TiO_2NPs); TiO_2-nanocomposite (NT), polyphenol oxidase (PPO); peroxidase (POD); phenylalanine ammonia lyase (PAL); chemically and green synthesized silver nanoparticles based glycerosomes (C/G-AgNPs); chemically synthesized polyvinylpyrrolidone-silver nanoparticles (C-PVP-AgNPs); green synthesized polyvinylpyrrolidone -silver nanoparticles (G-PVP-AgNPs); oregano essential oil (EO). Polymers: (PLA); Low density polyethylene (LDPE); Chitosan (CS); Poly(vinyl alcohol) (PVOH); Polyvinylpyrrolidone (PVP); Cellulose nanocrystals (CNC); Nano-fibrillated cellulose (NFC); Poly(acrylic acid) (PAA); Mahua oil-based polyurethane (PU); Poly (butylene adipate-co-terephthalate) (PBAT) and leading to cell death. Some studies suggested that ZnO can be more efficient and attractive than silver because of its lower toxicity and cost-effectiveness (Duncan 2011; Kim et al. 2020). In addition to the antibacterial capacity, ZnONPs enhance packaging properties comprising mechanical strength, barrier properties, and stability (Indumathi et al. 2019; Kumar et al. 2019; Li et al. 2017a). ZnONPs have been incorporated in different materials, including PVC (Li et al. 2011), CS/Cellulose acetate phthalate (CAP) (Indumathi et al. 2019), LDPE (Li et al. 2017a; Emamifar and Mohammadizadeh 2015), Agar (Kumar et al. 2019), PLA (Li et al. 2017b), among others, to extend the shelf life of fruits. An LDPE film, coated with a nanocomposite of CS and ZnONPs, was used to reduced fungal and bacterial growth and extend the shelf life of okra at room temperature from 5-days to 12 days (Al-Naamani et al. 2018). The authors also determined the Zn^{2+} release, after 12 days of storage, was around 1.8% of the initial amount in the coating. Despite, the material was developed to intentionally increase the concentration of Zn^{2+}, the ion release from the coating was very low due to the high stability of ZnONPs in the chitosan. Sportelli et al. (2019) developed an active film of Polyvinyl acetate (PVA) coated with a nanocomposite formed by copper oxide nanoparticles and polyethylene oxide (PEO) to maintain the quality of a fruit salad composed of apple and pear. The active films preserved color and texture of fruit up to 6 days more than the control sample, maintained sensorial properties and microbiological activity for 18 days of storage, and browning was delayed up to 14 days. Further, the authors also performed the controlled release kinetics in physiological and acidic conditions (to simulate lemon juice) of the active film and observed that the release of Cu^{2+} was five-fold higher at low pHs. In the last years, multimetallic NPs have played an important role in food packaging applications due to their improved physical and chemical properties, respect to monometallic NPs. The most usual are the bimetallic and trimetallic nanoparticles that result from the combination of two and three different metals and metal oxides, respectively. Compared with monometallic NPs, multimetallic NPs have diverse shapes, sizes, high surface/volume ratios, physicochemical stability, activity, and a greater degree of selectivity that enhances barrier properties, quality, safety, stability, and mechanical and thermal strength of packaging materials (Duncan 2011). Cao et al. (2020) developed a Poly (butylene adipate-co-terephthalate) (PBAT) film by casting with bimetallic NPs of TiO_2/AgNPs to preserve cherry tomato up to 25 days of storage. TiO_2/AgNPs enhanced the physical properties of the PBTA matrix, such as mechanical, UV blocking, barrier properties, thermal stability, food preservation properties, and antibacterial activity against E. coli and S. aureus.

A novel composite formed by Ag^+/Zn^{2+} and permutite (zeolite) and Cinnamon essential oil (CEO), with a particle size of 10 nm, was dispersed in ethyl cellulose pads to be used in the preservation of Chinese bayberries by Niu et al. (2018). Ag^+/Zn^{2+}-permutite/CEO assured a controlled release effect and inhibited mold growth up to 5 days of storage. These authors performed the kinetics of CEO release from Ag^+/Zn^{2+}-permutite at different temperatures (4, 15, and 25 °C) and observed that the Ag^+/Zn^{2+}-permutite/CEO had a sustainable controlled-release effect. With Ag^+/Zn^{2+}-permutite the CEO released was low (1%) in comparison to that observed without the nanocomposite that was up to ten-fold higher at 25 °C.

A nanocomposite formed by cellulose nanocrystals (CNC) and oregano essential oil (OEO) incorporated into PLA films by casting, was developed to packed mixed vegetables. The authors observed that the nanocomposite improved material properties and antimicrobial activity against L. monocytogenes after 14 days of storage at 4 °C. The kinetics tests performed with PLA–CNC films showed a slow controlled release of the total phenolic compounds that increased continuously from initial time up to 16.6% at the 14th day of storage (Salmieri et al. 2014).

Electrospinning has been extensively used to produce nanofibers with high specific surface area and porosity to encapsulate active additives, including essential oils and polyphenols. Two examples of nanofibers used for strawberries preservation are shown in Table 15.5. The production of Pullulan-Carboxymethylcellulose sodium (Pul-CMC) nanofibers to encapsulate tea polyphenols at different concentration was described by Shao et al. (2018). Besides, electrospinning was also used to produce electrospun nanofibrous Poly(vinyl alcohol) (PVOH) film with a nanocomposite of cinnamon essential oil (CEO) and β-cyclodextrin (β-CD) (Wen et al. 2016). The authors observed that the mild electrospinning process maintains CEO (volatile antimicrobial agent) in the film, improving its antimicrobial activity against *S. aureus* and *E. coli* compared to the casting film.

Other examples of the use of essential oils and natural extracts in combination with different inorganic nanofillers (Kumar et al. 2018; Kowsalya et al. 2019; Kumar et al. 2019; Li et al. 2017b; Chi et al. 2019), and organic glycerosomes (Saravanakumar et al. 2020), and nanotubes (Shemesh et al. 2016) may also be found in Table 15.6.

Inorganic nanoparticles such as silver, zinc oxide, copper, titanium oxide, and selenium have a high antimicrobial efficacy at the lowest levels and may substitute common chemical antimicrobial materials to preserve the fruits and vegetables from spoilage. The combination of these inorganic NPs between them or with organic ones, including natural extracts, improves the active packaging materials and leads to a significant extension of fruits and vegetable shelf life. Active food packaging systems contribute to food quality and stability by emitting antimicrobial agents into food. However, some gaps need to be fulfilled, and therefore, future research should be dedicated to the interactions between antimicrobial-loaded nanocarriers and food components (Bahrami et al. 2020). Besides some studies that tested the controlled release by nanocarriers, scientific data concerning the migration of nanostructures onto food and their levels of toxicity to humans and the environment remain incomplete, demanding special attention from researchers and regulators (Azeredo et al. 2019).

5 Conclusions

When focusing on fruits and vegetables, the most important releasers active systems are those intended to release antimicrobial compounds. A great variety of EOs, that have strong broad-spectrum antibacterial properties, are used inside sachets or lately incorporated into biodegradable packaging as an eco-friend alternative.

Oregano and cinnamon are the most described in the literature for use with vegetables and fruits respectively, but some limitations, as its organoleptic properties, need to be overcome. Ethanol release systems were also investigated to use mainly with fresh-cut vegetables and fruits since ethanol has shown an effective reduction of ripening. Other volatile and non-volatile compounds have been also tested, mainly incorporated in sachets in the packaging and/or combined with EMAP but also encapsulated or as a coating. Each combination of active compound/format is specifically designed to preserve fresh foods (generally fruits) by inhiving the growth of target microorganisms.

The field of nanomaterials for fruits and vegetables packaging is the most innovative and offers many solutions to extend the shelf life of fruits and vegetables. AgNPs, ZnONPs in different polymeric matrices have been extensively studied, but more recent applications include the use of multimetallic NPs that have shown better performance than monometallic NPs. The combination of EOs and natural extracts with nanofillers improves the active material extending the shelf life of fruits and vegetables.

However, some aspects still require special attention such as the interactions among antimicrobial-loaded nanocarriers and food components, the migration of nanostructures into foodstuffs, and the levels of toxicity for humans and the environment.

The great variety of release systems found in the literature with proven efficacy provides many starting points to develop commercial systems, although there seem to be critical points in its implementation due to its limited presence in the market of fruits and vegetables.

References

Abugoukh AA, Elamin MA. Effect of polyethylene film lining and potassium permanganate on quality and shelf-life of Banana fruits. Gezira J Agric Sci. 2009;7(2):217–30.

Almenar E, Auras R, Rubino M, Harte B. A new technique to prevent the main post harvest diseases in berries during storage: inclusion complexes β-cyclodextrin-hexanal. Int J Food Microbiol. 2007;118(2):164–72.

Almenar E, Catala R, Hernandez-Muñoz P, Gavara R. Optimization of an active package for wild strawberries based on the release of 2-nonanone. LWT Food Sci Technol. 2009;42(2):587–93.

Al-Naamani L, Dutta J, Dobretsov S. Nanocomposite zinc oxide-chitosan coatings on polyethylene films for extending storage life of okra (Abelmoschus esculentus). Nanomaterials. 2018;8(7):479.

Aloui H, Khwaldia K, Sánchez-González L, Muneret L, Jeandel C, Hamdi M, Desobry S. Alginate coatings containing grapefruit essential oil or grapefruit seed extract for grapes preservation. Int J Food Sci Technol. 2014;49(4):952–9.

Awalgaonkar G, Beaudry R, Almenar E. Ethylene-removing packaging: basis for development and latest advances. Compr Rev Food Sci Food Saf. 2020;2020:1–28.

Azeredo HM, Otoni CG, Corrêa DS, Assis OB, de Moura MR, Mattoso LHC. Nanostructured antimicrobials in food packaging—recent advances. Biotechnol J. 2019;14(12):1900068.

Aziz M, Karboune S. Natural antimicrobial/antioxidant agents in meat and poultry products as well as fruits and vegetables: a review. Crit Rev Food Sci Nutr. 2018;58(3):486–511.

Bahrami A, Delshadi R, Assadpour E, Jafari SM, Williams L. Antimicrobial-loaded nanocarriers for food packaging applications. In: Advances in colloid and Interface science; 2020. p. 102140.

Bai J, Baldwin EA, Soliva Fortuny RC, Mattheis JP, Stanley R, Perera C, Brecht JK. Effect of pretreatment of intact 'Gala' apple with ethanol vapor, heat, or 1-methylcyclopropene on quality and shelf life of fresh-cut slices. J Am Soc Hortic Sci. 2004;129(4):583–93.

Bai J, Plotto A, Spotts R, Rattanapanone N. Ethanol vapor and saprophytic yeast treatments reduce decay and maintain quality of intact and fresh-cut sweet cherries. Postharvest Biol Technol. 2011;62:204–12.

Bajpai VK, Baek KH. Biological efficacy and application of essential oils in foods-a review. J Essent Oil Bearing Plants. 2016;19(1):1–19.

Beaudry RM. Effect of O2 and CO2 partial pressure on selected phenomena affecting fruit and vegetable quality. Postharvest Biol Technol. 1999;15:293–303.

Becaro AA, Puti FC, Panosso AR, Gern JC, Brandão HM, Correa DS, Ferreira MD. Postharvest quality of fresh-cut carrots packaged in plastic films containing silver nanoparticles. Food Bioprocess Technol. 2016;9(4):637–49.

Beltran CAU, Florez IF. Preparation of bio-packages from native starches and essential oils prolonging shelf life in strawberries. Revista Colombiana de Investigadores. 2018;5:71–86.

Bhople S, Gaikwad S, Deshmukh S, Bonde S, Gade A, Sen S, et al. Myxobacteria-mediated synthesis of silver nanoparticles and their impregnation in wrapping paper used for enhancing shelf life of apples. IET Nanobiotechnol. 2016;10(6):389–94.

Bodaghi H, Hagh ZG. Application of clay-TiO 2 nanocomposite packaging films on pears (Prunus communis L. cv. Williams) under cold storage. J Food Measur Charact. 2019;13(3):2377–88.

Bodaghi H, Mostofi Y, Oromiehie A, Zamani Z, Ghanbarzadeh B, Costa C, et al. Evaluation of the photocatalytic antimicrobial effects of a TiO2 nanocomposite food packaging film by in vitro and in vivo tests. LWT Food Sci Technol. 2013;50(2):702–6.

Bugatti V, Cefola M, Montemurro N, Palumbo M, Quintieri L, Pace B, Gorrasi G. Combined effect of active packaging of polyethylene filled with a nano-carrier of salicylate and modified atmosphere to improve the shelf life of fresh blueberries. Nanomaterials. 2020;10(12):2513.

Campos-Requena VH, Rivas BL, Pérez MA, Figueroa CR, Figueroa NE, Sanfuentes EA. Thermoplastic starch/clay nanocomposites loaded with essential oil constituents as packaging for strawberries– in vivo antimicrobial synergy over Botrytis cinerea. Postharvest Biol Technol. 2017;129:29–36.

Cao C, Wang Y, Zheng S, Zhang J, Li W, Li B, et al. Poly (butylene adipate-co-terephthalate)/titanium dioxide/silver composite biofilms for food packaging application. LWT Food Sci Technol. 2020;132:109874.

Chang Y, Choi I, Cho AR, Han J. Reduction of Dickeya chrysanthemi on fresh-cut iceberg lettuce using antimicrobial sachet containing microencapsulated oregano essential oil. LWT Food Sci Technol. 2017;82:361–8.

Chervin C, Westercamp P, Monteils G. Ethanol vapours limit Botrytis development over the postharvest life of table grapes. Postharvest Biol Technol. 2005;36:319–22.

Chi H, Song S, Luo M, Zhang C, Li W, Li L, Qin Y. Effect of PLA nanocomposite films containing bergamot essential oil, TiO2 nanoparticles, and Ag nanoparticles on shelf life of mangoes. Sci Hortic. 2019;249:192–8.

Dao T, Dantigny P. Control of food spoilage fungi by ethanol. Food Control. 2011;22:360–8.

Deng J, Chen QJ, Peng ZY, Wang JH, Li W, Ding QM, et al. Nano-silver-containing polyvinyl alcohol composite film for grape fresh-keeping. Mater Express. 2019;9(9):985–92.

Deng LZ, Mujumdar AS, Pan Z, Vidyarthi SK, Xu J, Zielinska M, Xiao HW. Emerging chemical and physical disinfection technologies of fruits and vegetables: a comprehensive review. Crit Rev Food Sci Nutr. 2020;60(15):2481–508.

Duncan TV. Applications of nanotechnology in food packaging and food safety: barrier materials, antimicrobials, and sensors. J Colloid Interface Sci. 2011;363:1–24.

Emamifar A, Mohammadizadeh M. Preparation and application of LDPE/ZnO nanocomposites for extending shelf life of fresh strawberries. Food Technol Biotechnol. 2015;53(4):488–95.

Enescu D, Cerqueira MA, Fucinos P, Pastrana LM. Recent advances and challenges on applications of nanotechnology in food packaging. A literature review. Food Chem Toxicol. 2019;134:110814.

Espitia PJP, Soares NDFF, Botti LCM, Melo NRD, Pereira OL, Silva WAD. Assessment of the efficiency of essential oils in the preservation of postharvest papaya in an antimicrobial packaging system. Brazilian J Food Technol. 2012;15(4):333–42.

FAOSTAT. Food and Agriculture Organization of the United Nations. FAOSTAT statistical database. Rome: FAO; 2021.

Food and Drug Administration (FDA). Code of Federal Regulations – Title 21: Foods and drugs; 2021. Volume 3, Part 182, Sec. 182.20.

Gaikwad KK, Singh S, Negi YS. Ethylene scavengers for active packaging of fresh food produce. Environ Chem Lett. 2020;18:269–84.

Gorrasi G, Bugatti V, Vertuccio L, Vittoria V, Pace B, Cefola M, Quintieri L, Bernardo P, Clarizia G. Active packaging for table grapes: evaluation of antimicrobial performances of packaging for shelf life of the grapes under thermal stress. Food Packag Shelf Life. 2020;25:100545.

Hadi A, Mahsa JO, Ayda S. A review on techniques utilized for design of controlled release food active packaging. Crit Rev Food Sci Nutr. 2020;

Han Y, Yu M, Wang L. Physical and antimicrobial properties of sodium alginate/carboxymethyl cellulose films incorporated with cinnamon essential oil. Food Packag Shelf Life. 2018;15: 35–42.

Hasan HK, Fugang X, Majher S, Jin TZ, Sousa AMM, Cheng-Kung L, Peggy MT, LinShu L. Antibacterial poly (lactic acid) (PLA) films grafted with electrospun PLA/allyl isothiocyanate fibers for food packaging. J Appl Polym Sci. 2015;

He Y, Li H, Fei X, Peng L. Carboxymethyl cellulose/cellulose nanocrystals immobilized silver nanoparticles as an effective coating to improve barrier and antibacterial properties of paper for food packaging applications. Carbohydr Polym. 2020;252:117156.

Hu Q, Fang Y, Yang Y, Ma N, Zhao L. Effect of nanocomposite-based packaging on postharvest quality of ethylene-treated kiwifruit (Actinidia deliciosa) during cold storage. Food Res Int. 2011;44(6):1589–96.

Hyun-Sun S, Jihyun B, Hoikyung K, Larry RB, Seung Yong C, Jee-Hoon R. Development of an antimicrobial sachet containing encapsulated allyl isothiocyanate to inactivate Escherichia coli O157:H7 on spinach leaves. Int J Food Microbiol. 2012;159(2):136–43.

Indumathi MP, Rajarajeswari GR. Mahua oil-based polyurethane/chitosan/nano ZnO composite films for biodegradable food packaging applications. Int J Biol Macromol. 2019;124:163–74.

Indumathi MP, Sarojini KS, Rajarajeswari GR. Antimicrobial and biodegradable chitosan/cellulose acetate phthalate/ZnO nano composite films with optimal oxygen permeability and hydrophobicity for extending the shelf life of black grape fruits. Int J Biol Macromol. 2019;132:1112–20.

Issa A, Ibrahim SA, Tahergorabi R. Impact of sweet potato starch-based nanocomposite films activated with thyme essential oil on the shelf-life of baby spinach leaves. Foods. 2017;6(6):43.

Jamróz E, Kopel P, Juszczak L, Kawecka A, Bytesnikova Z, Milosavljevic V, Makarewicz M. Development of furcellaran-gelatin films with Se-AgNPs as an active packaging system for extension of mini kiwi shelf life. Food Packag Shelf Life. 2019;21:100339.

Jaya S, Jitendra P, Marcus JG, Sudhagar M, Hitesh H. Antimicrobial and physicochemical characterization of biodegradable, nitric oxide-releasing nanocellulose–chitosan packaging membranes. J Agric Food Chem. 2016;64(25):5260–6.

Jin YZ, Lv DQ, Liu WW, Qi HY, Bai XH. Ethanol vapor treatment maintains postharvest storage quality and inhibits internal ethylene biosynthesis during storage of oriental sweet melons. Postharvest Biol Technol. 2013;86:372–80.

Ju J, Chen X, Xie Y, Yu H, Guo Y, Cheng Y, Qian H, Yao W. Application of essential oil as a sustained release preparation in food packaging. Trends Food Sci Technol. 2019;92:22–32.

Jung J, Zhao Y. Antimicrobial packaging for fresh and minimally processed fruits and vegetables. In: Antimicrobial food packaging. Academic; 2016. p. 243–56.

Kaewklin P, Siripatrawan U, Suwanagul A, Lee YS. Active packaging from chitosan-titanium dioxide nanocomposite film for prolonging storage life of tomato fruit. Int J Biol Macromol. 2018;112:523–9.

Kapetanakou AE, Skandamis PN. Applications of active packaging for increasing microbial stability in foods: natural volatile antimicrobial compounds. Curr Opin Food Sci. 2016;12:1–12.

Kim I, Viswanathan K, Kasi G, Thanakkasaranee S, Sadeghi K, Seo J. ZnO nanostructures in active antibacterial food packaging: preparation methods, antimicrobial mechanisms, safety issues, future prospects, and challenges. Food Rev Intl. 2020:1–29.

Kowsalya E, MosaChristas K, Balashanmugam P, Rani JC. Biocompatible silver nanoparticles/poly (vinyl alcohol) electrospun nanofibers for potential antimicrobial food packaging applications. Food Packag Shelf Life. 2019;21:100379.

Kumar S, Shukla A, Baul PP, Mitra A, Halder D. Biodegradable hybrid nanocomposites of chitosan/gelatin and silver nanoparticles for active food packaging applications. Food Packag Shelf Life. 2018;16:178–84.

Kumar S, Boro JC, Ray D, Mukherjee A, Dutta J. Bionanocomposite films of agar incorporated with ZnO nanoparticles as an active packaging material for shelf life extension of green grape. Heliyon. 2019;5(6):e01867.

Kuorwel KK, Cran MJ, Sonneveld K, Miltz J. Antimicrobial activity of biodegradable polysaccharide and protein-based films containing active agents. J Food Sci. 2011;

Kwon SJ, Chang Y, Han J. Oregano essential oil-based natural antimicrobial packaging film to inactivate Salmonella enterica and yeasts/molds in the atmosphere surrounding cherry tomatoes. Food Microbiol. 2017;65:114–21.

Li X, Li W, Jiang Y, Ding Y, Yun J, Tang Y, Zhang P. Effect of nano-ZnO-coated active packaging on quality of fresh-cut 'Fuji' apple. Int J Food Sci Technol. 2011;46(9):1947–55.

Li W, Li L, Cao Y, Lan T, Chen H, Qin Y. Effects of PLA film incorporated with ZnO nanoparticle on the quality attributes of fresh-cut apple. Nanomaterials. 2017a;7(8):207.

Li D, Li L, Luo Z, Lu H, Yue Y. Effect of nano-ZnO-packaging on chilling tolerance and pectin metabolism of peaches during cold storage. Sci Hortic. 2017b;225:128–33.

Li D, Ye Q, Jiang L, Luo Z. Effects of nano-TiO2-LDPE packaging on postharvest quality and antioxidant capacity of strawberry (Fragaria ananassa Duch.) stored at refrigeration temperature. J Sci Food Agric. 2017c;97(4):1116–23.

Lichter A, Zutkhy Y, Sonego L, Dvir O, Kaplunov T, Sarig P, Ben-Arie R. Ethanol controls postharvest decay of table grapes. Postharvest Biol Technol. 2002;24:301–8.

Lichter A, Zutahy Y, Kaplunov T, Aharoni N, Lurie S. The effect of ethanol dip and modified atmosphere on prevention of Botrytis rot of table grapes. HortTechnology. 2005;15(2):284–91.

Linchun M, Guoze W, Fei Q. Application of 1-methylcyclopropene prior to cutting reduces wound responses and maintains quality in cut kiwifruit. J Food Eng. 2007;78(1):361–5.

Luo Z, Wang Y, Wang H, Feng S. Impact of nano-CaCO 3-LDPE packaging on quality of fresh-cut sugarcane. J Sci Food Agric. 2014;94(15):3273–80.

Luo Z, Wang Y, Jiang L, Xu X. Effect of nano-CaCO3-LDPE packaging on quality and browning of fresh-cut yam. LWT Food Sci Technol. 2015;60(2):1155–61.

Ma J, Gu J, Wang X, Peng H, Wang Q, Zhang R, Bao, J. Effects of nano-zerovalent iron on antibiotic resistance genes during the anaerobic digestion of cattle manure. Biores Technol. 2019;289:121688.

Ma L, Zhang M, Bhandari B, Gao Z. Recent developments in novel shelf life extension technologies of fresh-cut fruits and vegetables. Trends Food Sci Technol. 2017;64:23–38.

Marques CS, Grillo RP, Bravim DG, Pereira PV, Villanova JCO, Pinheiro PF, Carneiro JCS, Bernardes PC. Preservation of ready-to-eat salad: a study with combination of sanitizers, ultrasound, and essential oil-containing β-cyclodextrin inclusion complex. LWT Food Sci Technol. 2019;115:108433.

Martínez-Romero D, Bailén G, Serrano M, Guillén F, Valverde JM, Zapata P, Castillo S, Valero D. Tools to maintain postharvest fruit and vegetable quality through the inhibition of ethylene action: a review. Crit Rev Food Sci Nutr. 2007;47:543–60.

Mastromatteo M, Mastromatteo M, Conte A, Alessandro Del Nobile M. Advances in controlled release devices for food packaging applications. Trends Food Sci Technol. 2010;21(12):591–8.

Mehmet SA, Cengiz C. The applications of 'active packaging and chlorine dioxide' for extended shelf life of fresh strawberries; 2010.

Mehyar GF, Han JH. Chapter 14 Active packaging for fresh-cut fruits and vegetables. In: Modified atmosphere packaging for fresh-cut fruits and vegetables. Wiley-Blackwell; 2011. p. 267.

Merve D, Mehmet SA, Nükhet NDZ, Riza T, Mehmet BB, Cengiz C. Potential of antimicrobial active packaging 'containing natamycin, nisin, pomegranate and grape seed extract in chitosan coating' to extend shelf life of fresh strawberry. Food Bioprod Process. 2016;98:354–63.

Mu H, Gao H, Chen H, Fang X, Han Q. A novel controlled release ethanol emitter: preparation and effect on some postharvest quality parameters of Chinese bayberry during storage. J Sci Food Agric. 2017;97:4929–36.

Muriel-Galet V, Cerisuelo JP, López-Carballo G, Lara M, Gavara R, Hernández-Muñoz P. Development of antimicrobial films for microbiological control of packaged salad. Int J Food Microbiol. 2012;157(2):195–201.

Muriel-Galet V, Cerisuelo JP, López-Carballo G, Aucejo S, Gavara R, Hernández-Muñoz P. Evaluation of EVOH-coated PP films with oregano essential oil and citral to improve the shelf-life of packaged salad. Food Control. 2013;30(1):137–43.

Niu B, Yan Z, Shao P, Kang J, Chen H. Encapsulation of cinnamon essential oil for active food packaging film with synergistic antimicrobial activity. Nanomaterials. 2018;8(8):598.

Ozdemir M, Floros JD. Active food packaging technologies. Crit Rev Food Sci Nutr. 2004;44(3):185–93.

Patrignani F, Siroli L, Serrazanetti DI, Gardini F, Lanciotti R. Innovative strategies based on the use of essential oils and their components to improve safety, shelf-life and quality of minimally processed fruits and vegetables. Trends Food Sci Technol. 2015;46(2):311–9.

Pereira de Abreu DA, Cruz JM, Paseiro Losada P. Active and intelligent packaging for the food industry. Food Rev Intl. 2012;28:146–87.

Piercey MJ, Mazzanti G, Budge SM, Delaquis PJ, Paulson AT, Truelstrup HL. Antimicrobial activity of cyclodextrin entrapped allyl isothiocyanate in a model system and packaged fresh-cut onions. Food Microbiol. 2012;30(1):213–8.

Plotto A, Bai J, Narciso JA, Brecht JK, Baldwin EA. Ethanol vapor prior to processing extends fresh-cut mango storage by decreasing spoilage, but does not always delay ripening. Postharvest Biol Technol. 2006;39:134–45.

Ponni P, Subramanian KS, Janavi GJ, Subramanian J. Synthesis of nano-film from nanofibrillated cellulose of banana pseudostem (Musa spp.) to extend the shelf life of tomato. Bioresources. 2020;15(2):2882–905.

Popa I, Hanson EJ, Todd ECD, Schilder AC, Ryser ET. Efficacy of chlorine dioxide gas sachets for enhancing the microbiological quality and safety of blueberries. J Food Prot. 2007;70:2084–8.

Ritenour MA, Mangrich ME, Beaulieu JC, Rab A, Saltveit ME. Ethanol effects on the ripening of climacteric fruit. Postharvest Biol Technol. 1997;12:35–42.

Salmieri S, Islam F, Khan RA, Hossain FM, Ibrahim HM, Miao C, Hamad WY, Lacroix M. Antimicrobial nanocomposite films made of poly (lactic acid)–cellulose nanocrystals (PLA–CNC) in food applications—part B: effect of oregano essential oil release on the inactivation of Listeria monocytogenes in mixed vegetables. Cellulose. 2014;21(6):4271–85.

Saravanakumar K, Hu X, Chelliah R, Oh DH, Kathiresan K, Wang MH. Biogenic silver nanoparticles-polyvinylpyrrolidone based glycerosomes coating to expand the shelf life of fresh-cut bell pepper (Capsicum annuum L. var. grossum (L.) Sendt). Postharvest Biol Technol. 2020;160:111039.

Serrano M, Martinez-Romero D, Castillo S, Guillén F, Valero D. The use of natural antifungal compounds improves the beneficial effect of MAP in sweet cherry storage. Innovative Food Sci Emerg Technol. 2005;6(1):115–23.

Shao P, Niu B, Chen H, Sun P. Fabrication and characterization of tea polyphenols loaded pullulan-CMC electrospun nanofiber for fruit preservation. Int J Biol Macromol. 2018;107:1908–14.

Shemesh R, Krepker M, Nitzan N, Vaxman A, Segal E. Active packaging containing encapsulated carvacrol for control of postharvest decay. Postharvest Biol Technol. 2016;118:175–82.

Shinde R, Rodov V, Krishnakumar S, Subramanian J. Active and intelligent packaging for reducing postharvest losses of fruits and vegetables. Postharvest Biol Nanotechnol. 2018:171–89.

Sportelli MC, Izzi M, Volpe A, Lacivita V, Clemente M, Di Franco C, et al. A new nanocomposite based on LASiS-generated CuNPs as a preservation system for fruit salads. Food Packag Shelf Life. 2019;22:100422.

Sung SY, Sin LT, Tee TT, Bee ST, Rahmat AR, Rahman WAWA, Vikhraman M. Antimicrobial agents for food packaging applications. Trends Food Sci Technol. 2013;33:110–23.

Suzuki Y, Uji T, Terai H. Inhibition of senescence in broccoli florets with ethanol vapor from alcohol powder. Postharvest Biol Technol. 2004;31:177–82.

Synowiec A, Gniewosz M, Kraśniewska K, Leon PJ, Bączek K, Węglarz Z. Antimicrobial and antioxidant properties of pullulan film containing sweet basil extract and an evaluation of coating effectiveness in the prolongation of the shelf life of apples stored in refrigeration conditions. Innovative Food Sci Emerg Technol. 2014;23:171–81.

Takala PN, Vu KD, Salmieri S, Khan RA, Lacroix M. Antibacterial effect of biodegradable active packaging on the growth of Escherichia coli, Salmonella typhimurium and Listeria monocytogenes in fresh broccoli stored at 4 C. LWT Food Sci Technol. 2013;53(2):499–506.

Ueda Y, Bai JH. Effect of short term exposure of elevated CO2 on flesh firmness and ester production of strawberry. J Jpn Soc Hortic Sci. 1993;62(2):457–64.

Utto W, Noomhorm A, Bronlund JE. Modeling of ethanol vapor controlled release active packaging for fresh peeled shallots. Ital J Food Sci. 2018;Supl. (SLIM 2017):157–61.

Valero D, Valverde JM, Martínez-Romero D, Guillén F, Castillo S, Serrano M. The combination of modified atmosphere packaging with eugenol or thymol to maintain quality, safety and functional properties of table grapes. Postharvest Biol Technol. 2006;41(3):317–27.

Valverde JM, Guillén F, Salvador CD, Serrano M, Valero D. Improvement of table grapes quality and safety by the combination of modified atmosphere packaging (MAP) and eugenol, menthol, or thymol. J Agric Food Chem. 2005;53(19):7458–64.

Wang L, Shao S, Madebo MP, Hou Y, Zheng Y, Jin P. Effect of nano-SiO2 packing on postharvest quality and antioxidant capacity of loquat fruit under ambient temperature storage. Food Chem. 2020;315:126295.

Weerawate U, Mawson AJ, Bronlund JE. Hexanal reduces infection of tomatoes by Botrytis cinerea whilst maintaining quality. Postharvest Biol Technol. 2008;47(3):434–7.

Wei H, Seidi F, Zhang T, Jin Y, Xiao H. Ethylene scavengers for the preservation of fruits and vegetables: a review. Food Chem. 2021;337(127750):1–12.

Wen P, Zhu DH, Wu H, Zong MH, Jing YR, Han SY. Encapsulation of cinnamon essential oil in electrospun nanofibrous film for active food packaging. Food Control. 2016;59:366–76.

Wieczyńska J, Cavoski I. Antimicrobial, antioxidant and sensory features of eugenol, carvacrol and trans-anethole in active packaging for organic ready-to-eat iceberg lettuce. Food Chem. 2018;259:251–60.

Zanetti M, Carniel TK, Dalcanton F, dos Anjos RS, Riella HG, de Araujo PH, et al. Use of encapsulated natural compounds as antimicrobial additives in food packaging: a brief review. Trends Food Sci Technol. 2018;81:51–60.

Zhang C, Li W, Zhu B, Chen H, Chi H, Li L, Qin Y, Xue J. The quality evaluation of postharvest strawberries stored in nano-Ag packages at refrigeration temperature. Polymers. 2018;10(8):894.

Zhang W, Shu C, Chen Q, Cao J, Jiang W. The multi-layer film system improved the release and retention properties of cinnamon essential oil and its application as coating in inhibition to penicillium expansion of apple fruit. Food Chem. 2019;299:125109.

Zhijun H, Chunxia T, Zhibin H, Jiang L, Yonghao N. 1-Methylcyclopropene (MCP)-containing cellulose paper packaging for fresh fruit and vegetable preservation: a review. Bioresources. 2017;12(1)

Chapter 16
Application of Releasing Active Packaging in Oils and Fats

M. Fraga-Corral, C. Lourenço-Lopes, C. Jimenez-Lopez, J. Echave, F. Chamorro, M. A. Prieto, and J. Simal-Gandara

Abstract The oil industry is continuously evolving by the application of innovative tools and ingredients towards more effective, safe, natural and eco-friendly solutions to satisfy the claims of customers. Currently, packaging involves the use of a great diversity of materials ranging from different types of plastics to very diverse kinds of biodegradable fibers. The current trend is the incorporation of molecules with recognized bioactivities into these materials for their controlled release. The development of active packaging is aimed to prevent oils and fats oxidation/spoilage. Compounds with bioactivities can be freely added into oil/fats, or they can be applied as encapsulated molecules. In fact, this last option has proven to be very promising since the encapsulation has proved to enhance the biological qualities of the compounds. The most relevant bioactivities that the encapsulated molecules can provide to the product are antioxidant or antimicrobial capacities which eventually extend the shelf-life of the product by preventing food spoilage. Therefore, the incorporation of natural biomolecules to oily/fatty products through biodegradable active packaging may represent multiple benefits for human health and reduce the use of petroleum-derivate plastics. In the present work, current issues and solutions to oils and fats preservation through the use of active packaging systems will be reviewed.

M. Fraga-Corral (✉) · C. Jimenez-Lopez
Nutrition and Bromatology Group, Analytical and Food Chemistry Department. Faculty of Food Science and Technology, University of Vigo, Ourense, Spain

Centro de Investigação de Montanha (CIMO), Instituto Politécnico de Bragança, Bragança, Portugal
e-mail: mfraga@uvigo.es

C. Lourenço-Lopes · J. Echave · F. Chamorro · M. A. Prieto · J. Simal-Gandara
Nutrition and Bromatology Group, Analytical and Food Chemistry Department. Faculty of Food Science and Technology, University of Vigo, Ourense, Spain
e-mail: jsimal@uvigo.es

S. M. Jafari, A. S. Silva (eds.), *Releasing Systems in Active Food Packaging*, Food Bioactive Ingredients, https://doi.org/10.1007/978-3-030-90299-5_16

Keywords Active food packaging · Encapsulation · Oil · Fat · Preservation

Abbreviations

Film Materials

BOPP	Biaxially Oriented Polypropylene
CPP	Cast Polypropylene,
EVA	Ethylene Vinyl Acetate
EVOH	Ethylene Vinyl Alcohol Copolymer
EVOO	Extra Virgin Olive Oil
HDPE	High Density Polyethylene
LDPE	Low Density Polyethylene
PET	Polyethylene Terephthalate
PLA	Polylactic Acid
PP	Polypropylene
PVC	Polyvinyl Chloride

Compounds

BHT	Butylated Hydroxytoluene
PUFAs	Polyunsaturated Fatty Acids
TBHQ	Tert-butylhydroquinone
BHA	Butylated Hydroxyanisole
PG	Propyl Gallate
DHA	Docosahexaenoic Acid
EPA	Eicosapentaenoic
4-HNE	4-hydroxynonenal
MDA	Malondialdehyde

Others

PV	Peroxide value
TBAR	2-thiobarbituric acid reductive value assay
ppm	Parts Per Million
CFU	Colony Forming Units
MAP	Modified Atmosphere Packaging
ROS	Reactive Oxygen Species
UV	Ultraviolet Light
CIE	Commission Internationale de l'Eclairage

1 Introduction

Edible oils obtained from olives, sunflowers, soybeans or rapeseed for example, are used worldwide for cooking, preserving or marinating foods. They are considered a healthy and nutritious product that can provide flavor, aroma and even color when used for cooking, but their utilization has been also extended to the preservation of foods (Choe and Min 2006).

Packaging and storing are highly important processes in the preservation of edible oils. Indeed, lipid oxidation, which is the main cause of fatty products spoilage, is the major issue for preserving their nutritional value. A higher fat content has been correlated with higher oxidation rates in foods. As the unsaturation grade of the fatty acids (abundance of double bonds between linear carbons) increases, the fatty acids will be more unstable and prone to oxidation. Hence, polyunsaturated fats (i.e. fish oils) are more likely to present earlier oxidative damage than monounsaturated fats (vegetal oils) or other animal fats (Tena et al. 2019). In the case of vegetable oils obtained from seeds, the three main unsaturated fatty acids, ordered by decreasing unsaturation grade, are linolenic acid (C18:3), linoleic acid (C18:2) and oleic acid (C18:1). This means that oils with higher oleic acid content (i.e. olive, sunflower oil) will generally have a greater shelf-life than those with more abundant content of polyunsaturated fats (Kráčmar et al. 2019).

However, not just the nature of the fatty acids affects lipid oxidation, it is strongly influenced and induced by several environmental factors such as ultraviolet (UV) light, heat, enzymatic activity, metal ions, or microorganisms (Fig. 16.1). Storage is a key factor in the development of lipid oxidation reactions. The main parameters that need to be controlled during storage are temperature, humidity and light exposure since they notably influence the oxidation rate of the stored oil or fat (Tena

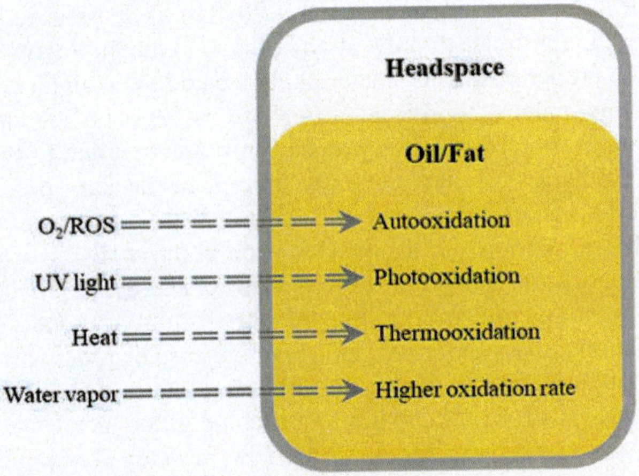

Fig. 16.1 Main factors responsible for triggering the lipid oxidation of fatty products when stored for long periods of time

et al. 2019). Depending on the main inducer that acts to trigger the lipid oxidation, it can be mainly categorized as autooxidation, photooxidation or thermooxidation, which may also take place simultaneously (Fereidoon and Ying 2010).

Autooxidation is the most common process among all and is defined as the spontaneous reaction of lipids with atmospheric oxygen through a chain reaction. Lipid oxidation is frequently induced by free radicals and/or other reactive species such as reactive oxygen species (ROS) acting as intermediate molecules in the process (Romani et al. 2020). Water vapor, for instance, may alter the oxidation rate of fats with the addition of higher levels of free hydroxyl (OH^-) radicals, which leads to a faster chain reaction of the oxidation process (Lloret et al. 2016). Either way, the presence of any levels of O_2 can trigger the oxidation process that essentially consists of the initial peroxidation of long chain fatty acids (Koontz 2016). The resulting oxidized lipids will cause further chain reactions throughout the lipid matrix that generates a great variety of secondary oxidation products, including alcohols, hydrocarbons, volatiles, epoxy compounds, among others (Frankel 2012). Some of these have a strong rancid odor with very low thresholds, but they also reduce the nutritional value and quality properties of the product (Ghorbani Gorji et al. 2019). Additionally, some of these sub-products (i.e. unsaturated aldehydes, ketones) have been correlated to the development of several ailments and diseases (Fereidoon and Ying 2010).

The selection of the adequate packaging material has been revealed as a key tool. The characteristic of the material and the packaging design, have been proven to be a useful strategy to prevent or delay oil spoilage to a greater extent (Rababah et al. 2011). Given the need to avoid oxidation and contamination of the products, packaging must act as a barrier against external hazards, both physical and chemical. When selecting an appropriate material for food packaging, some key factors must be taken into account such as: O_2 permeability, water-vapor permeability, durability, thermostability and transparency (Robertson 2013). However, storage conditions are another key point for preventing lipid oxidation. Optimal conditions include the maintenance of low storing temperatures (15–20 °C), while limiting light exposure and careful handling in order to prevent any breach in the packaging (Sionek et al. 2013). Thus, the proper packaging material choice and the maintenance of optimal storing conditions make a significant difference in the shelf-life of the product. However, the oxidation status of fatty acids can be broken due to the presence of metal ions or oxidative enzymes naturally present in the fatty products (Trapani et al. 2017). In this scenario the use of antioxidants that avoid the premature initialization of oxidative reaction permits the extension of the shelf-life of fatty foods by preserving its quality and safety (Lozano-Sánchez et al. 2013).

Traditional solutions to avoid lipid degradation in fatty products and oils have been focused on effective barrier packaging. In the case of food with high fat content should be packaged with poor porous materials (Haas et al. 2019). In fact, the most utilized materials for packaging oil/fat include glass, metals and thermoplastics. Currently, other biodegradable films are being developed to replace single-use plastics and improve packaging quality. Independently of the final selected material, most of the times, the process is carried out under a modified atmosphere packaging

(MAP). It has been demonstrated to minimize microbial contamination and limit oxygen concentration in the container headspace to the maximum extent possible. Thus, the application of MAP prevents lipid oxidation induced by oxygen presence. Nitrogen is usually employed as the filler gas and preservation may be furthered by employing hermetic seals (Lloret et al. 2016; Smeriglio et al. 2019). Using MAP conditions allows for minimum-processed foodstuffs to last for extended periods of time (50–100% more) until its seal is broken (Karaman et al. 2015).

Other strategies used to reduce lipid oxidation of fatty products and extend their shelf-life have consisted of the addition of free molecules with recognized bioactivities into the matrix, mainly antioxidants and antimicrobials. The direct addition of antioxidant compounds into the food product has been demonstrated to delay both the lipid oxidation and, by extension, the microbial growth. Among the synthetic options BHT or TBHQ has usually been selected to prevent lipid oxidation. However, many works have pointed to the efficiency of incorporating natural extracts into the oil or the fatty foods. Among the variability of molecules employed for fortifying fatty products pigments, phenolic compounds, lipids or amino acids are most widely utilized for their preservation (Xu et al. 2020; Stoll et al. 2017; Talón et al. 2019; López-de-Dicastillo et al. 2012; Tammineni et al. 2012; Mohammad Zadeh et al. 2019; Carpiné et al. 2015; Fayaz Dastgerdi et al. 2016; Moraes et al. 2007; São José et al. 2019; Ribeiro-Santos et al. 2017; Graciano-Verdugo et al. 2010; Wang et al. 2019; Manzanarez-López et al. 2011; Otero-Pazos et al. 2016; Torrieri et al. 2011; Mirzaei-Mohkam et al. 2019; Kchaou et al. 2020; Jiang et al. 2019; Chavoshizadeh et al. 2020; Colín-Chávez et al. 2013, 2014a, b; Samsudin et al. 2014; Stoll et al. 2019; Gaikwad et al. 2017). Nevertheless, this approach may alter the original organoleptic properties of the final product which ultimately reduce consumers' acceptance. To prevent these organoleptic modifications, molecules can be encapsulated to mask their flavor, odor and/or color. Encapsulation allows the protection of biomolecules from degradation factors extending their capacity to act and it also allows the organoleptic characteristics of the food matrix to go unaltered. Nowadays, one of the most innovative strategies in the food industry aimed to prolong oily and fatty foods is the incorporation of natural molecules, free or encapsulated, within different types of films to create active packaging. This system may eliminate undesirable by-products that can shorten the commercial life of food or it can slowly release biomolecules that can exert their effect, directly by exerting their action into the matrix or indirectly by providing to the package physical features that reduce the lipid oxidation and microbial spoilage. When biofilms are used for creating active packaging, they may provide a real alternative to the utilization of single-use plastics. Therefore, active packaging based on the use of biodegradable materials can reduce the amount of worldwide food waste, single-use plastics, and reduce the risk of food-borne diseases through a better preservation of the final product.

Hence, in this chapter, different types of active packaging with diverse bioactivities will be reviewed and assessed in terms of safety, nutritional value and quality of edible oils and fatty products.

2 Innovative Materials for Packaging for Oils and Fats

Over the past few decades, packaging has been revealed as a key elementfor extending the storage of oils/fats (Koontz 2016; Pristouri et al. 2010). The importance of the characteristics of the different kinds of package, the parameters under which packaging is developed and the storage conditions of the product have been disclosed. Colored packages and diverse materials have been shown to be involved in the shelf-life extension of lipophilic molecules (Koontz 2016; Rizzo et al. 2014; Sanmartin et al. 2018). A few environmental factors, atmospheric O_2, light irradiation, water vapor and high temperatures have been demonstrated to accelerate the rancidity process and reduce the quality and safety of fatty foods (Pristouri et al. 2010; Sanmartin et al. 2018; Iqdiam et al. 2020).

Traditionally, the most utilized materials for packaging oil/fat include glass, metals and thermoplastics. Silica-based glass presents some specific advantages, such as high durability, recyclability and an absolute impermeability to both gas and moisture if the seal is adequate. Glass also accounts for being chemically inert, which is also desirable for further preservation (Frankel 2012). In some countries, like Italy or Spain, glass is the most commonly used material for olive oil packaging (Piscopo and Poiana 2012). Glass bottles may present different shapes and sizes, but the defining property is its transparency grade since UV light is one of the main inducers of lipid oxidation, thus darkened and opaque glass bottles provide a better oxidation protection (Frankel 2012). It has been observed that under equal storage conditions, oils contained in glass showed less degradation in comparison with steel or thermoplastics. Even though glass is considered the best material regarding oil packaging, its fragility and higher cost pose as inconvenient for its use that has led many oil production companies to switch to thermoplastics (Pristouri et al. 2010; Kouveli et al. 2017; Tsimis and Karakasides 2002).

Although metals are impermeable to light and oxygen when paired with a proper seal; metal containers are not widely used nowadays. As mentioned, ionized metals present in the oil and containers can induce lipid oxidation. This reaction may take place through surface contact between the oil and the container walls, yet it may be prevented by coating the inner walls with an inert enamel (Deshwal and Panjagari 2020). It has been observed that oil storage in tin-plate containers show a higher oxidation degree in comparison with glass or thermoplastics, which explains the lower trend of use (Rababah et al. 2011). Nevertheless, bulk storage (10–20 tons) in the oil industry is carried out in metal containers prior to retail packaging. Stainless steel and tinplate are the most commonly utilized metals, while aluminum is not, due to its relatively high cost and ease of ionization (Kontominas 2017).

Among thermoplastics, polyethylene terephthalate (PET) is the most popular material due to its many advantages including clarity, chemical inertness, low oxygen permeability, and excellent mechanical properties. Besides PET, polyethylene (PE), the most prominent material, is mainly utilized in the form of low-density polyethylene (LDPE) or high-density polyethylene (HDPE), which differ in its oxygen permeability (Tena et al. 2019). Other widely used thermoplastic would be polypropylene (PP), ethylene vinyl acetate (EVA) or ethylene vinyl alcohol (EVOH) (Pristouri et al. 2010). All of these thermoplastics are the most common and extensively used materials for almost every type of food packaging worldwide, mainly due to their affordability, durability, barrier properties, and light weight (Robertson 2013). However, they exhibit some disadvantages. As in the same case as glass, transparent plastic bottles offer less protection against photooxidation. Also, plastics show high, but not complete impermeability to gases, like glass and metals and also allow the passing of small molecules that can alter its flavor, degradation status or acidity (Hu et al. 2020). In order to improve their packaging properties, combinations of these materials as multilayered packages, like bag-in-box or Tetra Brik, have also been studied for oil packaging. These films showed high protection to light and oxygen paired with chemical stability and mechanical properties. Even though they are not the most trendy choice regarding oil, they have demonstrated to prevent its degradation for long storage periods ($\approx$ 12 months) (Kontominas 2017; Mailer et al. 2012). Similarly, the most popular packaging materials for packaging butter, are plastics, (usually PET) or tin-plate, while in the industry, it is bulk stored in plastic coated cardboard (Karaman et al. 2015).

The current materials utilized for the development of alternative packages for oils and fatty food products can be mainly divided into non-biodegradable, petroleum-derived, and biodegradable ones (Table 16.1).

(a) Non bio-degradable materials: biaxially orientated polypropylene/vacuum metalized polyester/cast polypropylene (BOPP/VMPET/CPP), LDPE, HDPE, EVA or EVOH (López-de-Dicastillo et al. 2012; Fayaz Dastgerdi et al. 2016; Graciano-Verdugo et al. 2010; Torrieri et al. 2011; Colín-Chávez et al. 2013, 2014a, b; Pereira de Abreu et al. 2010; Liu et al. 2018a; Zhu et al. 2013; Otero-Pazos et al. 2018).

(b) Biodegradable ingredients:

- Polysaccharides: extracted from different natural sources, easily handled, and economic.

  (i) Polysaccharides from plants: starch, mostly extracted from cassava or corn, pectin or cellulose have been widely selected for extending the shelf-life of fatty foods, such as EVOO, sunflower oil, coconut oil or butter (Stoll et al. 2017; Talón et al. 2019; Moraes et al. 2007; São José et al. 2019; Mirzaei-Mohkam et al. 2019; Malherbi et al. 2019; Perazzo et al. 2014; Gautam and Mishra 2017).

  (ii) Polysaccharides from animals: chitosan and its derivatives, used in combination with other molecules such as gelatin, have been assayed for the conservation of corn oil, tuna or butter (Xu et al. 2020; Otero-Pazos et al. 2016; Jiang et al. 2019; Apjok et al. 2019).

  (iii) Polysaccharides from algae: alginate- or carrageenan-based films have been described to store coconut and sesame oils (Gautam and Mishra 2017; Ganesan et al. 2019)

Table 16.1 Active packaging approaches for oils and fatty food products

Packaging material	Active ingredient (source)	Incorporation	Properties	Oil/Fat	References
Casein and sodium alginate–pectin	Cooper and/or BHT	Freely incorporated into the film	Antioxidant, mechanical, barrier, and antimicrobial	Coconut oil	Gautam and Mishra (2017)
Chitosan/gelatin	ε-polylysine and astaxanthin (*Litopenaeus vannamei* by-products)	Freely incorporated into the film	Antioxidant and antimicrobial	Corn oil	Xu et al. (2020)
LDPE	α-Tocopherol	Freely incorporated into the film	Antioxidant (up to 16 weeks, 30 °C)	Corn oil	Graciano-Verdugo et al. (2010)
Cassava starch	Anthocyanins (wine grape pomace)	Microcapsules included into the film	Antioxidant	EVOO	Stoll et al. (2017)
Gelatin and corn starch	Guabiroba pulp (*Campomanesia xanthocarpa*)	Freely incorporated into the sachet	Antioxidant	EVOO	Graciano-Verdugo et al. (2010)
Fish gelatin	Glucose+lysine by Maillard reaction	Freely incorporated into the pouches	Antioxidant, UV and oxygen barrier	Flaxseed oil	Kchaou et al. (2020)
HDPE, LDPE or EVA	Sesamol	Freely incorporated into the film	Antioxidant	Linoleic acid and oat cereal	Zhu et al. (2013)
Soy protein	Brazilian pine(*Araucaria angustifolia* (Bertol.) Kuntze)	Freely incorporated to the film	Antioxidant (oxidative stability)	Linseed oil	de Souza et al. (2020)
Soy protein	Virgin coconut oil and soy lecithin	Freely incorporated into the film	Slow peroxides increment up to 28 days of storage	Olive oil	Carpiné et al. (2015)
Bovine gelatin	Anthocyanins (cranberry)	Nanocomplexes (chitosan-based) included into the film	Antioxidant and light barrier	Olive oil (applicable to fatty food)	Wang et al. (2019)
Durian rind cellulose reinforced PLA	α-Tocopherol or BHT	Freely incorporated into the film	Antioxidant (low PV)	Palm oil/Food simulant	Penjumras et al. (2018)
Semi- & refined carrageenan	Konjac glucomannan	Freely incorporated into the film	Antioxidant (low oxidation, PV and iodine value)	Sesame oil	Ganesan et al. (2019)

Wheat gluten	Chlorophyll	Freely incorporated for coating the film	Antioxidant (change of color with oxidation)	Sesame oil	Chavoshizadeh et al. (2020)
PLA	TBHQ	Freely incorporated into the film	Antioxidant	Soybean oil	Almasi et al. (2014)
LDPE	Pyrogallol	Freely incorporated for coating the film	Antioxidant (O_2 scavenging)	Soybean oil	Gaikwad et al. (2017)
LDPE and LDPE+HDPE	Astaxanthin	Freely incorporated into the film	Antioxidant potential	Soybean oil	Colín-Chávez et al. (2014a)
HDPE+TiO_2	Carotenoids (marigold flower, *Tagetes erecta*)	Freely incorporated into the film	Antioxidant and opacity	Soybean oil	Colín-Chávez et al. (2014b)
Bilayer PLA	Astaxanthin (Marigold flower extract)	Freely incorporated into the film	Antioxidant (10 days oxidation retard)	Soybean oil	Samsudin et al. (2014)
PLA	α-Tocopherol	Freely incorporated into the film	Antioxidant (up to 30 °C)	Soybean oil	Manzanarez-López et al. (2011)
HDPE	BHT and peppermint essential oil	Freely incorporated into the film	Antioxidant	Soybean oil	Fayaz Dastgerdi et al. (2016)
Chitosan	D-α-tocopheryl PEGS and/ or SiO2 nanoparticles	Freely incorporated into the film	Antioxidant and antimicrobial	Soybean oil	Bi et al. (2020)
Chitosan	Polyphenols from *Garcinia mangostana* L. rind extract	Freely incorporated into the film	Antioxidant and antimicrobial	Soybean oil	Zhang et al. (2020)
TPS-LLDPE	Polyphenols from green tea extract powder	Freely incorporated into microstructures of the film	Antioxidant (reduced lipid oxidation)	Soybean oil	Panrong et al. (2019)
Chitosan	*Urtica dioica* L.	Freely or nanoliposomal incorporated into the film	Antioxidant (60 days oxidation induction delay)	Soybean oil	Almasi et al. (2016)

(continued)

Table 16.1 (continued)

Packaging material	Active ingredient (source)	Incorporation	Properties	Oil/Fat	References
Methylcellulose	Cinnamon, clove, ginger, green tea and thyme extracts	Freely incorporated into the film	Antioxidant (radical scavenging activity and inhibition of lipid oxidation)	Soybean oil	Phooprritham et al. (2012)
Soybean protein enzymatically modified	Alkali lignin and lignosulfonate	Freely incorporated into the film	Antioxidant, prevent migration of chemicals, UV-blocker	Soybean oil and fish fatty acid ethyl ester	Mohammad Zadeh et al. (2019)
PLA	Carotenoids: β-carotene, lycopene, and bixin (carrots and tomatoes)	Freely incorporated into the film	Antioxidant (O_2 and light barrier)	Sunflower oil	Stoll et al. (2019)
Residues of gelatin	Beet root residue powder (*Beta vulgaris* L. var. Conditiva)	Freely incorporated into the film	Antioxidant (radical scavenging)	Sunflower oil	Iahnke et al. (2016)
Corn starch	Eugenol and oleic acid (free whey protein or lecithin encapsulated)	Freely or encapsulated incorporated into the film	Antioxidant (up to 53 days, at 30 °C)	Sunflower oil	Talón et al. (2019)
Cellulose paper	Sorbic acid	Freely incorporated into the paper	Antimicrobial	Butter	Moraes et al. (2007)
Cellulose paper	Omega-3(fish oil) and oregano essential oil	Freely incorporated into the paper	Reduction in the permeability to water vapor	Butter	São José et al. (2019)
Paper+PP (ATCO OS200 and LH210)	Oxide iron	Freely incorporated	Reduction on oxygen levels	Butter	Otero-Pazos et al. (2018)
Cassava starch	Green tea extract and carotenoids from oil palm colorant	Freely incorporated into the film	Antioxidant (PV reduction) and water vapor barrier	Butter/fatty products	Perazzo et al. (2014)
Chitosan and polycaprolactone	α-Tocopherol	Freely incorporated into the film	Antioxidant (longer than 20 days)	Butter/food simulants	Otero-Pazos et al. (2016)

Chitosan coating cellulose paper	TiO_2 or Ag/TiO_2	Freely incorporated into the film coating the paper	Humidity, O_2, grease and light barrier, antioxidant, antimicrobial	Clarified butter	Apjok et al. (2019)
Multilayer OPP25 μm (bags and trays)	Green tea extract	Grafted in the internal layer of the film	Antioxidant (hexanal reduction and higher fatty acids values)	Dark chocolate roasted peanuts and milk chocolate cereals	Carrizo et al. (2016)
Laminated film (OPP, PET, MET, PE, STC)	*Satureja thymbra* extract	Sprayed or rolled on for coating the film	Antioxidant	Fried potato chips	Choulitoudi et al. (2020)
Cassava starch–carboxymethyl cellulose	Quercetin or TBHQ	Freely incorporated into the film	Antioxidant (stable phenolic content and retard of lipid oxidation 35-70 days)	Lard	Tongdeesoontorn et al. (2020)
Gelatin	Fungal melamine	Freely incorporated into the film	Antioxidant (low PV)	Lard	Łopusiewicz et al. (2018)
EVOH	Ascorbic acid, ferulic acid, quercetin, and green tea extract	Freely incorporated into the film	Antioxidant (PV/ malondialdehyde reduction)	Brined *Sardina pilchardus*	López-de-Dicastillo et al. (2012)
Oncorhynchus mykiss skin gelatin	Epigallocatechin gallate and green tea powder	Freely incorporated into the film	Oxidation delay and O_2 barrier	Cod-liver oil	Tammineni et al. (2012)
LDPE	Phenolic compounds from barley husks	Freely incorporated for coating the film	Antioxidant (slow lipid hydrolysis and oxidative stability)	*Salmo salar* L.	Pereira de Abreu et al. (2010)
Chicken feather protein/ pork gelatin	Clove oil	Freely incorporated into the film	Antioxidant and antimicrobial	Smoked salmon	Song et al. (2014)
Gelatin and carboxymethyl chitosan	Squid Maillard peptides	Freely incorporated into the film	Antioxidant and antimicrobial (4 days shelf-life extension)	*Thunnus thynnus*	Jiang et al. (2019)

(continued)

Table 16.1 (continued)

Packaging material	Active ingredient (source)	Incorporation	Properties	Oil/Fat	References
LDPE	α-Tocopherol (modified atmosphere packaging)	Freely incorporated into the film	Antioxidant (hemoglobin protection, fat oxidation reduction)	*Thunnus thynnus*	Torrieri et al. (2011)
Carboxymethyl cellulose	α-Tocopherol and lecithin (polycaprolactone nanocapsules)	Nanocapsules included into the film	Antioxidant and controlled release	Fatty food simulant	Mirzaei-Mohkam et al. (2019)
Bovine gelatin	Tea polyphenols (chitosan-cyclodextrin nanocapsules)	Nanocapsules included into the film	Antioxidant and controlled release	Fatty food simulant	Liu et al. (2017)
Whey protein	Cinnamon, rosemary and basil essential oils	Freely incorporated into the film	Antioxidant and antimicrobial	Fatty food simulant	Ribeiro-Santos et al. (2017)
LDPE and LDPE+HDPE	Astaxanthin	Freely incorporated into the film	Antioxidant potential	Fatty food simulant	Colín-Chávez et al. (2013)

Different packaging materials with diverse active ingredients, free or encapsulated incorporated, have been tested to provide shelf-life extending properties in multiple fatty matrices

LDPE Low Density Polyethylene, *LLDPE* Linear Low-Density Polyethylene, *HDPE* High Density Polyethylene, *EVA* Ethylene-Vinyl Acetate, *PLA* Poly(lactic Acid), *EVOH* Ethylene Vinyl Alcohol, *BHT* Butylated Hydroxytoluene, *TBHQ* Tert-butylhydroquinone, *PEGS* Polyethylene Glycol Succinate, *TPS* Thermoplastic Starch, *PP* Polypropylene, *OPP* Oriented Polypropylene, *PET* Polyethylene terephthalate, *MET* Metal Foil, *PE* Polyethylene, *STC* Single Trip Container

- Proteins are a useful ingredient for extending shelf-life packages for fatty foods.

(i) Animal proteins: bovine or fish gelatin, whey or casein are the most employed ones for preventing lipid oxidation in olive, coconut, corn, flaxseed, sunflower or cod-liver oils or tuna (Xu et al. 2020; Tammineni et al. 2012; Ribeiro-Santos et al. 2017; Wang et al. 2019; Kchaou et al. 2020; Jiang et al. 2019; Malherbi et al. 2019; Gautam and Mishra 2017; Iahnke et al. 2016; Liu et al. 2017).
(ii) Vegetal proteins: soybean proteins or wheat gluten had been studied for delaying the oxidation of linseed, olive, sesame, soybean oils or fish fatty acids (Mohammad Zadeh et al. 2019; Carpiné et al. 2015; Chavoshizadeh et al. 2020; de Souza et al. 2020).

- Polymers: the polylactic acid (PLA), a polymer of lactic acid monomers obtained using bacterial fermentation of corn, sugarcane, potatoes, and another biomass. PLA has been repeatedly evaluated for preserving edible oils or fatty foods like palm, soybean, sunflower oils (Manzanarez-López et al. 2011; Stoll et al. 2019; Penjumras et al. 2018; Almasi et al. 2014).

3 Active Packaging: Improved Properties Conferred to Oil/Fat

3.1 *Antioxidant Capacity*

Oil and fats represent the main target of oxidation processes in food industry. Lipid oxidation begins as an initial peroxidation of long chain fatty acids that generates hydroperoxides. They subsequently get oxidized and result in several aldehydes, ketones or furans (Fig. 16.2). Some of these compounds, as in the case of aldehydes, can be quantified after reacting with thiobarbituric acid. The result of the thiobarbituric acid reactive acid substances (TBARS) assay allows the extent of lipid degradation in the sample to be determined (Kapchie et al. 2013). Another common measure of the degree of lipid oxidation in a sample is performed through the quantification of peroxidation value (PV) that measures the reactivity of free hydrogen atoms (Repetto et al. 2012). Depending on the fatty acid composition and saturation grade, oxidation products will vary in concentration. Aldehydes are the most common lipid oxidation termination product, of which 4-hydroxynonenal (4-HNE) and malondialdehyde (MDA) are the best studied (Guéraud et al. 2010). Among ketones, vinyl ketones are the most undesired as they give a strong off-flavor (Rababah et al. 2011). Furans, however, are prone to appear mostly by thermal-induced oxidation, as furans already present in the extracted oil react and bond with short chain fatty acids. The most commonly found furans are 2-butyl and 2-pentylfuran (Fan 2015). Other products derived from the lysis of fatty acid chains are alkanes and alkenes. These may be further oxidized to aldehydes like pentanal or hexanal (Joseph 2016).

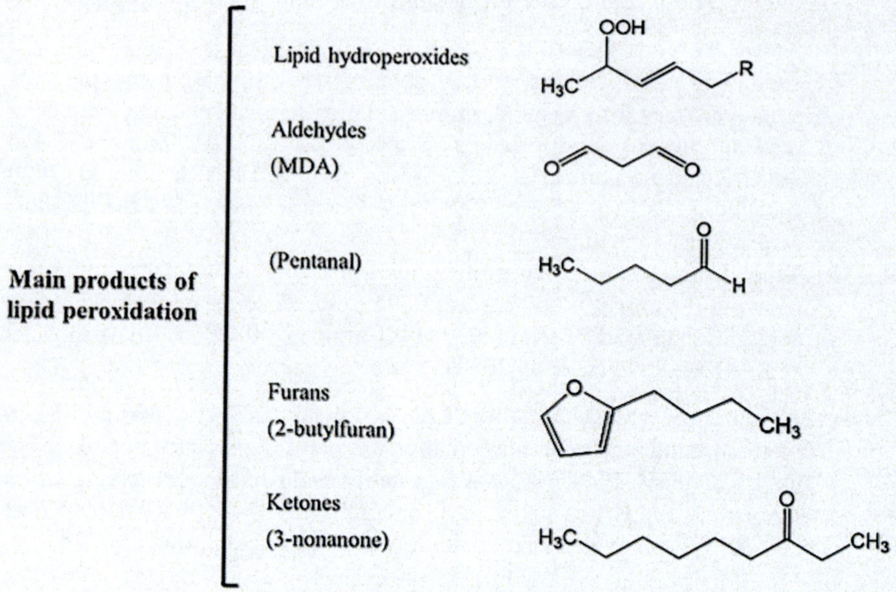

Fig. 16.2 Main products derived from the lipid oxidation of fatty products

The result of all these oxidative reactions involves huge economical loses in terms of waste of product and also because the sub-products produced during the lipid oxidation have been considered to threaten human health (Tian et al. 2013). Therefore, in order minimize the damage that lipid oxidation generates, the food industry has considered the application of different approaches that have evolved over the years.

Traditionally, oils and fats had been fortified by the addition of antioxidant molecules. The use of natural antioxidants in fatty products, where the lipid oxidation induces rancidity, has been proven to delay these reactions and extend their shelf-life. Antioxidants can delay the oxidation reaction rate by some of these mechanisms: scavenging free radicals, chelating pro-oxidative metals, quenching singlet oxygen/photosensitizers, and/or inactivating lipoxygenase (Choe and Min 2006). In food industry, synthetic antioxidants, such as BHA, BHT, PG or TBHQ, have been commonly used for their chemical stability, antioxidant and cost-efficiency. However, their safety has been questioned for its association with severe side effects related with toxicity and cancerogenic capacity. Additionally, their application requires strict statutory controls (Restuccia et al. 2010; İnanç and Maskan 2012). Therefore, the food industry has shown huge interest in the biological properties that natural molecules provide. Natural ingredients had not been associated with side effects; thus, they represent an alternative to replace the use of synthetic ones.

Pigments and phenolic compounds present in nature have been extensively demonstrated to possess antioxidant capacities which have been proved to improve shelf-life of edible oils. It has been known for more than 30 years that adding β-carotene and citric acid to soybean oil prevents its flavor's deterioration while preserving its color quality (Warner and Frankel 1987). Flavor preservation can be also achieved through the delay of lipid oxidation such as in the case of soybean oil fortified with curcumin, extracted from turmeric, which showed a reduction in the oxidation rate and low peroxide values (Eshghi et al. 2014). Similarly, sunflower oil has been blended with different extracts obtained from garlic or peels from bananas or tomatoes. All these approaches displayed have stabilized the oxidative status of the oil due to the presence of antioxidant molecules, like the lycopene recovered from tomatoes (Kehili et al. 2018; Bravi et al. 2017; Ling et al. 2015). The fortification of rapeseed oil with phenolic compounds obtained from two different berries, *Sorbus aucuparia* L. and *Malus baccata* L., showed antioxidant capacity for storing and frying (Aladedunye and Matthäus 2014). However, the direct application of antioxidants can get limited activity when incorporated into complex matrixes, besides it has low selectivity to target the food surface, where most oxidative reactions occur (Gemili et al. 2010). Besides, the direct addition of antioxidants is not allowed for fresh or raw foodstuffs and when permitted, it can modify the organoleptic features of the end product (Tovar et al. 2005). All these drawbacks have prompted the advancement of other alternatives such as the incorporation of the natural ingredients into micro- or nanocapsules, the design of different active packaging models and/or the application of a suitable MAP.

Nowadays, the trend is the development of packages into which molecules with recognized bioactivities that improve the shelf-life of the fatty products have been added, which in some scenarios has been improved when combined with suitable MAP. In the active packaging field, the most innovative tendency is the encapsulation of biomolecules since it avoids the modification of the organoleptic properties of the fatty food matrix (preventing consumer rejection) and enhances the efficiency of the bioactivities (Granata et al. 2018; Asensio et al. 2013). As for the direct application, several molecules have been evaluated for their incorporation into packages with the aim of preserving fatty foods. They can be classified into synthetic compounds (like BHT or TBHQ) and natural ones, being the variability of this last group large. Among the natural pigments, phenolic compounds, PUFAS, etc. have been evaluated for their incorporation into packages to create active packaging. Some of these molecules were recovered from vegetal or animal residues as essential oils or purified extracts (Table 16.1).

Among the pigments used for extending the shelf-life of fatty products astaxanthin, anthocyanins, chlorophyll, β-carotene, lycopene, and bixin, are some of examples that had displayed successful results as antioxidants for corn, olive, sesame, soybean and sunflower oils (Xu et al. 2020; Stoll et al. 2017; Wang et al. 2019; Chavoshizadeh et al. 2020; Colín-Chávez et al. 2014a, b; Samsudin et al. 2014; Stoll et al. 2019; Colín-Chávez et al. 2013). The antioxidant action of pigments included into different packaging materials has been suggested to exert their action by two different protective mechanisms. Pigments can act by direct contact with the matrix by migrating from the active packaging into the fatty food, as demonstrated by different studies (Colín-Chávez et al. 2014a, b; Samsudin et al. 2014; Stoll et al. 2019; Colín-Chávez et al. 2013). Besides, pigments reduce the lipid oxidation by indirect action since their presence in the packaging material offer light protection to the food content (Colín-Chávez et al. 2014b; Stoll et al. 2019).

The composition of the phenolic compounds obtained from tea or form barley husks and applied as antioxidants for fatty foods models and salmon was chemically undetermined (Pereira de Abreu et al. 2010; Liu et al. 2017). Few specific phenols like sesamol, pyrogallol, eugenol, ferulic acid, quercetin or lignin (and its derivatives), used by itself or in combination with other molecules were demonstrated to prevent lipid oxidation in linoleic acid, soybean, sunflower, cod-liver oils, brined sardines and fish fatty acid ethyl ester (Talón et al. 2019; López-de-Dicastillo et al. 2012; Tammineni et al. 2012; Mohammad Zadeh et al. 2019; Gaikwad et al. 2017). The antioxidant activity that polyphenols provide to fatty acids can be by direct contact with the matrix since the diffusion of the molecules from the packages into the matrix has been repeatedly described (Talón et al. 2019; López-de-Dicastillo et al. 2012; Zhu et al. 2013; Liu et al. 2017; Penjumras et al. 2018; Panrong et al. 2019; Almasi et al. 2016). However, a work pointed out that the free radicals produced in the food matrix can migrate into the active packaging. In this case, they get trapped in the inner layer of a multilayer OPP film where the active molecules, polyphenols extracted from green tea, were incorporated. Thus, polyphenols exerted their antioxidant activity by indirect contact with the food matrix (Carrizo et al. 2016).

The addition of lipophilic molecules (α-tocopherol, lecithin, oleic acid, PUFAS like sorbic acid or omega-3 from fish oil, essential oils or virgin coconut oil) into fatty products also improved the oxidative stability of corn, olive, soybean, sunflower oils, butter and tuna (Talón et al. 2019; Carpiné et al. 2015; Fayaz Dastgerdi et al. 2016; Moraes et al. 2007; São José et al. 2019; Ribeiro-Santos et al. 2017; Graciano-Verdugo et al. 2010; Manzanarez-López et al. 2011; Otero-Pazos et al. 2016; Torrieri et al. 2011; Mirzaei-Mohkam et al. 2019). The diffusion of lipophilic compounds indicates the direct contact as a potential mechanism for developing their antioxidant protection (Manzanarez-López et al. 2011; Otero-Pazos et al. 2016; Torrieri et al. 2011; Mirzaei-Mohkam et al. 2019; AbouZid and Ahmed 2013).

The antioxidant efficiency of several peptides or amino acids, such as ε-polylysine, lysine, squid Maillard peptides, was tested in corn, flaxseed, olive oils and tuna (Xu et al. 2020; Carpiné et al. 2015; Kchaou et al. 2020; Jiang et al. 2019). Saccharides have not been so commonly used as active ingredients but glucose (mixed with lysine) or Konjac glucomannan were positively evaluated for sesame and flaxseed oils (Kchaou et al. 2020; Ganesan et al. 2019). Amino acids have been demonstrated to diffuse from the package to the food matrix, thus this is a likely mechanism of action (Gemili et al. 2010; Liu et al. 2018b).

Independently of the kind of packaging material or the core antioxidant, active packaging has been extensively demonstrated to improve product shelf-life. The antioxidant efficiency of active packaging films was mainly evaluated through PV data. Most of them could maintain PV under the regulatory limits (10–20 mEq/kg) for longer periods of time than non-activated films. The improved antioxidant capacity was always related to the presence of a core ingredient. Even though most of them were naturally obtained molecules, some studies still selected synthetic compounds. Among the mechanisms of action of the biomolecules, the direct one, apart from preventing food lipid oxidation, implies, in most of the cases, the incorporation of the antioxidant into the food matrix which may have additional health benefits for the consumer. Thus, the application of active packaging with antioxidant activity represents a useful tool to extend fatty foods shelf-life, to keep their quality, and to provide to the consumer a safer product enriched with antioxidant molecules.

3.2 *Antimicrobial Capacity*

The most common bioactivity associated with active packaging is the antioxidant followed by the antimicrobial. This is because lipid deterioration is mostly caused by oxidation or microorganisms and represents one of the major causes of food loss. Thus, the control of the growth of these microorganisms and of the lipid oxidation is necessary for the foods to maintain a high level of quality, thus increasing the shelf life of the product and the profitability of the industry (Moraes et al. 2007). Indeed, the antioxidant capacity is strongly related with the antimicrobial, since along the microbial development diverse gases are synthesized such as hydrogen sulfide (H_2S) and hydrogen peroxide (H_2O_2) (Panrong et al. 2019; Fields et al. 1968). These sulfur and oxygen volatiles have been described to possess oxidant and reactive capacity, especially against proteins and lipids. Thus, the presence of antioxidant molecules may prevent the indirect deterioration of fatty products induced by the effect of said volatiles. Even though the use of active packaging with antioxidant capacity reveals antimicrobial benefits, the best approach is the application of molecules that possess double effect, both antioxidant and antimicrobial. Active packaging has been described to be especially effective when applied with solid and semi-solid foods, since microbial growth is superficial; hence the greater is the contact between the product and the antimicrobial agent, greater is the efficiency. Among the variability of materials used for the development of innovative packaging, different natural polymers have been recognized with antimicrobial capacity, such as chitosan. These biofilms can become active and thus reinforced their antimicrobial activity by the additional integration of natural or synthetic compounds. Active packaging additives' release increases consumer safety, since these compounds, instead of being directly added to the food, are delivered in a controlled manner. Therefore, they are present in smaller quantities, and mostly on the product surface, where most of the deterioration reactions occur (Gontard 1997). Some of the most successful approaches included as part of active packaging are oxygen and ethylene absorbers, CO_2 liberators or the immobilization of inert molecules, such as TiO_2/Ag, or biomolecules such as pigments, polyphenols, amino acids or enzymes (Xu et al. 2020; Moraes et al. 2007; Apjok et al. 2019; Liu et al. 2018c; Zhang et al. 2020; Bi et al. 2020). These preservation techniques are aimed to overall extend the shelf-life of fatty products but also to improve their quality and safety.

Different examples of active packaging showing both antioxidant and antimicrobial ability have been demonstrated to delay the degradation of vegetable edible oils. Several works have profited the chitosan properties for developing biodegradable films. The chitosan antimicrobial ability has been suggested to be due to two potential mechanisms: a) the interaction of its protonated groups with the microbial cell membrane that triggers an increment in the permeability that ultimately induce the inhibition of the food-borne pathogen; b) the capacity for chelating compounds that may result as essential for microbial growth (Zhang et al. 2020; Bi et al. 2020). For instance, an antimicrobial active packaging was developed for protecting corn oil using a chitosan and gelatin film later incorporated with polylysine and astaxanthin, both extracted from *Litopenaeus vannamei*. This film displayed inhibitory effect on spoilage bacteria derived from *L. vannamei*. It also presented antioxidant activity quantified in less than 5 nM/mL of MDA after the incubation of the corn oil at 60 °C for 59 h. As mentioned above, chitosan has been demonstrated to possess antimicrobial activity itself which may be reinforced by the antioxidant effect associated to the incorporation of polylysine and astaxanthin (Xu et al. 2020). Another antimicrobial active packaging created with a chitosan base was enriched with the polyphenolic compounds obtained from Mangosteen (*Garcinia mangostana* L.) rind. The additional content of polyphenols improved the antimicrobial of chitosan by itself. Previously, *G. mangostana* had been characterized for its content in polyphenols and tannins, both kind of molecules suggested to possess antibacterial capacity, especially against Gram-positive bacteria (Samprasit et al. 2014; Fraga-Corral et al. 2020). In fact, these films were less effective against Gram-negative bacteria probably due to the presence of the external membrane that minimized the impact on the cell permeability. Besides, this film was able to inhibit the increase of the peroxide value and thiobarbituric acid reactive substances on soybean oil during storage which likely hold up the antimicrobial activity (Zhang et al. 2020). Other work developed new chitosan-based films with antioxidant and antimicrobial properties to improve the oxidative stability of soybean oil and increase its shelf-life. This film was created by adding D-α-tocopheryl polyethylene glycol succinate (PEGS) and/or silicon dioxide (SiO_2) nanoparticles into chitosan films. While only addition of the D-α-tocopheryl PEGS did not improve the antimicrobial effect of chitosan, the inclusion of SiO_2 by itself enhanced it and in combination with D-α-tocopheryl PEGS showed the highest effect against all the four food-borne pathogens tested: as Gram-negative *Escherichia coli* and *Salmonella typhimurium* and as Gram-positive *Staphylococcus aureus* and *Listeria monocytogenes*. Among the bacteria, the film was again more effective against Gram-positive than against Gram-negative (Bi et al. 2020).

Other approaches of active packaging aimed to protect edible oils had been designed using alternative protein-based biodegradable materials. This is the case of a copper nanocomposite fabricated as a bilayer pouch from a heat sealable casein protein later laminated with sodium alginate–pectin and tested for preserving coconut oil. This nanocomposite showed antibacterial activity when tested against *Escherichia coli*. This ability was suggested to be due to the presence of the copper, besides the heat treatment of the casein seems to preserve the antibacterial activity even after washing it. This new pouch also reduced the rate of oxidation of coconut oil during storage which may support the antibacterial property (Gautam and Mishra 2017). Another option is the inclusion of essential oils into the films. Essential oils had been repeatedly demonstrated to be very effective natural ingredients for food preservation and thus for creating active packaging. The incorporation of essential oils into a whey protein-based film was evaluated through a fatty food stimulant, ethanol 95%. The final concentration of essential oils added to the film was very relevant. But also, the modification in the proportions of the essential oils showed different results. In terms of antimicrobial activity, the best film was that contained just a 5% of the following combination of essential oils: 51% *Cinnamomum cassia*, 34% *C. zeylanicum* and 15% of *Rosmarinus officinalis*. In fact, *C. cassia* displayed anti-bacterial and anti-fungal capacity when evaluated itself. Thus, the antimicrobial assays performed in the fatty food simulant demonstrated that this specific film was the most effective for inhibiting the growth of fungus *Penicillium* spp. and bacteria, *Escherichia coli* and *Staphylococcus aureus* (Ribeiro-Santos et al. 2017).

Active packaging for protecting butter has been also deeply studied. The traditional packages used for butter preservation usually present several disadvantages such as light transmission, oxidized off-flavor, microorganism and metallic contaminations. Thus, the use of biopolymers like chitosan, starch and cellulose and its derivatives are an interesting approach to solve these drawbacks. A study focused on the protection of butters by testing three different cellulose-based papers coated with chitosan, chitosan-TiO_2 and chitosan-Ag/TiO_2. These films were aged for 6 months at 55% relative humidity, 4 °C, 15 h light/9 h dark cycle and characterized in relation to plain paper. Physical-chemical-microbial measurements of the three active papers proved that the chitosan and active agent properties were lost during the storage time. This process also explains the deterioration of the physical barrier as well as the antimicrobial capacity reduction which leads to the depletion of the efficacy of papers preservation in some of the cases. The best results seemed to be the ones obtained with the chitosan-Ag/TiO_2 paper since after 6 months of storage they presented the lowest colony forming unit counts for molds and yeasts (5.8 CFU/g) as well as for *E. coli* (6.12 CFU/g). The antioxidant capacity detected for this paper was also the highest with low levels of peroxide of 2.72 mEq O_2/kg. Therefore, this chitosan-Ag/TiO_2 paper significantly improved the shelf-life of butter while keeping a clean appearance without greasy transference spots after the long storage period, which confirms its self-cleaning properties (Apjok et al. 2019). Another approach aimed to preserve butter developed a cellulosic film infused with a 7% sorbic acid. In this experiment, butter was sliced and inoculated with both yeast and filamentous fungi previously isolated from a different butter. In this scenario, the active film showed an inhibition halo of 3.4 cm for yeast and filamentous fungi. Additional experiments were performed using samples wrapped with the active film, that were then wrapped with an aluminum foil and then stored at 7 °C for 0, 10 and 20 days. The microbiological analyses performed from day 0 with an initial counting of 3×10^6 FCU/g and after 10 and 20 days the reductions were 1 log cycle (9×10^5 FCU/g) and 2 log cycles (8×10^4 FCU/g), respectively. These results were better than the ones obtained for the control film and also inhibited filamentous fungi and yeast growth in butter (Moraes et al. 2007).

Therefore, the use of active films designed with antimicrobial purposes might be a good alternative to avoid the use of antibiotics for food preservation that had been proved to have strong side effects, such as bacterial resistance, that directly affect human and animal health and the environment.

3.3 *Flavor/Aroma Properties*

Lipid oxidation is a complex chain of reactions which can be grossly divided into two main stages. During the primary oxidation unsaturated fatty acids in the presence of oxygen and some catalyst agent (iron, copper, enzymes, heat or light) produces lipid hydroperoxides. These initial molecules possess no color, odor or taste. In a second phase, the auto-oxidation step, peroxides compounds which are unstable and thus very reactive oxidize other food matrix components such as fats. The secondary products created are a complex mixture of low molecular weight compounds. The volatile ones, alkanes, alkenes, aldehydes and ketones, are responsible for the rancid odor and taste of fatty products. Hexanal, for example, is a commonly volatile aldehyde formed during the lipid oxidation of linoleic, gamma-linolenic and arachidonic acids and increases during storage. Concentrations about 5–10 μg/mL have been considered to induce a rancid odor which make them sensorially unacceptable (Robards et al. 1988; Pastorelli et al. 2006). These products continue reacting with other molecules triggering structural and organoleptic alterations. Aldehydes that are able to react with proteins and DNA molecules affecting their structural and functional properties (Pereira de Abreu et al. 2010; Tian et al. 2013).

Hence, lipid oxidation negatively reduces the nutritional value of products due to the chemical modification of fatty acids and other relevant lipophilic molecules such as the liposoluble vitamins A, D, E and K. Besides, the secondary oxidative products are directly related with rancidity which alters the organoleptic characteristics by producing off-flavors and modifying the original pigmentation of the product (both by darkening fats or lightening pigments (Otero-Pazos et al. 2018; Nerín et al. 2008). The flavor degradation of fatty products can be measured by the peroxide quantification since it indicates the lipid oxidation stage. Indeed, the peroxide index is usually considered for establishing its expiration date in oil (Chavoshizadeh et al. 2020). In other kinds of fatty foods such as fish, the presence of some biogenic amines, like histamine, tyramine, cadaverine, putrescine, spermidine and spermine permit to determine the oxidation status of the matrix (Jiang et al. 2019).

In order to prevent this chain of oxidative reactions different antioxidant molecules can be included into the food matrix. The most common synthetic compounds used as antioxidants are butylated hydroxyanisole (BHA), butylatedhydroxytoluene (BHT) and propyl gallate (PG) (Fayaz Dastgerdi et al. 2016). Nevertheless, nowadays, consumers claim the replacement of synthetic molecules with natural ones. Some aromatic herbs are a natural source of molecules such as β-carotene, astaxanthin, limonene, menthol, eugenol, allicin, carvacrol, thymol or terpineol, among others. Some of these molecules, apart from providing antioxidant protection, are recognized as flavoring and aromatic. Flavoring substances are defined as chemicals with flavoring properties which can be added into food matrixes in order to impart or modify odor and/or taste (EC EC 2008). A natural extract obtained from garlic, using supercritical fluids, was added into sunflower oil. Allicin, a thiosulfinate compound present in garlic, which had been previously demonstrated to have antioxidant activity, showed preservation properties when applied to sunflower oil (Bravi et al. 2017). Rosemary and oregano added to fried potatoes or used for frying improved the oxidative stability (Choulitoudi et al. 2020). Essential oils obtained from mint, laurel and myrtle leaves have been demonstrated to prevent the oxidation of edible oils and oily products. In the case of mint extracts, with identified major molecules being l-menthol, menthone and isomenthone, their direct application into the oil at concentrations of 200–400 ppm extended its shelf-life, but they required much higher concentrations, 3700–8400 ppm, when included into the HDPE bottle to perform similar results (Fayaz Dastgerdi et al. 2016). Hence, the addition of natural compounds for preventing lipid oxidation may modify its organoleptic properties. Sometimes, the combination of natural products may reveal a new taste with successful reception by consumers such as in the case of fried potato chips using oil treated with extracts from rosemary, sage and citric acid (Choulitoudi et al. 2020). Other combinations such as the olive oil flavored with oregano essential oils that displayed low lipid oxidation indicators, induced a negative impact in the sensorial panel indicating a potential negative consumer acceptance (Asensio et al. 2013). However, the inclusion of antioxidants becomes essential to maintain the fresh flavor of oils and fatty products in order to prevent the modification of their organoleptic properties. Several antioxidants had been included into the packaging material either free or encapsulated with the purpose of preserving the aroma and flavor of the fresh product by reducing the rancidity (antioxidant) and in some cases they also prevent the microbial spoilage (Table 16.1).

An active film was designed using LDPE into which different concentrations of α-tocopherol (0–40 mg/g) were incorporated. Their evaluation was performed through the quantification of hexanal production in corn oil stored with the active films at a temperature of 30 °C. The active film without α-tocopherol showed a concentration of hexanal 3–4 times higher than the films with α-tocopherol. The value hexanal for these films was 6–7 μg/mL after 16 weeks of storage. This concentration indicates the beginning of the rancid odor production however the lack of antioxidant in the active packaging over passed the upper limit for rancidity (10 μg/mL). Thus, the addition of antioxidants is relevant for preserving the corn oil flavor (Graciano-Verdugo et al. 2010).

Another experimental approach considered the use of a soybean protein isolate enzymatically modified into which two different lignin derivatives were added: alkali lignin and lignosulfonate. These biofilms were tested in soybean oil and fish. Fresh soybean oil consists of a high amount of PUFAs including α-linolenic acid and linoleic acid, and the monounsaturated oleic acid while fish oil contains long chain omega-3 fatty acids like eicosapentaenoic (EPA) and docosahexaenoic acids (DHA). Both kind of fats, vegetal and animal, are a main target of the lipid oxidation that reduces their flavor quality. The active biofilms evaluated reduced up to

50% the production of volatile compounds caused during autooxidation that produce off-flavors. In soybean oil the concentration of 2-heptanal, pentanal, hexanal, and nonanal was reduced when using the lignin-based biofilms. Similar results were displayed for fish packaged with the active films where the production of hexanal, heptanal, and nonanal was lower than in the control. Besides, this work described another kind of off flavor derived from contact with the plastic, named "aroma scalping", which can be avoided by the utilization of biopolymeric films as those based on lignin (Mohammad Zadeh et al. 2019).

In fish, there are other types of molecules that can also generate unpleasant flavors, the biogenic amines (histamine, tyramine, cadaverine, putrescine, spermidine and spermine). A research group developed a gelatin-water soluble chitosan film incorporated with squid Maillard peptides to prevent the production of said amines. The main highlight of this work is the reduction in the amount of histamine, tyramine, and cadaverine present in the packaged tuna after 6 days, when compared to other wrapping alternatives (Jiang et al. 2019).

The fresh butter aroma contains diacetyl and 3-hydroxy-2-butanone as responsible for the buttery odor; (Z)-4-heptenal, 2-pentanone and furfural for caramel or cream-like smell; 2-heptanone for dairy-like, 2-nonanone for a hot milk-like odor, acetic acid for acidic stages, and butanoic acid for cheese-like aromas (Mallia et al. 2008). When lipid oxidation happens, the presence of hydroxyl acid products reduces the flavor and aroma characteristic of fresh butter. Different approaches had been developed by using active biodegradable films. A cellulose-based paper coated in chitosan-Ag/TiO_2 reinforced the humidity, oxygen, grease and light barrier that reduced the fat lipolysis, extended the flavor of butter and preserved its quality longer (Apjok et al. 2019). Another active cellulosic film was developed by the incorporation of sorbic acid to evaluate its preserving capacity for butter. Sorbic acid and its salts (calcium, potassium, and sodium) are used in the food industry with preserving purposes for their antioxidant and anti-microbial capacities. They are commonly utilized molecules due to their solubility, low taste interference and low toxicity (Moraes et al. 2007). Other molecules that induce scarce organoleptic alterations were added to create an active experimental film developed by using ethylene vinyl alcohol copolymer (EVOH). Four different core ingredients were included into the EVOH film: ascorbic acid, ferulic acid, quercetin, and a green tea extract, mostly composed by catechins. These molecules are non-volatile and non-aromatic antioxidants and thus minimize the sensorial effect on the final product while preserving its original organoleptic properties. The active films tested along this study, especially the one containing the green tea extract, enhanced sardine stability by reducing the peroxide index and the MDA content, which in last term preserved the fresh flavor longer (López-de-Dicastillo et al. 2012).

A multilayer oriented polypropylene (OPP) film was biologically activated by grafting green tea extract in the inner layer. Catechins were described to exert the antioxidant capacity up to 16 months in both food products by indirect contact. The study suggested that free radicals get trapped into the layer where catechins had been grafted. This mechanism was described to occur by the migration of the free radical through the polymeric layers. Thus, the molecules contained into the green tea extract utilized as an active ingredient does not affect the organoleptic properties of the products since it does not migrate (Carrizo et al. 2016). Nevertheless, not all active packaging are capable of masking the aroma or flavor of the extracts used as core ingredients. In a work where rosemary and cinnamon essential oils were incorporated into a whey protein-based film to act as antimicrobials, the addition of concentrations from 5% were sensorially detectable (Ribeiro-Santos et al. 2017). Another study further analyzed the antimicrobial capacity of a PP film into which different combinations of natural compounds were incorporated. These mixtures were made using some of the main compounds identified in essential oils (hydrocinnamaldehyde, cinnamaldehyde, thymol, and carvacrol) and three different aromas (banana, strawberry and vanilla). The results showed that thymol, carvacrol, and cinnamaldehyde possess strong antimicrobial activity against several bacteria, yeast and molds. However their combination with the aromas was not successful, with the exception of thymol-strawberry that was the only mixture organoleptically acceptable (Gutiérrez et al. 2009).

A research group reported the beneficial effects of active packaging from an aroma-preserving point of view. The work analyzed the evolution of the aroma along the storage of three different food products: oil, potato chips and coffee. To this aim they monitored their lipid oxidation, especially by measuring hexanal. They compared the results obtained through different devices, an electronic nose, a gas chromatography coupled to mass spectrometry, a static headspace extraction and solid phase microextraction. Data displayed that active packaging allowed the decrease of lipid oxidation and even absorbed components from the headspace which definitely extend products shelf-life (Strathmann et al. 2005).

Therefore, the main purpose of active packaging, from the point of view of aroma and flavors, is to preserve the chemical properties of the volatile molecules characteristic of fresh products associated with a higher consumer acceptance but also with high quality products and nutritional values.

3.4 Color Enhancer

Color is an important quality attribute related to the presence of different pigments, some of them with beneficial bioactivities, such as carotenoids and chlorophylls that provide antioxidant capacities. The Commission Internationale de l'Eclairage (CIE) established in 1976 the CIEL*a*b* color space that permits the quantitative determination of colors through three chromatic coordinates L* (lightness/brightness), a* (redness/greenness), and b* (yellowness/blueness). From these, cylindrical Cartesian coordinates can be obtained : C* (chroma, indicates the relative saturation), h° (hue angle) and L* (lightness/brightness) (CIE Colorimetry Committee 1974; McLaren 1976). Based on this, in the 90s, a work correlated the concentration of pigments in virgin olive oil and the chromatic coordinates. The study concluded that b* can be correlated with the carotenoid concentration while a* refers to the pheophytins evolution. For olive oils that possess similar a* and b* values, L* permits to distinguished them (Isabel Minguez-Mosquera et al. 1991). The alteration of these values is usually due to the degeneration of the pigments which tends to be induced by oxidant processes. Lipid oxidation is mainly responsible for inducing color changes and off-flavors that shorten the shelf-life of fatty products. In animal products the blood presence implies the hemoglobin provides purplish-red colors that when oxidized forms deoxy-myoglobin and metmyoglobin which turn original colors into brownish ones. These oxidation-related color modifications implies a loss of organoleptic properties that reduces the consumer acceptance since they are associated with a loss of the product nutritional value (Torrieri et al. 2011). By the end of the 80s, β-carotene was suggested as a useful yellow colorant convenient for coloring salad oil and shortening, frying oil, and corn popping oil. The addition of low concentrations, 15–20 ppm, to soybean oils was described to avoid off-flavors and poor color quality (Warner and Frankel 1987). Nowadays, the trend is the incorporation of key ingredients to develop active packaging to create a useful tool to prevent color alterations as result of oxidation processes. The main molecules used are antioxidants that can protect the natural color of the food product by acting in different ways: (a) they can counteract free radicals; (b) some antioxidants had been described to possess antimicrobial capacity by scavenging free radicals produced as consequence of microbial growth, thus they may also prevent discoloration induced by the presence of bacteria, yeasts or molds; (c) they may migrate to get incorporated into and colored the food matrix, and (d) their inclusion provides color to the package that offers light protection reducing food photobleaching.

The antioxidant capacity of active packaging has been extensively discussed above; however, we will present few specific works in this section where its relevance was measured in terms of color protection. A study analyzed the pigment concentration when EVOO was packaged in an active film (starch cassava-based with anthocyanins included) or in a PP film. The anthocyanins film possesses the strongest antioxidant capacity and offered and slightly better carotenoid protection, even though it still showed a 77.5% loss. Thus these active films did not prevent the color degradation mediated by the light incidence which may be improved by modifying the amount of microencapsulated anthocyanins per unit of film surface (29 malvidin-3-glucoside equivalent mg/cm^2) (Stoll et al. 2017). Indeed, previous works suggested that lower concentrations of antioxidants, α-tocopherol about 50 ppm were more effective in other edible oils since higher antioxidant concentration may lead to pro-oxidation (Zuta et al. 2007). In another work, a wheat gluten film coated with chlorophyll (0–5 M) was tested for its antioxidant capacity and its ability to conserve the color of sesame oil. The oxidation status of the oil was intentionally induced by adding HNO_3, which displayed changes in all the three-color parameters a*, b*, and L*. The presence of the active film did not enhance the parameters b* and L* but it could slow down the color alteration of the coordinate a*, and thus the yellow index. Besides, when the oxidation status of the sesame oil was higher, the package modified its color to a more yellow one. Thus, this material extends the shelf life of the oil through the antioxidant molecules that protect the oil color and provides color sensor of the oil expire date (Chavoshizadeh et al. 2020). In the case of butter, the color alterations may be due to the lipid oxidation but also to the microorganism growth. The presence of microorganisms on its surface alters the typical yellow tone because of the colors of the spores they produce. Two different approaches of active packaging were described to possess antioxidant and antimicrobial activities. The simplest active packaging consisted of a cellulosic based film with sorbic acid and was capable of inhibiting the microbial growth that contributed to protect the fresh color of butter (Moraes et al. 2007). The other active packaging also used a cellulose-based paper coated in chitosan, Ag and TiO_2. The light exposure of the film made it to turn itself into dark brown but no fatty or dark spots were found on its surface. Besides, after 6-months, butter still showed low peroxide values, low count of yeasts, molds and bacteria which prevent discoloration processes (Graciano-Verdugo et al. 2010). Another example of color preservation had been observed in bacon packaged with a LLDPE/cassava starch/green tea extract. Meat color behavior differently depending on the film used, LLDPE reduced the redness color (a*) over time, while all the active films containing green tea extract improved a* parameter for at least 5 days. The active films with higher proportions of green tea extracts maintained and even improved the redness color of the fatty meat probably due to the higher content in antioxidants that may speed up the oxy-myoglobin formation. Additionally, antioxidants were suggested to act as antimicrobial agents too. During microbial growth, different compounds (H_2S and H_2O_2) have been described to react with myoglobin and shift its color into greenish tones. The change of coloration has been also associated to the creation of free radicals that may have been scavenged by the antioxidants contained into the film (Panrong et al. 2019). Fish color goes through similar color modification more so than meat since the main oxidized component is the myoglobin. Myoglobin possesses a bright red pigment when it initially makes contact with oxygen and gets transformed into oxymyoglobin. However, a long oxygen exposition provokes the oxidation of the ferrous ions present in the oxymyoglobin (providing reddish tones) into ferric ones (brown color). Tuna was packaged using LDPE film containing α-tocopherol and applying different MAP. Tuna packed in MAP with 100% N_2 keep constant a* parameter and slightly increase b* (still providing the best results for b*). This package was suggested to reduce the amount of oxygen present in the sample and the antioxidant capacity to scavenge free radical. Thus, this combination of active packaging and MAP preserved the bright red color of tuna that maintained the hygienic and sensory characteristics of bluefin tuna fillets (Torrieri et al. 2011). Fish spoilage implies water loss and microbial growth which are accompanied by discoloration and loss of texture features, like what happens in meat products. A study analyzed the capacity of a gelatin-chitosan film infused with squid Maillard peptides to prevent this process. Fish unwrapped, wrapped with a control (gelatin-chitosan film), and PE films did not avoid the coloration change from bright red to brown more than 4 days whereas the active film extended the red color up to 8 days of storage probably due to the additional antioxidant and antibacterial activities that provides the presence of the squid Maillard peptides (Jiang et al. 2019).

The migration of pigments from active packaging into food matrixes has been suggested to provide direct antioxidant capacity. A HDPE resin containing 2% of marigold extract and 2% of TiO_2 was tested with soybean oil. This film released up to a 45% of the included carotenoids when the storage temperature was elevated up to 40 °C. However, at this condition, the pigments are likely to be totally degraded. At lower temperatures (10–25 °C) the film showed lower migration rates (13% and 42%, respectively) but it probably improved the shelf life of the pigments. Even though, this work does not present quantitative data about the color modification that the carotenoid migration generated, it reported a visual color change that might be desirable for oil commercialization (Colín-Chávez et al. 2014b). Similarly, LDPE and HDPE/LDPE films were incorporated with carotenoids from marigold extracts and evaluated in terms of pigments release. LDPE films released them at 40 and 30 °C (59% and 34%, respectively) reaching the equilibrium in short times (8–9 h) and none release was achieved at 25 and 10 °C. While HDPE/LDPE films showed 47%, 66%, 41% and 19% of pigment release, respectively at 40, 30, 25 and 10 °C but required longer equilibrium times (22, 191, 357, and 936 h). Hence, even though the presence of HDPE delayed the pigment diffusion, it allowed it to be at lower temperatures that reduced the degradation rate of the carotenoids. As in the previous work, authors did not provide quantitative color values, but the diffusion of the pigments into the oil was stated (Colín-Chávez et al. 2014a). Other experiments performed with different core ingredients, and films did not show pigment migration. This is the case of a film created with beet root extracts incorporated into gelatin for storing sunflower oil. The oil was exposed to severe storage conditions that modified its color. However, it was not possible to demonstrate that this color change was due to the migration of betalains, even though they were identified as mainly responsible for the antioxidant activity of the film (Iahnke et al. 2016).

Finally, active packaging has been also evaluated as a physical light barrier that the addition of some biomolecules induces into the final film. Polymers have been described to possess higher stability when fortified with carotenoids since they reduce the oxidant status and prevent the damage induced by UV radiation. The antioxidant capacity may avoid the biomolecules degradation while the color may provide light protection of the product by preventing photobleaching. Several pigments, such as astaxanthin, β-carotene, bixin, lycopene, or tocopherol, have been incorporated into different polymer material and evaluated through their antioxidant capacity. PLA films were added with β-carotene, bixin or lycopene and their stability when packaging oil was evaluated under dark and light conditions. PLA containing carotenoids provided a UV and visible light barrier when compared with pure PLA films. Among the tested pigments, β-carotene was the one that presented better light protection results. However, the storage increased the transparency of all the films, especially for the β-carotene-PLA one likely due to the low chemical stability of this pigment (Stoll et al. 2019). Active films created with marigold extracts (rich in carotenoids) and HDPE with or without TiO_2 where evaluated in terms of chrome and astaxanthin (as representative of carotenoids content) decrease. The films with TiO_2 were demonstrated to slow down both losses. Astaxanthin degradation was quantified in 13 and 91% after 16 days of storage for the films with and without TiO_2. Similarly, the film with TiO_2 lost 80% of chrome after 48 storage days while that without TiO_2 reached same value in just 20 days (turning into orange or red dulls, respectively). Thus the presence of TiO_2 in HDPE prevents the color deterioration of the film and avoids carotenoids degradation (Colín-Chávez et al. 2014b). The incorporation of different molecules provides different packaging colors. In fact, other compounds such as phenols have been also explored in terms of film coloring agent. The inclusion of quercetin into EVOH films gave it a slight yellow tone while it was slightly brown with green tea extract (López-de-Dicastillo et al. 2012). Hence, color that biomolecules endow to the packaging materials protects the natural color of fresh products packaged with the active films extending their shelf-life and thus consumer acceptance.

3.5 Vitamin Fortification

The development of active packaging accounts for a vast variability in part due to the huge diversity of compounds used in the manufacture process. Those aimed to improve the properties of oils or fats mainly involve a conservation mechanism led through antioxidant or antimicrobial pathways. These protective functions also avoid the deterioration or degradation of key nutrients constituting the product, such as vitamins. Fat-soluble vitamins, namely, vitamins A, D, E, and K, are considered essential micronutrients. Fat-soluble vitamins possess an isoprenoid chemical structure and they are derived from apolar compounds. They are required in small amounts since the excessive consume of fat-soluble vitamins A and D may lead to toxic effects. Even though minimum amounts are enough to provide health benefits, they cannot be endogenously synthesized, thus they must be ingested through diet. The presence of these vitamins in the body is essential for its proper functioning and for the prevention of certain diseases, since their deficiency is a trigger for disorders such as type 2 diabetes, blindness, osteoporosis, immune diseases or even cancer (Van Wayenbergh et al. 2020; Pedrali et al. 2020). Fat-soluble vitamins can be naturally found in some fatty foods, like liver and fish oil which are rich in vitamin A. Besides, in order to enhance the nutritional value of some fatty foods, lipophilic vitamins can be added to fatty matrices such as oils or margarines which can be fortify with vitamins A or E (Van Wayenbergh et al. 2020). However, it has been demonstrated that fat-soluble vitamins undergo degradation processes that deteriorate them. Vitamins are compounds sensitive to several factors such as light, temperature, exposure to oxygen or acidic environments. The degradation mechanism is mainly triggered by isomerization, oxidation or polymerization reactions (Pedrali et al. 2020). There are other intrinsic factors of the matrix that surrounds these molecules that also play a fundamental role in their deterioration such as its concentration or the presence of other chemical compounds. For instance, when embedded into a matrix characterized for unsaturated fatty acids their degradation also favors that of vitamins; while the presence of triglycerides increases their stability (Van Wayenbergh et al. 2020; Moccand et al. 2016). Vitamins are normally broken down into by-products that reduce the nutritional value of the food and can alter its taste and aroma. However, it has been stated that the use of high quality oils and the prevention of its oxidation are two main conditions to protect both the inherent and fortified vitamins (Diosady and Krishnaswamy 2018). In fact, packaging that reduces the concentration of oxygen in the headspace also prevents the degradation of vitamins (Otero-Pazos et al. 2018). Therefore, all the active packaging alternatives discussed in previous sections and summarized in Table 16.1 can mitigate vitamin degradation that also favors the conservation of food's organoleptic characteristics. Thus, active packaging can prevent chemical and biological deterioration of fat-soluble vitamins that will also ultimately preserve their necessary and beneficial health properties (Ertugrul et al. 2020; Clark et al. 2004).

On the other hand, in some cases, the molecules included in active packaging, especially for those with antioxidant purposes, are precisely vitamins, such as α-tocopherol, or substances that behave like provitamins (compounds that lead to the formation of vitamins). Such is the case of β-carotene, which, in addition to its functions as a pigment and antioxidant, has provitamin A capacity. Once introduced into the body, β-carotene can undergo a series of structural modifications that transform it into vitamin A (Van Wayenbergh et al. 2020). Therefore, the inclusion of these molecules into active packaging provides a double effect: oxidative protection and indirect fortification. Hence, vitamin fortification of fatty foods can be done indirectly through the release of vitamins or provitamins from the package. Vitamin E has been demonstrated to be efficiently included into a chitosan biofilm with good characteristics to contain vitamin E which may provide an alternative system for fortifying food. In fact, chitosan has been repeatedly displayed as a good film-forming agent suitable for being enriched with this vitamin, among other molecules (Table 16.1), although its formulation has to be specifically adjusted according to its use (Il and Zhao 2004). Other researches focused on the inclusion of carotenoids in active packaging, which also entails vitamin fortification, since molecules with provitamin actions are included in this group of pigments, as mentioned above (Colín-Chávez et al. 2014a, b; Nderitu et al. 2018). However, most of the studies carried out so far are based on the incorporation of α-tocopherol in the packaging (Table 16.1) (Otero-Pazos et al. 2016). The presence of α-tocopherol in the packaging material also increases its antioxidant, antimicrobial and light resistance capacity, as well as its migration to food whose consecutive intake may prevent diseases such as cardiovascular or certain skin conditions (Torrieri et al. 2011; Bi et al. 2020). Another work evaluated the incorporation of α-tocopherol in LDPE films and its corresponding diffusion coefficient to corn oil. It was determined that the incorporation of higher concentrations of α-tocopherol (19–30 mg/g) in the film achieved a greater migration rate to the oil. Another important parameter was the temperature that proportionally influenced the migration, since at 30 °C the diffusion coefficient was greater than at 5 °C or 20 °C (Graciano-Verdugo et al. 2010). This fact can also be advantageous, since the degradation of vitamins is favored with an increase in temperature, at which moment the presence of antioxidant compounds is most necessary to prevent this degradation and favor the fortification. The migration of α-tocopherol from packaging to food is also affected by other factors, such as the matrix in which it is embedded, the presence of other natural antioxidants, or even the incorporation of additional synthetic antioxidants, such as BHT, which achieves a faster release (Manzanarez-López et al. 2011; Penjumras et al. 2018). In fact, α-tocopherol can be unstable under certain conditions, thus the use of stabilizing techniques becomes necessary for its inclusion in active packaging, such as encapsulation. This system manages to preserve and extend its antioxidant effects by offering a physical barrier that protects it from external factors but it also provides a controlled release. (Mirzaei-Mohkam et al. 2019). Therefore, the development of active packaging based on the inclusion of vitamins/pro-vitamins represents a double advantage, for being able to exert protective functions on fatty foods and for increasing their vitamin content when transferred into food. Thus, the preservative properties offered by the application of active packaging may ensure that the vitamin content of the final product is the same or even higher than the original contained in its composition. For instance, the vitamin fortification mediated by active packaging systems can potentially be used to prevent or treat vitamin deficiencies, a global sanitary concern, that affect millions of people around the world (Moccand et al. 2016; Il and Zhao 2004).

4 Health and Environmental Benefits of Using Active Food Packaging for Oil/Fats

Currently, the trend in active food packaging in the field of oils and fatty products includes the improvement of plastic material barriers, but mostly the development of new biodegradable biopolymers which represent an alternative to replace the use of traditional one-use plastics (Becerril et al. 2020; Neethirajan and Jayas 2011). For example, chitosan has been demonstrated as an alternative functional polymer due to its low toxicity, biodegradability, film formation, antioxidant and antimicrobial capacities (Liu et al. 2018c, 2020). The incorporation of molecules with functional attributes such as antimicrobial and antioxidant properties endows benefits to the fatty food and also to consumers' health. One of the mechanisms of action of the active package, especially in oils and fatty products, is the migration of biocomponents by direct contact. Biomolecules mainly exert their function in the matrix, secondarily they may also translate their beneficial properties by their direct consumption when released from the active packages (Jugreet et al. 2020). The principal bioactivity associated to active packaging for fatty food is the antioxidant capacity that prevents lipid oxidation that ultimately may reduce the antimicrobial load. Therefore, health benefits of active packaging for oils and fats are strongly related to the preservation of the nutritional, microbiological and organoleptic properties of the final product by avoiding lipid oxidation (Farhoodi 2016; Gómez-Estaca et al. 2014).

Along the lipid oxidation process aldehydes, ketones, hydrocarbons, among others, are produced, which are responsible for the deterioration of the product, but also the accumulation of these substances are related to the damage of biological tissues and the appearance of degenerative diseases (Rehman et al. 2020). The ingestion of food containing products derived from the lipid oxidation, as well as, the endogenous generation of these products, has been demonstrated to represent a health risk when evaluated through animal models. Indeed, free radicals produced by the presence of products and sub-products created during lipid oxidation are considered responsible for inducing inflammation processes which may derive into different kinds of diseases such as cancer, immunodepression, heart affections, Alzheimer's or Parkinson's disease, among others (Guillen and Goicoechea 2008; Aruoma 1998). Even though the oral toxicity of oxidized lipids is low, their chronic uptake in large amounts increases the tumor frequency and the incidence of atherosclerosis in animals (Esterbauer 1996). In fact two aldehydes that can be present in food, 4-HNE and MDA, both by-products of the decomposition of peroxides, are considered as cytotoxic agents and very reactive compounds that may alter the normal functioning of the cell (Zarkovic 2003). A work stated that after ingestion of oxidized fats in animals and humans, increased amounts of MDA are excreted in urine and lipophilic carbonyl compounds that are absorbed from the intestine into the circulatory system. These derivatives act as harmful chemical substances that activate an inflammatory response to organs such as the liver, kidneys, lungs, and intestine (Kanner 2007). Another study provided the analysis of the composition of the headspace in sunflower oil. The chemical profile was characterized for the presence of α, β-unsaturated aldehydes, such as 4-hydroxy-(E)-2-nonenal, and oxygenates derivatives, such as 4 -hydroxy-2-nonenal, 4-oxo-2-nonenal or 4,5-epoxy-2-decennial which have been established as genotoxic and cytotoxic agents. In summary, lipid oxidation negatively reduces the nutritional value of products due to the chemical modification of fatty acids and other relevant lipophilic molecules such as fat-soluble vitamins A, D, E and K which ultimately might produce undesirable sensory and biological effects (Otero-Pazos et al. 2018). The use of bioactive substances, such as those extracted from natural sources, i.e. essential oils, with antioxidant capacity are being considered for control or preferably preventing these secondary oxidation products. The uses of these biomolecules apart from reducing oxidative stress have been described to possess anti-mutagenic, anticancer, anti-inflammatory, immunomodulatory or anti-protozoal agents. Therefore, natural molecules, with no side effects associated, are deeply considered to provide alternatives to the prevention and/or treatment of various diseases (Mahomoodally et al. 2019).

The direct consumption of antioxidants through the diet, or indirectly through the release from active packaging, might prevent the consumption of harmful substances derived from the lipid oxidation produced during their storage (Gorelik et al. 2005). The use of some natural antioxidants is known to delay lipid oxidation and rancidity (Jugreet et al. 2020). In fact, in the 80s, the use of β-carotene and citric acid in oils for preventing the deterioration of their organoleptic features was reported (Warner and Frankel 1987). Over the last decades, several components of essential oils have been described to possess redox properties which play a role in the breakdown of peroxides and a neutralization of free radicals (Burt 2004). Cinnamon, clove, basil, parsley or oregano essential oils have been chemically characterized with thymol and carvacrol as major components in some of them. These molecules recognized as antioxidants which capacity has been related to their phenolic structure. As an example, an edible coating developed with chitosan combined with kojic acid and clove essential oil significantly lengthened the shrimp shelf life. This biofilm inhibited the bacterial growth, pH change, and reduced total volatile basic nitrogen which serves as an indicator of the chemical spoilage. It also conserved shrimp color and avoided sample melanosis which increased the consumer acceptance and improved the final nutritional quality of the product (Liu et al. 2020). Thus, the use of essential oils can be an alternative to the use of synthetic preservatives (Kumar et al. 2020). However, they can alter the organoleptic properties of the final product; therefore, their utilization is preferable when encapsulated to mask their odor and flavor. Other natural molecules further used in active packaging are pigments, among others like phenolic compounds, PUFAS, or amino acids/peptides. For instance, a work stated the antioxidant and antimicrobial capacity of ε-polylysine, astaxanthin, and tocopherol which delay lipid oxidation corn oil inhibiting the formation and decomposition of hydroperoxides making this product safer (Wang et al. 2019; Huang et al. 1994). Another example of the inclusion of peptides, is the use of nisin that was integrated within layer-by-layer structures to develop antimicrobial films and thus reduce the health risk associated to food microbial ingestion (Haynie et al. 2006). Regarding the benefit of biocomposites fortified with phenolic compounds, they had been demonstrated to protect sardines from lipid oxidation which reduces the free radicals load to be consumed (López-de-Dicastillo et al. 2012).Other kinds of molecules used in active packaging are metals, such as iron powder that resulted very efficient for absorbing oxygen from fish oil. This mechanism prevented the deterioration of its high content of unsaturated fatty acids which preserved the nutritional value of the product while reducing the harmful presence of peroxide derivatives (Hidayah et al. 2020).

Worldwide, a third of all food produced for human consumption is wasted yearly. A huge percentage of food loss has been attributed to microbiological, fungal or mold spoilage. High microbial loads shorten the useful product life and risen the risk of food-borne diseases that may lead to serious threats to public health. Besides, this food loss creates a negative impact from both economic and ecological points of view (Becerril et al. 2020; Gómez-Estaca et al. 2014). Therefore, there is a huge interest focused on the design of active packaging aimed to maintain the physical, chemical and microbiological characteristics of the food that will improve their stability and safety, especially when using compounds obtained from natural sources for both film development and activation (Carpiné et al. 2015; Sapper et al. 2018; Figueroa-Lopez et al. 2020). Nowadays, the food industry has designed diverse packaging strategies for oil and fatty food using biopolymers (polysaccharides, proteins, etc.) in combination with several active compounds (pigments, phenols, amino acids, etc.) that have been proved to possess antioxidant and antimicrobial activity (Jamróz and Kopel 2020; Luangapai et al. 2019). Biodegradable active packaging can represent a true alternative to conventional packaging based on petroleum polymers aimed to single use. Biodegradable films will reduce the utilization of not friendly environmentally materials and their associated bioactivities will decrease food losses by extending shelf-life products and improving food quality. Therefore, active packaging can offer a green package solution that conserves the nutritional benefits of the food and avoids the consumption of harmful products as those derived from the lipid oxidation or microbial spoilage.

5 Limitations, Future Trends and Conclusions

The application of active packaging is an innovative solution for improving the storage time of oily foods, such as oils, butters, fatty fishes or fried products, among others. Since the main preservative issue concerning these products is the lipid oxidation, the main bioactivity expected to accomplish by active packaging is the antioxidant one. Among the multiple core ingredients that have been included with antioxidant capacity into the film materials, those with natural origins are mostly preferred, since the use of those that are synthetic have been associated to different harmful side effects. Natural molecules, especially when utilizing non-purified extracts, such as essential oils, present the advantage of providing additional properties such as being antimicrobial. The presence of antioxidants and antimicrobial molecules offers synergy and positive feedback inasmuch it permits the delay of oxidation reactions which imply the delay of the growth of microorganisms. This antioxidant environment protects the antimicrobial molecules that can exert their activity longer. The extended antimicrobial capacity further prevents the creation of free radicals related with the microorganism growth and thus, it ultimately extends the oxidative stability of the product. The addition of natural products into packaging films also provides color features and less common flavors and aromas. These features can influence it both negatively and positively. The inclusion of color to the package can be positive because it can prevent photodegradation and can represent a quality sensor of the product. However, if it results in opaque and/or a non-attractive appearance, it will not have consumer acceptance. The same occurs with flavors and odors; the inclusion of new organoleptic properties to traditional products may have negative impact for marketing purposes. For these scenarios, the food industry has already provided a solution: the encapsulation of core ingredients. This innovative approach enhances the bioactivity and bioavailability of the molecules by protecting them from environmental factors. Few examples of this technique have been described in the field of active packaging aimed for fatty products. However, it has to be considered that this encapsulation system may increase the final price of the product since it may require the use of expensive equipments and trained personnel.

Therefore, active packaging has been demonstrated to be a useful tool aimed to extend the shelf-life of fatty products while preserving their nutritional value, organoleptic properties, and food safety of the final product. Active packaging based on the use of biodegradable materials and natural molecules are the most attractive approach. This design may have a positive environmental impact by reducing the utilization of single-use plastics, but it would also diminish the current food waste rate. This option may be also translated into economic benefits since the final product would not contain synthetic compounds as additives and would be packaged with sustainable materials, which would ultimately increase consumer acceptance. Additionally, this preservation system would reduce the risk of food-borne diseases by decreasing the amount of oxidative products and the presence of microorganism colonies, while it could represent a fortification tool for enriching food matrices with essential antioxidant molecules, such as vitamins.

Therefore, the application of active films designed with biodegradable materials represents a green alternative to single-use plastic wrapping materials that preserves the nutritional benefits of the food and avoids the consumption of harmful products as those derived from the lipid oxidation or microbial spoilage.

Acknowledgements The research leading to these results received financial support from Programa de Cooperación Interreg V-A España—Portugal (POCTEP) 2014–2020 (projects Ref. 0181_NANOEATERS_01_E and Ref. 0377_IBERPHENOL_6_E) that supports the pre-doctoral grant for C. Jimenez-Lopez; from MICINN supporting the Ramón&Cajal grant for M.A. Prieto (RYC-2017-22891); from Xunta de Galicia and University of Vigo supporting the post-doctoral grant for M. Fraga-Corral (ED481B-2019/096) and the pre-doctoral grants for A.G. Pereira (ED481A-2019/0228) and P. García-Oliveira (ED481A-2019/295); from Axudas Conecta Peme (Xunta de Galicia) supporting the IN852A 2018/58 NeuroFood Project; AlgaMar (www.algamar.com) that supports the pre-doctoral grant for C. Lourenço-Lopes; from EcoChestnut Project (Erasmus+ KA202) that supports the work of M. Carpena; from Ibero-American Program on Science and Technology (CYTED - AQUA-CIBUS, P317RT0003); and from Bio Based Industries Joint Undertaking (JU) under grant agreement No 888003 UP4HEALTH Project (H2020-BBI-JTI-2019).

References

AbouZid S, Ahmed OM. Chapter 14 – Silymarin Flavonolignans: structure–activity relationship and biosynthesis. In: Atta-ur-Rahman BT-S in NPC, editors. Studies in natural products chemistry. Elsevier; 2013. p. 469–484.

Aladedunye F, Matthäus B. Phenolic extracts from *Sorbus aucuparia* (L.) and *Malus baccata* (L.) berries: Antioxidant activity and performance in rapeseed oil during frying and storage. Food Chem. 2014;159:273–81. https://doi.org/10.1016/j.foodchem.2014.02.139.

Almasi H, Ghanbarzadeh B, Dehghannya J, Entezami AA, Khosrowshahi Asl A. Development of a novel controlled-release nanocomposite based on poly(lactic acid) to increase the oxidative stability of soybean oil. Food Addit Contam A. 2014;31:1586–97. https://doi.org/10.1080/19440049.2014.935962.

Almasi H, Zandi M, Beigzadeh S, Haghju S, Mehrnow N. Chitosan films incorporated with nettle (*Urtica dioica* L.) extract-loaded nanoliposomes: II. Antioxidant activity and release properties. J Microencapsul. 2016;33:449–59. https://doi.org/10.1080/02652048.2016.1208295.

Apjok R, Mihaly Cozmuta A, Peter A, Mihaly Cozmuta L, Nicula C, Baia M, Vulpoi A. Active packaging based on cellulose-chitosan-Ag/TiO_2 nanocomposite for storage of clarified butter. Cellulose. 2019;26:1923–46. https://doi.org/10.1007/s10570-018-02226-7.

Aruoma OI. Free radicals, oxidative stress, and antioxidants in human health and disease. J Am Oil Chem Soc. 1998;75:199–212. https://doi.org/10.1007/s11746-998-0032-9.

Asensio CM, Nepote V, Grosso NR. Consumers' acceptance and quality stability of olive oil flavoured with essential oils of different oregano species. Int J Food Sci Technol. 2013;48:2417–28. https://doi.org/10.1111/ijfs.12233.

Becerril R, Nerín C, Silva F. Encapsulation systems for antimicrobial food packaging components: an update. Molecules. 2020;25 https://doi.org/10.3390/molecules25051134.

Bi F, Zhang X, Liu J, Yong H, Gao L, Liu J. Development of antioxidant and antimicrobial packaging films based on chitosan, D-α-tocopheryl polyethylene glycol 1000 succinate and silicon dioxide nanoparticles. Food Packag Shelf Life. 2020;24:100503. https://doi.org/10.1016/j.fpsl.2020.100503.

Bravi E, Perretti G, Falconi C, Marconi O, Fantozzi P. Antioxidant effects of supercritical fluid garlic extracts in sunflower oil. J Sci Food Agric. 2017;97:102–7. https://doi.org/10.1002/jsfa.7690.

Burt S. Essential oils: their antibacterial properties and potential applications in foods – a review. Int J Food Microbiol. 2004;94:223–53.

Carpiné D, Dagostin JLA, Bertan LC, Mafra MR. Development and characterization of soy protein isolate emulsion-based edible films with added coconut oil for olive oil packaging: barrier, mechanical, and thermal properties. Food Bioprocess Technol. 2015;8:1811–23. https://doi.org/10.1007/s11947-015-1538-4.

Carrizo D, Taborda G, Nerín C, Bosetti O. Extension of shelf life of two fatty foods using a new antioxidant multilayer packaging containing green tea extract. Innov Food Sci Emerg Technol. 2016;33:534–41. https://doi.org/10.1016/j.ifset.2015.10.018.

Chavoshizadeh S, Pirsa S, Mohtarami F. Sesame oil oxidation control by active and smart packaging system using wheat gluten/chlorophyll film to increase shelf life and detecting expiration date. Eur J Lipid Sci Technol. 2020;122:1900385. https://doi.org/10.1002/ejlt.201900385.

Choe E, Min DB. Mechanisms and factors for edible oil oxidation. Compr Rev Food Sci Food Saf. 2006;5:169–86. https://doi.org/10.1111/j.1541-4337.2006.00009.x.

Choulitoudi E, Velliopoulou A, Tsimogiannis D, Oreopoulou V. Effect of active packaging with *Satureja thymbra* extracts on the oxidative stability of fried potato chips. Food Packag Shelf Life. 2020;23:100455. https://doi.org/10.1016/j.fpsl.2019.100455.

CIE Colorimetry Committee. Technical notes. J Opt Soc Am. 1974;64:896–7. https://doi.org/10.1364/JOSA.64.000896.

Clark S, Youngman LD, Chukwurah B, Palmer A, Parish S, Peto R, Collins R. Effect of temperature and light on the stability of fat-soluble vitamins in whole blood over several days: Implications for epidemiological studies. Int J Epidemiol. 2004;33:518–25. https://doi.org/10.1093/ije/dyh064.

Colín-Chávez C, Soto-Valdez H, Peralta E, Lizardi-Mendoza J, Balandrán-Quintana R. Diffusion of natural astaxanthin from polyethylene active packaging films into a fatty food simulant. Food Res Int. 2013;54:873–80. https://doi.org/10.1016/j.foodres.2013.08.021.

Colín-Chávez C, Soto-Valdez H, Peralta E. Diffusion of carotenoids from mono and bilayer polyethylene active packaging into soybean oil. Food Packag Shelf Life. 2014a;1:170–8. https://doi.org/10.1016/j.fpsl.2014.01.008.

Colín-Chávez C, Vicente-Ramírez EB, Soto-Valdez H, Peralta E, Auras R. The release of carotenoids from a light-protected antioxidant active packaging designed to improve the stability of soybean oil. Food Bioprocess Technol. 2014b;7:3504–15. https://doi.org/10.1007/s11947-014-1359-x.

de Souza KC, Correa LG, da Silva TBV, Moreira TFM, de Oliveira A, Sakanaka LS, Dias MI, Barros L, Ferreira ICFR, Valderrama P, Leimann FV, Shirai MA. Soy Protein Isolate Films Incorporated with Pinhão (*Araucaria angustifolia* (Bertol.) Kuntze) Extract for Potential Use as Edible Oil Active Packaging. Food Bioprocess Technol. 2020;13:998–1008. https://doi.org/10.1007/s11947-020-02454-5.

Deshwal GK, Panjagari NR. Review on metal packaging: materials, forms, food applications, safety and recyclability. J Food Sci Technol. 2020;57:2377–92. https://doi.org/10.1007/s13197-019-04172-z.

Diosady LL, Krishnaswamy K. Micronutrient fortification of edible oils. In: Food fortification in a globalized world. Elsevier; 2018. p. 167–74.

EC EC. Regulation (EC) No 1334/2008 of the European Parliament and of the Council of 16 December 2008 on flavourings and certain food ingredients with flavouring properties for use in and on foods and amending Council Regulation (EEC) No 1601/91. Off J Eur Communities. 2008;354:34–50.

Ertugrul S, Yucel C, Sertoglu E, Ozkan Y, Ozgurtas T. Development and optimization of simultaneous determination of fat soluble vitamins by liquid chromatography tandem mass spectrometry. Chem Phys Lipids. 2020;230:104932. https://doi.org/10.1016/j.chemphyslip.2020.104932.

Eshghi N, Asnaashari M, Khodaparast MHH, Hosseini F. Evaluating the potential of natural curcumin for oxidative stability of soybean oil. Nat Prod Res. 2014;28:1375–8. https://doi.org/10.1080/14786419.2014.901319.

Esterbauer H. Estimation of peroxidative damage. A critical review. Pathol Biol. 1996;44:25–8.

Fan X. Furan formation from fatty acids as a result of storage, gamma irradiation, UV-C and heat treatments. Food Chem. 2015;175:439–44. https://doi.org/10.1016/j.foodchem.2014.12.002.

Farhoodi M. Nanocomposite materials for food packaging applications: characterization and safety evaluation. Food Eng Rev. 2016;8:35–51. https://doi.org/10.1007/s12393-015-9114-2.

Fayaz Dastgerdi G, Goli SAH, Kadivar M. A new antioxidant active film based on HDPE and peppermint essential oil for packaging soybean oil. J Am Oil Chem Soc. 2016;93:657–64. https://doi.org/10.1007/s11746-016-2806-9.

Fereidoon S, Ying Z. Lipid oxidation and improving the oxidative stability. Chem Soc Rev. 2010;39:4067–79. https://doi.org/10.1039/b922183m.

Fields ML, Richmond BS, Baldwin RE. Food quality as determined by metabolic by-products of microorganisms. In: Chichester CO, Mrak EM, Stewart GFBT-A in FR, editors. Academic; 1968. p. 161–229.

Figueroa-Lopez KJ, Enescu D, Torres-Giner S, Cabedo L, Cerqueira MA, Pastrana L, Fuciños P, Lagaron JM. Development of electrospun active films of poly(3-hydroxybutyrate-co-3-hydroxyvalerate) by the incorporation of cyclodextrin inclusion complexes containing oregano essential oil. Food Hydrocoll. 2020;108. https://doi.org/10.1016/j.foodhyd.2020.106013

Fraga-Corral M, García-Oliveira P, Pereira AG, Lourenço-Lopes C, Jimenez-Lopez C, Prieto MA, Simal-Gandara J. Technological application of tannin-based extracts. Molecules. 2020;25:1–27.

Frankel EN. Chapter 3 - Photooxidation of unsaturated fats. In: Lipid oxidation; 2012. p. 51–66.

Gaikwad KK, Singh S, Lee YS. A new pyrogallol coated oxygen scavenging film and their effect on oxidative stability of soybean oil under different storage conditions. Food Sci Biotechnol. 2017;26:1535–43. https://doi.org/10.1007/s10068-017-0232-x.

Ganesan AR, Shanmugam M, Ilansuriyan P, Anandhakumar R, Balasubramanian B. Composite film for edible oil packaging from carrageenan derivative and konjac glucomannan: application and quality evaluation. Polym Test. 2019;78:105936. https://doi.org/10.1016/j.polymertesting.2019.105936.

Gautam G, Mishra P. Development and characterization of copper nanocomposite containing bilayer film for coconut oil packaging. J Food Process Preserv. 2017;41:e13243. https://doi.org/10.1111/jfpp.13243.

Gemili S, Yemenicioğlu A, Altinkaya SA. Development of antioxidant food packaging materials with controlled release properties. J Food Eng. 2010;96:325–32. https://doi.org/10.1016/j.jfoodeng.2009.08.020.

Ghorbani Gorji S, Calingacion M, Smyth HE, Fitzgerald M. Comprehensive profiling of lipid oxidation volatile compounds during storage of mayonnaise. J Food Sci Technol. 2019;56:4076–90. https://doi.org/10.1007/s13197-019-03876-6.

Gómez-Estaca J, López-de-Dicastillo C, Hernández-Muñoz P, Catalá R, Gavara R. Advances in antioxidant active food packaging. Trends Food Sci. Technol. 2014;35:42–51.

Gontard N. Active packaging. Workshop sobre biopolímeros. Pirassununga – SP. FZEA, 1997.

Gorelik S, Lapidot T, Shaham I, Granit R, Ligumsky M, Kohen R, Kanner J. Lipid peroxidation and coupled vitamin oxidation in simulated and human gastric fluid inhibited by dietary polyphenols: Health implications. J Agric Food Chem. 2005;53:3397–402. https://doi.org/10.1021/jf040401o.

Graciano-Verdugo AZ, Soto-Valdez H, Peralta E, Cruz-Zárate P, Islas-Rubio AR, Sánchez-Valdes S, Sánchez-Escalante A, González-Méndez N, González-Ríos H. Migration of α-tocopherol from LDPE films to corn oil and its effect on the oxidative stability. Food Res Int. 2010;43:1073–8. https://doi.org/10.1016/j.foodres.2010.01.019.

Granata G, Stracquadanio S, Leonardi M, Napoli E, Consoli GML, Cafiso V, Stefani S, Geraci C. Essential oils encapsulated in polymer-based nanocapsules as potential candidates for application in food preservation. Food Chem. 2018;269:286–92. https://doi.org/10.1016/j.foodchem.2018.06.140.

Guéraud F, Atalay M, Bresgen N, Cipak A, Eckl PM, Huc L, Jouanin I, Siems W, Uchida K. Chemistry and biochemistry of lipid peroxidation products. Free Radic Res. 2010;44:1098–124. https://doi.org/10.3109/10715762.2010.498477.

Guillen MD, Goicoechea E. Formation of oxygenated α,β-unsaturated aldehydes and other toxic compounds in sunflower oil oxidation at room temperature in closed receptacles. Food Chem. 2008;111:157–64. https://doi.org/10.1016/j.foodchem.2008.03.052.

Gutiérrez L, Escudero A, Batlle R, Nerín C. Effect of mixed antimicrobial agents and flavors in active packaging films. J Agric Food Chem. 2009;57:8564–71. https://doi.org/10.1021/jf901459e.

Haas K, Obernberger J, Zehetner E, Kiesslich A, Volkert M, Jaeger H. Impact of powder particle structure on the oxidation stability and color of encapsulated crystalline and emulsified carotenoids in carrot concentrate powders. J Food Eng. 2019;263:398–408. https://doi.org/10.1016/j.jfoodeng.2019.07.025.

Haynie DT, Zhang L, Zhao W, Rudra JS. Protein-inspired multilayer nanofilms: science, technology and medicine. Nanomed Nanotechnol Biol Med. 2006;2:150–7. https://doi.org/10.1016/j.nano.2006.07.008.

Hidayah EN, Triastuti RRJ, Abdillah AA. The effect of iron powder as oxygen absorber active packaging on fish oil total oxidation value. In: IOP conference series: earth and environmental science; 2020. p. 241–67.

Hu K, Huyan Z, Ding S, Dong Y, Yu X. Investigation on food packaging polymers: effects on vegetable oil oxidation. Food Chem. 2020;315:126299. https://doi.org/10.1016/j.foodchem.2020.126299.

Huang SW, Frankel EN, German JB. Antioxidant activity of α-and γ-tocopherols in bulk oils and in oil-in-water emulsions. J Agric Food Chem. 1994;42:2108–14. https://doi.org/10.1021/jf00046a007.

Iahnke AOeS, Costa TMH, de Oliveira Rios A, Flôres SH. Antioxidant films based on gelatin capsules and minimally processed beet root (*Beta vulgaris* L. var. Conditiva) residues. J Appl Polym Sci. 2016;133. https://doi.org/10.1002/app.43094

Il PS, Zhao Y. Incorporation of a high concentration of mineral or vitamin into chitosan-based films. J Agric Food Chem. 2004;52:1933–9. https://doi.org/10.1021/jf034612p.

İnanç T, Maskan M. The potential application of plant essential oils/extracts as natural preservatives in oils during processing: a review. J Food Sci Eng. 2012;2:1–9.

Iqdiam BM, Welt BA, Goodrich-Schneider R, Sims CA, Baker GL, Marshall MR. Influence of headspace oxygen on quality and shelf life of extra virgin olive oil during storage. Food Packag Shelf Life. 2020;23:100433. https://doi.org/10.1016/j.fpsl.2019.100433.

Isabel Minguez-Mosquera M, Rejano-Navarro L, Gandul-Rojas B, SanchezGomez AH, Garrido-Fernandez J. Color-pigment correlation in virgin olive oil. J Am Oil Chem Soc. 1991;68:332–6. https://doi.org/10.1007/BF02657688.

Jamróz E, Kopel P. Polysaccharide and protein films with antimicrobial/antioxidant activity in the food industry: a review. Polymers (Basel). 2020;12:1289. https://doi.org/10.3390/polym12061289.

Jiang W, Hu S, Li S, Liu Y. Evaluation of the preservation effect of gelatin-water soluble chitosan film incorporated with maillard peptides on bluefin tuna (*Thunnus thynnus*) slices packaging. LWT. 2019;113:108294. https://doi.org/10.1016/j.lwt.2019.108294.

Joseph P. Oxidative stability and shelf life of bulk animal fats and poultry fats. Elsevier Inc.; 2016.

Jugreet BS, Suroowan S, Rengasamy RRK, Mahomoodally MF. Chemistry, bioactivities, mode of action and industrial applications of essential oils. Trends Food Sci Technol. 2020;101:89–105.

Kanner J. Dietary advanced lipid oxidation endproducts are risk factors to human health. Mol Nutr Food Res. 2007;51:1094–101. https://doi.org/10.1002/mnfr.200600303.

Kapchie VN, Yao L, Hauck CC, Wang T, Murphy PA. Oxidative stability of soybean oil in oleosomes as affected by pH and iron. Food Chem. 2013;141:2286–93. https://doi.org/10.1016/j.foodchem.2013.05.018.

Karaman AD, Özer B, Pascall MA, Alvarez V. Recent advances in dairy packaging. Food Rev Int. 2015;31:295–318. https://doi.org/10.1080/87559129.2015.1015138.

Kchaou H, Jridi M, Nasri M, Debeaufort F. Design of gelatin pouches for the preservation of flaxseed oil during storage. Coatings. 2020;10. https://doi.org/10.3390/coatings10020150

Kehili M, Choura S, Zammel A, Allouche N, Sayadi S. Oxidative stability of refined olive and sunflower oils supplemented with lycopene-rich oleoresin from tomato peels industrial by-product, during accelerated shelf-life storage. Food Chem. 2018;246:295–304. https://doi.org/10.1016/j.foodchem.2017.11.034.

Kontominas MG. Olive oil packaging: recent developments. Olives Olive Oil Funct Foods. 2017:279–94. https://doi.org/10.1002/9781119135340.ch14.

Koontz JL. Chapter 13 - Packaging technologies to control lipid oxidation. In: Hu M, Jacobsen CBT-OS and SL of FCO and F, editors. Oxidative stability and shelf life of foods containing oils and fats. AOCS Press; 2016. p. 479–517

Kouveli A, Tzetzis D, Kyratsis P. Packaging design in today's industries: the case of olive oil packaging design. J Packag Technol Res. 2017;1:13–23. https://doi.org/10.1007/s41783-017-0008-z.

Kráčmar S, Fišera M, Přikrylová V, Fišerová L, Málek Z, Tvrzník P. Storage of extra virgin olive oil and its impact on fatty acid levels. J Microbiol Biotechnol Food Sci. 2019;8:1228–30. https://doi.org/10.15414/jmbfs.2019.8.5.1228-1230.

Kumar A, Singh P, Gupta V, Prakash B. Application of nanotechnology to boost the functional and preservative properties of essential oils. In: Functional and preservative properties of phytochemicals. Elsevier Inc.; 2020. p. 241–67.

Ling SSC, Chang SK, Sia WCM, Yim HS. Antioxidant efficacy of unripe banana (*Musa acuminata* Colla) peel extracts in sunflower oil during accelerated storage. Acta Sci Pol Technol Aliment. 2015;14:343–56.

Liu F, Avena-Bustillos RJ, Chiou B-S, Li Y, Ma Y, Williams TG, Wood DF, McHugh TH, Zhong F. Controlled-release of tea polyphenol from gelatin films incorporated with different ratios of free/nanoencapsulated tea polyphenols into fatty food simulants. Food Hydrocoll. 2017;62:212–21. https://doi.org/10.1016/j.foodhyd.2016.08.004.

Liu Y, Chen Q, Chen G. Study on shelf life prediction of edible oil in BOPP/VMPET/CPP flexible packaging BT. In: Zhao P, Ouyang Y, Xu M, Yang L, Ren Y, editors. Applied sciences in graphic communication and packaging. Singapore: Springer; 2018a. p. 517–23.

Liu J, Xiao J, Li F, Shi Y, Li D, Huang Q. Chitosan-sodium alginate nanoparticle as a delivery system for ε-polylysine: preparation, characterization and antimicrobial activity. Food Control. 2018b;91:302–10. https://doi.org/10.1016/j.foodcont.2018.04.020.

Liu X, Xia W, Jiang Q, Yu P, Yue L. Chitosan oligosaccharide-N-chlorokojic acid mannich base polymer as a potential antibacterial material. Carbohydr Polym. 2018c;182:225–34. https://doi.org/10.1016/j.carbpol.2017.11.019.

Liu X, Zhang C, Liu S, Gao J, Cui SW, Xia W. Coating white shrimp (*Litopenaeus vannamei*) with edible fully deacetylated chitosan incorporated with clove essential oil and kojic acid improves preservation during cold storage. Int J Biol Macromol. 2020;162:1276–82. https://doi.org/10.1016/j.ijbiomac.2020.06.248.

Lloret E, Picouet PA, Trbojevich R, Fernández A. Colour stability of cooked ham packed under modified atmospheres in polyamide nanocomposite blends. LWT Food Sci Technol. 2016;66:582–9. https://doi.org/10.1016/j.lwt.2015.11.012.

López-de-Dicastillo C, Gómez-Estaca J, Catalá R, Gavara R, Hernández-Muñoz P. Active antioxidant packaging films: development and effect on lipid stability of brined sardines. Food Chem. 2012;131:1376–84. https://doi.org/10.1016/j.foodchem.2011.10.002.

Łopusiewicz Ł, Jędra F, Bartkowiak A. The application of melanin modified gelatin coatings for packaging and the oxidative stability of pork lard. World Sci News. 2018;101:108–19.

Lozano-Sánchez J, Bendini A, Quirantes-Piné R, Cerretani L, Segura-Carretero A, Fernández-Gutiérrez A. Monitoring the bioactive compounds status of extra-virgin olive oil and storage by-products over the shelf life. Food Control. 2013;30:606–15. https://doi.org/10.1016/j.foodcont.2012.06.036.

Luangapai F, Peanparkdee M, Iwamoto S. Biopolymer films for food industries: properties, applications, and future aspects based on chitosan. Rev Agric Sci. 2019;7:59–67. https://doi.org/10.7831/ras.7.0_59.

Mahomoodally F, Aumeeruddy-Elalfi Z, Venugopala KN, Hosenally M. Antiglycation, comparative antioxidant potential, phenolic content and yield variation of essential oils from 19 exotic and endemic medicinal plants. Saudi J Biol Sci. 2019;26:1779–88. https://doi.org/10.1016/j.sjbs.2018.05.002.

Mailer RJ, Graham K, Ayton J. The effect of storage in collapsible containers on olive oil quality. Rural Industries Research and Development Corporation (Australian Government), 2012.

Malherbi NM, Schmitz AC, Grando RC, Bilck AP, Yamashita F, Tormen L, Fakhouri FM, Velasco JI, Bertan LC. Corn starch and gelatin-based films added with guabiroba pulp for application in food packaging. Food Packag Shelf Life. 2019;19:140–6. https://doi.org/10.1016/j.fpsl.2018.12.008.

Mallia S, Escher F, Schlichtherle-Cerny H. Aroma-active compounds of butter: a review. Eur Food Res Technol. 2008;226:315–25. https://doi.org/10.1007/s00217-006-0555-y.

Manzanarez-López F, Soto-Valdez H, Auras R, Peralta E. Release of α-Tocopherol from Poly(lactic acid) films, and its effect on the oxidative stability of soybean oil. J Food Eng. 2011;104:508–17. https://doi.org/10.1016/j.jfoodeng.2010.12.029.

McLaren K. XIII—the development of the CIE 1976 (L* a* b*) uniform colour space and colour-difference formula. J Soc Dye Colour. 1976;92:338–41. https://doi.org/10.1111/j.1478-4408.1976.tb03301.x.

Mirzaei-Mohkam A, Garavand F, Dehnad D, Keramat J, Nasirpour A. Optimisation, antioxidant attributes, stability and release behaviour of carboxymethyl cellulose films incorporated with nanoencapsulated vitamin E. Prog Org Coatings. 2019;134:333–41. https://doi.org/10.1016/j.porgcoat.2019.05.026.

Moccand C, Martin F, Martiel I, Gancel C, Michel M, Fries L, Sagalowicz L. Vitamin A degradation in triglycerides varying by their saturation levels. Food Res Int. 2016;88:3–9. https://doi.org/10.1016/j.foodres.2016.06.001.

Mohammad Zadeh E, O'Keefe SF, Kim Y-T. Lignin-based biopolymeric active packaging system for oil products. J Food Sci. 2019;84:1420–6. https://doi.org/10.1111/1750-3841.14632.

Moraes ARF, LER G, Soares N d FF, Santos MMDS, MPJC G. Desenvolvimento e avaliação de filme antimicrobiano na conservação de manteiga. Food Sci Technol. 2007;27:33–6.

Nderitu SM, Nawiri MP, Nyambaka HN. Fortification of sunflower and palm oils using beta carotene extracted from *Asystasia mysorensis* and *Solanum nigrum*. Food Res. 2018;2:437–42. https://doi.org/10.26656/fr.2017.2(5).080.

Neethirajan S, Jayas DS. Nanotechnology for the food and bioprocessing industries. Food Bioprocess Technol. 2011;4:39–47. https://doi.org/10.1007/s11947-010-0328-2.

Nerín C, Tovar L, Salafranca J. Behaviour of a new antioxidant active film versus oxidizable model compounds. J Food Eng. 2008;84:313–20. https://doi.org/10.1016/j.jfoodeng.2007.05.027.

Otero-Pazos P, Sendón R, Blanco-Fernandez B, Blanco-Dorado S, Alvarez-Lorenzo C, Concheiro A, Angulo I, Paseiro-Losada P, Rodríguez-Bernaldo de Quirós A. Preparation of antioxidant active films based on chitosan: diffusivity study of α-tocopherol into food simulants. J Food Sci Technol. 2016;53:2817–26. https://doi.org/10.1007/s13197-016-2256-2.

Otero-Pazos P, Sendón R, Martínez IP, Aurrekoetxea G, Angulo I, Rodríguez-Bernaldo de Quirós A. Evaluation of oxygen absorber system effectiveness in butter containers. CyTA-J Food. 2018;16:205–12. https://doi.org/10.1080/19476337.2017.1379561.

Panrong T, Karbowiak T, Harnkarnsujarit N. Thermoplastic starch and green tea blends with LLDPE films for active packaging of meat and oil-based products. Food Packag Shelf Life. 2019;21:100331. https://doi.org/10.1016/j.fpsl.2019.100331.

Pastorelli S, Valzacchi S, Rodriguez A, Simoneau C. Solid-phase microextraction method for the determination of hexanal in hazelnuts as an indicator of the interaction of active packaging materials with food aroma compounds. Food Addit Contam. 2006;23:1236–41. https://doi.org/10.1080/02652030600778744.

Pedrali D, Gallotti F, Proserpio C, Pagliarini E, Lavelli V. Kinetic study of vitamin D2 degradation in mushroom powder to improve its applications in fortified foods. LWT. 2020;125:109248. https://doi.org/10.1016/j.lwt.2020.109248.

Penjumras P, Rahman RA, Thammakulkrajang R. Migration study of antioxidant in durian rind cellulose reinforced poly(lactic acid) (PLA) biocomposites and its effect on oxidative stability of edible oil. Solid State Phenom. 2018;278:89–95. https://doi.org/10.4028/www.scientific.net/SSP.278.89.

Perazzo KKNCL, Conceição AC d V, Santos JCP d, Assis D d J, Souza CO, Druzian JI. Properties and antioxidant action of actives cassava starch films incorporated with green tea and palm oil extracts. PLoS One. 2014;9:e105199.

Pereira de Abreu DA, Losada PP, Maroto J, Cruz JM. Evaluation of the effectiveness of a new active packaging film containing natural antioxidants (from barley husks) that retard lipid damage in frozen Atlantic salmon (*Salmo salar* L.). Food Res Int. 2010;43:1277–82. https://doi.org/10.1016/j.foodres.2010.03.019.

Phoopuritham P, Thongngam M, Yoksan R, Suppakul P. Antioxidant properties of selected plant extracts and application in packaging as antioxidant cellulose-based films for vegetable oil. Packag Technol Sci. 2012;25:125–36. https://doi.org/10.1002/pts.963.

Piscopo A, Poiana M. Packaging and storage of olive oil. IntechOpen. 2012;23. https://doi.org/10.5772/51827

Pristouri G, Badeka A, Kontominas MG. Effect of packaging material headspace, oxygen and light transmission, temperature and storage time on quality characteristics of extra virgin olive oil. Food Control. 2010;21:412–8. https://doi.org/10.1016/j.foodcont.2009.06.019.

Rababah TM, Feng H, Yang W, Eriefej K, Al-Omoush M. Effects of type of packaging material on physicochemical and sensory properties of olive oil. Int J Agric Biol Eng. 2011;4:66–72. https://doi.org/10.3965/j.issn.1934-6344.2011.04.066-072.

Rehman A, Jafari SM, Aadil RM, Assadpour E, Randhawa MA, Mahmood S. Development of active food packaging via incorporation of biopolymeric nanocarriers containing essential oils. Trends Food Sci Technol. 2020;101:106–21.

Repetto M, Semprine J, Boveris A. Lipid peroxidation: chemical mechanism, biological implications and analytical determination. IntechOpen; 2012. p. 3–30.

Restuccia D, Spizzirri UG, Parisi OI, Cirillo G, Curcio M, Iemma F, Puoci F, Vinci G, Picci N. New EU regulation aspects and global market of active and intelligent packaging for food industry applications. Food Control. 2010;21:1425–35. https://doi.org/10.1016/j.foodcont.2010.04.028.

Ribeiro-Santos R, Sanches-Silva A, Motta JFG, Andrade M, Neves I d A, Teófilo RF, Carvalho MG d, Melo NR d. Combined use of essential oils applied to protein base active food packaging: study in vitro and in a food simulant. Eur Polym J. 2017;93:75–86. https://doi.org/10.1016/j.eurpolymj.2017.03.055.

Rizzo V, Torri L, Licciardello F, Piergiovanni L, Muratore G. Quality changes of extra virgin olive oil packaged in coloured polyethylene terephthalate bottles stored under different lighting conditions. Packag Technol Sci. 2014;27:437–48. https://doi.org/10.1002/pts.2044.

Robards K, Kerr AF, Patsalides E, Korth J. Headspace gas analysis as a measure of rancidity in corn chips. J Am Oil Chem Soc. 1988;65:1621–6.

Robertson GL. Food packaging: principles and practice. 3rd ed; 2013.

Romani VP, Martins VG, Goddard JM. Radical scavenging polyethylene films as antioxidant active packaging materials. Food Control. 2020;109:106946. https://doi.org/10.1016/j.foodcont.2019.106946.

Samprasit W, Opanasopit P, Sukma M, Kaomongkolgit R. Antibacterial activity of *Garcinia mangostana* extracts on oral pathogens. Minerva Stomatol. 2014;63:249–57.

Samsudin H, Soto-Valdez H, Auras R. Poly(lactic acid) film incorporated with marigold flower extract (*Tagetes erecta*) intended for fatty-food application. Food Control. 2014;46:55–66. https://doi.org/10.1016/j.foodcont.2014.04.045.

Sanmartin C, Venturi F, Sgherri C, Nari A, Macaluso M, Flamini G, Quartacci MF, Taglieri I, Andrich G, Zinnai A. The effects of packaging and storage temperature on the shelf-life of extra virgin olive oil. Heliyon. 2018;4:e00888. https://doi.org/10.1016/j.heliyon.2018.e00888.

São José JFB d, Medeiros HS, Oliveira FCE d, Filhao e Moares AR, Oliveira D d S, ÉAA M, Soares N d FF. Development and characterization of active film with omega-3 as a proposal for enrichment of butter. Food Sci Technol. 2019;39:304–8.

Sapper M, Wilcaso P, Santamarina MP, Roselló J, Chiralt A. Antifungal and functional properties of starch-gellan films containing thyme (Thymus zygis) essential oil. Food Control. 2018;92:505–15. https://doi.org/10.1016/j.foodcont.2018.05.004.

Sionek B, Krygier K, Ukalski K, Ukalska J, Amarowicz R. The influence of nitrogen and carbon dioxide on the oxidative stability of fully refined rapeseed oil. Eur J Lipid Sci Technol. 2013;115:1426–33. https://doi.org/10.1002/ejlt.201200328.

Smeriglio A, Toscano G, Denaro M, De Francesco C, Agozzino S, Trombetta D. Nitrogen headspace improves the extra virgin olive oil shelf-life, preserving its functional properties. Antioxidants. 2019;8:1–12. https://doi.org/10.3390/antiox8090331.

Song N-B, Lee J-H, Al Mijan M, Bin SK. Development of a chicken feather protein film containing clove oil and its application in smoked salmon packaging. LWT Food Sci Technol. 2014;57:453–60. https://doi.org/10.1016/j.lwt.2014.02.009.

Stoll L, da Silva AM, Iahnke AOS, TMH C, Flôres SH, Rios A d O. Active biodegradable film with encapsulated anthocyanins: effect on the quality attributes of extra-virgin olive oil during storage. J Food Process Preserv. 2017;41:e13218. https://doi.org/10.1111/jfpp.13218.

Stoll L, Rech R, Flôres SH, Nachtigall SMB, de Oliveira RA. Poly(acid lactic) films with carotenoids extracts: release study and effect on sunflower oil preservation. Food Chem. 2019;281:213–21. https://doi.org/10.1016/j.foodchem.2018.12.100.

Strathmann S, Pastorelli S, Simoneau C. Investigation of the interaction of active packaging material with food aroma compounds. Sensors Actuators B Chem. 2005;106:83–7. https://doi.org/10.1016/j.snb.2004.05.040.

Talón E, Vargas M, Chiralt A, González-Martínez C. Antioxidant starch-based films with encapsulated eugenol. Application to sunflower oil preservation. LWT. 2019;113:108290. https://doi.org/10.1016/j.lwt.2019.108290.

Tammineni N, Ünlü G, Rasco B, Powers J, Sablani S, Nindo C. Trout-skin gelatin-based edible films containing phenolic antioxidants: effect on physical properties and oxidative stability of Cod-liver oil model food. J Food Sci. 2012;77:E342–7. https://doi.org/10.1111/j.1750-3841.2012.02964.x.

Tena N, Lobo-Prieto A, Aparicio R, García-González DL. Storage and preservation of fats and oils. In: Encyclopedia of food security and sustainability; 2019. p. 605–18.

Tian F, Decker EA, Goddard JM. Controlling lipid oxidation of food by active packaging technologies. Food Funct. 2013;4:669–80. https://doi.org/10.1039/C3FO30360H.

Tongdeesoontorn W, Mauer LJ, Wongruong S, Sriburi P, Rachtanapun P. Physical and antioxidant properties of cassava starch–carboxymethyl cellulose incorporated with quercetin and TBHQ as active food packaging. Polymers (Basel). 2020;12:366. https://doi.org/10.3390/polym12020366.

Torrieri E, Carlino PA, Cavella S, Fogliano V, Attianese I, Buonocore GG, Masi P. Effect of modified atmosphere and active packaging on the shelf-life of fresh bluefin tuna fillets. J Food Eng. 2011;105:429–35. https://doi.org/10.1016/j.jfoodeng.2011.02.038.

Tovar L, Salafranca J, Sánchez C, Nerín C. Migration studies to assess the safety in use of a new antioxidant active packaging. J Agric Food Chem. 2005;53:5270–5. https://doi.org/10.1021/jf050076k.

Trapani S, Breschi C, Cecchi L, Guerrini L, Mulinacci N, Parenti A, Canuti V, Picchi M, Caruso G, Gucci R, Zanoni B. Indirect indices of oxidative damage to phenolic compounds for the implementation of olive paste malaxation optimization charts. J Food Eng. 2017;207:24–34. https://doi.org/10.1016/j.jfoodeng.2017.03.012.

Tsimis DA, Karakasides NG. How the choice of container affects olive oil quality – a review. Packag Technol Sci. 2002;15:147–54. https://doi.org/10.1002/pts.581.

Van Wayenbergh E, Struyf N, Rezaei MN, Sagalowicz L, Bel-Rhlid R, Moccand C, Courtin CM. Cereal bran protects vitamin A from degradation during simmering and storage. Food Chem. 2020;331:127292. https://doi.org/10.1016/j.foodchem.2020.127292.

Wang S, Xia P, Wang S, Liang J, Sun Y, Yue P, Gao X. Packaging films formulated with gelatin and anthocyanins nanocomplexes: physical properties, antioxidant activity and its application for olive oil protection. Food Hydrocoll. 2019;96:617–24. https://doi.org/10.1016/j.foodhyd.2019.06.004.

Warner K, Frankel EN. Effects of β-carotene on light stability of soybean oil. J Am Oil Chem Soc. 1987;64:213–8. https://doi.org/10.1007/BF02542004.

Xu J, Wei R, Jia Z, Song R. Characteristics and bioactive functions of chitosan/gelatin-based film incorporated with ε-polylysine and astaxanthin extracts derived from by-products of shrimp (*Litopenaeus vannamei*). Food Hydrocoll. 2020;100:105436. https://doi.org/10.1016/j.foodhyd.2019.105436.

Zarkovic N. 4-hydroxynonenal as a bioactive marker of pathophysiological processes. Mol Aspects Med. 2003;24:281–91. https://doi.org/10.1016/S0098-2997(03)00023-2.

Zhang X, Liu J, Yong H, Qin Y, Liu J, Jin C. Development of antioxidant and antimicrobial packaging films based on chitosan and mangosteen (*Garcinia mangostana* L.) rind powder. Int J Biol Macromol. 2020;145:1129–39. https://doi.org/10.1016/j.ijbiomac.2019.10.038.

Zhu X, Schaich K, Chen X, Yam K. Antioxidant effects of sesamol released from polymeric films on lipid oxidation in linoleic acid and oat cereal. Packag Technol Sci. 2013;26:31–8. https://doi.org/10.1002/pts.1964.

Zuta PC, Simpson BK, Zhao X, Leclerc L. The effect of α-tocopherol on the oxidation of mackerel oil. Food Chem. 2007;100:800–7. https://doi.org/10.1016/j.foodchem.2005.11.003.

Index

S. M. Jafari, A. S. Silva (eds.), *Releasing Systems in Active Food Packaging*, Food Bioactive Ingredients, https://doi.org/10.1007/978-3-030-90299-5